算法设计与分析基础

(第 3 版　影印版)

(美) Anany Levitin　著

清華大學出版社
北京

图书在版编目(CIP)数据

算法设计与分析基础 (第 3 版 影印版) = Introduction to the Design and Analysis of Algorithms, Third Edition / (美)莱维丁(Levitin, A.)著. --影印本. --北京：清华大学出版社，2013（2022.9 重印）
ISBN 978-7-302-31185-0

Ⅰ. ①算…　Ⅱ. ①莱…　Ⅲ. ①电子计算机一算法设计一英文 ②电子计算机一算法分析一英文　Ⅳ. ①TP301.6

中国版本图书馆 CIP 数据核字(2013)第 002529 号

责任编辑：文开琪　汤涌涛
封面设计：杨玉兰
责任校对：周剑云
责任印制：杨　艳

出版发行：清华大学出版社
网　　址：http://www.tup.com.cn，http://www.wqbook.com
地　　址：北京清华大学学研大厦 A 座　　**邮　　编**：100084
社 总 机：010-83470000　　**邮　　购**：010-62786544
投稿与读者服务：010-62776969，c-service@tup.tsinghua.edu.cn
质 量 反 馈：010-62772015，zhiliang@tup.tsinghua.edu.cn

印 装 者：三河市龙大印装有限公司
经　　销：全国新华书店
开　　本：185mm×230mm　　**印　张**：37.25　　**字　　数**：1047 千字
版　　次：2013 年 5 月第 1 版　　**印　　次**：2022 年 9月第 8 次印刷
定　　价：129.00 元

产品编号：050764-02

前　言

"一个人接受科技教育时所能获得的最珍贵的收获，是那些能够受用终身的通用智能工具[①]。"

——George Forsythe，*What to do till the computer scientist comes*，1968

无论是计算科学还是计算实践，算法都在其中扮演着重要角色。因此，这门学科中出现了大量的教材。它们在介绍算法的时候，基本上都选择了以下两种方案中的一种。第一种方案是按照问题的类型对算法分类。这类教材安排了不同的章节分别讨论排序、查找、图等算法。这种做法的优点是，对于解决同一问题的不同算法，它能够立即比较这些算法的效率。其缺点在于，由于过于强调问题的类型，它忽略了对算法设计技术的讨论。

第二种方案围绕着算法设计技术来组织章节。在这种结构中，即使算法来自于不同的计算领域，如果它们采用了相同的设计技术，就会被编成一组。从各方(例如[BaY95])获得的信心使我相信，这种结构更适合于算法设计与分析的基础课程。强调算法设计技术有三个主要原因。第一，学生们在解决新问题时，可以运用这些技术设计出新的算法。从实用的角度看，这使得学习算法设计技术颇有价值。第二，学生们会试图按照算法的内在设计方法对已知的众多算法进行分类。计算机科学教育的一个主要目的，就是让学生们知道如何发掘不同应用领域的算法间的共性。毕竟，每门学科都会倾向于把它的重要主题归纳为几个甚至一个规则。第三，依我看来，算法设计技术作为问题求解的一般性策略，在解决计算机领域以外的问题时，也能发挥相当大的作用。

遗憾的是，无论是从理论还是从教学的角度，传统的算法设计技术分类法都存在一些严重的缺陷。其中最显著的缺陷就是无法对许多重要的算法进行分类。由于这种局限性，这些书的作者不得不在按照设计技术进行分类的同时，另外增加一些章节来讨论特殊的问题类型。但这种改变将导致课程缺乏一致性，而且很可能会使学生感到迷惑。

算法设计技术的新分类法

传统算法设计技术分类法的缺陷令我感到失望，它激发我开发一套新的分类法[Lev99]，这套分类法就是本书的基础。以下是这套新分类法的几个主要优势。

- 新分类法比传统分类法更容易理解。它包含的某些设计策略，例如蛮力法、减治法、变治法、时空权衡和迭代改进，几乎从不曾被看作重要的设计范例。
- 新分类法很自然地覆盖了许多传统方法无法分类的经典算法(欧几里得算法、堆排序、查找树、散列法、拓扑排序、高斯消去法、霍纳法则等，不胜枚举)。所以，新分类法能够以一种连贯的、一致的方式表达这些经典算法的标准内容。
- 新分类法很自然地容纳了某些设计技术的重要变种(例如，它能涵盖减治法的 3 个变种

① 译注：George Forsythe 认为，在这些工具当中，最重要的三项依次是自然语言、数学和计算机科学。

和变治法的 3 个变种)。

- 在分析算法效率时，新分类法与分析方法结合得更好(参见附录 B)。

设计技术作为问题求解的一般性策略

在本书中，主要将设计技术应用于计算机科学中的经典问题(这里唯一的创新是引入了一些数值算法的内容，我们也是用同样的通用框架来表述这些算法的)。但把这些设计技术看作问题求解的一般性工具时，它们的应用就不仅限于传统的计算问题和数学问题了。有两个因素令这一点变得尤其重要。第一，越来越多的计算类应用超越了它们的传统领域，并且有足够的理由使人相信，这种趋势会愈演愈烈。第二，人们渐渐认识到，提高学生们的问题求解能力是高等教育的一个主要目标。为了满足这个目标，在计算机科学课程体系中安排一门算法设计和分析课程是非常合适的，因为它会告诉学生如何应用一些特定的策略来解决问题。

虽然我并不建议将算法设计和分析课程变成一门教授一般性问题求解方法的课程，但我的确认为，我们不应错过算法设计和分析课程提供的这样一个独一无二的机会。为了这个目标，本书包含了一些和谜题相关的应用。虽然利用谜题来教授算法课程绝不是我的创新，但本书打算通过引进一些全新的谜题来系统地实现这个思路。

如何使用本书

我的目标是写一本既不泛泛而谈，又可供学生们独立阅读的教材。为了实现这个目标，本书做了如下努力。

- 根据 George Forsythe 的观点(参见引言)，我试图着重强调那些隐藏在算法设计和分析背后的主要思想。在选择特定的算法来阐述这些思想的时候，我并不倾向于涉及大量的算法，而是选择那些最能揭示其内在设计技术或分析方法的算法。幸运的是，大多数经典算法满足了这个要求。
- 第 2 章主要分析算法的效率，该章将分析非递归算法的方法和分析递归算法的典型方法区别开来。这一章还花了一些篇幅介绍算法经验分析和算法可视化。
- 书中系统地穿插着一些面向读者的提问。其中有些问题是经过精心设计的，而且答案紧随其后，目的是引起读者的注意或引发疑问。其余问题的用意是防止读者走马观花，不能充分理解本书的内容。
- 每一章结束时都会对本章最重要的概念和结论做一个总结。
- 本书包含 600 多道习题。有些习题是为了给大家练习，另外一些则是为了指出书中正文部分所涉及内容的重要意义，或是为了介绍一些书中没有涉及的算法。有一些习题利用了因特网上的资源。较难的习题数量不多，会在教师用书中用一种特殊的记号标注出来(因为有些学生可能没有勇气做那些标有难度的习题，所以本书没有对习题标注难度)。谜题类的习题用一种特殊的图标做标注。
- 本书所有的习题都附有提示。除了编程练习，习题的详细解法都能够在教师用书中找到，符合条件的教师可以填写书后的教师证明表，发传真到 010-62791865，以获得教师用书(也可联系培生公司的当地销售代表，或者访问 *www. pearsonhighered.com/irc*)。本书的任何读者都可以在 CS 支持网站 *http://cssupport.pearsoncmg.com* 上找到 PowerPoint 格式的幻灯片文件。

第 3 版的变化

第 3 版有若干变化。其中最重要的变化是介绍减治法和分治法的先后顺序。第 3 版会先介绍减治法，后介绍分治法，这样做有以下几个优点。

- 较之分治法，减治法更简单。
- 在求解问题方面，减治法应用更广。
- 这样的编排顺序便于先介绍插入排序，后介绍合并排序和快速排序。
- 数组划分的概念通过选择性问题引入，这次利用 Lomuto 算法的单向扫描来实现，而将 Hoare 划分方法的双向扫描留至后文与快速排序一并介绍。
- 折半查找归入介绍减常量算法的章节。

另一重要变化是重新编排第 8 章关于动态规划的内容，具体如下所述。

- 导述部分的内容是全新的。在前两版中用计算二项式系数的例子来引入动态规划这一重要技术，但在第 3 版中会介绍 3 个基础性示例，这样介绍的效果更好。
- 8.1 节的习题是全新的，包括一些在前两版中没有涉及的流行的应用。
- 第 8 章其他小节的顺序也做了调整，以便达到由浅入深、循序渐进的效果。

此外，还有其他一些变化。增加了不少与本书所述算法相关的应用。遍历图算法不再随减治法介绍，而是纳入蛮力算法和穷举查找的范畴，我认为这样更合理。在介绍生成组合对象的算法时，会新增格雷码算法。对求解最近对问题的分治法会有更深入的探讨。改进的内容包括算法可视化和求解旅行商问题的近似算法，当然参考文献也相应更新。

第 3 版新增约 70 道习题，其中涉及算法谜题和面试问题。

读者所需的知识背景

本书假定读者已经学习了离散数学的标准课程和一门基础性的编程课程。有了这样的知识背景，读者应该能够掌握本书的内容而不会遇到太大的困难。尽管如此，1.4 节、附录 A 和附录 B 仍然对基本的数据结构，必须用到的求和公式和递推关系分别进行了复习和回顾。只有 3 个小节(2.2 节、11.4 节和 12.4 节)会用到一些简单的微积分知识，如果读者缺少必要的微积分知识，完全可以跳过这 3 个涉及微积分的小节，这并不会妨碍对本书其余部分的理解。

进度安排

如果打算开设一门围绕算法设计技术来讲解算法设计和分析理论的基础课程，可以采用本书作为教材。但要想在一个学期内完成该课程，本书涵盖的内容可能过于丰富了。大体上来说，跳过第 3～12 章的部分内容不会影响读者对后面部分的理解。本书的任何一个部分都可以安排学生自学。尤其是 2.6 节和 2.7 节，它们分别介绍了经验分析和算法可视化，这两小节的内容可以结合练习[①]布置给学生。

① 译注："练习"的原文为"project"，一般应该翻译成"项目"，但国外一般将布置在课后完成的、较大型的、要求实际演练的习题称为 project，国内没有相应的称呼，所以姑且译为"练习"。

下面给出了一种针对一个学期课程的教学计划，这是按照 40 课时的集中教学来设计的。

课　次	主　题	小　节
1	课程简介	1.1～1.3
2，3	分析框架；O、Θ 和 Ω 符号	2.1，2.2
4	非递归算法的数学分析	2.3
5，6	递归算法的数学分析	2.4，2.5(+附录 B)
7	蛮力算法	3.1，3.2(+3.3)
8	穷举查找	3.4
9	深度优先查找和广度优先查找	3.5
10～11	减一算法：插入排序、拓扑排序	4.1，4.2
12	折半查找和其他减常量算法	4.4
13	减变量算法	4.5
14～15	分治法：合并排序、快速排序	5.1～5.2
16	其他分治法示例	5.3、5.4 或 5.5
16	减变量算法	5.6
17～19	实例化简：预排序、高斯消去法、平衡查找树	6.1～6.3
20	改变表现：堆和堆排序 或者霍纳法则和二进制幂	6.4 或 6.5
21	问题化简	6.6
22～24	时空权衡：串匹配、散列法、B 树	7.2～7.4
25～27	动态规划算法	8.1～8.4(选 3 节)
28～30	贪婪算法：Prim 算法、Kruskal 算法、Dijkstra 算法、哈夫曼算法	9.1～9.4
31～33	迭代改进算法	10.1～10.4(选 3 节)
34	下界的参数	11.1
35	决策树	11.2
36	*P*、*NP* 和 *NP* 完全问题	11.3
37	数值算法	11.4(+12.4)
38	回溯法	12.1
39	分支界限法	12.2
40	*NP* 困难问题的近似算法	12.3

Anany Levitin
anany.levitin@villanova.edu

Brief Contents

Contents

New to the Third Edition

- Reordering of chapters to introduce decrease-and-conquer before divide-and-conquer
- Restructuring of chapter 8 on dynamic programming, including all new introductory material and new exercises focusing on well-known applications
- More coverage of the applications of the algorithms discussed
- Reordering of select sections throughout the book to achieve a better alignment of specific algorithms and general algorithm design techniques
- Addition of the Lomuto partition and Gray code algorithms
- Seventy new problems added to the end-of-chapter exercises, including algorithmic puzzles and questions asked during job interviews

Preface

The most valuable acquisitions in a scientific or technical education are the general-purpose mental tools which remain serviceable for a life-time.
—George Forsythe, "What to do till the computer scientist comes." (1968)

Algorithms play the central role both in the science and practice of computing. Recognition of this fact has led to the appearance of a considerable number of textbooks on the subject. By and large, they follow one of two alternatives in presenting algorithms. One classifies algorithms according to a problem type. Such a book would have separate chapters on algorithms for sorting, searching, graphs, and so on. The advantage of this approach is that it allows an immediate comparison of, say, the efficiency of different algorithms for the same problem. The drawback of this approach is that it emphasizes problem types at the expense of algorithm design techniques.

The second alternative organizes the presentation around algorithm design techniques. In this organization, algorithms from different areas of computing are grouped together if they have the same design approach. I share the belief of many (e.g., [BaY95]) that this organization is more appropriate for a basic course on the design and analysis of algorithms. There are three principal reasons for emphasis on algorithm design techniques. First, these techniques provide a student with tools for designing algorithms for new problems. This makes learning algorithm design techniques a very valuable endeavor from a practical standpoint. Second, they seek to classify multitudes of known algorithms according to an underlying design idea. Learning to see such commonality among algorithms from different application areas should be a major goal of computer science education. After all, every science considers classification of its principal subject as a major if not the central point of its discipline. Third, in my opinion, algorithm design techniques have utility as general problem solving strategies, applicable to problems beyond computing.

Unfortunately, the traditional classification of algorithm design techniques has several serious shortcomings, from both theoretical and educational points of view. The most significant of these shortcomings is the failure to classify many important algorithms. This limitation has forced the authors of other textbooks to depart from the design technique organization and to include chapters dealing with specific problem types. Such a switch leads to a loss of course coherence and almost unavoidably creates a confusion in students' minds.

New taxonomy of algorithm design techniques

My frustration with the shortcomings of the traditional classification of algorithm design techniques has motivated me to develop a new taxonomy of them [Lev99], which is the basis of this book. Here are the principal advantages of the new taxonomy:

- The new taxonomy is more comprehensive than the traditional one. It includes several strategies—brute-force, decrease-and-conquer, transform-and-conquer, space and time trade-offs, and iterative improvement—that are rarely if ever recognized as important design paradigms.
- The new taxonomy covers naturally many classic algorithms (Euclid's algorithm, heapsort, search trees, hashing, topological sorting, Gaussian elimination, Horner's rule—to name a few) that the traditional taxonomy cannot classify. As a result, the new taxonomy makes it possible to present the standard body of classic algorithms in a unified and coherent fashion.
- It naturally accommodates the existence of important varieties of several design techniques. For example, it recognizes three variations of decrease-and-conquer and three variations of transform-and-conquer.
- It is better aligned with analytical methods for the efficiency analysis (see Appendix B).

Design techniques as general problem solving strategies

Most applications of the design techniques in the book are to classic problems of computer science. (The only innovation here is an inclusion of some material on numerical algorithms, which are covered within the same general framework.) But these design techniques can be considered general problem solving tools, whose applications are not limited to traditional computing and mathematical problems. Two factors make this point particularly important. First, more and more computing applications go beyond the traditional domain, and there are reasons to believe that this trend will strengthen in the future. Second, developing students' problem solving skills has come to be recognized as a major goal of college education. Among all the courses in a computer science curriculum, a course on the design and analysis of algorithms is uniquely suitable for this task because it can offer a student specific strategies for solving problems.

I am not proposing that a course on the design and analysis of algorithms should become a course on general problem solving. But I do believe that the unique opportunity provided by studying the design and analysis of algorithms should not be missed. Toward this goal, the book includes applications to puzzles and puzzle-like games. Although using puzzles in teaching algorithms is certainly not a new idea, the book tries to do this systematically by going well beyond a few standard examples.

Textbook pedagogy

My goal was to write a text that would not trivialize the subject but would still be readable by most students on their own. Here are some of the things done toward this objective.

- Sharing the opinion of George Forsythe expressed in the epigraph, I have sought to stress major ideas underlying the design and analysis of algorithms. In choosing specific algorithms to illustrate these ideas, I limited the number of covered algorithms to those that demonstrate an underlying design technique or an analysis method most clearly. Fortunately, most classic algorithms satisfy this criterion.
- In Chapter 2, which is devoted to efficiency analysis, the methods used for analyzing nonrecursive algorithms are separated from those typically used for analyzing recursive algorithms. The chapter also includes sections devoted to empirical analysis and algorithm visualization.
- The narrative is systematically interrupted by questions to the reader. Some of them are asked rhetorically, in anticipation of a concern or doubt, and are answered immediately. The goal of the others is to prevent the reader from drifting through the text without a satisfactory level of comprehension.
- Each chapter ends with a summary recapping the most important concepts and results discussed in the chapter.
- The book contains over 600 exercises. Some of them are drills; others make important points about the material covered in the body of the text or introduce algorithms not covered there at all. A few exercises take advantage of Internet resources. More difficult problems—there are not many of them—are marked by special symbols in the Instructor's Manual. (Because marking problems as difficult may discourage some students from trying to tackle them, problems are not marked in the book itself.) Puzzles, games, and puzzle-like questions are marked in the exercises with a special icon.
- The book provides hints to all the exercises. Detailed solutions, except for programming projects, are provided in the Instructor's Manual, available to qualified adopters through Pearson's Instructor Resource Center. (Please contact your local Pearson sales representative or go to www.pearsonhighered.com/irc to access this material.) Slides in PowerPoint are available to all readers of this book via anonymous ftp at the CS Support site: http://cssupport.pearsoncmg.com/.

Changes for the third edition

There are a few changes in the third edition. The most important is the new order of the chapters on decrease-and-conquer and divide-and-conquer. There are several advantages in introducing decrease-and-conquer before divide-and-conquer:

- Decrease-and-conquer is a simpler strategy than divide-and-conquer.
- Decrease-and-conquer is applicable to more problems than divide-and-conquer.
- The new order makes it possible to discuss insertion sort before mergesort and quicksort.
- The idea of array partitioning is now introduced in conjunction with the selection problem. I took advantage of an opportunity to do this via the one-directional scan employed by Lomuto's algorithm, leaving the two-directional scan used by Hoare's partitioning to a later discussion in conjunction with quicksort.
- Binary search is now considered in the section devoted to decrease-by-a-constant-factor algorithms, where it belongs.

The second important change is restructuring of Chapter 8 on dynamic programming. Specifically:

- The introductory section is completely new. It contains three basic examples that provide a much better introduction to this important technique than computing a binomial coefficient, the example used in the first two editions.
- All the exercises for Section 8.1 are new as well; they include well-known applications not available in the previous editions.
- I also changed the order of the other sections in this chapter to get a smoother progression from the simpler applications to the more advanced ones.

The other changes include the following. More applications of the algorithms discussed are included. The section on the graph-traversal algorithms is moved from the decrease-and-conquer chapter to the brute-force and exhaustive-search chapter, where it fits better, in my opinion. The Gray code algorithm is added to the section dealing with algorithms for generating combinatorial objects. The divide-and-conquer algorithm for the closest-pair problem is discussed in more detail. Updates include the section on algorithm visualization, approximation algorithms for the traveling salesman problem, and, of course, the bibliography.

I also added about 70 new problems to the exercises. Some of them are algorithmic puzzles and questions asked during job interviews.

Prerequisites

The book assumes that a reader has gone through an introductory programming course and a standard course on discrete structures. With such a background, he or she should be able to handle the book's material without undue difficulty.

Still, fundamental data structures, necessary summation formulas, and recurrence relations are reviewed in Section 1.4, Appendix A, and Appendix B, respectively. Calculus is used in only three sections (Section 2.2, 11.4, and 12.4), and to a very limited degree; if students lack calculus as an assured part of their background, the relevant portions of these three sections can be omitted without hindering their understanding of the rest of the material.

Use in the curriculum

The book can serve as a textbook for a basic course on design and analysis of algorithms organized around algorithm design techniques. It might contain slightly more material than can be covered in a typical one-semester course. By and large, portions of Chapters 3 through 12 can be skipped without the danger of making later parts of the book incomprehensible to the reader. Any portion of the book can be assigned for self-study. In particular, Sections 2.6 and 2.7 on empirical analysis and algorithm visualization, respectively, can be assigned in conjunction with projects.

Here is a possible plan for a one-semester course; it assumes a 40-class meeting format.

Lecture	Topic	Sections
1	Introduction	1.1–1.3
2, 3	Analysis framework; O, Ω, Θ notations	2.1, 2.2
4	Mathematical analysis of nonrecursive algorithms	2.3
5, 6	Mathematical analysis of recursive algorithms	2.4, 2.5 (+ App. B)
7	Brute-force algorithms	3.1, 3.2 (+ 3.3)
8	Exhaustive search	3.4
9	Depth-first search and breadth-first search	3.5
10, 11	Decrease-by-one: insertion sort, topological sorting	4.1, 4.2
12	Binary search and other decrease-by-a-constant-factor algorithms	4.4
13	Variable-size-decrease algorithms	4.5
14, 15	Divide-and-conquer: mergesort, quicksort	5.1–5.2
16	Other divide-and-conquer examples	5.3 or 5.4 or 5.5
17–19	Instance simplification: presorting, Gaussian elimination, balanced search trees	6.1–6.3
20	Representation change: heaps and heapsort or Horner's rule and binary exponentiation	6.4 or 6.5
21	Problem reduction	6.6
22–24	Space-time trade-offs: string matching, hashing, B-trees	7.2–7.4
25–27	Dynamic programming algorithms	3 from 8.1–8.4

28–30	Greedy algorithms: Prim's, Kruskal's, Dijkstra's, Huffman's	9.1–9.4
31–33	Iterative improvement algorithms	3 from 10.1–10.4
34	Lower-bound arguments	11.1
35	Decision trees	11.2
36	*P*, *NP*, and *NP*-complete problems	11.3
37	Numerical algorithms	11.4 (+ 12.4)
38	Backtracking	12.1
39	Branch-and-bound	12.2
40	Approximation algorithms for *NP*-hard problems	12.3

Acknowledgments

I would like to express my gratitude to the reviewers and many readers who have shared with me their opinions about the first two editions of the book and suggested improvements and corrections. The third edition has certainly benefited from the reviews by Andrew Harrington (Loyola University Chicago), David Levine (Saint Bonaventure University), Stefano Lombardi (UC Riverside), Daniel McKee (Mansfield University), Susan Brilliant (Virginia Commonwealth University), David Akers (University of Puget Sound), and two anonymous reviewers.

My thanks go to all the people at Pearson and their associates who worked on my book. I am especially grateful to my editor, Matt Goldstein; the editorial assistant, Chelsea Bell; the marketing manager, Yez Alayan; and the production supervisor, Kayla Smith-Tarbox. I am also grateful to Richard Camp for copyediting the book, Paul Anagnostopoulos of Windfall Software and Jacqui Scarlott for its project management and typesetting, and MaryEllen Oliver for proofreading the book.

Finally, I am indebted to two members of my family. Living with a spouse writing a book is probably more trying than doing the actual writing. My wife, Maria, lived through several years of this, helping me any way she could. And help she did: over 400 figures in the book and the Instructor's Manual were created by her. My daughter Miriam has been my English prose guru over many years. She read large portions of the book and was instrumental in finding the chapter epigraphs.

Anany Levitin
anany.levitin@villanova.edu
June 2011

1

Introduction

Two ideas lie gleaming on the jeweler's velvet. The first is the calculus, the second, the algorithm. The calculus and the rich body of mathematical analysis to which it gave rise made modern science possible; but it has been the algorithm that has made possible the modern world.

—David Berlinski, *The Advent of the Algorithm*, 2000

Why do you need to study algorithms? If you are going to be a computer professional, there are both practical and theoretical reasons to study algorithms. From a practical standpoint, you have to know a standard set of important algorithms from different areas of computing; in addition, you should be able to design new algorithms and analyze their efficiency. From the theoretical standpoint, the study of algorithms, sometimes called ***algorithmics***, has come to be recognized as the cornerstone of computer science. David Harel, in his delightful book pointedly titled *Algorithmics: the Spirit of Computing*, put it as follows:

> Algorithmics is more than a branch of computer science. It is the core of computer science, and, in all fairness, can be said to be relevant to most of science, business, and technology. [Har92, p. 6]

But even if you are not a student in a computing-related program, there are compelling reasons to study algorithms. To put it bluntly, computer programs would not exist without algorithms. And with computer applications becoming indispensable in almost all aspects of our professional and personal lives, studying algorithms becomes a necessity for more and more people.

Another reason for studying algorithms is their usefulness in developing analytical skills. After all, algorithms can be seen as special kinds of solutions to problems—not just answers but precisely defined procedures for getting answers. Consequently, specific algorithm design techniques can be interpreted as problem-solving strategies that can be useful regardless of whether a computer is involved. Of course, the precision inherently imposed by algorithmic thinking limits the kinds of problems that can be solved with an algorithm. You will not find, for example, an algorithm for living a happy life or becoming rich and famous. On the other hand, this required precision has an important educational advantage. Donald Knuth, one of the most prominent computer scientists in the history of algorithmics, put it as follows:

> A person well-trained in computer science knows how to deal with algorithms: how to construct them, manipulate them, understand them, analyze them. This knowledge is preparation for much more than writing good computer programs; it is a general-purpose mental tool that will be a definite aid to the understanding of other subjects, whether they be chemistry, linguistics, or music, etc. The reason for this may be understood in the following way: It has often been said that a person does not really understand something until after teaching it to someone else. Actually, a person does not *really* understand something until after teaching it to a *computer*, i.e., expressing it as an algorithm . . . An attempt to formalize things as algorithms leads to a much deeper understanding than if we simply try to comprehend things in the traditional way. [Knu96, p. 9]

We take up the notion of algorithm in Section 1.1. As examples, we use three algorithms for the same problem: computing the greatest common divisor. There are several reasons for this choice. First, it deals with a problem familiar to everybody from their middle-school days. Second, it makes the important point that the same problem can often be solved by several algorithms. Quite typically, these algorithms differ in their idea, level of sophistication, and efficiency. Third, one of these algorithms deserves to be introduced first, both because of its age—it appeared in Euclid's famous treatise more than two thousand years ago—and its enduring power and importance. Finally, investigation of these three algorithms leads to some general observations about several important properties of algorithms in general.

Section 1.2 deals with algorithmic problem solving. There we discuss several important issues related to the design and analysis of algorithms. The different aspects of algorithmic problem solving range from analysis of the problem and the means of expressing an algorithm to establishing its correctness and analyzing its efficiency. The section does not contain a magic recipe for designing an algorithm for an arbitrary problem. It is a well-established fact that such a recipe does not exist. Still, the material of Section 1.2 should be useful for organizing your work on designing and analyzing algorithms.

Section 1.3 is devoted to a few problem types that have proven to be particularly important to the study of algorithms and their application. In fact, there are textbooks (e.g., [Sed11]) organized around such problem types. I hold the view—shared by many others—that an organization based on algorithm design techniques is superior. In any case, it is very important to be aware of the principal problem types. Not only are they the most commonly encountered problem types in real-life applications, they are used throughout the book to demonstrate particular algorithm design techniques.

Section 1.4 contains a review of fundamental data structures. It is meant to serve as a reference rather than a deliberate discussion of this topic. If you need a more detailed exposition, there is a wealth of good books on the subject, most of them tailored to a particular programming language.

1.1 What Is an Algorithm?

Although there is no universally agreed-on wording to describe this notion, there is general agreement about what the concept means:

> An ***algorithm*** is a sequence of unambiguous instructions for solving a problem, i.e., for obtaining a required output for any legitimate input in a finite amount of time.

This definition can be illustrated by a simple diagram (Figure 1.1).

The reference to "instructions" in the definition implies that there is something or someone capable of understanding and following the instructions given. We call this a "computer," keeping in mind that before the electronic computer was invented, the word "computer" meant a human being involved in performing numeric calculations. Nowadays, of course, "computers" are those ubiquitous electronic devices that have become indispensable in almost everything we do. Note, however, that although the majority of algorithms are indeed intended for eventual computer implementation, the notion of algorithm does not depend on such an assumption.

As examples illustrating the notion of the algorithm, we consider in this section three methods for solving the same problem: computing the greatest common divisor of two integers. These examples will help us to illustrate several important points:

- The nonambiguity requirement for each step of an algorithm cannot be compromised.
- The range of inputs for which an algorithm works has to be specified carefully.
- The same algorithm can be represented in several different ways.
- There may exist several algorithms for solving the same problem.

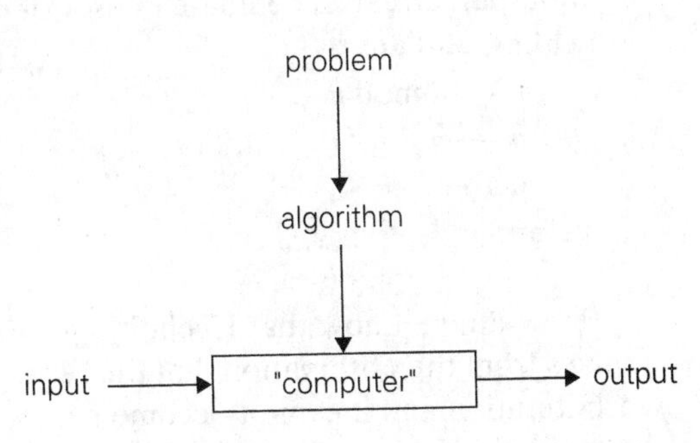

FIGURE 1.1 The notion of the algorithm.

- Algorithms for the same problem can be based on very different ideas and can solve the problem with dramatically different speeds.

Recall that the greatest common divisor of two nonnegative, not-both-zero integers m and n, denoted gcd(m, n), is defined as the largest integer that divides both m and n evenly, i.e., with a remainder of zero. Euclid of Alexandria (third century B.C.) outlined an algorithm for solving this problem in one of the volumes of his *Elements* most famous for its systematic exposition of geometry. In modern terms, ***Euclid's algorithm*** is based on applying repeatedly the equality

$$\gcd(m, n) = \gcd(n, m \bmod n),$$

where $m \bmod n$ is the remainder of the division of m by n, until $m \bmod n$ is equal to 0. Since $\gcd(m, 0) = m$ (why?), the last value of m is also the greatest common divisor of the initial m and n.

For example, gcd(60, 24) can be computed as follows:

$$\gcd(60, 24) = \gcd(24, 12) = \gcd(12, 0) = 12.$$

(If you are not impressed by this algorithm, try finding the greatest common divisor of larger numbers, such as those in Problem 6 in this section's exercises.)

Here is a more structured description of this algorithm:

Euclid's algorithm for computing gcd(m, n)

Step 1 If $n = 0$, return the value of m as the answer and stop; otherwise, proceed to Step 2.

Step 2 Divide m by n and assign the value of the remainder to r.

Step 3 Assign the value of n to m and the value of r to n. Go to Step 1.

Alternatively, we can express the same algorithm in pseudocode:

ALGORITHM *Euclid*(m, n)

```
//Computes gcd(m, n) by Euclid's algorithm
//Input: Two nonnegative, not-both-zero integers m and n
//Output: Greatest common divisor of m and n
while n ≠ 0 do
    r ← m mod n
    m ← n
    n ← r
return m
```

How do we know that Euclid's algorithm eventually comes to a stop? This follows from the observation that the second integer of the pair gets smaller with each iteration and it cannot become negative. Indeed, the new value of n on the next iteration is $m \bmod n$, which is always smaller than n (why?). Hence, the value of the second integer eventually becomes 0, and the algorithm stops.

Just as with many other problems, there are several algorithms for computing the greatest common divisor. Let us look at the other two methods for this problem. The first is simply based on the definition of the greatest common divisor of m and n as the largest integer that divides both numbers evenly. Obviously, such a common divisor cannot be greater than the smaller of these numbers, which we will denote by $t = \min\{m, n\}$. So we can start by checking whether t divides both m and n: if it does, t is the answer; if it does not, we simply decrease t by 1 and try again. (How do we know that the process will eventually stop?) For example, for numbers 60 and 24, the algorithm will try first 24, then 23, and so on, until it reaches 12, where it stops.

Consecutive integer checking algorithm for computing gcd(m, n)

Step 1 Assign the value of $\min\{m, n\}$ to t.

Step 2 Divide m by t. If the remainder of this division is 0, go to Step 3; otherwise, go to Step 4.

Step 3 Divide n by t. If the remainder of this division is 0, return the value of t as the answer and stop; otherwise, proceed to Step 4.

Step 4 Decrease the value of t by 1. Go to Step 2.

Note that unlike Euclid's algorithm, this algorithm, in the form presented, does not work correctly when one of its input numbers is zero. This example illustrates why it is so important to specify the set of an algorithm's inputs explicitly and carefully.

The third procedure for finding the greatest common divisor should be familiar to you from middle school.

Middle-school procedure for computing gcd(m, n)

Step 1 Find the prime factors of m.

Step 2 Find the prime factors of n.

Step 3 Identify all the common factors in the two prime expansions found in Step 1 and Step 2. (If p is a common factor occurring p_m and p_n times in m and n, respectively, it should be repeated $\min\{p_m, p_n\}$ times.)

Step 4 Compute the product of all the common factors and return it as the greatest common divisor of the numbers given.

Thus, for the numbers 60 and 24, we get

$$60 = 2 \cdot 2 \cdot 3 \cdot 5$$
$$24 = 2 \cdot 2 \cdot 2 \cdot 3$$
$$\gcd(60, 24) = 2 \cdot 2 \cdot 3 = 12.$$

Nostalgia for the days when we learned this method should not prevent us from noting that the last procedure is much more complex and slower than Euclid's algorithm. (We will discuss methods for finding and comparing running times of algorithms in the next chapter.) In addition to inferior efficiency, the middle-school procedure does not qualify, in the form presented, as a legitimate algorithm. Why? Because the prime factorization steps are not defined unambiguously: they require a list of prime numbers, and I strongly suspect that your middle-school math teacher did not explain how to obtain such a list. This is not a matter of unnecessary nitpicking. Unless this issue is resolved, we cannot, say, write a program implementing this procedure. Incidentally, Step 3 is also not defined clearly enough. Its ambiguity is much easier to rectify than that of the factorization steps, however. How would you find common elements in two sorted lists?

So, let us introduce a simple algorithm for generating consecutive primes not exceeding any given integer $n > 1$. It was probably invented in ancient Greece and is known as the ***sieve of Eratosthenes*** (ca. 200 B.C.). The algorithm starts by initializing a list of prime candidates with consecutive integers from 2 to n. Then, on its first iteration, the algorithm eliminates from the list all multiples of 2, i.e., 4, 6, and so on. Then it moves to the next item on the list, which is 3, and eliminates its multiples. (In this straightforward version, there is an overhead because some numbers, such as 6, are eliminated more than once.) No pass for number 4 is needed: since 4 itself and all its multiples are also multiples of 2, they were already eliminated on a previous pass. The next remaining number on the list, which is used on the third pass, is 5. The algorithm continues in this fashion until no more numbers can be eliminated from the list. The remaining integers of the list are the primes needed.

As an example, consider the application of the algorithm to finding the list of primes not exceeding $n = 25$:

2	3	4	5	6	7	8	9	10	11	12	13	14	15	16	17	18	19	20	21	22	23	24	25
2	3		5		7		9		11		13		15		17		19		21		23		25
2	**3**		5		7				11		13				17		19				23		25
2	3		**5**		7				11		13				17		19				23		

For this example, no more passes are needed because they would eliminate numbers already eliminated on previous iterations of the algorithm. The remaining numbers on the list are the consecutive primes less than or equal to 25.

What is the largest number p whose multiples can still remain on the list to make further iterations of the algorithm necessary? Before we answer this question, let us first note that if p is a number whose multiples are being eliminated on the current pass, then the first multiple we should consider is $p \cdot p$ because all its smaller multiples $2p, \ldots, (p-1)p$ have been eliminated on earlier passes through the list. This observation helps to avoid eliminating the same number more than once. Obviously, $p \cdot p$ should not be greater than n, and therefore p cannot exceed $\sqrt{n}$ rounded down (denoted $\lfloor \sqrt{n} \rfloor$ using the so-called ***floor function***). We assume in the following pseudocode that there is a function available for computing $\lfloor \sqrt{n} \rfloor$ alternatively, we could check the inequality $p \cdot p \le n$ as the loop continuation condition there.

ALGORITHM *Sieve*(n)

//Implements the sieve of Eratosthenes
//Input: A positive integer $n > 1$
//Output: Array L of all prime numbers less than or equal to n

```
for p ← 2 to n do A[p] ← p
for p ← 2 to ⌊√n⌋ do    //see note before pseudocode
    if A[p] ≠ 0          //p hasn't been eliminated on previous passes
        j ← p * p
        while j ≤ n do
            A[j] ← 0     //mark element as eliminated
            j ← j + p
//copy the remaining elements of A to array L of the primes
i ← 0
for p ← 2 to n do
    if A[p] ≠ 0
        L[i] ← A[p]
        i ← i + 1
return L
```

So now we can incorporate the sieve of Eratosthenes into the middle-school procedure to get a legitimate algorithm for computing the greatest common divisor of two positive integers. Note that special care needs to be exercised if one or both input numbers are equal to 1: because mathematicians do not consider 1 to be a prime number, strictly speaking, the method does not work for such inputs.

Before we leave this section, one more comment is in order. The examples considered in this section notwithstanding, the majority of algorithms in use today—even those that are implemented as computer programs—do not deal with mathematical problems. Look around for algorithms helping us through our daily routines, both professional and personal. May this ubiquity of algorithms in today's world strengthen your resolve to learn more about these fascinating engines of the information age.

Exercises 1.1

1. Do some research on al-Khorezmi (also al-Khwarizmi), the man from whose name the word "algorithm" is derived. In particular, you should learn what the origins of the words "algorithm" and "algebra" have in common.
2. Given that the official purpose of the U.S. patent system is the promotion of the "useful arts," do you think algorithms are patentable in this country? Should they be?
3. **a.** Write down driving directions for going from your school to your home with the precision required from an algorithm's description.

   b. Write down a recipe for cooking your favorite dish with the precision required by an algorithm.
4. Design an algorithm for computing $\lfloor\sqrt{n}\rfloor$ for any positive integer n. Besides assignment and comparison, your algorithm may only use the four basic arithmetical operations.

5. Design an algorithm to find all the common elements in two sorted lists of numbers. For example, for the lists 2, 5, 5, 5 and 2, 2, 3, 5, 5, 7, the output should be 2, 5, 5. What is the maximum number of comparisons your algorithm makes if the lengths of the two given lists are m and n, respectively?

6. **a.** Find gcd(31415, 14142) by applying Euclid's algorithm.
   b. Estimate how many times faster it will be to find gcd(31415, 14142) by Euclid's algorithm compared with the algorithm based on checking consecutive integers from $\min\{m, n\}$ down to $\gcd(m, n)$.

7. Prove the equality $\gcd(m, n) = \gcd(n, m \bmod n)$ for every pair of positive integers m and n.

8. What does Euclid's algorithm do for a pair of integers in which the first is smaller than the second? What is the maximum number of times this can happen during the algorithm's execution on such an input?

9. **a.** What is the minimum number of divisions made by Euclid's algorithm among all inputs $1 \le m, n \le 10$?
   b. What is the maximum number of divisions made by Euclid's algorithm among all inputs $1 \le m, n \le 10$?

10. **a.** Euclid's algorithm, as presented in Euclid's treatise, uses subtractions rather than integer divisions. Write pseudocode for this version of Euclid's algorithm.

   b. *Euclid's game* (see [Bog]) starts with two unequal positive integers on the board. Two players move in turn. On each move, a player has to write on the board a positive number equal to the difference of two numbers already on the board; this number must be new, i.e., different from all the numbers already on the board. The player who cannot move loses the game. Should you choose to move first or second in this game?

11. The ***extended Euclid's algorithm*** determines not only the greatest common divisor d of two positive integers m and n but also integers (not necessarily positive) x and y, such that $mx + ny = d$.
   a. Look up a description of the extended Euclid's algorithm (see, e.g., [KnuI, p. 13]) and implement it in the language of your choice.
   b. Modify your program to find integer solutions to the Diophantine equation $ax + by = c$ with any set of integer coefficients a, b, and c.

12. *Locker doors* There are n lockers in a hallway, numbered sequentially from 1 to n. Initially, all the locker doors are closed. You make n passes by the lockers, each time starting with locker #1. On the ith pass, $i = 1, 2, \ldots, n$, you toggle the door of every ith locker: if the door is closed, you open it; if it is open, you close it. After the last pass, which locker doors are open and which are closed? How many of them are open?

1.2 Fundamentals of Algorithmic Problem Solving

Let us start by reiterating an important point made in the introduction to this chapter:

> We can consider algorithms to be procedural solutions to problems.

These solutions are not answers but specific instructions for getting answers. It is this emphasis on precisely defined constructive procedures that makes computer science distinct from other disciplines. In particular, this distinguishes it from theoretical mathematics, whose practitioners are typically satisfied with just proving the existence of a solution to a problem and, possibly, investigating the solution's properties.

We now list and briefly discuss a sequence of steps one typically goes through in designing and analyzing an algorithm (Figure 1.2).

Understanding the Problem

From a practical perspective, the first thing you need to do before designing an algorithm is to understand completely the problem given. Read the problem's description carefully and ask questions if you have any doubts about the problem, do a few small examples by hand, think about special cases, and ask questions again if needed.

There are a few types of problems that arise in computing applications quite often. We review them in the next section. If the problem in question is one of them, you might be able to use a known algorithm for solving it. Of course, it helps to understand how such an algorithm works and to know its strengths and weaknesses, especially if you have to choose among several available algorithms. But often you will not find a readily available algorithm and will have to design your own. The sequence of steps outlined in this section should help you in this exciting but not always easy task.

An input to an algorithm specifies an ***instance*** of the problem the algorithm solves. It is very important to specify exactly the set of instances the algorithm needs to handle. (As an example, recall the variations in the set of instances for the three greatest common divisor algorithms discussed in the previous section.) If you fail to do this, your algorithm may work correctly for a majority of inputs but crash on some "boundary" value. Remember that a correct algorithm is not one that works most of the time, but one that works correctly for *all* legitimate inputs.

Do not skimp on this first step of the algorithmic problem-solving process; otherwise, you will run the risk of unnecessary rework.

Ascertaining the Capabilities of the Computational Device

Once you completely understand a problem, you need to ascertain the capabilities of the computational device the algorithm is intended for. The vast majority of

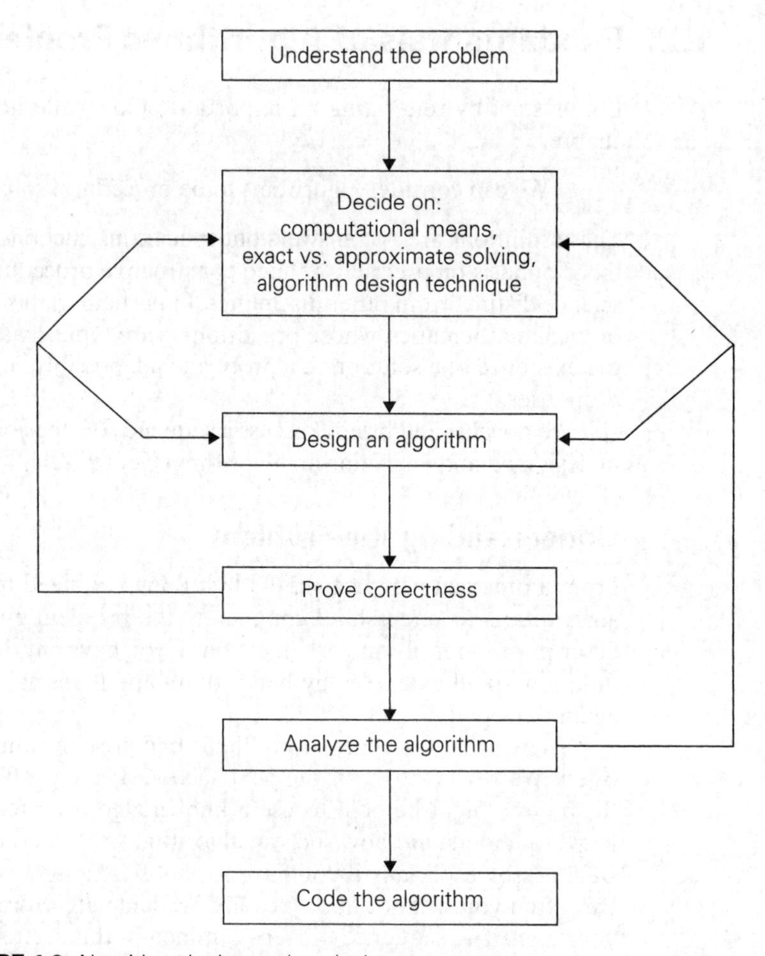

FIGURE 1.2 Algorithm design and analysis process.

algorithms in use today are still destined to be programmed for a computer closely resembling the von Neumann machine—a computer architecture outlined by the prominent Hungarian-American mathematician John von Neumann (1903–1957), in collaboration with A. Burks and H. Goldstine, in 1946. The essence of this architecture is captured by the so-called ***random-access machine*** (***RAM***). Its central assumption is that instructions are executed one after another, one operation at a time. Accordingly, algorithms designed to be executed on such machines are called ***sequential algorithms***.

The central assumption of the RAM model does not hold for some newer computers that can execute operations concurrently, i.e., in parallel. Algorithms that take advantage of this capability are called ***parallel algorithms***. Still, studying the classic techniques for design and analysis of algorithms under the RAM model remains the cornerstone of algorithmics for the foreseeable future.

Should you worry about the speed and amount of memory of a computer at your disposal? If you are designing an algorithm as a scientific exercise, the answer is a qualified no. As you will see in Section 2.1, most computer scientists prefer to study algorithms in terms independent of specification parameters for a particular computer. If you are designing an algorithm as a practical tool, the answer may depend on a problem you need to solve. Even the "slow" computers of today are almost unimaginably fast. Consequently, in many situations you need not worry about a computer being too slow for the task. There are important problems, however, that are very complex by their nature, or have to process huge volumes of data, or deal with applications where the time is critical. In such situations, it is imperative to be aware of the speed and memory available on a particular computer system.

Choosing between Exact and Approximate Problem Solving

The next principal decision is to choose between solving the problem exactly or solving it approximately. In the former case, an algorithm is called an ***exact algorithm***; in the latter case, an algorithm is called an ***approximation algorithm***. Why would one opt for an approximation algorithm? First, there are important problems that simply cannot be solved exactly for most of their instances; examples include extracting square roots, solving nonlinear equations, and evaluating definite integrals. Second, available algorithms for solving a problem exactly can be unacceptably slow because of the problem's intrinsic complexity. This happens, in particular, for many problems involving a very large number of choices; you will see examples of such difficult problems in Chapters 3, 11, and 12. Third, an approximation algorithm can be a part of a more sophisticated algorithm that solves a problem exactly.

Algorithm Design Techniques

Now, with all the components of the algorithmic problem solving in place, how do you design an algorithm to solve a given problem? This is the main question this book seeks to answer by teaching you several general design techniques.

What is an algorithm design technique?

> An ***algorithm design technique*** (or "strategy" or "paradigm") is a general approach to solving problems algorithmically that is applicable to a variety of problems from different areas of computing.

Check this book's table of contents and you will see that a majority of its chapters are devoted to individual design techniques. They distill a few key ideas that have proven to be useful in designing algorithms. Learning these techniques is of utmost importance for the following reasons.

First, they provide guidance for designing algorithms for new problems, i.e., problems for which there is no known satisfactory algorithm. Therefore—to use the language of a famous proverb—learning such techniques is akin to learning to fish as opposed to being given a fish caught by somebody else. It is not true, of course, that each of these general techniques will be necessarily applicable to every problem you may encounter. But taken together, they do constitute a powerful collection of tools that you will find quite handy in your studies and work.

Second, algorithms are the cornerstone of computer science. Every science is interested in classifying its principal subject, and computer science is no exception. Algorithm design techniques make it possible to classify algorithms according to an underlying design idea; therefore, they can serve as a natural way to both categorize and study algorithms.

Designing an Algorithm and Data Structures

While the algorithm design techniques do provide a powerful set of general approaches to algorithmic problem solving, designing an algorithm for a particular problem may still be a challenging task. Some design techniques can be simply inapplicable to the problem in question. Sometimes, several techniques need to be combined, and there are algorithms that are hard to pinpoint as applications of the known design techniques. Even when a particular design technique is applicable, getting an algorithm often requires a nontrivial ingenuity on the part of the algorithm designer. With practice, both tasks—choosing among the general techniques and applying them—get easier, but they are rarely easy.

Of course, one should pay close attention to choosing data structures appropriate for the operations performed by the algorithm. For example, the sieve of Eratosthenes introduced in Section 1.1 would run longer if we used a linked list instead of an array in its implementation (why?). Also note that some of the algorithm design techniques discussed in Chapters 6 and 7 depend intimately on structuring or restructuring data specifying a problem's instance. Many years ago, an influential textbook proclaimed the fundamental importance of both algorithms and data structures for computer programming by its very title: *Algorithms + Data Structures = Programs* [Wir76]. In the new world of object-oriented programming, data structures remain crucially important for both design and analysis of algorithms. We review basic data structures in Section 1.4.

Methods of Specifying an Algorithm

Once you have designed an algorithm, you need to specify it in some fashion. In Section 1.1, to give you an example, Euclid's algorithm is described in words (in a free and also a step-by-step form) and in pseudocode. These are the two options that are most widely used nowadays for specifying algorithms.

Using a natural language has an obvious appeal; however, the inherent ambiguity of any natural language makes a succinct and clear description of algorithms surprisingly difficult. Nevertheless, being able to do this is an important skill that you should strive to develop in the process of learning algorithms.

Pseudocode is a mixture of a natural language and programming language-like constructs. Pseudocode is usually more precise than natural language, and its usage often yields more succinct algorithm descriptions. Surprisingly, computer scientists have never agreed on a single form of pseudocode, leaving textbook authors with a need to design their own "dialects." Fortunately, these dialects are so close to each other that anyone familiar with a modern programming language should be able to understand them all.

This book's dialect was selected to cause minimal difficulty for a reader. For the sake of simplicity, we omit declarations of variables and use indentation to show the scope of such statements as **for**, **if**, and **while**. As you saw in the previous section, we use an arrow "$\leftarrow$" for the assignment operation and two slashes "//" for comments.

In the earlier days of computing, the dominant vehicle for specifying algorithms was a ***flowchart***, a method of expressing an algorithm by a collection of connected geometric shapes containing descriptions of the algorithm's steps. This representation technique has proved to be inconvenient for all but very simple algorithms; nowadays, it can be found only in old algorithm books.

The state of the art of computing has not yet reached a point where an algorithm's description—be it in a natural language or pseudocode—can be fed into an electronic computer directly. Instead, it needs to be converted into a computer program written in a particular computer language. We can look at such a program as yet another way of specifying the algorithm, although it is preferable to consider it as the algorithm's implementation.

Proving an Algorithm's Correctness

Once an algorithm has been specified, you have to prove its ***correctness***. That is, you have to prove that the algorithm yields a required result for every legitimate input in a finite amount of time. For example, the correctness of Euclid's algorithm for computing the greatest common divisor stems from the correctness of the equality $\gcd(m, n) = \gcd(n, m \bmod n)$ (which, in turn, needs a proof; see Problem 7 in Exercises 1.1), the simple observation that the second integer gets smaller on every iteration of the algorithm, and the fact that the algorithm stops when the second integer becomes 0.

For some algorithms, a proof of correctness is quite easy; for others, it can be quite complex. A common technique for proving correctness is to use mathematical induction because an algorithm's iterations provide a natural sequence of steps needed for such proofs. It might be worth mentioning that although tracing the algorithm's performance for a few specific inputs can be a very worthwhile activity, it cannot prove the algorithm's correctness conclusively. But in order to show that an algorithm is incorrect, you need just one instance of its input for which the algorithm fails.

The notion of correctness for approximation algorithms is less straightforward than it is for exact algorithms. For an approximation algorithm, we usually would like to be able to show that the error produced by the algorithm does not exceed a predefined limit. You can find examples of such investigations in Chapter 12.

Analyzing an Algorithm

We usually want our algorithms to possess several qualities. After correctness, by far the most important is ***efficiency***. In fact, there are two kinds of algorithm efficiency: ***time efficiency***, indicating how fast the algorithm runs, and ***space efficiency***, indicating how much extra memory it uses. A general framework and specific techniques for analyzing an algorithm's efficiency appear in Chapter 2.

Another desirable characteristic of an algorithm is ***simplicity***. Unlike efficiency, which can be precisely defined and investigated with mathematical rigor, simplicity, like beauty, is to a considerable degree in the eye of the beholder. For example, most people would agree that Euclid's algorithm is simpler than the middle-school procedure for computing gcd(m, n), but it is not clear whether Euclid's algorithm is simpler than the consecutive integer checking algorithm. Still, simplicity is an important algorithm characteristic to strive for. Why? Because simpler algorithms are easier to understand and easier to program; consequently, the resulting programs usually contain fewer bugs. There is also the undeniable aesthetic appeal of simplicity. Sometimes simpler algorithms are also more efficient than more complicated alternatives. Unfortunately, it is not always true, in which case a judicious compromise needs to be made.

Yet another desirable characteristic of an algorithm is ***generality***. There are, in fact, two issues here: generality of the problem the algorithm solves and the set of inputs it accepts. On the first issue, note that it is sometimes easier to design an algorithm for a problem posed in more general terms. Consider, for example, the problem of determining whether two integers are relatively prime, i.e., whether their only common divisor is equal to 1. It is easier to design an algorithm for a more general problem of computing the greatest common divisor of two integers and, to solve the former problem, check whether the gcd is 1 or not. There are situations, however, where designing a more general algorithm is unnecessary or difficult or even impossible. For example, it is unnecessary to sort a list of n numbers to find its median, which is its $\lceil n/2 \rceil$th smallest element. To give another example, the standard formula for roots of a quadratic equation cannot be generalized to handle polynomials of arbitrary degrees.

As to the set of inputs, your main concern should be designing an algorithm that can handle a set of inputs that is natural for the problem at hand. For example, excluding integers equal to 1 as possible inputs for a greatest common divisor algorithm would be quite unnatural. On the other hand, although the standard formula for the roots of a quadratic equation holds for complex coefficients, we would normally not implement it on this level of generality unless this capability is explicitly required.

If you are not satisfied with the algorithm's efficiency, simplicity, or generality, you must return to the drawing board and redesign the algorithm. In fact, even if your evaluation is positive, it is still worth searching for other algorithmic solutions. Recall the three different algorithms in the previous section for computing the greatest common divisor: generally, you should not expect to get the best algorithm on the first try. At the very least, you should try to fine-tune the algorithm you already have. For example, we made several improvements in our implementation of the sieve of Eratosthenes compared with its initial outline in Section 1.1. (Can you identify them?) You will do well if you keep in mind the following observation of Antoine de Saint-Exupéry, the French writer, pilot, and aircraft designer: "A designer knows he has arrived at perfection not when there is no longer anything to add, but when there is no longer anything to take away."[1]

Coding an Algorithm

Most algorithms are destined to be ultimately implemented as computer programs. Programming an algorithm presents both a peril and an opportunity. The peril lies in the possibility of making the transition from an algorithm to a program either incorrectly or very inefficiently. Some influential computer scientists strongly believe that unless the correctness of a computer program is proven with full mathematical rigor, the program cannot be considered correct. They have developed special techniques for doing such proofs (see [Gri81]), but the power of these techniques of formal verification is limited so far to very small programs.

As a practical matter, the validity of programs is still established by testing. Testing of computer programs is an art rather than a science, but that does not mean that there is nothing in it to learn. Look up books devoted to testing and debugging; even more important, test and debug your program thoroughly whenever you implement an algorithm.

Also note that throughout the book, we assume that inputs to algorithms belong to the specified sets and hence require no verification. When implementing algorithms as programs to be used in actual applications, you should provide such verifications.

Of course, implementing an algorithm correctly is necessary but not sufficient: you would not like to diminish your algorithm's power by an inefficient implementation. Modern compilers do provide a certain safety net in this regard, especially when they are used in their code optimization mode. Still, you need to be aware of such standard tricks as computing a loop's invariant (an expression that does not change its value) outside the loop, collecting common subexpressions, replacing expensive operations by cheap ones, and so on. (See [Ker99] and [Ben00] for a good discussion of code tuning and other issues related to algorithm programming.) Typically, such improvements can speed up a program only by a constant factor, whereas a better algorithm can make a difference in running time by orders of magnitude. But once an algorithm is selected, a 10–50% speedup may be worth an effort.

---

1. I found this call for design simplicity in an essay collection by Jon Bentley [Ben00]; the essays deal with a variety of issues in algorithm design and implementation and are justifiably titled *Programming Pearls*. I wholeheartedly recommend the writings of both Jon Bentley and Antoine de Saint-Exupéry.

A working program provides an additional opportunity in allowing an empirical analysis of the underlying algorithm. Such an analysis is based on timing the program on several inputs and then analyzing the results obtained. We discuss the advantages and disadvantages of this approach to analyzing algorithms in Section 2.6.

In conclusion, let us emphasize again the main lesson of the process depicted in Figure 1.2:

> As a rule, a good algorithm is a result of repeated effort and rework.

Even if you have been fortunate enough to get an algorithmic idea that seems perfect, you should still try to see whether it can be improved.

Actually, this is good news since it makes the ultimate result so much more enjoyable. (Yes, I did think of naming this book *The Joy of Algorithms.*) On the other hand, how does one know when to stop? In the real world, more often than not a project's schedule or the impatience of your boss will stop you. And so it should be: perfection is expensive and in fact not always called for. Designing an algorithm is an engineering-like activity that calls for compromises among competing goals under the constraints of available resources, with the designer's time being one of the resources.

In the academic world, the question leads to an interesting but usually difficult investigation of an algorithm's ***optimality***. Actually, this question is not about the efficiency of an algorithm but about the complexity of the problem it solves: What is the minimum amount of effort *any* algorithm will need to exert to solve the problem? For some problems, the answer to this question is known. For example, any algorithm that sorts an array by comparing values of its elements needs about $n \log_2 n$ comparisons for some arrays of size n (see Section 11.2). But for many seemingly easy problems such as integer multiplication, computer scientists do not yet have a final answer.

Another important issue of algorithmic problem solving is the question of whether or not every problem can be solved by an algorithm. We are not talking here about problems that do not have a solution, such as finding real roots of a quadratic equation with a negative discriminant. For such cases, an output indicating that the problem does not have a solution is all we can and should expect from an algorithm. Nor are we talking about ambiguously stated problems. Even some unambiguous problems that must have a simple yes or no answer are "undecidable," i.e., unsolvable by any algorithm. An important example of such a problem appears in Section 11.3. Fortunately, a vast majority of problems in practical computing *can* be solved by an algorithm.

Before leaving this section, let us be sure that you do not have the misconception—possibly caused by the somewhat mechanical nature of the diagram of Figure 1.2—that designing an algorithm is a dull activity. There is nothing further from the truth: inventing (or discovering?) algorithms is a very creative and rewarding process. This book is designed to convince you that this is the case.

Exercises 1.2

1. *Old World puzzle* A peasant finds himself on a riverbank with a wolf, a goat, and a head of cabbage. He needs to transport all three to the other side of the river in his boat. However, the boat has room for only the peasant himself and one other item (either the wolf, the goat, or the cabbage). In his absence, the wolf would eat the goat, and the goat would eat the cabbage. Solve this problem for the peasant or prove it has no solution. (Note: The peasant is a vegetarian but does not like cabbage and hence can eat neither the goat nor the cabbage to help him solve the problem. And it goes without saying that the wolf is a protected species.)

2. *New World puzzle* There are four people who want to cross a rickety bridge; they all begin on the same side. You have 17 minutes to get them all across to the other side. It is night, and they have one flashlight. A maximum of two people can cross the bridge at one time. Any party that crosses, either one or two people, must have the flashlight with them. The flashlight must be walked back and forth; it cannot be thrown, for example. Person 1 takes 1 minute to cross the bridge, person 2 takes 2 minutes, person 3 takes 5 minutes, and person 4 takes 10 minutes. A pair must walk together at the rate of the slower person's pace. (Note: According to a rumor on the Internet, interviewers at a well-known software company located near Seattle have given this problem to interviewees.)

3. Which of the following formulas can be considered an algorithm for computing the area of a triangle whose side lengths are given positive numbers a, b, and c?
   a. $S = \sqrt{p(p-a)(p-b)(p-c)}$, where $p = (a+b+c)/2$
   b. $S = \frac{1}{2}bc \sin A$, where A is the angle between sides b and c
   c. $S = \frac{1}{2}ah_a$, where h_a is the height to base a

4. Write pseudocode for an algorithm for finding real roots of equation $ax^2 + bx + c = 0$ for arbitrary real coefficients a, b, and c. (You may assume the availability of the square root function $sqrt(x)$.)

5. Describe the standard algorithm for finding the binary representation of a positive decimal integer
   a. in English.
   b. in pseudocode.

6. Describe the algorithm used by your favorite ATM machine in dispensing cash. (You may give your description in either English or pseudocode, whichever you find more convenient.)

7. **a.** Can the problem of computing the number π be solved exactly?
   b. How many instances does this problem have?
   c. Look up an algorithm for this problem on the Internet.

8. Give an example of a problem other than computing the greatest common divisor for which you know more than one algorithm. Which of them is simpler? Which is more efficient?

9. Consider the following algorithm for finding the distance between the two closest elements in an array of numbers.

ALGORITHM *MinDistance*($A[0..n-1]$)
//Input: Array $A[0..n-1]$ of numbers
//Output: Minimum distance between two of its elements
$dmin \leftarrow \infty$
for $i \leftarrow 0$ **to** $n-1$ **do**
for $j \leftarrow 0$ **to** $n-1$ **do**
if $i \neq j$ **and** $|A[i] - A[j]| < dmin$
$dmin \leftarrow |A[i] - A[j]|$
return *dmin*

Make as many improvements as you can in this algorithmic solution to the problem. If you need to, you may change the algorithm altogether; if not, improve the implementation given.

10. One of the most influential books on problem solving, titled *How To Solve It* [Pol57], was written by the Hungarian-American mathematician George Pólya (1887–1985). Pólya summarized his ideas in a four-point summary. Find this summary on the Internet or, better yet, in his book, and compare it with the plan outlined in Section 1.2. What do they have in common? How are they different?

1.3 Important Problem Types

In the limitless sea of problems one encounters in computing, there are a few areas that have attracted particular attention from researchers. By and large, their interest has been driven either by the problem's practical importance or by some specific characteristics making the problem an interesting research subject; fortunately, these two motivating forces reinforce each other in most cases.

In this section, we are going to introduce the most important problem types:

- Sorting
- Searching
- String processing
- Graph problems
- Combinatorial problems
- Geometric problems
- Numerical problems

These problems are used in subsequent chapters of the book to illustrate different algorithm design techniques and methods of algorithm analysis.

Sorting

The ***sorting problem*** is to rearrange the items of a given list in nondecreasing order. Of course, for this problem to be meaningful, the nature of the list items must allow such an ordering. (Mathematicians would say that there must exist a relation of total ordering.) As a practical matter, we usually need to sort lists of numbers, characters from an alphabet, character strings, and, most important, records similar to those maintained by schools about their students, libraries about their holdings, and companies about their employees. In the case of records, we need to choose a piece of information to guide sorting. For example, we can choose to sort student records in alphabetical order of names or by student number or by student grade-point average. Such a specially chosen piece of information is called a ***key***. Computer scientists often talk about sorting a list of keys even when the list's items are not records but, say, just integers.

Why would we want a sorted list? To begin with, a sorted list can be a required output of a task such as ranking Internet search results or ranking students by their GPA scores. Further, sorting makes many questions about the list easier to answer. The most important of them is searching: it is why dictionaries, telephone books, class lists, and so on are sorted. You will see other examples of the usefulness of list presorting in Section 6.1. In a similar vein, sorting is used as an auxiliary step in several important algorithms in other areas, e.g., geometric algorithms and data compression. The greedy approach—an important algorithm design technique discussed later in the book—requires a sorted input.

By now, computer scientists have discovered dozens of different sorting algorithms. In fact, inventing a new sorting algorithm has been likened to designing the proverbial mousetrap. And I am happy to report that the hunt for a better sorting mousetrap continues. This perseverance is admirable in view of the following facts. On the one hand, there are a few good sorting algorithms that sort an arbitrary array of size n using about $n \log_2 n$ comparisons. On the other hand, no algorithm that sorts by key comparisons (as opposed to, say, comparing small pieces of keys) can do substantially better than that.

There is a reason for this embarrassment of algorithmic riches in the land of sorting. Although some algorithms are indeed better than others, there is no algorithm that would be the best solution in all situations. Some of the algorithms are simple but relatively slow, while others are faster but more complex; some work better on randomly ordered inputs, while others do better on almost-sorted lists; some are suitable only for lists residing in the fast memory, while others can be adapted for sorting large files stored on a disk; and so on.

Two properties of sorting algorithms deserve special mention. A sorting algorithm is called ***stable*** if it preserves the relative order of any two equal elements in its input. In other words, if an input list contains two equal elements in positions i and j where $i < j$, then in the sorted list they have to be in positions i' and j', respectively, such that $i' < j'$. This property can be desirable if, for example, we have a list of students sorted alphabetically and we want to sort it according to student GPA: a stable algorithm will yield a list in which students with the same GPA will still be sorted alphabetically. Generally speaking, algorithms that can exchange keys located far apart are not stable, but they usually work faster; you will see how this general comment applies to important sorting algorithms later in the book.

The second notable feature of a sorting algorithm is the amount of extra memory the algorithm requires. An algorithm is said to be ***in-place*** if it does not require extra memory, except, possibly, for a few memory units. There are important sorting algorithms that are in-place and those that are not.

Searching

The ***searching problem*** deals with finding a given value, called a ***search key***, in a given set (or a multiset, which permits several elements to have the same value). There are plenty of searching algorithms to choose from. They range from the straightforward sequential search to a spectacularly efficient but limited binary search and algorithms based on representing the underlying set in a different form more conducive to searching. The latter algorithms are of particular importance for real-world applications because they are indispensable for storing and retrieving information from large databases.

For searching, too, there is no single algorithm that fits all situations best. Some algorithms work faster than others but require more memory; some are very fast but applicable only to sorted arrays; and so on. Unlike with sorting algorithms, there is no stability problem, but different issues arise. Specifically, in applications where the underlying data may change frequently relative to the number of searches, searching has to be considered in conjunction with two other operations: an addition to and deletion from the data set of an item. In such situations, data structures and algorithms should be chosen to strike a balance among the requirements of each operation. Also, organizing very large data sets for efficient searching poses special challenges with important implications for real-world applications.

String Processing

In recent decades, the rapid proliferation of applications dealing with nonnumerical data has intensified the interest of researchers and computing practitioners in string-handling algorithms. A ***string*** is a sequence of characters from an alphabet. Strings of particular interest are text strings, which comprise letters, numbers, and special characters; bit strings, which comprise zeros and ones; and gene sequences, which can be modeled by strings of characters from the four-character alphabet {A, C, G, T}. It should be pointed out, however, that string-processing algorithms have been important for computer science for a long time in conjunction with computer languages and compiling issues.

One particular problem—that of searching for a given word in a text—has attracted special attention from researchers. They call it ***string matching***. Several algorithms that exploit the special nature of this type of searching have been invented. We introduce one very simple algorithm in Chapter 3 and discuss two algorithms based on a remarkable idea by R. Boyer and J. Moore in Chapter 7.

Graph Problems

One of the oldest and most interesting areas in algorithmics is graph algorithms. Informally, a ***graph*** can be thought of as a collection of points called vertices, some of which are connected by line segments called edges. (A more formal definition is given in the next section.) Graphs are an interesting subject to study, for both theoretical and practical reasons. Graphs can be used for modeling a wide variety of applications, including transportation, communication, social and economic networks, project scheduling, and games. Studying different technical and social aspects of the Internet in particular is one of the active areas of current research involving computer scientists, economists, and social scientists (see, e.g., [Eas10]).

Basic graph algorithms include graph-traversal algorithms (how can one reach all the points in a network?), shortest-path algorithms (what is the best route between two cities?), and topological sorting for graphs with directed edges (is a set of courses with their prerequisites consistent or self-contradictory?). Fortunately, these algorithms can be considered illustrations of general design techniques; accordingly, you will find them in corresponding chapters of the book.

Some graph problems are computationally very hard; the most well-known examples are the traveling salesman problem and the graph-coloring problem. The ***traveling salesman problem (TSP)*** is the problem of finding the shortest tour through n cities that visits every city exactly once. In addition to obvious applications involving route planning, it arises in such modern applications as circuit board and VLSI chip fabrication, X-ray crystallography, and genetic engineering. The ***graph-coloring problem*** seeks to assign the smallest number of colors to the vertices of a graph so that no two adjacent vertices are the same color. This problem arises in several applications, such as event scheduling: if the events are represented by vertices that are connected by an edge if and only if the corresponding events cannot be scheduled at the same time, a solution to the graph-coloring problem yields an optimal schedule.

Combinatorial Problems

From a more abstract perspective, the traveling salesman problem and the graph-coloring problem are examples of ***combinatorial problems***. These are problems that ask, explicitly or implicitly, to find a combinatorial object—such as a permutation, a combination, or a subset—that satisfies certain constraints. A desired combinatorial object may also be required to have some additional property such as a maximum value or a minimum cost.

Generally speaking, combinatorial problems are the most difficult problems in computing, from both a theoretical and practical standpoint. Their difficulty stems from the following facts. First, the number of combinatorial objects typically grows extremely fast with a problem's size, reaching unimaginable magnitudes even for moderate-sized instances. Second, there are no known algorithms for solving most such problems exactly in an acceptable amount of time. Moreover, most computer scientists believe that such algorithms do not exist. This conjecture has been neither proved nor disproved, and it remains the most important unresolved issue in theoretical computer science. We discuss this topic in more detail in Section 11.3.

Some combinatorial problems can be solved by efficient algorithms, but they should be considered fortunate exceptions to the rule. The shortest-path problem mentioned earlier is among such exceptions.

Geometric Problems

Geometric algorithms deal with geometric objects such as points, lines, and polygons. The ancient Greeks were very much interested in developing procedures (they did not call them algorithms, of course) for solving a variety of geometric problems, including problems of constructing simple geometric shapes—triangles, circles, and so on—with an unmarked ruler and a compass. Then, for about 2000 years, intense interest in geometric algorithms disappeared, to be resurrected in the age of computers—no more rulers and compasses, just bits, bytes, and good old human ingenuity. Of course, today people are interested in geometric algorithms with quite different applications in mind, such as computer graphics, robotics, and tomography.

We will discuss algorithms for only two classic problems of computational geometry: the closest-pair problem and the convex-hull problem. The ***closest-pair problem*** is self-explanatory: given n points in the plane, find the closest pair among them. The ***convex-hull problem*** asks to find the smallest convex polygon that would include all the points of a given set. If you are interested in other geometric algorithms, you will find a wealth of material in such specialized monographs as [deB10], [ORo98], and [Pre85].

Numerical Problems

Numerical problems, another large special area of applications, are problems that involve mathematical objects of continuous nature: solving equations and systems of equations, computing definite integrals, evaluating functions, and so on. The majority of such mathematical problems can be solved only approximately. Another principal difficulty stems from the fact that such problems typically require manipulating real numbers, which can be represented in a computer only approximately. Moreover, a large number of arithmetic operations performed on approximately represented numbers can lead to an accumulation of the round-off error to a point where it can drastically distort an output produced by a seemingly sound algorithm.

Many sophisticated algorithms have been developed over the years in this area, and they continue to play a critical role in many scientific and engineering applications. But in the last 30 years or so, the computing industry has shifted its focus to business applications. These new applications require primarily algorithms for information storage, retrieval, transportation through networks, and presentation to users. As a result of this revolutionary change, numerical analysis has lost its formerly dominating position in both industry and computer science programs. Still, it is important for any computer-literate person to have at least a rudimentary idea about numerical algorithms. We discuss several classical numerical algorithms in Sections 6.2, 11.4, and 12.4.

Exercises 1.3

1. Consider the algorithm for the sorting problem that sorts an array by counting, for each of its elements, the number of smaller elements and then uses this information to put the element in its appropriate position in the sorted array:

ALGORITHM *ComparisonCountingSort*($A[0..n-1]$)

```
//Sorts an array by comparison counting
//Input: Array A[0..n − 1] of orderable values
//Output: Array S[0..n − 1] of A's elements sorted
//   in nondecreasing order
for i ← 0 to n − 1 do
    Count[i] ← 0
for i ← 0 to n − 2 do
    for j ← i + 1 to n − 1 do
        if A[i] < A[j]
            Count[j] ← Count[j] + 1
        else Count[i] ← Count[i] + 1
for i ← 0 to n − 1 do
    S[Count[i]] ← A[i]
return S
```

a. Apply this algorithm to sorting the list 60, 35, 81, 98, 14, 47.

b. Is this algorithm stable?

c. Is it in-place?

2. Name the algorithms for the searching problem that you already know. Give a good succinct description of each algorithm in English. If you know no such algorithms, use this opportunity to design one.

3. Design a simple algorithm for the string-matching problem.

4. *Königsberg bridges* The Königsberg bridge puzzle is universally accepted as the problem that gave birth to graph theory. It was solved by the great Swiss-born mathematician Leonhard Euler (1707–1783). The problem asked whether one could, in a single stroll, cross all seven bridges of the city of Königsberg exactly once and return to a starting point. Following is a sketch of the river with its two islands and seven bridges:

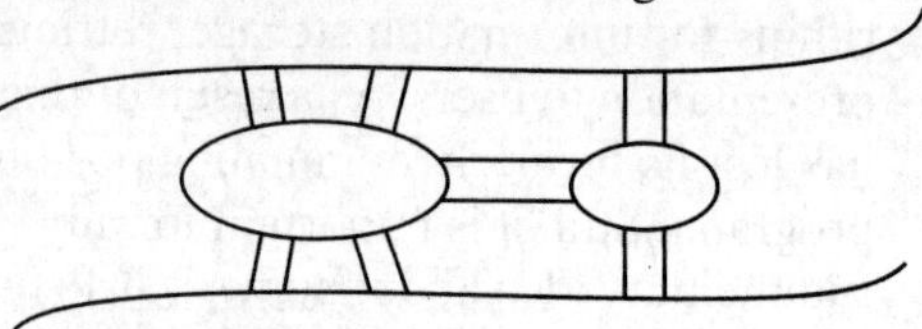

a. State the problem as a graph problem.

b. Does this problem have a solution? If you believe it does, draw such a stroll; if you believe it does not, explain why and indicate the smallest number of new bridges that would be required to make such a stroll possible.

5. *Icosian Game* A century after Euler's discovery (see Problem 4), another famous puzzle—this one invented by the renowned Irish mathematician Sir William Hamilton (1805–1865)—was presented to the world under the name of the Icosian Game. The game's board was a circular wooden board on which the following graph was carved:

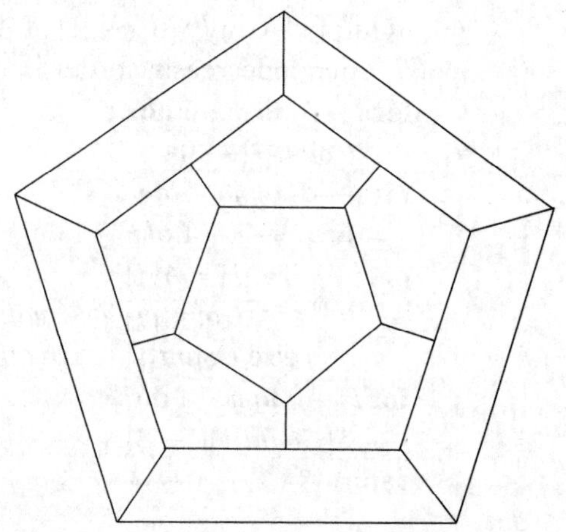

Find a ***Hamiltonian circuit***—a path that visits all the graph's vertices exactly once before returning to the starting vertex—for this graph.

6. Consider the following problem: Design an algorithm to determine the best route for a subway passenger to take from one designated station to another in a well-developed subway system similar to those in such cities as Washington, D.C., and London, UK.

a. The problem's statement is somewhat vague, which is typical of real-life problems. In particular, what reasonable criterion can be used for defining the "best" route?

b. How would you model this problem by a graph?

7. a. Rephrase the traveling-salesman problem in combinatorial object terms.

b. Rephrase the graph-coloring problem in combinatorial object terms.

8. Consider the following map:

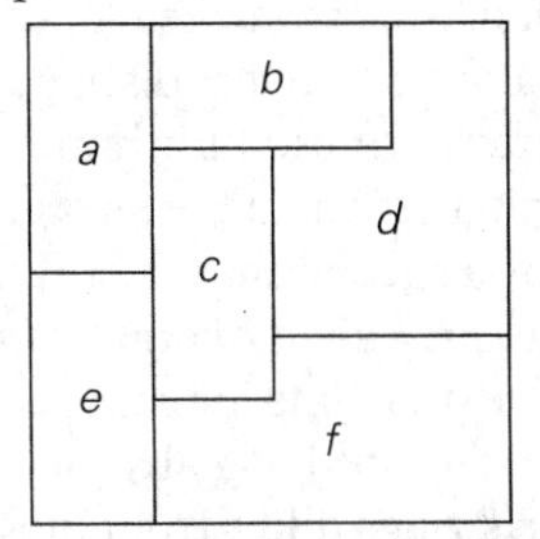

a. Explain how we can use the graph-coloring problem to color the map so that no two neighboring regions are colored the same.

b. Use your answer to part (a) to color the map with the smallest number of colors.

9. Design an algorithm for the following problem: Given a set of n points in the Cartesian plane, determine whether all of them lie on the same circumference.

10. Write a program that reads as its inputs the (x, y) coordinates of the endpoints of two line segments P_1Q_1 and P_2Q_2 and determines whether the segments have a common point.

1.4 Fundamental Data Structures

Since the vast majority of algorithms of interest operate on data, particular ways of organizing data play a critical role in the design and analysis of algorithms. A ***data structure*** can be defined as a particular scheme of organizing related data items. The nature of the data items is dictated by the problem at hand; they can range from elementary data types (e.g., integers or characters) to data structures (e.g., a one-dimensional array of one-dimensional arrays is often used for implementing matrices). There are a few data structures that have proved to be particularly important for computer algorithms. Since you are undoubtedly familiar with most if not all of them, just a quick review is provided here.

Linear Data Structures

The two most important elementary data structures are the array and the linked list. A (one-dimensional) ***array*** is a sequence of n items of the same data type that are stored contiguously in computer memory and made accessible by specifying a value of the array's ***index*** (Figure 1.3).

In the majority of cases, the index is an integer either between 0 and $n-1$ (as shown in Figure 1.3) or between 1 and n. Some computer languages allow an array index to range between any two integer bounds low and $high$, and some even permit nonnumerical indices to specify, for example, data items corresponding to the 12 months of the year by the month names.

Each and every element of an array can be accessed in the same constant amount of time regardless of where in the array the element in question is located. This feature positively distinguishes arrays from linked lists, discussed below.

Arrays are used for implementing a variety of other data structures. Prominent among them is the ***string***, a sequence of characters from an alphabet terminated by a special character indicating the string's end. Strings composed of zeros and ones are called ***binary strings*** or ***bit strings***. Strings are indispensable for processing textual data, defining computer languages and compiling programs written in them, and studying abstract computational models. Operations we usually perform on strings differ from those we typically perform on other arrays (say, arrays of numbers). They include computing the string length, comparing two strings to determine which one precedes the other in ***lexicographic*** (i.e., alphabetical) ***order***, and concatenating two strings (forming one string from two given strings by appending the second to the end of the first).

A ***linked list*** is a sequence of zero or more elements called ***nodes***, each containing two kinds of information: some data and one or more links called ***pointers*** to other nodes of the linked list. (A special pointer called "null" is used to indicate the absence of a node's successor.) In a ***singly linked list***, each node except the last one contains a single pointer to the next element (Figure 1.4).

To access a particular node of a linked list, one starts with the list's first node and traverses the pointer chain until the particular node is reached. Thus, the time needed to access an element of a singly linked list, unlike that of an array, depends on where in the list the element is located. On the positive side, linked lists do

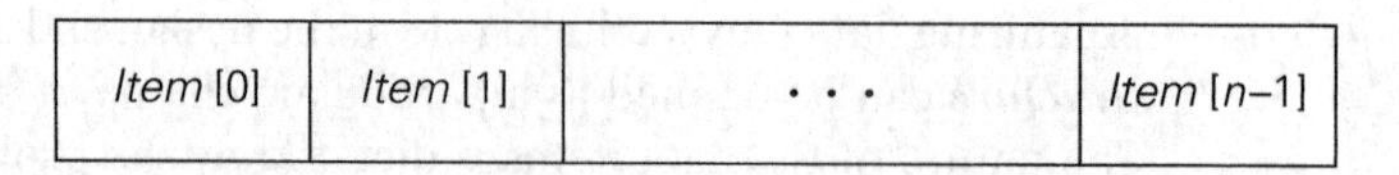

FIGURE 1.3 Array of n elements.

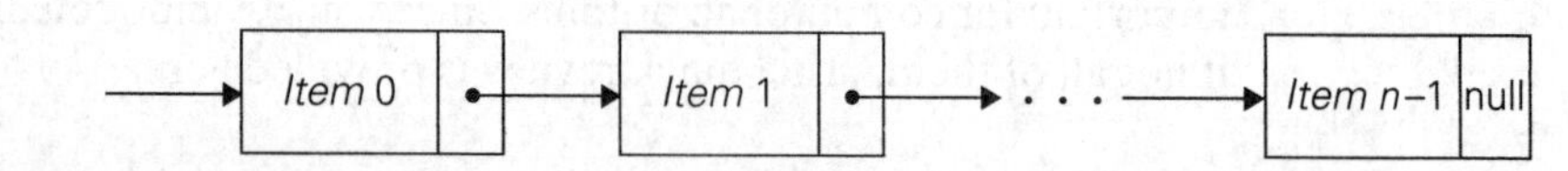

FIGURE 1.4 Singly linked list of n elements.

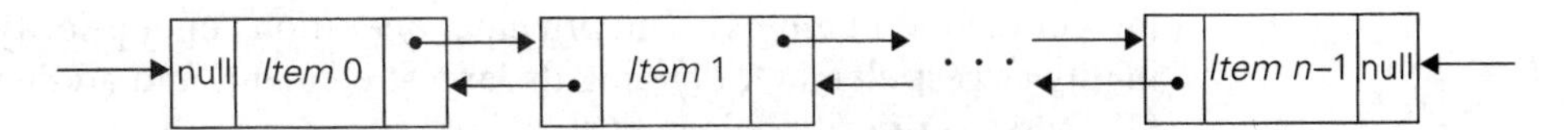

FIGURE 1.5 Doubly linked list of n elements.

not require any preliminary reservation of the computer memory, and insertions and deletions can be made quite efficiently in a linked list by reconnecting a few appropriate pointers.

We can exploit flexibility of the linked list structure in a variety of ways. For example, it is often convenient to start a linked list with a special node called the ***header***. This node may contain information about the linked list itself, such as its current length; it may also contain, in addition to a pointer to the first element, a pointer to the linked list's last element.

Another extension is the structure called the ***doubly linked list***, in which every node, except the first and the last, contains pointers to both its successor and its predecessor (Figure 1.5).

The array and linked list are two principal choices in representing a more abstract data structure called a linear list or simply a list. A ***list*** is a finite sequence of data items, i.e., a collection of data items arranged in a certain linear order. The basic operations performed on this data structure are searching for, inserting, and deleting an element.

Two special types of lists, stacks and queues, are particularly important. A ***stack*** is a list in which insertions and deletions can be done only at the end. This end is called the ***top*** because a stack is usually visualized not horizontally but vertically—akin to a stack of plates whose "operations" it mimics very closely. As a result, when elements are added to (pushed onto) a stack and deleted from (popped off) it, the structure operates in a "last-in–first-out" (LIFO) fashion—exactly like a stack of plates if we can add or remove a plate only from the top. Stacks have a multitude of applications; in particular, they are indispensable for implementing recursive algorithms.

A ***queue***, on the other hand, is a list from which elements are deleted from one end of the structure, called the ***front*** (this operation is called ***dequeue***), and new elements are added to the other end, called the ***rear*** (this operation is called ***enqueue***). Consequently, a queue operates in a "first-in–first-out" (FIFO) fashion—akin to a queue of customers served by a single teller in a bank. Queues also have many important applications, including several algorithms for graph problems.

Many important applications require selection of an item of the highest priority among a dynamically changing set of candidates. A data structure that seeks to satisfy the needs of such applications is called a priority queue. A ***priority queue*** is a collection of data items from a totally ordered universe (most often, integer or real numbers). The principal operations on a priority queue are finding its largest element, deleting its largest element, and adding a new element. Of course, a priority queue must be implemented so that the last two operations yield another priority queue. Straightforward implementations of this data structure can be based on either an array or a sorted array, but neither of these options yields the most efficient solution possible. A better implementation of a priority queue is based on an ingenious data structure called the ***heap***. We discuss heaps and an important sorting algorithm based on them in Section 6.4.

Graphs

As we mentioned in the previous section, a graph is informally thought of as a collection of points in the plane called "vertices" or "nodes," some of them connected by line segments called "edges" or "arcs." Formally, a ***graph*** $G = \langle V, E \rangle$ is defined by a pair of two sets: a finite nonempty set V of items called ***vertices*** and a set E of pairs of these items called ***edges***. If these pairs of vertices are unordered, i.e., a pair of vertices (u, v) is the same as the pair (v, u), we say that the vertices u and v are ***adjacent*** to each other and that they are connected by the ***undirected edge*** (u, v). We call the vertices u and v ***endpoints*** of the edge (u, v) and say that u and v are ***incident*** to this edge; we also say that the edge (u, v) is incident to its endpoints u and v. A graph G is called ***undirected*** if every edge in it is undirected.

If a pair of vertices (u, v) is not the same as the pair (v, u), we say that the edge (u, v) is ***directed*** from the vertex u, called the edge's ***tail***, to the vertex v, called the edge's ***head***. We also say that the edge (u, v) leaves u and enters v. A graph whose every edge is directed is called ***directed***. Directed graphs are also called ***digraphs***.

It is normally convenient to label vertices of a graph or a digraph with letters, integer numbers, or, if an application calls for it, character strings (Figure 1.6). The graph depicted in Figure 1.6a has six vertices and seven undirected edges:

$$V = \{a, b, c, d, e, f\}, \; E = \{(a, c), (a, d), (b, c), (b, f), (c, e), (d, e), (e, f)\}.$$

The digraph depicted in Figure 1.6b has six vertices and eight directed edges:

$$V = \{a, b, c, d, e, f\}, \; E = \{(a, c), (b, c), (b, f), (c, e), (d, a), (d, e), (e, c), (e, f)\}.$$

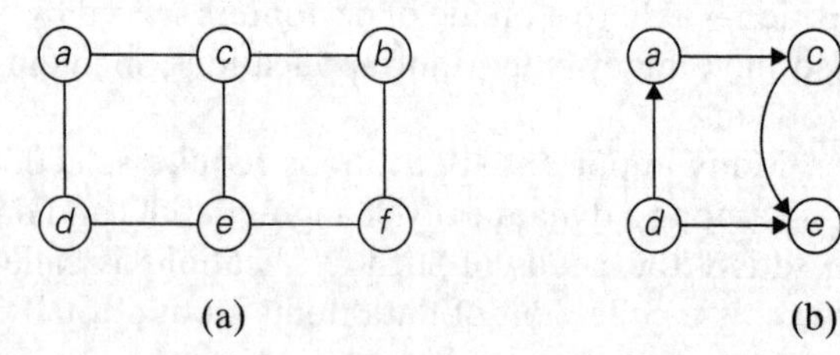

FIGURE 1.6 (a) Undirected graph. (b) Digraph.

Our definition of a graph does not forbid ***loops***, or edges connecting vertices to themselves. Unless explicitly stated otherwise, we will consider graphs without loops. Since our definition disallows multiple edges between the same vertices of an undirected graph, we have the following inequality for the number of edges $|E|$ possible in an undirected graph with $|V|$ vertices and no loops:

$$0 \le |E| \le |V|(|V|-1)/2.$$

(We get the largest number of edges in a graph if there is an edge connecting each of its $|V|$ vertices with all $|V|-1$ other vertices. We have to divide product $|V|(|V|-1)$ by 2, however, because it includes every edge twice.)

A graph with every pair of its vertices connected by an edge is called ***complete***. A standard notation for the complete graph with $|V|$ vertices is $K_{|V|}$. A graph with relatively few possible edges missing is called ***dense***; a graph with few edges relative to the number of its vertices is called ***sparse***. Whether we are dealing with a dense or sparse graph may influence how we choose to represent the graph and, consequently, the running time of an algorithm being designed or used.

Graph Representations Graphs for computer algorithms are usually represented in one of two ways: the adjacency matrix and adjacency lists. The ***adjacency matrix*** of a graph with n vertices is an $n \times n$ boolean matrix with one row and one column for each of the graph's vertices, in which the element in the ith row and the jth column is equal to 1 if there is an edge from the ith vertex to the jth vertex, and equal to 0 if there is no such edge. For example, the adjacency matrix for the graph of Figure 1.6a is given in Figure 1.7a.

Note that the adjacency matrix of an undirected graph is always symmetric, i.e., $A[i, j] = A[j, i]$ for every $0 \le i, j \le n-1$ (why?).

The ***adjacency lists*** of a graph or a digraph is a collection of linked lists, one for each vertex, that contain all the vertices adjacent to the list's vertex (i.e., all the vertices connected to it by an edge). Usually, such lists start with a header identifying a vertex for which the list is compiled. For example, Figure 1.7b represents the graph in Figure 1.6a via its adjacency lists. To put it another way,

$$\begin{array}{c} \\ a \\ b \\ c \\ d \\ e \\ f \end{array}\begin{array}{c} \begin{array}{cccccc} a & b & c & d & e & f \end{array} \\ \begin{bmatrix} 0 & 0 & 1 & 1 & 0 & 0 \\ 0 & 0 & 1 & 0 & 0 & 1 \\ 1 & 1 & 0 & 0 & 1 & 0 \\ 1 & 0 & 0 & 0 & 1 & 0 \\ 0 & 0 & 1 & 1 & 0 & 1 \\ 0 & 1 & 0 & 0 & 1 & 0 \end{bmatrix} \end{array}$$

(a)

a	→	c	→	d		
b	→	c	→	f		
c	→	a	→	b	→	e
d	→	a	→	e		
e	→	c	→	d	→	f
f	→	b	→	e		

(b)

FIGURE 1.7 (a) Adjacency matrix and (b) adjacency lists of the graph in Figure 1.6a.

adjacency lists indicate columns of the adjacency matrix that, for a given vertex, contain 1's.

If a graph is sparse, the adjacency list representation may use less space than the corresponding adjacency matrix despite the extra storage consumed by pointers of the linked lists; the situation is exactly opposite for dense graphs. In general, which of the two representations is more convenient depends on the nature of the problem, on the algorithm used for solving it, and, possibly, on the type of input graph (sparse or dense).

Weighted Graphs A ***weighted graph*** (or weighted digraph) is a graph (or digraph) with numbers assigned to its edges. These numbers are called ***weights*** or ***costs***. An interest in such graphs is motivated by numerous real-world applications, such as finding the shortest path between two points in a transportation or communication network or the traveling salesman problem mentioned earlier.

Both principal representations of a graph can be easily adopted to accommodate weighted graphs. If a weighted graph is represented by its adjacency matrix, then its element $A[i, j]$ will simply contain the weight of the edge from the ith to the jth vertex if there is such an edge and a special symbol, e.g., ∞, if there is no such edge. Such a matrix is called the ***weight matrix*** or ***cost matrix***. This approach is illustrated in Figure 1.8b for the weighted graph in Figure 1.8a. (For some applications, it is more convenient to put 0's on the main diagonal of the adjacency matrix.) Adjacency lists for a weighted graph have to include in their nodes not only the name of an adjacent vertex but also the weight of the corresponding edge (Figure 1.8c).

Paths and Cycles Among the many properties of graphs, two are important for a great number of applications: ***connectivity*** and ***acyclicity***. Both are based on the notion of a path. A ***path*** from vertex u to vertex v of a graph G can be defined as a sequence of adjacent (connected by an edge) vertices that starts with u and ends with v. If all vertices of a path are distinct, the path is said to be ***simple***. The ***length*** of a path is the total number of vertices in the vertex sequence defining the path minus 1, which is the same as the number of edges in the path. For example, a, c, b, f is a simple path of length 3 from a to f in the graph in Figure 1.6a, whereas a, c, e, c, b, f is a path (not simple) of length 5 from a to f.

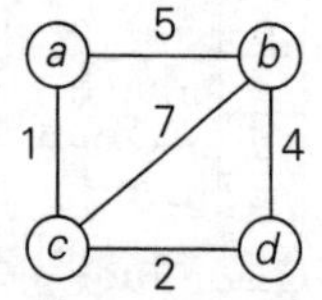

	a	b	c	d
a	∞	5	1	∞
b	5	∞	7	4
c	1	7	∞	2
d	∞	4	2	∞

a → b, 5 → c, 1
b → a, 5 → c, 7 → d, 4
c → a, 1 → b, 7 → d, 2
d → b, 4 → c, 2

(a) (b) (c)

FIGURE 1.8 (a) Weighted graph. (b) Its weight matrix. (c) Its adjacency lists.

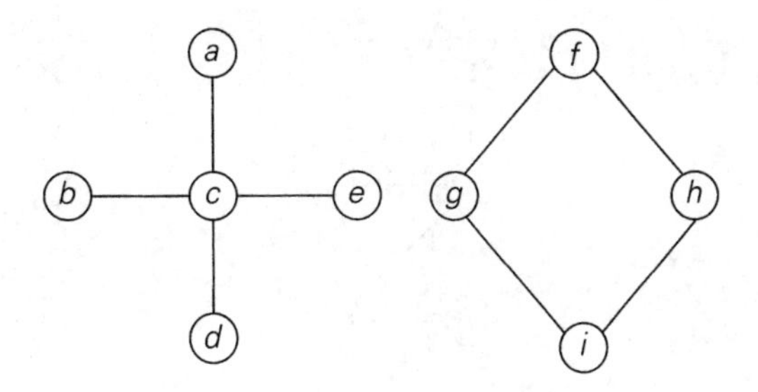

FIGURE 1.9 Graph that is not connected.

In the case of a directed graph, we are usually interested in directed paths. A ***directed path*** is a sequence of vertices in which every consecutive pair of the vertices is connected by an edge directed from the vertex listed first to the vertex listed next. For example, a, c, e, f is a directed path from a to f in the graph in Figure 1.6b.

A graph is said to be ***connected*** if for every pair of its vertices u and v there is a path from u to v. If we make a model of a connected graph by connecting some balls representing the graph's vertices with strings representing the edges, it will be a single piece. If a graph is not connected, such a model will consist of several connected pieces that are called connected components of the graph. Formally, a ***connected component*** is a maximal (not expandable by including another vertex and an edge) connected subgraph[2] of a given graph. For example, the graphs in Figures 1.6a and 1.8a are connected, whereas the graph in Figure 1.9 is not, because there is no path, for example, from a to f. The graph in Figure 1.9 has two connected components with vertices $\{a, b, c, d, e\}$ and $\{f, g, h, i\}$, respectively.

Graphs with several connected components do happen in real-world applications. A graph representing the Interstate highway system of the United States would be an example (why?).

It is important to know for many applications whether or not a graph under consideration has cycles. A ***cycle*** is a path of a positive length that starts and ends at the same vertex and does not traverse the same edge more than once. For example, f, h, i, g, f is a cycle in the graph in Figure 1.9. A graph with no cycles is said to be ***acyclic***. We discuss acyclic graphs in the next subsection.

Trees

A ***tree*** (more accurately, a ***free tree***) is a connected acyclic graph (Figure 1.10a). A graph that has no cycles but is not necessarily connected is called a ***forest***: each of its connected components is a tree (Figure 1.10b).

---

2. A ***subgraph*** of a given graph $G = \langle V, E\rangle$ is a graph $G' = \langle V', E'\rangle$ such that $V' \subseteq V$ and $E' \subseteq E$.

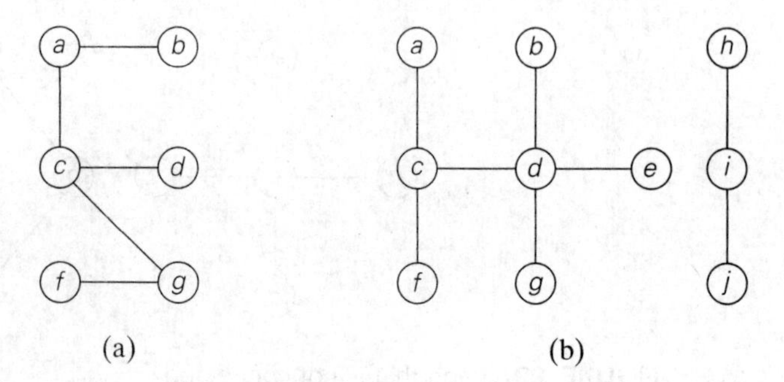

FIGURE 1.10 (a) Tree. (b) Forest.

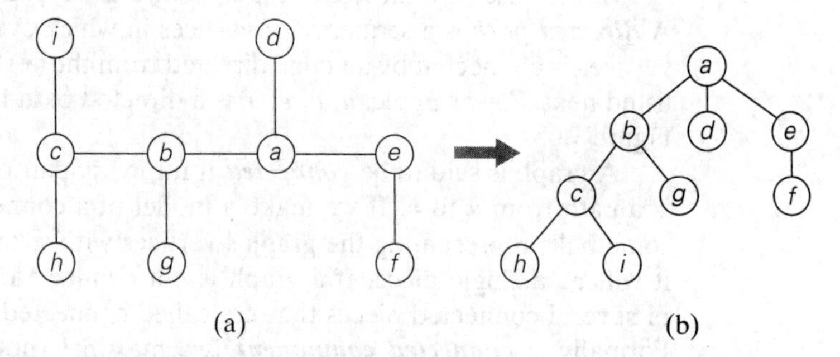

FIGURE 1.11 (a) Free tree. (b) Its transformation into a rooted tree.

Trees have several important properties other graphs do not have. In particular, the number of edges in a tree is always one less than the number of its vertices:

$$|E| = |V| - 1.$$

As the graph in Figure 1.9 demonstrates, this property is necessary but not sufficient for a graph to be a tree. However, for connected graphs it is sufficient and hence provides a convenient way of checking whether a connected graph has a cycle.

Rooted Trees Another very important property of trees is the fact that for every two vertices in a tree, there always exists exactly one simple path from one of these vertices to the other. This property makes it possible to select an arbitrary vertex in a free tree and consider it as the ***root*** of the so-called ***rooted tree***. A rooted tree is usually depicted by placing its root on the top (level 0 of the tree), the vertices adjacent to the root below it (level 1), the vertices two edges apart from the root still below (level 2), and so on. Figure 1.11 presents such a transformation from a free tree to a rooted tree.

Rooted trees play a very important role in computer science, a much more important one than free trees do; in fact, for the sake of brevity, they are often referred to as simply "trees." An obvious application of trees is for describing hierarchies, from file directories to organizational charts of enterprises. There are many less obvious applications, such as implementing dictionaries (see below), efficient access to very large data sets (Section 7.4), and data encoding (Section 9.4). As we discuss in Chapter 2, trees also are helpful in analysis of recursive algorithms. To finish this far-from-complete list of tree applications, we should mention the so-called ***state-space trees*** that underline two important algorithm design techniques: backtracking and branch-and-bound (Sections 12.1 and 12.2).

For any vertex v in a tree T, all the vertices on the simple path from the root to that vertex are called ***ancestors*** of v. The vertex itself is usually considered its own ancestor; the set of ancestors that excludes the vertex itself is referred to as the set of ***proper ancestors***. If (u, v) is the last edge of the simple path from the root to vertex v (and $u \neq v$), u is said to be the ***parent*** of v and v is called a ***child*** of u; vertices that have the same parent are said to be ***siblings***. A vertex with no children is called a ***leaf***; a vertex with at least one child is called ***parental***. All the vertices for which a vertex v is an ancestor are said to be ***descendants*** of v; the ***proper descendants*** exclude the vertex v itself. All the descendants of a vertex v with all the edges connecting them form the ***subtree*** of T rooted at that vertex. Thus, for the tree in Figure 1.11b, the root of the tree is a; vertices d, g, f, h, and i are leaves, and vertices a, b, e, and c are parental; the parent of b is a; the children of b are c and g; the siblings of b are d and e; and the vertices of the subtree rooted at b are $\{b, c, g, h, i\}$.

The ***depth*** of a vertex v is the length of the simple path from the root to v. The ***height*** of a tree is the length of the longest simple path from the root to a leaf. For example, the depth of vertex c of the tree in Figure 1.11b is 2, and the height of the tree is 3. Thus, if we count tree levels top down starting with 0 for the root's level, the depth of a vertex is simply its level in the tree, and the tree's height is the maximum level of its vertices. (You should be alert to the fact that some authors define the height of a tree as the number of levels in it; this makes the height of a tree larger by 1 than the height defined as the length of the longest simple path from the root to a leaf.)

Ordered Trees An ***ordered tree*** is a rooted tree in which all the children of each vertex are ordered. It is convenient to assume that in a tree's diagram, all the children are ordered left to right.

A ***binary tree*** can be defined as an ordered tree in which every vertex has no more than two children and each child is designated as either a ***left child*** or a ***right child*** of its parent; a binary tree may also be empty. An example of a binary tree is given in Figure 1.12a. The binary tree with its root at the left (right) child of a vertex in a binary tree is called the ***left*** (***right***) ***subtree*** of that vertex. Since left and right subtrees are binary trees as well, a binary tree can also be defined recursively. This makes it possible to solve many problems involving binary trees by recursive algorithms.

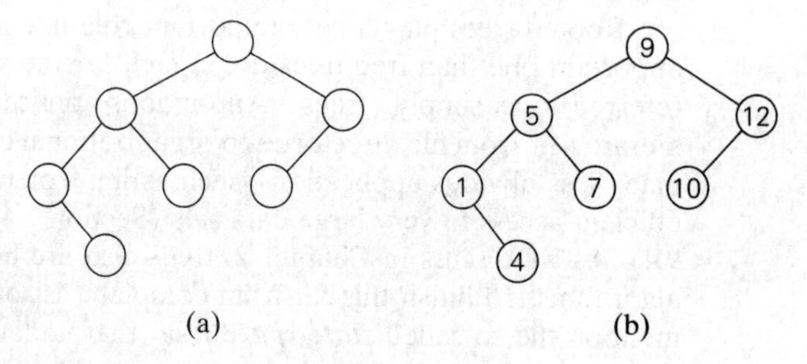

FIGURE 1.12 (a) Binary tree. (b) Binary search tree.

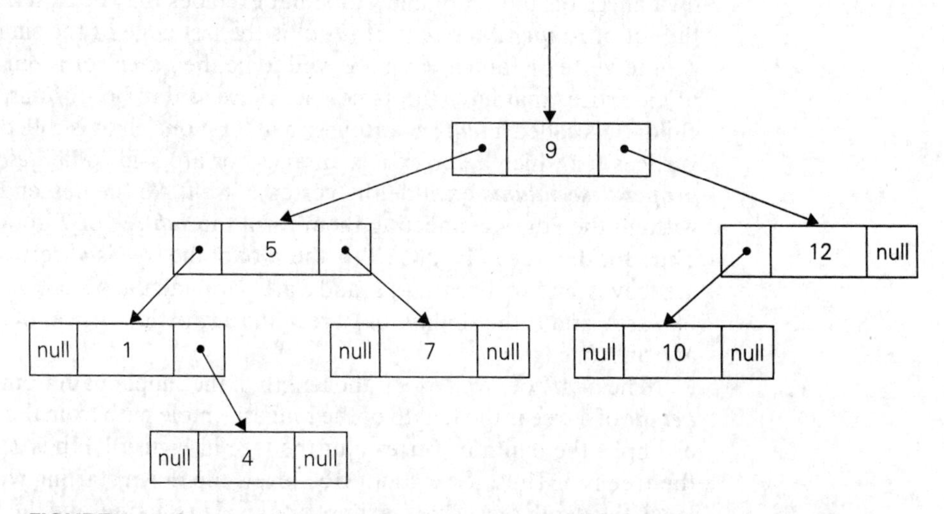

FIGURE 1.13 Standard implementation of the binary search tree in Figure 1.12b.

In Figure 1.12b, some numbers are assigned to vertices of the binary tree in Figure 1.12a. Note that a number assigned to each parental vertex is larger than all the numbers in its left subtree and smaller than all the numbers in its right subtree. Such trees are called ***binary search trees***. Binary trees and binary search trees have a wide variety of applications in computer science; you will encounter some of them throughout the book. In particular, binary search trees can be generalized to more general types of search trees called ***multiway search trees***, which are indispensable for efficient access to very large data sets.

As you will see later in the book, the efficiency of most important algorithms for binary search trees and their extensions depends on the tree's height. Therefore, the following inequalities for the height h of a binary tree with n nodes are especially important for analysis of such algorithms:

$$\lfloor \log_2 n \rfloor \le h \le n - 1.$$

A binary tree is usually implemented for computing purposes by a collection of nodes corresponding to vertices of the tree. Each node contains some information associated with the vertex (its name or some value assigned to it) and two pointers to the nodes representing the left child and right child of the vertex, respectively. Figure 1.13 illustrates such an implementation for the binary search tree in Figure 1.12b.

A computer representation of an arbitrary ordered tree can be done by simply providing a parental vertex with the number of pointers equal to the number of its children. This representation may prove to be inconvenient if the number of children varies widely among the nodes. We can avoid this inconvenience by using nodes with just two pointers, as we did for binary trees. Here, however, the left pointer will point to the first child of the vertex, and the right pointer will point to its next sibling. Accordingly, this representation is called the ***first child–next sibling representation***. Thus, all the siblings of a vertex are linked via the nodes' right pointers in a singly linked list, with the first element of the list pointed to by the left pointer of their parent. Figure 1.14a illustrates this representation for the tree in Figure 1.11b. It is not difficult to see that this representation effectively transforms an ordered tree into a binary tree said to be associated with the ordered tree. We get this representation by "rotating" the pointers about 45 degrees clockwise (see Figure 1.14b).

Sets and Dictionaries

The notion of a set plays a central role in mathematics. A ***set*** can be described as an unordered collection (possibly empty) of distinct items called ***elements*** of the

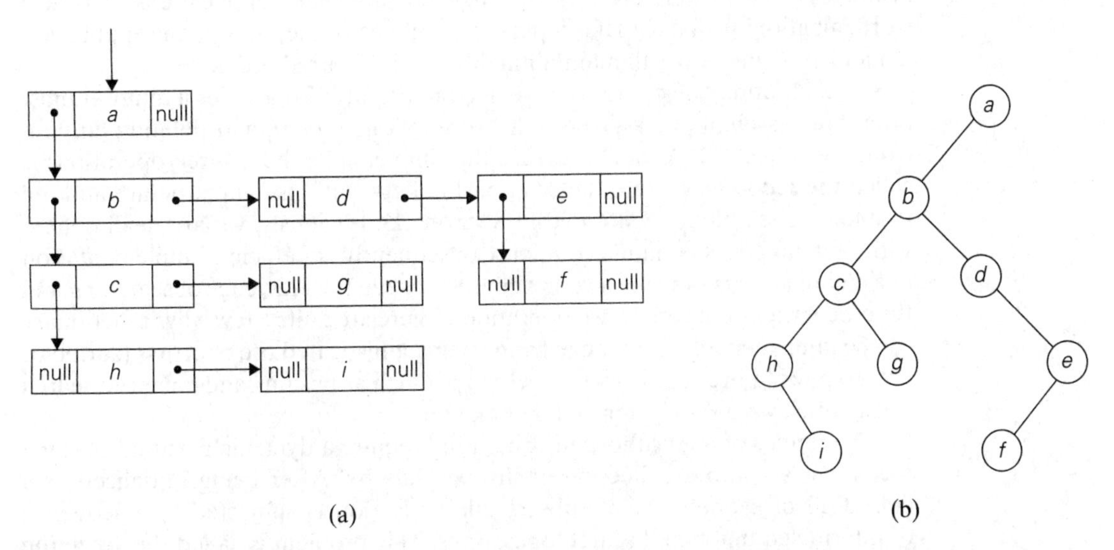

FIGURE 1.14 (a) First child–next sibling representation of the tree in Figure 1.11b. (b) Its binary tree representation.

set. A specific set is defined either by an explicit listing of its elements (e.g., $S = \{2, 3, 5, 7\}$) or by specifying a property that all the set's elements and only they must satisfy (e.g., $S = \{n: n \text{ is a prime number smaller than } 10\}$). The most important set operations are: checking membership of a given item in a given set; finding the union of two sets, which comprises all the elements in either or both of them; and finding the intersection of two sets, which comprises all the common elements in the sets.

Sets can be implemented in computer applications in two ways. The first considers only sets that are subsets of some large set U, called the ***universal set***. If set U has n elements, then any subset S of U can be represented by a bit string of size n, called a ***bit vector***, in which the ith element is 1 if and only if the ith element of U is included in set S. Thus, to continue with our example, if $U = \{1, 2, 3, 4, 5, 6, 7, 8, 9\}$, then $S = \{2, 3, 5, 7\}$ is represented by the bit string 011010100. This way of representing sets makes it possible to implement the standard set operations very fast, but at the expense of potentially using a large amount of storage.

The second and more common way to represent a set for computing purposes is to use the list structure to indicate the set's elements. Of course, this option, too, is feasible only for finite sets; fortunately, unlike mathematics, this is the kind of sets most computer applications need. Note, however, the two principal points of distinction between sets and lists. First, a set cannot contain identical elements; a list can. This requirement for uniqueness is sometimes circumvented by the introduction of a ***multiset***, or ***bag***, an unordered collection of items that are not necessarily distinct. Second, a set is an unordered collection of items; therefore, changing the order of its elements does not change the set. A list, defined as an ordered collection of items, is exactly the opposite. This is an important theoretical distinction, but fortunately it is not important for many applications. It is also worth mentioning that if a set is represented by a list, depending on the application at hand, it might be worth maintaining the list in a sorted order.

In computing, the operations we need to perform for a set or a multiset most often are searching for a given item, adding a new item, and deleting an item from the collection. A data structure that implements these three operations is called the ***dictionary***. Note the relationship between this data structure and the problem of searching mentioned in Section 1.3; obviously, we are dealing here with searching in a dynamic context. Consequently, an efficient implementation of a dictionary has to strike a compromise between the efficiency of searching and the efficiencies of the other two operations. There are quite a few ways a dictionary can be implemented. They range from an unsophisticated use of arrays (sorted or not) to much more sophisticated techniques such as hashing and balanced search trees, which we discuss later in the book.

A number of applications in computing require a dynamic partition of some n-element set into a collection of disjoint subsets. After being initialized as a collection of n one-element subsets, the collection is subjected to a sequence of intermixed union and search operations. This problem is called the ***set union problem***. We discuss efficient algorithmic solutions to this problem in Section 9.2, in conjunction with one of its important applications.

You may have noticed that in our review of basic data structures we almost always mentioned specific operations that are typically performed for the structure in question. This intimate relationship between the data and operations has been recognized by computer scientists for a long time. It has led them in particular to the idea of an ***abstract data type*** (***ADT***): a set of abstract objects representing data items with a collection of operations that can be performed on them. As illustrations of this notion, reread, say, our definitions of the priority queue and dictionary. Although abstract data types could be implemented in older procedural languages such as Pascal (see, e.g., [Aho83]), it is much more convenient to do this in object-oriented languages such as C++ and Java, which support abstract data types by means of ***classes***.

Exercises 1.4

1. Describe how one can implement each of the following operations on an array so that the time it takes does not depend on the array's size n.
   a. Delete the ith element of an array ($1 \le i \le n$).
   b. Delete the ith element of a sorted array (the remaining array has to stay sorted, of course).
2. If you have to solve the searching problem for a list of n numbers, how can you take advantage of the fact that the list is known to be sorted? Give separate answers for
   a. lists represented as arrays.
   b. lists represented as linked lists.
3. **a.** Show the stack after each operation of the following sequence that starts with the empty stack:

   $$push(a), push(b), pop, push(c), push(d), pop$$

   b. Show the queue after each operation of the following sequence that starts with the empty queue:

   $$enqueue(a), enqueue(b), dequeue, enqueue(c), enqueue(d), dequeue$$

4. **a.** Let A be the adjacency matrix of an undirected graph. Explain what property of the matrix indicates that
   i. the graph is complete.
   ii. the graph has a loop, i.e., an edge connecting a vertex to itself.
   iii. the graph has an isolated vertex, i.e., a vertex with no edges incident to it.
   b. Answer the same questions for the adjacency list representation.
5. Give a detailed description of an algorithm for transforming a free tree into a tree rooted at a given vertex of the free tree.

6. Prove the inequalities that bracket the height of a binary tree with n vertices:

$$\lfloor \log_2 n \rfloor \le h \le n - 1.$$

7. Indicate how the ADT priority queue can be implemented as
   a. an (unsorted) array.
   b. a sorted array.
   c. a binary search tree.

8. How would you implement a dictionary of a reasonably small size n if you knew that all its elements are distinct (e.g., names of the 50 states of the United States)? Specify an implementation of each dictionary operation.

9. For each of the following applications, indicate the most appropriate data structure:
   a. answering telephone calls in the order of their known priorities
   b. sending backlog orders to customers in the order they have been received
   c. implementing a calculator for computing simple arithmetical expressions

10. *Anagram checking* Design an algorithm for checking whether two given words are anagrams, i.e., whether one word can be obtained by permuting the letters of the other. For example, the words *tea* and *eat* are anagrams.

SUMMARY

- An *algorithm* is a sequence of nonambiguous instructions for solving a problem in a finite amount of time. An input to an algorithm specifies an *instance* of the problem the algorithm solves.

- Algorithms can be specified in a natural language or pseudocode; they can also be implemented as computer programs.

- Among several ways to classify algorithms, the two principal alternatives are:
  - to group algorithms according to types of problems they solve
  - to group algorithms according to underlying design techniques they are based upon

- The important problem types are sorting, searching, string processing, graph problems, combinatorial problems, geometric problems, and numerical problems.

- Algorithm *design techniques* (or "strategies" or "paradigms") are general approaches to solving problems algorithmically, applicable to a variety of problems from different areas of computing.

- Although designing an algorithm is undoubtedly a creative activity, one can identify a sequence of interrelated actions involved in such a process. They are summarized in Figure 1.2.
- A good algorithm is usually the result of repeated efforts and rework.
- The same problem can often be solved by several algorithms. For example, three algorithms were given for computing the greatest common divisor of two integers: *Euclid's algorithm*, the consecutive integer checking algorithm, and the middle-school method enhanced by the *sieve of Eratosthenes* for generating a list of primes.
- Algorithms operate on data. This makes the issue of data structuring critical for efficient algorithmic problem solving. The most important elementary data structures are the *array* and the *linked list*. They are used for representing more abstract data structures such as the *list*, the *stack*, the *queue*, the *graph* (via its *adjacency matrix* or *adjacency lists*), the *binary tree*, and the *set*.
- An abstract collection of objects with several operations that can be performed on them is called an *abstract data type* (*ADT*). The *list*, the *stack*, the *queue*, the *priority queue*, and the *dictionary* are important examples of abstract data types. Modern object-oriented languages support implementation of ADTs by means of classes.

2

Fundamentals of the Analysis of Algorithm Efficiency

I often say that when you can measure what you are speaking about and express it in numbers you know something about it; but when you cannot express it in numbers your knowledge is a meagre and unsatisfactory kind: it may be the beginning of knowledge but you have scarcely, in your thoughts, advanced to the stage of science, whatever the matter may be.

—Lord Kelvin (1824–1907)

Not everything that can be counted counts, and not everything that counts can be counted.

—Albert Einstein (1879–1955)

This chapter is devoted to analysis of algorithms. The *American Heritage Dictionary* defines "analysis" as "the separation of an intellectual or substantial whole into its constituent parts for individual study." Accordingly, each of the principal dimensions of an algorithm pointed out in Section 1.2 is both a legitimate and desirable subject of study. But the term "analysis of algorithms" is usually used in a narrower, technical sense to mean an investigation of an algorithm's efficiency with respect to two resources: running time and memory space. This emphasis on efficiency is easy to explain. First, unlike such dimensions as simplicity and generality, efficiency can be studied in precise quantitative terms. Second, one can argue—although this is hardly always the case, given the speed and memory of today's computers—that the efficiency considerations are of primary importance from a practical point of view. In this chapter, we too will limit the discussion to an algorithm's efficiency.

We start with a general framework for analyzing algorithm efficiency in Section 2.1. This section is arguably the most important in the chapter; the fundamental nature of the topic makes it also one of the most important sections in the entire book.

In Section 2.2, we introduce three notations: O ("big oh"), Ω ("big omega"), and Θ ("big theta"). Borrowed from mathematics, these notations have become *the* language for discussing the efficiency of algorithms.

In Section 2.3, we show how the general framework outlined in Section 2.1 can be systematically applied to analyzing the efficiency of nonrecursive algorithms. The main tool of such an analysis is setting up a sum representing the algorithm's running time and then simplifying the sum by using standard sum manipulation techniques.

In Section 2.4, we show how the general framework outlined in Section 2.1 can be systematically applied to analyzing the efficiency of recursive algorithms. Here, the main tool is not a summation but a special kind of equation called a recurrence relation. We explain how such recurrence relations can be set up and then introduce a method for solving them.

Although we illustrate the analysis framework and the methods of its applications by a variety of examples in the first four sections of this chapter, Section 2.5 is devoted to yet another example—that of the Fibonacci numbers. Discovered 800 years ago, this remarkable sequence appears in a variety of applications both within and outside computer science. A discussion of the Fibonacci sequence serves as a natural vehicle for introducing an important class of recurrence relations not solvable by the method of Section 2.4. We also discuss several algorithms for computing the Fibonacci numbers, mostly for the sake of a few general observations about the efficiency of algorithms and methods of analyzing them.

The methods of Sections 2.3 and 2.4 provide a powerful technique for analyzing the efficiency of many algorithms with mathematical clarity and precision, but these methods are far from being foolproof. The last two sections of the chapter deal with two approaches—empirical analysis and algorithm visualization—that complement the pure mathematical techniques of Sections 2.3 and 2.4. Much newer and, hence, less developed than their mathematical counterparts, these approaches promise to play an important role among the tools available for analysis of algorithm efficiency.

2.1 The Analysis Framework

In this section, we outline a general framework for analyzing the efficiency of algorithms. We already mentioned in Section 1.2 that there are two kinds of efficiency: time efficiency and space efficiency. ***Time efficiency***, also called ***time complexity***, indicates how fast an algorithm in question runs. ***Space efficiency***, also called ***space complexity***, refers to the amount of memory units required by the algorithm in addition to the space needed for its input and output. In the early days of electronic computing, both resources—time and space—were at a premium. Half a century of relentless technological innovations have improved the computer's speed and memory size by many orders of magnitude. Now the amount of extra space required by an algorithm is typically not of as much concern, with the caveat that there is still, of course, a difference between the fast main memory, the slower secondary memory, and the cache. The time issue has not diminished quite to the same extent, however. In addition, the research experience has shown that for most problems, we can achieve much more spectacular progress in speed than in space. Therefore, following a well-established tradition of algorithm textbooks, we primarily concentrate on time efficiency, but the analytical framework introduced here is applicable to analyzing space efficiency as well.

Measuring an Input's Size

Let's start with the obvious observation that almost all algorithms run longer on larger inputs. For example, it takes longer to sort larger arrays, multiply larger matrices, and so on. Therefore, it is logical to investigate an algorithm's efficiency as a function of some parameter n indicating the algorithm's input size.[1] In most cases, selecting such a parameter is quite straightforward. For example, it will be the size of the list for problems of sorting, searching, finding the list's smallest element, and most other problems dealing with lists. For the problem of evaluating a polynomial $p(x) = a_n x^n + \cdots + a_0$ of degree n, it will be the polynomial's degree or the number of its coefficients, which is larger by 1 than its degree. You'll see from the discussion that such a minor difference is inconsequential for the efficiency analysis.

There are situations, of course, where the choice of a parameter indicating an input size does matter. One such example is computing the product of two $n \times n$ matrices. There are two natural measures of size for this problem. The first and more frequently used is the matrix order n. But the other natural contender is the total number of elements N in the matrices being multiplied. (The latter is also more general since it is applicable to matrices that are not necessarily square.) Since there is a simple formula relating these two measures, we can easily switch from one to the other, but the answer about an algorithm's efficiency will be qualitatively different depending on which of these two measures we use (see Problem 2 in this section's exercises).

The choice of an appropriate size metric can be influenced by operations of the algorithm in question. For example, how should we measure an input's size for a spell-checking algorithm? If the algorithm examines individual characters of its input, we should measure the size by the number of characters; if it works by processing words, we should count their number in the input.

We should make a special note about measuring input size for algorithms solving problems such as checking primality of a positive integer n. Here, the input is just one number, and it is this number's magnitude that determines the input

1. Some algorithms require more than one parameter to indicate the size of their inputs (e.g., the number of vertices and the number of edges for algorithms on graphs represented by their adjacency lists).

size. In such situations, it is preferable to measure size by the number b of bits in the n's binary representation:

$$b = \lfloor \log_2 n \rfloor + 1. \quad \textbf{(2.1)}$$

This metric usually gives a better idea about the efficiency of algorithms in question.

Units for Measuring Running Time

The next issue concerns units for measuring an algorithm's running time. Of course, we can simply use some standard unit of time measurement—a second, or millisecond, and so on—to measure the running time of a program implementing the algorithm. There are obvious drawbacks to such an approach, however: dependence on the speed of a particular computer, dependence on the quality of a program implementing the algorithm and of the compiler used in generating the machine code, and the difficulty of clocking the actual running time of the program. Since we are after a measure of an *algorithm*'s efficiency, we would like to have a metric that does not depend on these extraneous factors.

One possible approach is to count the number of times each of the algorithm's operations is executed. This approach is both excessively difficult and, as we shall see, usually unnecessary. The thing to do is to identify the most important operation of the algorithm, called the ***basic operation***, the operation contributing the most to the total running time, and compute the number of times the basic operation is executed.

As a rule, it is not difficult to identify the basic operation of an algorithm: it is usually the most time-consuming operation in the algorithm's innermost loop. For example, most sorting algorithms work by comparing elements (keys) of a list being sorted with each other; for such algorithms, the basic operation is a key comparison. As another example, algorithms for mathematical problems typically involve some or all of the four arithmetical operations: addition, subtraction, multiplication, and division. Of the four, the most time-consuming operation is division, followed by multiplication and then addition and subtraction, with the last two usually considered together.[2]

Thus, the established framework for the analysis of an algorithm's time efficiency suggests measuring it by counting the number of times the algorithm's basic operation is executed on inputs of size n. We will find out how to compute such a count for nonrecursive and recursive algorithms in Sections 2.3 and 2.4, respectively.

Here is an important application. Let c_{op} be the execution time of an algorithm's basic operation on a particular computer, and let $C(n)$ be the number of times this operation needs to be executed for this algorithm. Then we can estimate

---

2. On some computers, multiplication does not take longer than addition/subtraction (see, for example, the timing data provided by Kernighan and Pike in [Ker99, pp. 185–186]).

the running time $T(n)$ of a program implementing this algorithm on that computer by the formula

$$T(n) \approx c_{op}C(n).$$

Of course, this formula should be used with caution. The count $C(n)$ does not contain any information about operations that are not basic, and, in fact, the count itself is often computed only approximately. Further, the constant c_{op} is also an approximation whose reliability is not always easy to assess. Still, unless n is extremely large or very small, the formula can give a reasonable estimate of the algorithm's running time. It also makes it possible to answer such questions as "How much faster would this algorithm run on a machine that is 10 times faster than the one we have?" The answer is, obviously, 10 times. Or, assuming that $C(n) = \frac{1}{2}n(n-1)$, how much longer will the algorithm run if we double its input size? The answer is about four times longer. Indeed, for all but very small values of n,

$$C(n) = \frac{1}{2}n(n-1) = \frac{1}{2}n^2 - \frac{1}{2}n \approx \frac{1}{2}n^2$$

and therefore

$$\frac{T(2n)}{T(n)} \approx \frac{c_{op}C(2n)}{c_{op}C(n)} \approx \frac{\frac{1}{2}(2n)^2}{\frac{1}{2}n^2} = 4.$$

Note that we were able to answer the last question without actually knowing the value of c_{op}: it was neatly cancelled out in the ratio. Also note that $\frac{1}{2}$, the multiplicative constant in the formula for the count $C(n)$, was also cancelled out. It is for these reasons that the efficiency analysis framework ignores multiplicative constants and concentrates on the count's ***order of growth*** to within a constant multiple for large-size inputs.

Orders of Growth

Why this emphasis on the count's order of growth for large input sizes? A difference in running times on small inputs is not what really distinguishes efficient algorithms from inefficient ones. When we have to compute, for example, the greatest common divisor of two small numbers, it is not immediately clear how much more efficient Euclid's algorithm is compared to the other two algorithms discussed in Section 1.1 or even why we should care which of them is faster and by how much. It is only when we have to find the greatest common divisor of two large numbers that the difference in algorithm efficiencies becomes both clear and important. For large values of n, it is the function's order of growth that counts: just look at Table 2.1, which contains values of a few functions particularly important for analysis of algorithms.

The magnitude of the numbers in Table 2.1 has a profound significance for the analysis of algorithms. The function growing the slowest among these is the logarithmic function. It grows so slowly, in fact, that we should expect a program

TABLE 2.1 Values (some approximate) of several functions important for analysis of algorithms

n	$\log_2 n$	n	$n \log_2 n$	n^2	n^3	2^n	$n!$
10	3.3	10^1	$3.3 \cdot 10^1$	10^2	10^3	10^3	$3.6 \cdot 10^6$
10^2	6.6	10^2	$6.6 \cdot 10^2$	10^4	10^6	$1.3 \cdot 10^{30}$	$9.3 \cdot 10^{157}$
10^3	10	10^3	$1.0 \cdot 10^4$	10^6	10^9		
10^4	13	10^4	$1.3 \cdot 10^5$	10^8	10^{12}		
10^5	17	10^5	$1.7 \cdot 10^6$	10^{10}	10^{15}		
10^6	20	10^6	$2.0 \cdot 10^7$	10^{12}	10^{18}		

implementing an algorithm with a logarithmic basic-operation count to run practically instantaneously on inputs of all realistic sizes. Also note that although specific values of such a count depend, of course, on the logarithm's base, the formula

$$\log_a n = \log_a b \log_b n$$

makes it possible to switch from one base to another, leaving the count logarithmic but with a new multiplicative constant. This is why we omit a logarithm's base and write simply $\log n$ in situations where we are interested just in a function's order of growth to within a multiplicative constant.

On the other end of the spectrum are the exponential function 2^n and the factorial function $n!$ Both these functions grow so fast that their values become astronomically large even for rather small values of n. (This is the reason why we did not include their values for $n > 10^2$ in Table 2.1.) For example, it would take about $4 \cdot 10^{10}$ years for a computer making a trillion (10^{12}) operations per second to execute 2^{100} operations. Though this is incomparably faster than it would have taken to execute 100! operations, it is still longer than 4.5 billion ($4.5 \cdot 10^9$) years—the estimated age of the planet Earth. There is a tremendous difference between the orders of growth of the functions 2^n and $n!$, yet both are often referred to as "exponential-growth functions" (or simply "exponential") despite the fact that, strictly speaking, only the former should be referred to as such. The bottom line, which is important to remember, is this:

> Algorithms that require an exponential number of operations are practical for solving only problems of very small sizes.

Another way to appreciate the qualitative difference among the orders of growth of the functions in Table 2.1 is to consider how they react to, say, a twofold increase in the value of their argument n. The function $\log_2 n$ increases in value by just 1 (because $\log_2 2n = \log_2 2 + \log_2 n = 1 + \log_2 n$); the linear function increases twofold, the linearithmic function $n \log_2 n$ increases slightly more than twofold; the quadratic function n^2 and cubic function n^3 increase fourfold and eightfold, respectively (because $(2n)^2 = 4n^2$ and $(2n)^3 = 8n^3$); the value of 2^n gets squared (because $2^{2n} = (2^n)^2$); and $n!$ increases much more than that (yes, even mathematics refuses to cooperate to give a neat answer for $n!$).

Worst-Case, Best-Case, and Average-Case Efficiencies

In the beginning of this section, we established that it is reasonable to measure an algorithm's efficiency as a function of a parameter indicating the size of the algorithm's input. But there are many algorithms for which running time depends not only on an input size but also on the specifics of a particular input. Consider, as an example, sequential search. This is a straightforward algorithm that searches for a given item (some search key K) in a list of n elements by checking successive elements of the list until either a match with the search key is found or the list is exhausted. Here is the algorithm's pseudocode, in which, for simplicity, a list is implemented as an array. It also assumes that the second condition $A[i] \neq K$ will not be checked if the first one, which checks that the array's index does not exceed its upper bound, fails.

```
ALGORITHM  SequentialSearch(A[0..n − 1], K)
    //Searches for a given value in a given array by sequential search
    //Input: An array A[0..n − 1] and a search key K
    //Output: The index of the first element in A that matches K
    //        or −1 if there are no matching elements
    i ← 0
    while i < n and A[i] ≠ K do
        i ← i + 1
    if i < n return i
    else return −1
```

Clearly, the running time of this algorithm can be quite different for the same list size n. In the worst case, when there are no matching elements or the first matching element happens to be the last one on the list, the algorithm makes the largest number of key comparisons among all possible inputs of size n: $C_{worst}(n) = n$.

The ***worst-case efficiency*** of an algorithm is its efficiency for the worst-case input of size n, which is an input (or inputs) of size n for which the algorithm runs the longest among all possible inputs of that size. The way to determine the worst-case efficiency of an algorithm is, in principle, quite straightforward: analyze the algorithm to see what kind of inputs yield the largest value of the basic operation's count $C(n)$ among all possible inputs of size n and then compute this worst-case value $C_{worst}(n)$. (For sequential search, the answer was obvious. The methods for handling less trivial situations are explained in subsequent sections of this chapter.) Clearly, the worst-case analysis provides very important information about an algorithm's efficiency by bounding its running time from above. In other words, it guarantees that for any instance of size n, the running time will not exceed $C_{worst}(n)$, its running time on the worst-case inputs.

The ***best-case efficiency*** of an algorithm is its efficiency for the best-case input of size n, which is an input (or inputs) of size n for which the algorithm runs the fastest among all possible inputs of that size. Accordingly, we can analyze the best-case efficiency as follows. First, we determine the kind of inputs for which the count $C(n)$ will be the smallest among all possible inputs of size n. (Note that the best case does not mean the smallest input; it means the input of size n for which the algorithm runs the fastest.) Then we ascertain the value of $C(n)$ on these most convenient inputs. For example, the best-case inputs for sequential search are lists of size n with their first element equal to a search key; accordingly, $C_{best}(n) = 1$ for this algorithm.

The analysis of the best-case efficiency is not nearly as important as that of the worst-case efficiency. But it is not completely useless, either. Though we should not expect to get best-case inputs, we might be able to take advantage of the fact that for some algorithms a good best-case performance extends to some useful types of inputs close to being the best-case ones. For example, there is a sorting algorithm (insertion sort) for which the best-case inputs are already sorted arrays on which the algorithm works very fast. Moreover, the best-case efficiency deteriorates only slightly for almost-sorted arrays. Therefore, such an algorithm might well be the method of choice for applications dealing with almost-sorted arrays. And, of course, if the best-case efficiency of an algorithm is unsatisfactory, we can immediately discard it without further analysis.

It should be clear from our discussion, however, that neither the worst-case analysis nor its best-case counterpart yields the necessary information about an algorithm's behavior on a "typical" or "random" input. This is the information that the ***average-case efficiency*** seeks to provide. To analyze the algorithm's average-case efficiency, we must make some assumptions about possible inputs of size n.

Let's consider again sequential search. The standard assumptions are that (a) the probability of a successful search is equal to p $(0 \le p \le 1)$ and (b) the probability of the first match occurring in the ith position of the list is the same for every i. Under these assumptions—the validity of which is usually difficult to verify, their reasonableness notwithstanding—we can find the average number of key comparisons $C_{avg}(n)$ as follows. In the case of a successful search, the probability of the first match occurring in the ith position of the list is p/n for every i, and the number of comparisons made by the algorithm in such a situation is obviously i. In the case of an unsuccessful search, the number of comparisons will be n with the probability of such a search being $(1 - p)$. Therefore,

$$
\begin{aligned}
C_{avg}(n) &= [1 \cdot \frac{p}{n} + 2 \cdot \frac{p}{n} + \cdots + i \cdot \frac{p}{n} + \cdots + n \cdot \frac{p}{n}] + n \cdot (1 - p) \\
&= \frac{p}{n}[1 + 2 + \cdots + i + \cdots + n] + n(1 - p) \\
&= \frac{p}{n}\frac{n(n+1)}{2} + n(1 - p) = \frac{p(n+1)}{2} + n(1 - p).
\end{aligned}
$$

This general formula yields some quite reasonable answers. For example, if $p = 1$ (the search must be successful), the average number of key comparisons made by sequential search is $(n + 1)/2$; that is, the algorithm will inspect, on average, about half of the list's elements. If $p = 0$ (the search must be unsuccessful), the average number of key comparisons will be n because the algorithm will inspect all n elements on all such inputs.

As you can see from this very elementary example, investigation of the average-case efficiency is considerably more difficult than investigation of the worst-case and best-case efficiencies. The direct approach for doing this involves dividing all instances of size n into several classes so that for each instance of the class the number of times the algorithm's basic operation is executed is the same. (What were these classes for sequential search?) Then a probability distribution of inputs is obtained or assumed so that the expected value of the basic operation's count can be found.

The technical implementation of this plan is rarely easy, however, and probabilistic assumptions underlying it in each particular case are usually difficult to verify. Given our quest for simplicity, we will mostly quote known results about the average-case efficiency of algorithms under discussion. If you are interested in derivations of these results, consult such books as [Baa00], [Sed96], [KnuI], [KnuII], and [KnuIII].

It should be clear from the preceding discussion that the average-case efficiency cannot be obtained by taking the average of the worst-case and the best-case efficiencies. Even though this average does occasionally coincide with the average-case cost, it is not a legitimate way of performing the average-case analysis.

Does one really need the average-case efficiency information? The answer is unequivocally yes: there are many important algorithms for which the average-case efficiency is much better than the overly pessimistic worst-case efficiency would lead us to believe. So, without the average-case analysis, computer scientists could have missed many important algorithms.

Yet another type of efficiency is called ***amortized efficiency***. It applies not to a single run of an algorithm but rather to a sequence of operations performed on the same data structure. It turns out that in some situations a single operation can be expensive, but the total time for an entire sequence of n such operations is always significantly better than the worst-case efficiency of that single operation multiplied by n. So we can "amortize" the high cost of such a worst-case occurrence over the entire sequence in a manner similar to the way a business would amortize the cost of an expensive item over the years of the item's productive life. This sophisticated approach was discovered by the American computer scientist Robert Tarjan, who used it, among other applications, in developing an interesting variation of the classic binary search tree (see [Tar87] for a quite readable nontechnical discussion and [Tar85] for a technical account). We will see an example of the usefulness of amortized efficiency in Section 9.2, when we consider algorithms for finding unions of disjoint sets.

Recapitulation of the Analysis Framework

Before we leave this section, let us summarize the main points of the framework outlined above.

- Both time and space efficiencies are measured as functions of the algorithm's input size.
- Time efficiency is measured by counting the number of times the algorithm's basic operation is executed. Space efficiency is measured by counting the number of extra memory units consumed by the algorithm.
- The efficiencies of some algorithms may differ significantly for inputs of the same size. For such algorithms, we need to distinguish between the worst-case, average-case, and best-case efficiencies.
- The framework's primary interest lies in the order of growth of the algorithm's running time (extra memory units consumed) as its input size goes to infinity.

In the next section, we look at formal means to investigate orders of growth. In Sections 2.3 and 2.4, we discuss particular methods for investigating nonrecursive and recursive algorithms, respectively. It is there that you will see how the analysis framework outlined here can be applied to investigating the efficiency of specific algorithms. You will encounter many more examples throughout the rest of the book.

Exercises 2.1

1. For each of the following algorithms, indicate (i) a natural size metric for its inputs, (ii) its basic operation, and (iii) whether the basic operation count can be different for inputs of the same size:
   a. computing the sum of n numbers
   b. computing $n!$
   c. finding the largest element in a list of n numbers
   d. Euclid's algorithm
   e. sieve of Eratosthenes
   f. pen-and-pencil algorithm for multiplying two n-digit decimal integers
2. **a.** Consider the definition-based algorithm for adding two $n \times n$ matrices. What is its basic operation? How many times is it performed as a function of the matrix order n? As a function of the total number of elements in the input matrices?
   b. Answer the same questions for the definition-based algorithm for matrix multiplication.

3. Consider a variation of sequential search that scans a list to return the number of occurrences of a given search key in the list. Does its efficiency differ from the efficiency of classic sequential search?

4. **a.** *Glove selection* There are 22 gloves in a drawer: 5 pairs of red gloves, 4 pairs of yellow, and 2 pairs of green. You select the gloves in the dark and can check them only after a selection has been made. What is the smallest number of gloves you need to select to have at least one matching pair in the best case? In the worst case?

   b. *Missing socks* Imagine that after washing 5 distinct pairs of socks, you discover that two socks are missing. Of course, you would like to have the largest number of complete pairs remaining. Thus, you are left with 4 complete pairs in the best-case scenario and with 3 complete pairs in the worst case. Assuming that the probability of disappearance for each of the 10 socks is the same, find the probability of the best-case scenario; the probability of the worst-case scenario; the number of pairs you should expect in the average case.

5. **a.** Prove formula (2.1) for the number of bits in the binary representation of a positive decimal integer.

   b. Prove the alternative formula for the number of bits in the binary representation of a positive integer n:

$$b = \lceil \log_2(n+1) \rceil.$$

   c. What would be the analogous formulas for the number of decimal digits?

   d. Explain why, within the accepted analysis framework, it does not matter whether we use binary or decimal digits in measuring n's size.

6. Suggest how any sorting algorithm can be augmented in a way to make the best-case count of its key comparisons equal to just $n-1$ (n is a list's size, of course). Do you think it would be a worthwhile addition to any sorting algorithm?

7. Gaussian elimination, the classic algorithm for solving systems of n linear equations in n unknowns, requires about $\frac{1}{3}n^3$ multiplications, which is the algorithm's basic operation.

   a. How much longer should you expect Gaussian elimination to work on a system of 1000 equations versus a system of 500 equations?

   b. You are considering buying a computer that is 1000 times faster than the one you currently have. By what factor will the faster computer increase the sizes of systems solvable in the same amount of time as on the old computer?

8. For each of the following functions, indicate how much the function's value will change if its argument is increased fourfold.

   a. $\log_2 n$ **b.** $\sqrt{n}$ **c.** n **d.** n^2 **e.** n^3 **f.** 2^n

9. For each of the following pairs of functions, indicate whether the first function of each of the following pairs has a lower, same, or higher order of growth (to within a constant multiple) than the second function.

a. $n(n+1)$ and $2000n^2$ **b.** $100n^2$ and $0.01n^3$

c. $\log_2 n$ and $\ln n$ **d.** $\log_2^2 n$ and $\log_2 n^2$

e. 2^{n-1} and 2^n **f.** $(n-1)!$ and $n!$

10. *Invention of chess*

a. According to a well-known legend, the game of chess was invented many centuries ago in northwestern India by a certain sage. When he took his invention to his king, the king liked the game so much that he offered the inventor any reward he wanted. The inventor asked for some grain to be obtained as follows: just a single grain of wheat was to be placed on the first square of the chessboard, two on the second, four on the third, eight on the fourth, and so on, until all 64 squares had been filled. If it took just 1 second to count each grain, how long would it take to count all the grain due to him?

b. How long would it take if instead of doubling the number of grains for each square of the chessboard, the inventor asked for adding two grains?

2.2 Asymptotic Notations and Basic Efficiency Classes

As pointed out in the previous section, the efficiency analysis framework concentrates on the order of growth of an algorithm's basic operation count as the principal indicator of the algorithm's efficiency. To compare and rank such orders of growth, computer scientists use three notations: O (big oh), Ω (big omega), and Θ (big theta). First, we introduce these notations informally, and then, after several examples, formal definitions are given. In the following discussion, $t(n)$ and $g(n)$ can be any nonnegative functions defined on the set of natural numbers. In the context we are interested in, $t(n)$ will be an algorithm's running time (usually indicated by its basic operation count $C(n)$), and $g(n)$ will be some simple function to compare the count with.

Informal Introduction

Informally, $O(g(n))$ is the set of all functions with a lower or same order of growth as $g(n)$ (to within a constant multiple, as n goes to infinity). Thus, to give a few examples, the following assertions are all true:

$$n \in O(n^2), \qquad 100n + 5 \in O(n^2), \qquad \frac{1}{2}n(n-1) \in O(n^2).$$

Indeed, the first two functions are linear and hence have a lower order of growth than $g(n) = n^2$, while the last one is quadratic and hence has the same order of growth as n^2. On the other hand,

$$n^3 \notin O(n^2), \qquad 0.00001n^3 \notin O(n^2), \qquad n^4 + n + 1 \notin O(n^2).$$

Indeed, the functions n^3 and $0.00001n^3$ are both cubic and hence have a higher order of growth than n^2, and so has the fourth-degree polynomial $n^4 + n + 1$.

The second notation, $\Omega(g(n))$, stands for the set of all functions with a higher or same order of growth as $g(n)$ (to within a constant multiple, as n goes to infinity). For example,

$$n^3 \in \Omega(n^2), \qquad \frac{1}{2}n(n-1) \in \Omega(n^2), \qquad \text{but } 100n + 5 \notin \Omega(n^2).$$

Finally, $\Theta(g(n))$ is the set of all functions that have the same order of growth as $g(n)$ (to within a constant multiple, as n goes to infinity). Thus, every quadratic function $an^2 + bn + c$ with $a > 0$ is in $\Theta(n^2)$, but so are, among infinitely many others, $n^2 + \sin n$ and $n^2 + \log n$. (Can you explain why?)

Hopefully, this informal introduction has made you comfortable with the idea behind the three asymptotic notations. So now come the formal definitions.

O-notation

DEFINITION A function $t(n)$ is said to be in $O(g(n))$, denoted $t(n) \in O(g(n))$, if $t(n)$ is bounded above by some constant multiple of $g(n)$ for all large n, i.e., if there exist some positive constant c and some nonnegative integer n_0 such that

$$t(n) \le cg(n) \quad \text{for all } n \ge n_0.$$

The definition is illustrated in Figure 2.1 where, for the sake of visual clarity, n is extended to be a real number.

As an example, let us formally prove one of the assertions made in the introduction: $100n + 5 \in O(n^2)$. Indeed,

$$100n + 5 \le 100n + n \text{ (for all } n \ge 5) = 101n \le 101n^2.$$

Thus, as values of the constants c and n_0 required by the definition, we can take 101 and 5, respectively.

Note that the definition gives us a lot of freedom in choosing specific values for constants c and n_0. For example, we could also reason that

$$100n + 5 \le 100n + 5n \text{ (for all } n \ge 1) = 105n$$

to complete the proof with $c = 105$ and $n_0 = 1$.

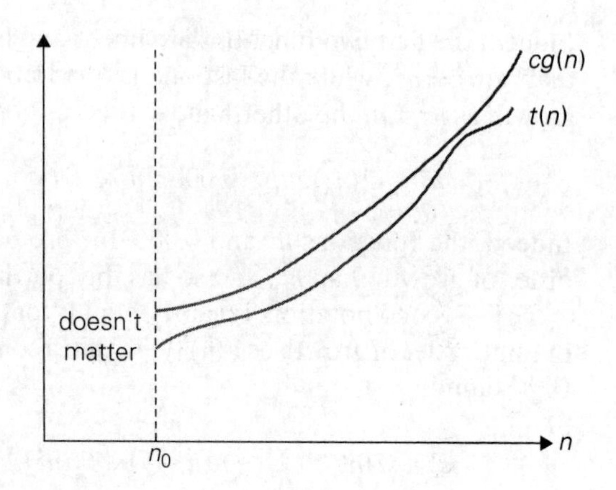

FIGURE 2.1 Big-oh notation: $t(n) \in O(g(n))$.

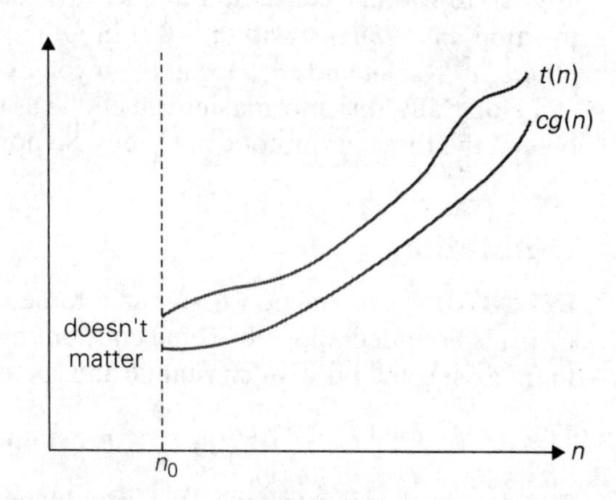

FIGURE 2.2 Big-omega notation: $t(n) \in \Omega(g(n))$.

Ω-notation

DEFINITION A function $t(n)$ is said to be in $\Omega(g(n))$, denoted $t(n) \in \Omega(g(n))$, if $t(n)$ is bounded below by some positive constant multiple of $g(n)$ for all large n, i.e., if there exist some positive constant c and some nonnegative integer n_0 such that

$$t(n) \geq cg(n) \quad \text{for all } n \geq n_0.$$

The definition is illustrated in Figure 2.2.

Here is an example of the formal proof that $n^3 \in \Omega(n^2)$:

$$n^3 \geq n^2 \quad \text{for all } n \geq 0,$$

i.e., we can select $c = 1$ and $n_0 = 0$.

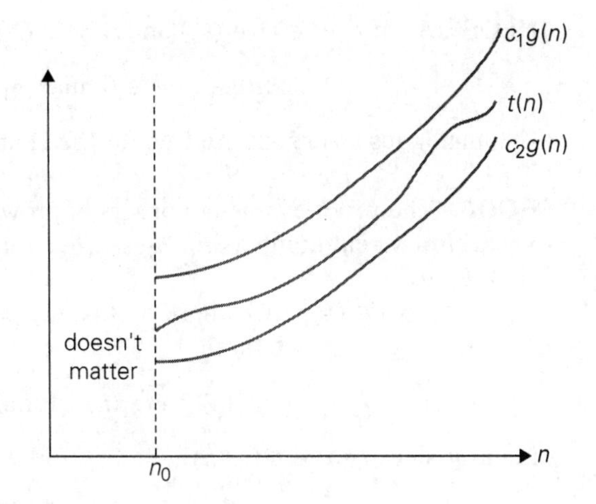

FIGURE 2.3 Big-theta notation: $t(n) \in \Theta(g(n))$.

Θ-notation

DEFINITION A function $t(n)$ is said to be in $\Theta(g(n))$, denoted $t(n) \in \Theta(g(n))$, if $t(n)$ is bounded both above and below by some positive constant multiples of $g(n)$ for all large n, i.e., if there exist some positive constants c_1 and c_2 and some nonnegative integer n_0 such that

$$c_2 g(n) \le t(n) \le c_1 g(n) \quad \text{for all } n \ge n_0.$$

The definition is illustrated in Figure 2.3.

For example, let us prove that $\frac{1}{2}n(n-1) \in \Theta(n^2)$. First, we prove the right inequality (the upper bound):

$$\frac{1}{2}n(n-1) = \frac{1}{2}n^2 - \frac{1}{2}n \le \frac{1}{2}n^2 \quad \text{for all } n \ge 0.$$

Second, we prove the left inequality (the lower bound):

$$\frac{1}{2}n(n-1) = \frac{1}{2}n^2 - \frac{1}{2}n \ge \frac{1}{2}n^2 - \frac{1}{2}n\frac{1}{2}n \text{ (for all } n \ge 2) = \frac{1}{4}n^2.$$

Hence, we can select $c_2 = \frac{1}{4}$, $c_1 = \frac{1}{2}$, and $n_0 = 2$.

Useful Property Involving the Asymptotic Notations

Using the formal definitions of the asymptotic notations, we can prove their general properties (see Problem 7 in this section's exercises for a few simple examples). The following property, in particular, is useful in analyzing algorithms that comprise two consecutively executed parts.

THEOREM If $t_1(n) \in O(g_1(n))$ and $t_2(n) \in O(g_2(n))$, then

$$t_1(n) + t_2(n) \in O(\max\{g_1(n), g_2(n)\}).$$

(The analogous assertions are true for the Ω and Θ notations as well.)

PROOF The proof extends to orders of growth the following simple fact about four arbitrary real numbers a_1, b_1, a_2, b_2: if $a_1 \le b_1$ and $a_2 \le b_2$, then $a_1 + a_2 \le 2\max\{b_1, b_2\}$.

Since $t_1(n) \in O(g_1(n))$, there exist some positive constant c_1 and some nonnegative integer n_1 such that

$$t_1(n) \le c_1 g_1(n) \quad \text{for all } n \ge n_1.$$

Similarly, since $t_2(n) \in O(g_2(n))$,

$$t_2(n) \le c_2 g_2(n) \quad \text{for all } n \ge n_2.$$

Let us denote $c_3 = \max\{c_1, c_2\}$ and consider $n \ge \max\{n_1, n_2\}$ so that we can use both inequalities. Adding them yields the following:

$$\begin{aligned} t_1(n) + t_2(n) &\le c_1 g_1(n) + c_2 g_2(n) \\ &\le c_3 g_1(n) + c_3 g_2(n) = c_3[g_1(n) + g_2(n)] \\ &\le c_3 2\max\{g_1(n), g_2(n)\}. \end{aligned}$$

Hence, $t_1(n) + t_2(n) \in O(\max\{g_1(n), g_2(n)\})$, with the constants c and n_0 required by the O definition being $2c_3 = 2\max\{c_1, c_2\}$ and $\max\{n_1, n_2\}$, respectively. ■

So what does this property imply for an algorithm that comprises two consecutively executed parts? It implies that the algorithm's overall efficiency is determined by the part with a higher order of growth, i.e., its least efficient part:

$$\left.\begin{array}{|c|}\hline t_1(n) \in O(g_1(n)) \\ \hline t_2(n) \in O(g_2(n)) \\ \hline \end{array}\right\} \quad t_1(n) + t_2(n) \in O(\max\{g_1(n), g_2(n)\}).$$

For example, we can check whether an array has equal elements by the following two-part algorithm: first, sort the array by applying some known sorting algorithm; second, scan the sorted array to check its consecutive elements for equality. If, for example, a sorting algorithm used in the first part makes no more than $\frac{1}{2}n(n-1)$ comparisons (and hence is in $O(n^2)$) while the second part makes no more than $n-1$ comparisons (and hence is in $O(n)$), the efficiency of the entire algorithm will be in $O(\max\{n^2, n\}) = O(n^2)$.

Using Limits for Comparing Orders of Growth

Though the formal definitions of O, Ω, and Θ are indispensable for proving their abstract properties, they are rarely used for comparing the orders of growth of two specific functions. A much more convenient method for doing so is based on computing the limit of the ratio of two functions in question. Three principal cases may arise:

$$\lim_{n\to\infty}\frac{t(n)}{g(n)} = \begin{cases} 0 & \text{implies that } t(n) \text{ has a smaller order of growth than } g(n),\\ c & \text{implies that } t(n) \text{ has the same order of growth as } g(n),\\ \infty & \text{implies that } t(n) \text{ has a larger order of growth than } g(n).^3 \end{cases}$$

Note that the first two cases mean that $t(n) \in O(g(n))$, the last two mean that $t(n) \in \Omega(g(n))$, and the second case means that $t(n) \in \Theta(g(n))$.

The limit-based approach is often more convenient than the one based on the definitions because it can take advantage of the powerful calculus techniques developed for computing limits, such as L'Hôpital's rule

$$\lim_{n\to\infty}\frac{t(n)}{g(n)} = \lim_{n\to\infty}\frac{t'(n)}{g'(n)}$$

and Stirling's formula

$$n! \approx \sqrt{2\pi n}\left(\frac{n}{e}\right)^n \quad \text{for large values of } n.$$

Here are three examples of using the limit-based approach to comparing orders of growth of two functions.

EXAMPLE 1 Compare the orders of growth of $\frac{1}{2}n(n-1)$ and n^2. (This is one of the examples we used at the beginning of this section to illustrate the definitions.)

$$\lim_{n\to\infty}\frac{\frac{1}{2}n(n-1)}{n^2} = \frac{1}{2}\lim_{n\to\infty}\frac{n^2-n}{n^2} = \frac{1}{2}\lim_{n\to\infty}(1-\frac{1}{n}) = \frac{1}{2}.$$

Since the limit is equal to a positive constant, the functions have the same order of growth or, symbolically, $\frac{1}{2}n(n-1) \in \Theta(n^2)$. ■

EXAMPLE 2 Compare the orders of growth of $\log_2 n$ and $\sqrt{n}$. (Unlike Example 1, the answer here is not immediately obvious.)

$$\lim_{n\to\infty}\frac{\log_2 n}{\sqrt{n}} = \lim_{n\to\infty}\frac{(\log_2 n)'}{(\sqrt{n})'} = \lim_{n\to\infty}\frac{(\log_2 e)\frac{1}{n}}{\frac{1}{2\sqrt{n}}} = 2\log_2 e \lim_{n\to\infty}\frac{1}{\sqrt{n}} = 0.$$

Since the limit is equal to zero, $\log_2 n$ has a smaller order of growth than $\sqrt{n}$. (Since $\lim_{n\to\infty}\frac{\log_2 n}{\sqrt{n}} = 0$, we can use the so-called ***little-oh notation***: $\log_2 n \in o(\sqrt{n})$. Unlike the big-Oh, the little-oh notation is rarely used in analysis of algorithms.) ■

---

3. The fourth case, in which such a limit does not exist, rarely happens in the actual practice of analyzing algorithms. Still, this possibility makes the limit-based approach to comparing orders of growth less general than the one based on the definitions of O, Ω, and Θ.

EXAMPLE 3 Compare the orders of growth of $n!$ and 2^n. (We discussed this informally in Section 2.1.) Taking advantage of Stirling's formula, we get

$$\lim_{n\to\infty}\frac{n!}{2^n}=\lim_{n\to\infty}\frac{\sqrt{2\pi n}\left(\frac{n}{e}\right)^n}{2^n}=\lim_{n\to\infty}\sqrt{2\pi n}\frac{n^n}{2^n e^n}=\lim_{n\to\infty}\sqrt{2\pi n}\left(\frac{n}{2e}\right)^n=\infty.$$

Thus, though 2^n grows very fast, $n!$ grows still faster. We can write symbolically that $n! \in \Omega(2^n)$; note, however, that while the big-Omega notation does not preclude the possibility that $n!$ and 2^n have the same order of growth, the limit computed here certainly does. ■

Basic Efficiency Classes

Even though the efficiency analysis framework puts together all the functions whose orders of growth differ by a constant multiple, there are still infinitely many such classes. (For example, the exponential functions a^n have different orders of growth for different values of base a.) Therefore, it may come as a surprise that the time efficiencies of a large number of algorithms fall into only a few classes. These classes are listed in Table 2.2 in increasing order of their orders of growth, along with their names and a few comments.

You could raise a concern that classifying algorithms by their asymptotic efficiency would be of little practical use since the values of multiplicative constants are usually left unspecified. This leaves open the possibility of an algorithm in a worse efficiency class running faster than an algorithm in a better efficiency class for inputs of realistic sizes. For example, if the running time of one algorithm is n^3 while the running time of the other is 10^6n^2, the cubic algorithm will outperform the quadratic algorithm unless n exceeds 10^6. A few such anomalies are indeed known. Fortunately, multiplicative constants usually do not differ that drastically. As a rule, you should expect an algorithm from a better asymptotic efficiency class to outperform an algorithm from a worse class even for moderately sized inputs. This observation is especially true for an algorithm with a better than exponential running time versus an exponential (or worse) algorithm.

Exercises 2.2

1. Use the most appropriate notation among O, Θ, and Ω to indicate the time efficiency class of sequential search (see Section 2.1)
   a. in the worst case.
   b. in the best case.
   c. in the average case.
2. Use the informal definitions of O, Θ, and Ω to determine whether the following assertions are true or false.

TABLE 2.2 Basic asymptotic efficiency classes

Class	Name	Comments
1	*constant*	Short of best-case efficiencies, very few reasonable examples can be given since an algorithm's running time typically goes to infinity when its input size grows infinitely large.
$\log n$	*logarithmic*	Typically, a result of cutting a problem's size by a constant factor on each iteration of the algorithm (see Section 4.4). Note that a logarithmic algorithm cannot take into account all its input or even a fixed fraction of it: any algorithm that does so will have at least linear running time.
n	*linear*	Algorithms that scan a list of size n (e.g., sequential search) belong to this class.
$n \log n$	*linearithmic*	Many divide-and-conquer algorithms (see Chapter 5), including mergesort and quicksort in the average case, fall into this category.
n^2	*quadratic*	Typically, characterizes efficiency of algorithms with two embedded loops (see the next section). Elementary sorting algorithms and certain operations on $n \times n$ matrices are standard examples.
n^3	*cubic*	Typically, characterizes efficiency of algorithms with three embedded loops (see the next section). Several nontrivial algorithms from linear algebra fall into this class.
2^n	*exponential*	Typical for algorithms that generate all subsets of an n-element set. Often, the term "exponential" is used in a broader sense to include this and larger orders of growth as well.
$n!$	*factorial*	Typical for algorithms that generate all permutations of an n-element set.

a. $n(n+1)/2 \in O(n^3)$ **b.** $n(n+1)/2 \in O(n^2)$

c. $n(n+1)/2 \in \Theta(n^3)$ **d.** $n(n+1)/2 \in \Omega(n)$

3. For each of the following functions, indicate the class $\Theta(g(n))$ the function belongs to. (Use the simplest $g(n)$ possible in your answers.) Prove your assertions.

   a. $(n^2+1)^{10}$ **b.** $\sqrt{10n^2+7n+3}$

   c. $2n\lg(n+2)^2 + (n+2)^2\lg\frac{n}{2}$ **d.** $2^{n+1}+3^{n-1}$

   e. $\lfloor \log_2 n \rfloor$

4. a. Table 2.1 contains values of several functions that often arise in the analysis of algorithms. These values certainly suggest that the functions

$$\log n, \quad n, \quad n\log_2 n, \quad n^2, \quad n^3, \quad 2^n, \quad n!$$

are listed in increasing order of their order of growth. Do these values prove this fact with mathematical certainty?

b. Prove that the functions are indeed listed in increasing order of their order of growth.

5. List the following functions according to their order of growth from the lowest to the highest:

$$(n-2)!, \quad 5\lg(n+100)^{10}, \quad 2^{2n}, \quad 0.001n^4 + 3n^3 + 1, \quad \ln^2 n, \quad \sqrt[3]{n}, \quad 3^n.$$

6. a. Prove that every polynomial of degree k, $p(n) = a_k n^k + a_{k-1}n^{k-1} + \cdots + a_0$ with $a_k > 0$, belongs to $\Theta(n^k)$.

b. Prove that exponential functions a^n have different orders of growth for different values of base $a > 0$.

7. Prove the following assertions by using the definitions of the notations involved, or disprove them by giving a specific counterexample.

a. If $t(n) \in O(g(n))$, then $g(n) \in \Omega(t(n))$.

b. $\Theta(\alpha g(n)) = \Theta(g(n))$, where $\alpha > 0$.

c. $\Theta(g(n)) = O(g(n)) \cap \Omega(g(n))$.

d. For any two nonnegative functions $t(n)$ and $g(n)$ defined on the set of nonnegative integers, either $t(n) \in O(g(n))$, or $t(n) \in \Omega(g(n))$, or both.

8. Prove the section's theorem for

a. Ω notation. **b.** Θ notation.

9. We mentioned in this section that one can check whether all elements of an array are distinct by a two-part algorithm based on the array's presorting.

a. If the presorting is done by an algorithm with a time efficiency in $\Theta(n \log n)$. what will be a time-efficiency class of the entire algorithm?

b. If the sorting algorithm used for presorting needs an extra array of size n, what will be the space-efficiency class of the entire algorithm?

10. The ***range*** of a finite nonempty set of n real numbers S is defined as the difference between the largest and smallest elements of S. For each representation of S given below, describe in English an algorithm to compute the range. Indicate the time efficiency classes of these algorithms using the most appropriate notation (O, Θ, or Ω).

a. An unsorted array

b. A sorted array

c. A sorted singly linked list

d. A binary search tree

11. *Lighter or heavier?* You have $n > 2$ identical-looking coins and a two-pan balance scale with no weights. One of the coins is a fake, but you do not know whether it is lighter or heavier than the genuine coins, which all weigh the same. Design a $\Theta(1)$ algorithm to determine whether the fake coin is lighter or heavier than the others.

12. *Door in a wall* You are facing a wall that stretches infinitely in both directions. There is a door in the wall, but you know neither how far away nor in which direction. You can see the door only when you are right next to it. Design an algorithm that enables you to reach the door by walking at most $O(n)$ steps where n is the (unknown to you) number of steps between your initial position and the door. [Par95]

2.3 Mathematical Analysis of Nonrecursive Algorithms

In this section, we systematically apply the general framework outlined in Section 2.1 to analyzing the time efficiency of nonrecursive algorithms. Let us start with a very simple example that demonstrates all the principal steps typically taken in analyzing such algorithms.

EXAMPLE 1 Consider the problem of finding the value of the largest element in a list of n numbers. For simplicity, we assume that the list is implemented as an array. The following is pseudocode of a standard algorithm for solving the problem.

```
ALGORITHM MaxElement(A[0..n − 1])
    //Determines the value of the largest element in a given array
    //Input: An array A[0..n − 1] of real numbers
    //Output: The value of the largest element in A
    maxval ← A[0]
    for i ← 1 to n − 1 do
        if A[i] > maxval
            maxval ← A[i]
    return maxval
```

The obvious measure of an input's size here is the number of elements in the array, i.e., n. The operations that are going to be executed most often are in the algorithm's **for** loop. There are two operations in the loop's body: the comparison $A[i] > maxval$ and the assignment $maxval \leftarrow A[i]$. Which of these two operations should we consider basic? Since the comparison is executed on each repetition of the loop and the assignment is not, we should consider the comparison to be the algorithm's basic operation. Note that the number of comparisons will be the same for all arrays of size n; therefore, in terms of this metric, there is no need to distinguish among the worst, average, and best cases here.

Let us denote $C(n)$ the number of times this comparison is executed and try to find a formula expressing it as a function of size n. The algorithm makes one comparison on each execution of the loop, which is repeated for each value of the loop's variable i within the bounds 1 and $n-1$, inclusive. Therefore, we get the following sum for $C(n)$:

$$C(n) = \sum_{i=1}^{n-1} 1.$$

This is an easy sum to compute because it is nothing other than 1 repeated $n-1$ times. Thus,

$$C(n) = \sum_{i=1}^{n-1} 1 = n - 1 \in \Theta(n). \qquad \blacksquare$$

Here is a general plan to follow in analyzing nonrecursive algorithms.

General Plan for Analyzing the Time Efficiency of Nonrecursive Algorithms

1. Decide on a parameter (or parameters) indicating an input's size.
2. Identify the algorithm's basic operation. (As a rule, it is located in the innermost loop.)
3. Check whether the number of times the basic operation is executed depends only on the size of an input. If it also depends on some additional property, the worst-case, average-case, and, if necessary, best-case efficiencies have to be investigated separately.
4. Set up a sum expressing the number of times the algorithm's basic operation is executed.[4]
5. Using standard formulas and rules of sum manipulation, either find a closed-form formula for the count or, at the very least, establish its order of growth.

Before proceeding with further examples, you may want to review Appendix A, which contains a list of summation formulas and rules that are often useful in analysis of algorithms. In particular, we use especially frequently two basic rules of sum manipulation

$$\sum_{i=l}^{u} ca_i = c\sum_{i=l}^{u} a_i, \tag{R1}$$

$$\sum_{i=l}^{u} (a_i \pm b_i) = \sum_{i=l}^{u} a_i \pm \sum_{i=l}^{u} b_i, \tag{R2}$$

4. Sometimes, an analysis of a nonrecursive algorithm requires setting up not a sum but a recurrence relation for the number of times its basic operation is executed. Using recurrence relations is much more typical for analyzing recursive algorithms (see Section 2.4).

and two summation formulas

$$\sum_{i=l}^{u} 1 = u - l + 1 \quad \text{where } l \le u \text{ are some lower and upper integer limits,} \tag{S1}$$

$$\sum_{i=0}^{n} i = \sum_{i=1}^{n} i = 1 + 2 + \cdots + n = \frac{n(n+1)}{2} \approx \frac{1}{2}n^2 \in \Theta(n^2). \tag{S2}$$

Note that the formula $\sum_{i=1}^{n-1} 1 = n - 1$, which we used in Example 1, is a special case of formula (S1) for $l = 1$ and $u = n - 1$.

EXAMPLE 2 Consider the ***element uniqueness problem***: check whether all the elements in a given array of n elements are distinct. This problem can be solved by the following straightforward algorithm.

ALGORITHM *UniqueElements*$(A[0..n-1])$

```
//Determines whether all the elements in a given array are distinct
//Input: An array A[0..n − 1]
//Output: Returns "true" if all the elements in A are distinct
//         and "false" otherwise
for i ← 0 to n − 2 do
    for j ← i + 1 to n − 1 do
        if A[i] = A[j] return false
return true
```

The natural measure of the input's size here is again n, the number of elements in the array. Since the innermost loop contains a single operation (the comparison of two elements), we should consider it as the algorithm's basic operation. Note, however, that the number of element comparisons depends not only on n but also on whether there are equal elements in the array and, if there are, which array positions they occupy. We will limit our investigation to the worst case only.

By definition, the worst case input is an array for which the number of element comparisons $C_{worst}(n)$ is the largest among all arrays of size n. An inspection of the innermost loop reveals that there are two kinds of worst-case inputs—inputs for which the algorithm does not exit the loop prematurely: arrays with no equal elements and arrays in which the last two elements are the only pair of equal elements. For such inputs, one comparison is made for each repetition of the innermost loop, i.e., for each value of the loop variable j between its limits $i + 1$ and $n - 1$; this is repeated for each value of the outer loop, i.e., for each value of the loop variable i between its limits 0 and $n - 2$. Accordingly, we get

$$C_{worst}(n) = \sum_{i=0}^{n-2} \sum_{j=i+1}^{n-1} 1 = \sum_{i=0}^{n-2} [(n-1) - (i+1) + 1] = \sum_{i=0}^{n-2} (n-1-i)$$

$$= \sum_{i=0}^{n-2} (n-1) - \sum_{i=0}^{n-2} i = (n-1) \sum_{i=0}^{n-2} 1 - \frac{(n-2)(n-1)}{2}$$

$$= (n-1)^2 - \frac{(n-2)(n-1)}{2} = \frac{(n-1)n}{2} \approx \frac{1}{2}n^2 \in \Theta(n^2).$$

We also could have computed the sum $\sum_{i=0}^{n-2}(n-1-i)$ faster as follows:

$$\sum_{i=0}^{n-2}(n-1-i) = (n-1) + (n-2) + \cdots + 1 = \frac{(n-1)n}{2},$$

where the last equality is obtained by applying summation formula (S2). Note that this result was perfectly predictable: in the worst case, the algorithm needs to compare all $n(n-1)/2$ distinct pairs of its n elements. ■

EXAMPLE 3 Given two $n \times n$ matrices A and B, find the time efficiency of the definition-based algorithm for computing their product $C = AB$. By definition, C is an $n \times n$ matrix whose elements are computed as the scalar (dot) products of the rows of matrix A and the columns of matrix B:

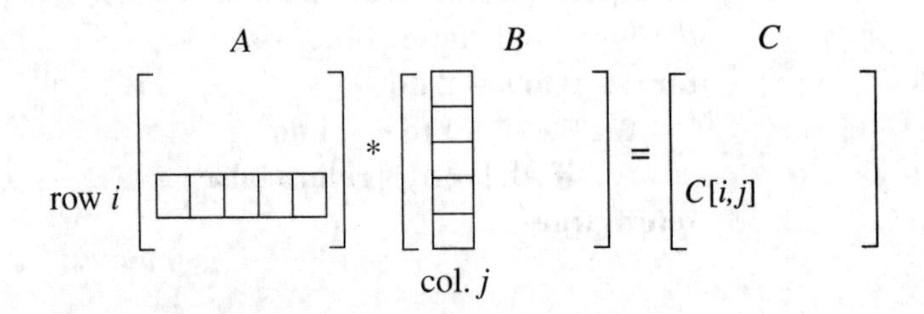

where $C[i, j] = A[i, 0]B[0, j] + \cdots + A[i, k]B[k, j] + \cdots + A[i, n-1]B[n-1, j]$ for every pair of indices $0 \le i, j \le n-1$.

```
ALGORITHM  MatrixMultiplication(A[0..n − 1, 0..n − 1], B[0..n − 1, 0..n − 1])
    //Multiplies two square matrices of order n by the definition-based algorithm
    //Input: Two n × n matrices A and B
    //Output: Matrix C = AB
    for i ← 0 to n − 1 do
        for j ← 0 to n − 1 do
            C[i, j] ← 0.0
            for k ← 0 to n − 1 do
                C[i, j] ← C[i, j] + A[i, k] ∗ B[k, j]
    return C
```

We measure an input's size by matrix order n. There are two arithmetical operations in the innermost loop here—multiplication and addition—that, in principle, can compete for designation as the algorithm's basic operation. Actually, we do not have to choose between them, because on each repetition of the innermost loop each of the two is executed exactly once. So by counting one we automatically count the other. Still, following a well-established tradition, we consider multiplication as the basic operation (see Section 2.1). Let us set up a sum for the total number of multiplications $M(n)$ executed by the algorithm. (Since this count depends only on the size of the input matrices, we do not have to investigate the worst-case, average-case, and best-case efficiencies separately.)

Obviously, there is just one multiplication executed on each repetition of the algorithm's innermost loop, which is governed by the variable k ranging from the lower bound 0 to the upper bound $n-1$. Therefore, the number of multiplications made for every pair of specific values of variables i and j is

$$\sum_{k=0}^{n-1} 1,$$

and the total number of multiplications $M(n)$ is expressed by the following triple sum:

$$M(n) = \sum_{i=0}^{n-1}\sum_{j=0}^{n-1}\sum_{k=0}^{n-1} 1.$$

Now, we can compute this sum by using formula (S1) and rule (R1) given above. Starting with the innermost sum $\sum_{k=0}^{n-1} 1$, which is equal to n (why?), we get

$$M(n) = \sum_{i=0}^{n-1}\sum_{j=0}^{n-1}\sum_{k=0}^{n-1} 1 = \sum_{i=0}^{n-1}\sum_{j=0}^{n-1} n = \sum_{i=0}^{n-1} n^2 = n^3.$$

This example is simple enough so that we could get this result without all the summation machinations. How? The algorithm computes n^2 elements of the product matrix. Each of the product's elements is computed as the scalar (dot) product of an n-element row of the first matrix and an n-element column of the second matrix, which takes n multiplications. So the total number of multiplications is $n \cdot n^2 = n^3$. (It is this kind of reasoning that we expected you to employ when answering this question in Problem 2 of Exercises 2.1.)

If we now want to estimate the running time of the algorithm on a particular machine, we can do it by the product

$$T(n) \approx c_m M(n) = c_m n^3,$$

where c_m is the time of one multiplication on the machine in question. We would get a more accurate estimate if we took into account the time spent on the additions, too:

$$T(n) \approx c_m M(n) + c_a A(n) = c_m n^3 + c_a n^3 = (c_m + c_a)n^3,$$

where c_a is the time of one addition. Note that the estimates differ only by their multiplicative constants and not by their order of growth. ■

You should not have the erroneous impression that the plan outlined above always succeeds in analyzing a nonrecursive algorithm. An irregular change in a loop variable, a sum too complicated to analyze, and the difficulties intrinsic to the average case analysis are just some of the obstacles that can prove to be insurmountable. These caveats notwithstanding, the plan does work for many simple nonrecursive algorithms, as you will see throughout the subsequent chapters of the book.

As a last example, let us consider an algorithm in which the loop's variable changes in a different manner from that of the previous examples.

EXAMPLE 4 The following algorithm finds the number of binary digits in the binary representation of a positive decimal integer.

ALGORITHM *Binary*(n)

//Input: A positive decimal integer n
//Output: The number of binary digits in n's binary representation
count ← 1
while $n > 1$ **do**
    count ← *count* + 1
    $n \leftarrow \lfloor n/2 \rfloor$
return *count*

First, notice that the most frequently executed operation here is not inside the **while** loop but rather the comparison $n > 1$ that determines whether the loop's body will be executed. Since the number of times the comparison will be executed is larger than the number of repetitions of the loop's body by exactly 1, the choice is not that important.

A more significant feature of this example is the fact that the loop variable takes on only a few values between its lower and upper limits; therefore, we have to use an alternative way of computing the number of times the loop is executed. Since the value of n is about halved on each repetition of the loop, the answer should be about $\log_2 n$. The exact formula for the number of times the comparison $n > 1$ will be executed is actually $\lfloor \log_2 n \rfloor + 1$—the number of bits in the binary representation of n according to formula (2.1). We could also get this answer by applying the analysis technique based on recurrence relations; we discuss this technique in the next section because it is more pertinent to the analysis of recursive algorithms. ■

Exercises 2.3

1. Compute the following sums.
   a. $1+3+5+7+\cdots+999$
   b. $2+4+8+16+\cdots+1024$
   c. $\sum_{i=3}^{n+1} 1$ **d.** $\sum_{i=3}^{n+1} i$ **e.** $\sum_{i=0}^{n-1} i(i+1)$
   f. $\sum_{j=1}^{n} 3^{j+1}$ **g.** $\sum_{i=1}^{n}\sum_{j=1}^{n} ij$ **h.** $\sum_{i=1}^{n} 1/i(i+1)$
2. Find the order of growth of the following sums. Use the $\Theta(g(n))$ notation with the simplest function $g(n)$ possible.
   a. $\sum_{i=0}^{n-1}(i^2+1)^2$ **b.** $\sum_{i=2}^{n-1} \lg i^2$
   c. $\sum_{i=1}^{n}(i+1)2^{i-1}$ **d.** $\sum_{i=0}^{n-1}\sum_{j=0}^{i-1}(i+j)$
3. The sample variance of n measurements $x_1, \ldots, x_n$ can be computed as either

$$\frac{\sum_{i=1}^{n}(x_i-\bar{x})^2}{n-1} \quad \text{where } \bar{x} = \frac{\sum_{i=1}^{n} x_i}{n}$$

   or

$$\frac{\sum_{i=1}^{n} x_i^2 - (\sum_{i=1}^{n} x_i)^2/n}{n-1}.$$

   Find and compare the number of divisions, multiplications, and additions/subtractions (additions and subtractions are usually bunched together) that are required for computing the variance according to each of these formulas.
4. Consider the following algorithm.

```
ALGORITHM Mystery(n)
    //Input: A nonnegative integer n
    S ← 0
    for i ← 1 to n do
        S ← S + i * i
    return S
```

   a. What does this algorithm compute?
   b. What is its basic operation?
   c. How many times is the basic operation executed?
   d. What is the efficiency class of this algorithm?
   e. Suggest an improvement, or a better algorithm altogether, and indicate its efficiency class. If you cannot do it, try to prove that, in fact, it cannot be done.

5. Consider the following algorithm.

ALGORITHM *Secret*($A[0..n-1]$)
//Input: An array $A[0..n-1]$ of n real numbers
minval ← $A[0]$; *maxval* ← $A[0]$
for $i \leftarrow 1$ **to** $n-1$ **do**
  if $A[i] < minval$
    minval ← $A[i]$
  if $A[i] > maxval$
    maxval ← $A[i]$
return *maxval* − *minval*

Answer questions (a)–(e) of Problem 4 about this algorithm.

6. Consider the following algorithm.

ALGORITHM *Enigma*($A[0..n-1, 0..n-1]$)
//Input: A matrix $A[0..n-1, 0..n-1]$ of real numbers
for $i \leftarrow 0$ **to** $n-2$ **do**
  for $j \leftarrow i+1$ **to** $n-1$ **do**
    if $A[i, j] \neq A[j, i]$
      return false
return true

Answer questions (a)–(e) of Problem 4 about this algorithm.

7. Improve the implementation of the matrix multiplication algorithm (see Example 3) by reducing the number of additions made by the algorithm. What effect will this change have on the algorithm's efficiency?

8. Determine the asymptotic order of growth for the total number of times all the doors are toggled in the locker doors puzzle (Problem 12 in Exercises 1.1).

9. Prove the formula

$$\sum_{i=1}^{n} i = 1 + 2 + \cdots + n = \frac{n(n+1)}{2}$$

either by mathematical induction or by following the insight of a 10-year-old school boy named Carl Friedrich Gauss (1777–1855) who grew up to become one of the greatest mathematicians of all times.

10. *Mental arithmetic* A 10×10 table is filled with repeating numbers on its diagonals as shown below. Calculate the total sum of the table's numbers in your head (after [Cra07, Question 1.33]).

1	2	3			…			9	10
2	3						9	10	11
3						9	10	11	
					9	10	11		
				9	10	11			
⋮			9	10	11				⋮
		9	10	11					
	9	10	11						17
9	10	11						17	18
10	11				…		17	18	19

11. Consider the following version of an important algorithm that we will study later in the book.

Algorithm $GE(A[0..n-1, 0..n])$
//Input: An n-by-$n+1$ matrix $A[0..n-1, 0..n]$ of real numbers
for $i \leftarrow 0$ **to** $n-2$ **do**
  for $j \leftarrow i+1$ **to** $n-1$ **do**
    for $k \leftarrow n$ **downto** i **do**
      $A[j, k] \leftarrow A[j, k] - A[i, k] * A[j, i] / A[i, i]$

a. Find the time efficiency class of this algorithm.

b. What glaring inefficiency does this pseudocode contain and how can it be eliminated to speed the algorithm up?

12. *von Neumann's neighborhood* Consider the algorithm that starts with a single square and on each of its n iterations adds new squares all around the outside. How many one-by-one squares are there after n iterations? [Gar99] (In the parlance of cellular automata theory, the answer is the number of cells in the von Neumann neighborhood of range n.) The results for $n = 0$, 1, and 2 are illustrated below.

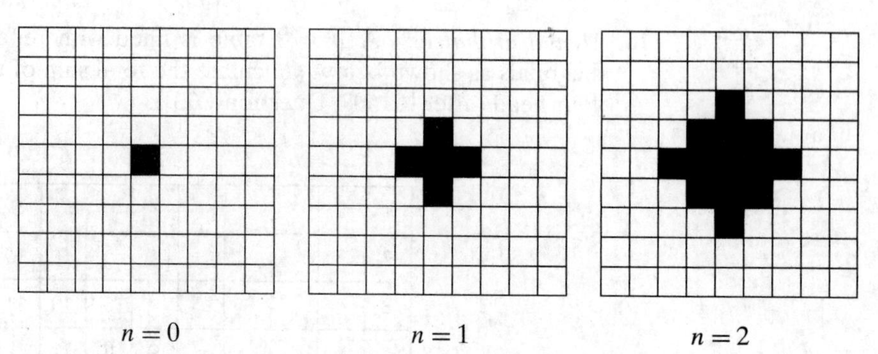

13. *Page numbering* Find the total number of decimal digits needed for numbering pages in a book of 1000 pages. Assume that the pages are numbered consecutively starting with 1.

2.4 Mathematical Analysis of Recursive Algorithms

In this section, we will see how to apply the general framework for analysis of algorithms to recursive algorithms. We start with an example often used to introduce novices to the idea of a recursive algorithm.

EXAMPLE 1 Compute the factorial function $F(n) = n!$ for an arbitrary nonnegative integer n. Since

$$n! = 1 \cdot \ldots \cdot (n-1) \cdot n = (n-1)! \cdot n \quad \text{for } n \geq 1$$

and $0! = 1$ by definition, we can compute $F(n) = F(n-1) \cdot n$ with the following recursive algorithm.

ALGORITHM $F(n)$

```
//Computes n! recursively
//Input: A nonnegative integer n
//Output: The value of n!
if n = 0 return 1
else return F(n − 1) ∗ n
```

For simplicity, we consider n itself as an indicator of this algorithm's input size (rather than the number of bits in its binary expansion). The basic operation of the algorithm is multiplication,[5] whose number of executions we denote $M(n)$. Since the function $F(n)$ is computed according to the formula

$$F(n) = F(n-1) \cdot n \quad \text{for } n > 0,$$

5. Alternatively, we could count the number of times the comparison $n = 0$ is executed, which is the same as counting the total number of calls made by the algorithm (see Problem 2 in this section's exercises).

the number of multiplications $M(n)$ needed to compute it must satisfy the equality

$$M(n) = \underbrace{M(n-1)}_{\substack{\text{to compute}\\ F(n-1)}} + \underbrace{1}_{\substack{\text{to multiply}\\ F(n-1)\text{ by } n}} \quad \text{for } n > 0.$$

Indeed, $M(n-1)$ multiplications are spent to compute $F(n-1)$, and one more multiplication is needed to multiply the result by n.

The last equation defines the sequence $M(n)$ that we need to find. This equation defines $M(n)$ not explicitly, i.e., as a function of n, but implicitly as a function of its value at another point, namely $n-1$. Such equations are called ***recurrence relations*** or, for brevity, ***recurrences***. Recurrence relations play an important role not only in analysis of algorithms but also in some areas of applied mathematics. They are usually studied in detail in courses on discrete mathematics or discrete structures; a very brief tutorial on them is provided in Appendix B. Our goal now is to solve the recurrence relation $M(n) = M(n-1) + 1$, i.e., to find an explicit formula for $M(n)$ in terms of n only.

Note, however, that there is not one but infinitely many sequences that satisfy this recurrence. (Can you give examples of, say, two of them?) To determine a solution uniquely, we need an ***initial condition*** that tells us the value with which the sequence starts. We can obtain this value by inspecting the condition that makes the algorithm stop its recursive calls:

if $n = 0$ **return** 1.

This tells us two things. First, since the calls stop when $n = 0$, the smallest value of n for which this algorithm is executed and hence $M(n)$ defined is 0. Second, by inspecting the pseudocode's exiting line, we can see that when $n = 0$, the algorithm performs no multiplications. Therefore, the initial condition we are after is

$$M(0) = 0.$$

the calls stop when $n = 0$ ——↑ ↑—— no multiplications when $n = 0$

Thus, we succeeded in setting up the recurrence relation and initial condition for the algorithm's number of multiplications $M(n)$:

$$\begin{aligned} M(n) &= M(n-1) + 1 \quad \text{for } n > 0, \\ M(0) &= 0. \end{aligned} \tag{2.2}$$

Before we embark on a discussion of how to solve this recurrence, let us pause to reiterate an important point. We are dealing here with two recursively defined functions. The first is the factorial function $F(n)$ itself; it is defined by the recurrence

$$\begin{aligned} F(n) &= F(n-1) \cdot n \quad \text{for every } n > 0, \\ F(0) &= 1. \end{aligned}$$

The second is the number of multiplications $M(n)$ needed to compute $F(n)$ by the recursive algorithm whose pseudocode was given at the beginning of the section.

As we just showed, $M(n)$ is defined by recurrence (2.2). And it is recurrence (2.2) that we need to solve now.

Though it is not difficult to "guess" the solution here (what sequence starts with 0 when $n = 0$ and increases by 1 on each step?), it will be more useful to arrive at it in a systematic fashion. From the several techniques available for solving recurrence relations, we use what can be called the ***method of backward substitutions***. The method's idea (and the reason for the name) is immediately clear from the way it applies to solving our particular recurrence:

$$\begin{aligned} M(n) &= M(n-1) + 1 && \text{substitute } M(n-1) = M(n-2) + 1 \\ &= [M(n-2) + 1] + 1 = M(n-2) + 2 && \text{substitute } M(n-2) = M(n-3) + 1 \\ &= [M(n-3) + 1] + 2 = M(n-3) + 3. \end{aligned}$$

After inspecting the first three lines, we see an emerging pattern, which makes it possible to predict not only the next line (what would it be?) but also a general formula for the pattern: $M(n) = M(n-i) + i$. Strictly speaking, the correctness of this formula should be proved by mathematical induction, but it is easier to get to the solution as follows and then verify its correctness.

What remains to be done is to take advantage of the initial condition given. Since it is specified for $n = 0$, we have to substitute $i = n$ in the pattern's formula to get the ultimate result of our backward substitutions:

$$M(n) = M(n-1) + 1 = \cdots = M(n-i) + i = \cdots = M(n-n) + n = n.$$

You should not be disappointed after exerting so much effort to get this "obvious" answer. The benefits of the method illustrated in this simple example will become clear very soon, when we have to solve more difficult recurrences. Also, note that the simple iterative algorithm that accumulates the product of n consecutive integers requires the same number of multiplications, and it does so without the overhead of time and space used for maintaining the recursion's stack.

The issue of time efficiency is actually not that important for the problem of computing $n!$, however. As we saw in Section 2.1, the function's values get so large so fast that we can realistically compute exact values of $n!$ only for very small n's. Again, we use this example just as a simple and convenient vehicle to introduce the standard approach to analyzing recursive algorithms. ■

Generalizing our experience with investigating the recursive algorithm for computing $n!$, we can now outline a general plan for investigating recursive algorithms.

General Plan for Analyzing the Time Efficiency of Recursive Algorithms

1. Decide on a parameter (or parameters) indicating an input's size.
2. Identify the algorithm's basic operation.

3. Check whether the number of times the basic operation is executed can vary on different inputs of the same size; if it can, the worst-case, average-case, and best-case efficiencies must be investigated separately.
4. Set up a recurrence relation, with an appropriate initial condition, for the number of times the basic operation is executed.
5. Solve the recurrence or, at least, ascertain the order of growth of its solution.

EXAMPLE 2 As our next example, we consider another educational workhorse of recursive algorithms: the ***Tower of Hanoi*** puzzle. In this puzzle, we (or mythical monks, if you do not like to move disks) have n disks of different sizes that can slide onto any of three pegs. Initially, all the disks are on the first peg in order of size, the largest on the bottom and the smallest on top. The goal is to move all the disks to the third peg, using the second one as an auxiliary, if necessary. We can move only one disk at a time, and it is forbidden to place a larger disk on top of a smaller one.

The problem has an elegant recursive solution, which is illustrated in Figure 2.4. To move $n > 1$ disks from peg 1 to peg 3 (with peg 2 as auxiliary), we first move recursively $n - 1$ disks from peg 1 to peg 2 (with peg 3 as auxiliary), then move the largest disk directly from peg 1 to peg 3, and, finally, move recursively $n - 1$ disks from peg 2 to peg 3 (using peg 1 as auxiliary). Of course, if $n = 1$, we simply move the single disk directly from the source peg to the destination peg.

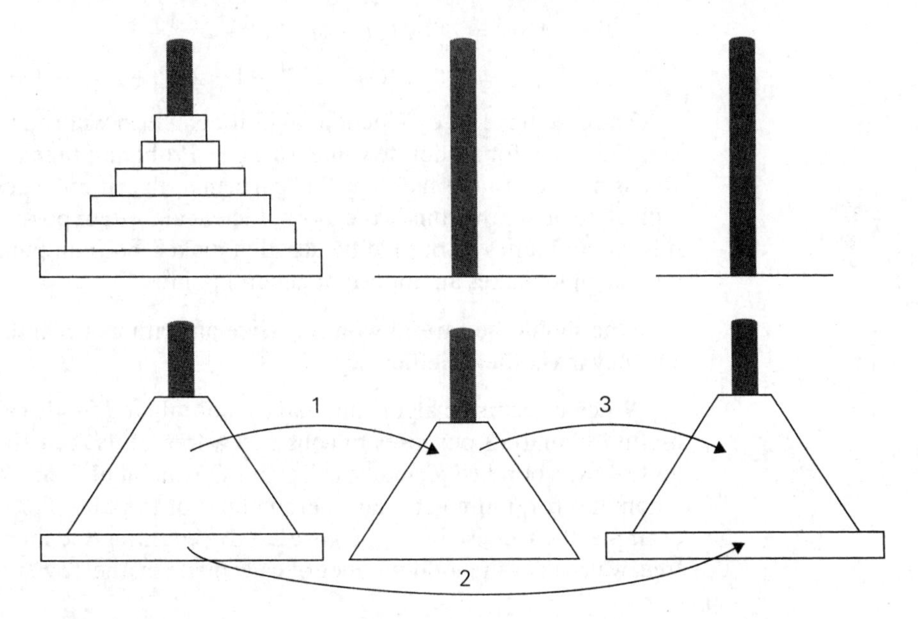

FIGURE 2.4 Recursive solution to the Tower of Hanoi puzzle.

Let us apply the general plan outlined above to the Tower of Hanoi problem. The number of disks n is the obvious choice for the input's size indicator, and so is moving one disk as the algorithm's basic operation. Clearly, the number of moves $M(n)$ depends on n only, and we get the following recurrence equation for it:

$$M(n) = M(n-1) + 1 + M(n-1) \quad \text{for } n > 1.$$

With the obvious initial condition $M(1) = 1$, we have the following recurrence relation for the number of moves $M(n)$:

$$\begin{aligned} M(n) &= 2M(n-1) + 1 \quad \text{for } n > 1, \\ M(1) &= 1. \end{aligned} \tag{2.3}$$

We solve this recurrence by the same method of backward substitutions:

$$\begin{aligned} M(n) &= 2M(n-1) + 1 && \text{sub. } M(n-1) = 2M(n-2) + 1 \\ &= 2[2M(n-2) + 1] + 1 = 2^2 M(n-2) + 2 + 1 && \text{sub. } M(n-2) = 2M(n-3) + 1 \\ &= 2^2[2M(n-3) + 1] + 2 + 1 = 2^3 M(n-3) + 2^2 + 2 + 1. \end{aligned}$$

The pattern of the first three sums on the left suggests that the next one will be $2^4 M(n-4) + 2^3 + 2^2 + 2 + 1$, and generally, after i substitutions, we get

$$M(n) = 2^i M(n-i) + 2^{i-1} + 2^{i-2} + \cdots + 2 + 1 = 2^i M(n-i) + 2^i - 1.$$

Since the initial condition is specified for $n = 1$, which is achieved for $i = n - 1$, we get the following formula for the solution to recurrence (2.3):

$$\begin{aligned} M(n) &= 2^{n-1} M(n-(n-1)) + 2^{n-1} - 1 \\ &= 2^{n-1} M(1) + 2^{n-1} - 1 = 2^{n-1} + 2^{n-1} - 1 = 2^n - 1. \end{aligned}$$

Thus, we have an exponential algorithm, which will run for an unimaginably long time even for moderate values of n (see Problem 5 in this section's exercises). This is not due to the fact that this particular algorithm is poor; in fact, it is not difficult to prove that this is the most efficient algorithm possible for this problem. It is the problem's intrinsic difficulty that makes it so computationally hard. Still, this example makes an important general point:

> One should be careful with recursive algorithms because their succinctness may mask their inefficiency.

When a recursive algorithm makes more than a single call to itself, it can be useful for analysis purposes to construct a tree of its recursive calls. In this tree, nodes correspond to recursive calls, and we can label them with the value of the parameter (or, more generally, parameters) of the calls. For the Tower of Hanoi example, the tree is given in Figure 2.5. By counting the number of nodes in the tree, we can get the total number of calls made by the Tower of Hanoi algorithm:

$$C(n) = \sum_{l=0}^{n-1} 2^l \text{ (where } l \text{ is the level in the tree in Figure 2.5)} = 2^n - 1.$$

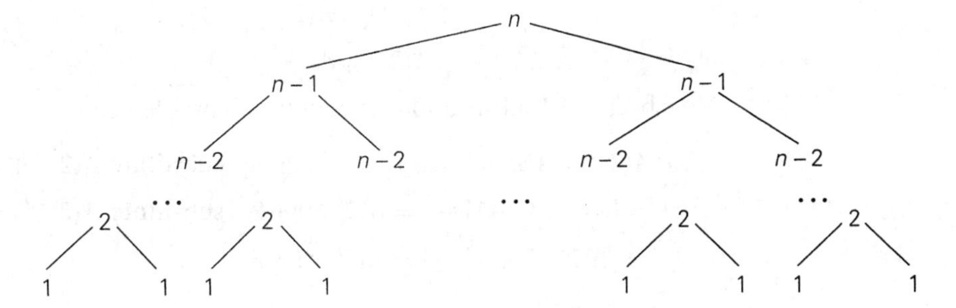

FIGURE 2.5 Tree of recursive calls made by the recursive algorithm for the Tower of Hanoi puzzle.

The number agrees, as it should, with the move count obtained earlier. ∎

EXAMPLE 3 As our next example, we investigate a recursive version of the algorithm discussed at the end of Section 2.3.

ALGORITHM *BinRec(n)*

```
//Input: A positive decimal integer n
//Output: The number of binary digits in n's binary representation
if n = 1 return 1
else return BinRec(⌊n/2⌋) + 1
```

Let us set up a recurrence and an initial condition for the number of additions $A(n)$ made by the algorithm. The number of additions made in computing $BinRec(\lfloor n/2 \rfloor)$ is $A(\lfloor n/2 \rfloor)$, plus one more addition is made by the algorithm to increase the returned value by 1. This leads to the recurrence

$$A(n) = A(\lfloor n/2 \rfloor) + 1 \quad \text{for } n > 1. \tag{2.4}$$

Since the recursive calls end when n is equal to 1 and there are no additions made then, the initial condition is

$$A(1) = 0.$$

The presence of $\lfloor n/2 \rfloor$ in the function's argument makes the method of backward substitutions stumble on values of n that are not powers of 2. Therefore, the standard approach to solving such a recurrence is to solve it only for $n = 2^k$ and then take advantage of the theorem called the ***smoothness rule*** (see Appendix B), which claims that under very broad assumptions the order of growth observed for $n = 2^k$ gives a correct answer about the order of growth for all values of n. (Alternatively, after getting a solution for powers of 2, we can sometimes fine-tune this solution to get a formula valid for an arbitrary n.) So let us apply this recipe to our recurrence, which for $n = 2^k$ takes the form

$$A(2^k) = A(2^{k-1}) + 1 \quad \text{for } k > 0,$$
$$A(2^0) = 0.$$

Now backward substitutions encounter no problems:

$$\begin{aligned} A(2^k) &= A(2^{k-1}) + 1 && \text{substitute } A(2^{k-1}) = A(2^{k-2}) + 1 \\ &= [A(2^{k-2}) + 1] + 1 = A(2^{k-2}) + 2 && \text{substitute } A(2^{k-2}) = A(2^{k-3}) + 1 \\ &= [A(2^{k-3}) + 1] + 2 = A(2^{k-3}) + 3 && \ldots \\ &\ldots \\ &= A(2^{k-i}) + i \\ &\ldots \\ &= A(2^{k-k}) + k. \end{aligned}$$

Thus, we end up with

$$A(2^k) = A(1) + k = k,$$

or, after returning to the original variable $n = 2^k$ and hence $k = \log_2 n$,

$$A(n) = \log_2 n \in \Theta(\log n).$$

In fact, one can prove (Problem 7 in this section's exercises) that the exact solution for an arbitrary value of n is given by just a slightly more refined formula $A(n) = \lfloor \log_2 n \rfloor$. ■

This section provides an introduction to the analysis of recursive algorithms. These techniques will be used throughout the book and expanded further as necessary. In the next section, we discuss the Fibonacci numbers; their analysis involves more difficult recurrence relations to be solved by a method different from backward substitutions.

Exercises 2.4

1. Solve the following recurrence relations.
   a. $x(n) = x(n-1) + 5$ for $n > 1$, $x(1) = 0$
   b. $x(n) = 3x(n-1)$ for $n > 1$, $x(1) = 4$
   c. $x(n) = x(n-1) + n$ for $n > 0$, $x(0) = 0$
   d. $x(n) = x(n/2) + n$ for $n > 1$, $x(1) = 1$ (solve for $n = 2^k$)
   e. $x(n) = x(n/3) + 1$ for $n > 1$, $x(1) = 1$ (solve for $n = 3^k$)
2. Set up and solve a recurrence relation for the number of calls made by $F(n)$, the recursive algorithm for computing $n!$.
3. Consider the following recursive algorithm for computing the sum of the first n cubes: $S(n) = 1^3 + 2^3 + \cdots + n^3$.

ALGORITHM $S(n)$

//Input: A positive integer n
//Output: The sum of the first n cubes
if $n = 1$ **return** 1
else return $S(n-1) + n * n * n$

a. Set up and solve a recurrence relation for the number of times the algorithm's basic operation is executed.

b. How does this algorithm compare with the straightforward nonrecursive algorithm for computing this sum?

4. Consider the following recursive algorithm.

ALGORITHM $Q(n)$

//Input: A positive integer n
if $n = 1$ **return** 1
else return $Q(n-1) + 2 * n - 1$

a. Set up a recurrence relation for this function's values and solve it to determine what this algorithm computes.

b. Set up a recurrence relation for the number of multiplications made by this algorithm and solve it.

c. Set up a recurrence relation for the number of additions/subtractions made by this algorithm and solve it.

5. *Tower of Hanoi*

a. In the original version of the Tower of Hanoi puzzle, as it was published in the 1890s by Édouard Lucas, a French mathematician, the world will end after 64 disks have been moved from a mystical Tower of Brahma. Estimate the number of years it will take if monks could move one disk per minute. (Assume that monks do not eat, sleep, or die.)

b. How many moves are made by the ith largest disk $(1 \le i \le n)$ in this algorithm?

c. Find a nonrecursive algorithm for the Tower of Hanoi puzzle and implement it in the language of your choice.

6. *Restricted Tower of Hanoi* Consider the version of the Tower of Hanoi puzzle in which n disks have to be moved from peg A to peg C using peg B so that any move should either place a disk on peg B or move a disk from that peg. (Of course, the prohibition of placing a larger disk on top of a smaller one remains in place, too.) Design a recursive algorithm for this problem and find the number of moves made by it.

7. a. Prove that the exact number of additions made by the recursive algorithm *BinRec*(n) for an arbitrary positive decimal integer n is $\lfloor \log_2 n \rfloor$.

b. Set up a recurrence relation for the number of additions made by the nonrecursive version of this algorithm (see Section 2.3, Example 4) and solve it.

8. a. Design a recursive algorithm for computing 2^n for any nonnegative integer n that is based on the formula $2^n = 2^{n-1} + 2^{n-1}$.

b. Set up a recurrence relation for the number of additions made by the algorithm and solve it.

c. Draw a tree of recursive calls for this algorithm and count the number of calls made by the algorithm.

d. Is it a good algorithm for solving this problem?

9. Consider the following recursive algorithm.

```
ALGORITHM  Riddle(A[0..n − 1])
    //Input: An array A[0..n − 1] of real numbers
    if n = 1 return A[0]
    else temp ← Riddle(A[0..n − 2])
        if temp ≤ A[n − 1] return temp
        else return A[n − 1]
```

a. What does this algorithm compute?

b. Set up a recurrence relation for the algorithm's basic operation count and solve it.

10. Consider the following algorithm to check whether a graph defined by its adjacency matrix is complete.

```
ALGORITHM  GraphComplete(A[0..n − 1, 0..n − 1])
    //Input: Adjacency matrix A[0..n − 1, 0..n − 1]) of an undirected graph G
    //Output: 1 (true) if G is complete and 0 (false) otherwise
    if n = 1 return 1  //one-vertex graph is complete by definition
    else
        if not GraphComplete(A[0..n − 2, 0..n − 2]) return 0
        else for j ← 0 to n − 2 do
                if A[n − 1, j] = 0 return 0
            return 1
```

What is the algorithm's efficiency class in the worst case?

11. The determinant of an $n \times n$ matrix

$$A = \begin{bmatrix} a_{0\,0} & \cdots & a_{0\,n-1} \\ a_{1\,0} & \cdots & a_{1\,n-1} \\ \vdots & & \vdots \\ a_{n-1\,0} & \cdots & a_{n-1\,n-1} \end{bmatrix},$$

denoted det A, can be defined as a_{00} for $n = 1$ and, for $n > 1$, by the recursive formula

$$\det A = \sum_{j=0}^{n-1} s_j a_{0\,j} \det A_j,$$

where s_j is +1 if j is even and −1 if j is odd, $a_{0\,j}$ is the element in row 0 and column j, and A_j is the $(n-1) \times (n-1)$ matrix obtained from matrix A by deleting its row 0 and column j.

a. Set up a recurrence relation for the number of multiplications made by the algorithm implementing this recursive definition.

b. Without solving the recurrence, what can you say about the solution's order of growth as compared to $n!$?

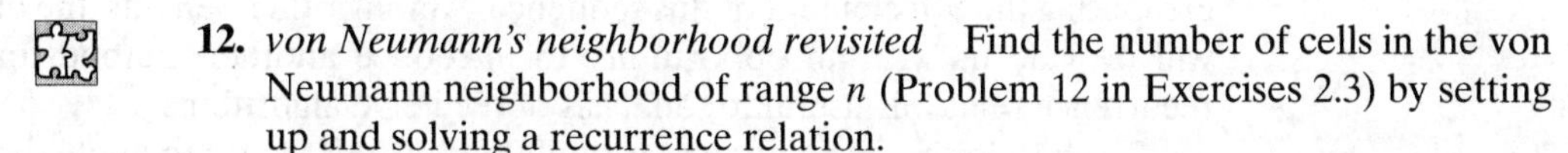

12. *von Neumann's neighborhood revisited* Find the number of cells in the von Neumann neighborhood of range n (Problem 12 in Exercises 2.3) by setting up and solving a recurrence relation.

13. *Frying hamburgers* There are n hamburgers to be fried on a small grill that can hold only two hamburgers at a time. Each hamburger has to be fried on both sides; frying one side of a hamburger takes 1 minute, regardless of whether one or two hamburgers are fried at the same time. Consider the following recursive algorithm for executing this task in the minimum amount of time. If $n \le 2$, fry the hamburger or the two hamburgers together on each side. If $n > 2$, fry any two hamburgers together on each side and then apply the same procedure recursively to the remaining $n - 2$ hamburgers.

a. Set up and solve the recurrence for the amount of time this algorithm needs to fry n hamburgers.

b. Explain why this algorithm does *not* fry the hamburgers in the minimum amount of time for all $n > 0$.

c. Give a correct recursive algorithm that executes the task in the minimum amount of time.

14. *Celebrity problem* A celebrity among a group of n people is a person who knows nobody but is known by everybody else. The task is to identify a celebrity by only asking questions to people of the form "Do you know him/her?" Design an efficient algorithm to identify a celebrity or determine that the group has no such person. How many questions does your algorithm need in the worst case?

2.5 Example: Computing the *n*th Fibonacci Number

In this section, we consider the ***Fibonacci numbers***, a famous sequence

$$0,\ 1,\ 1,\ 2,\ 3,\ 5,\ 8,\ 13,\ 21,\ 34, \ldots \tag{2.5}$$

that can be defined by the simple recurrence

$$F(n) = F(n-1) + F(n-2) \quad \text{for } n > 1 \tag{2.6}$$

and two initial conditions

$$F(0) = 0, \qquad F(1) = 1. \tag{2.7}$$

The Fibonacci numbers were introduced by Leonardo Fibonacci in 1202 as a solution to a problem about the size of a rabbit population (Problem 2 in this section's exercises). Many more examples of Fibonacci-like numbers have since been discovered in the natural world, and they have even been used in predicting the prices of stocks and commodities. There are some interesting applications of the Fibonacci numbers in computer science as well. For example, worst-case inputs for Euclid's algorithm discussed in Section 1.1 happen to be consecutive elements of the Fibonacci sequence. In this section, we briefly consider algorithms for computing the nth element of this sequence. Among other benefits, the discussion will provide us with an opportunity to introduce another method for solving recurrence relations useful for analysis of recursive algorithms.

To start, let us get an explicit formula for $F(n)$. If we try to apply the method of backward substitutions to solve recurrence (2.6), we will fail to get an easily discernible pattern. Instead, we can take advantage of a theorem that describes solutions to a ***homogeneous second-order linear recurrence with constant coefficients***

$$ax(n) + bx(n-1) + cx(n-2) = 0, \tag{2.8}$$

where a, b, and c are some fixed real numbers ($a \neq 0$) called the coefficients of the recurrence and $x(n)$ is the generic term of an unknown sequence to be found. Applying this theorem to our recurrence with the initial conditions given—see Appendix B—we obtain the formula

$$F(n) = \frac{1}{\sqrt{5}}(\phi^n - \hat{\phi}^n), \tag{2.9}$$

where $\phi = (1 + \sqrt{5})/2 \approx 1.61803$ and $\hat{\phi} = -1/\phi \approx -0.61803$.[6] It is hard to believe that formula (2.9), which includes arbitrary integer powers of irrational numbers, yields nothing else but all the elements of Fibonacci sequence (2.5), but it does!

One of the benefits of formula (2.9) is that it immediately implies that $F(n)$ grows exponentially (remember Fibonacci's rabbits?), i.e., $F(n) \in \Theta(\phi^n)$. This

6. Constant ϕ is known as the ***golden ratio***. Since antiquity, it has been considered the most pleasing ratio of a rectangle's two sides to the human eye and might have been consciously used by ancient architects and sculptors.

follows from the observation that $\hat{\phi}$ is a fraction between -1 and 0, and hence $\hat{\phi}^n$ gets infinitely small as n goes to infinity. In fact, one can prove that the impact of the second term $\frac{1}{\sqrt{5}}\hat{\phi}^n$ on the value of $F(n)$ can be obtained by rounding off the value of the first term to the nearest integer. In other words, for every nonnegative integer n,

$$F(n) = \frac{1}{\sqrt{5}}\phi^n \quad \text{rounded to the nearest integer.} \qquad \textbf{(2.10)}$$

In the algorithms that follow, we consider, for the sake of simplicity, such operations as additions and multiplications at unit cost. Since the Fibonacci numbers grow infinitely large (and grow very rapidly), a more detailed analysis than the one offered here is warranted. In fact, it is the size of the numbers rather than a time-efficient method for computing them that should be of primary concern here. Still, these caveats notwithstanding, the algorithms we outline and their analysis provide useful examples for a student of the design and analysis of algorithms.

To begin with, we can use recurrence (2.6) and initial conditions (2.7) for the obvious recursive algorithm for computing $F(n)$.

ALGORITHM $F(n)$

```
//Computes the nth Fibonacci number recursively by using its definition
//Input: A nonnegative integer n
//Output: The nth Fibonacci number
if n ≤ 1 return n
else return F(n − 1) + F(n − 2)
```

Before embarking on its formal analysis, can you tell whether this is an efficient algorithm? Well, we need to do a formal analysis anyway. The algorithm's basic operation is clearly addition, so let $A(n)$ be the number of additions performed by the algorithm in computing $F(n)$. Then the numbers of additions needed for computing $F(n-1)$ and $F(n-2)$ are $A(n-1)$ and $A(n-2)$, respectively, and the algorithm needs one more addition to compute their sum. Thus, we get the following recurrence for $A(n)$:

$$A(n) = A(n-1) + A(n-2) + 1 \quad \text{for } n > 1, \qquad \textbf{(2.11)}$$
$$A(0) = 0, \qquad A(1) = 0.$$

The recurrence $A(n) - A(n-1) - A(n-2) = 1$ is quite similar to recurrence $F(n) - F(n-1) - F(n-2) = 0$, but its right-hand side is not equal to zero. Such recurrences are called ***inhomogeneous***. There are general techniques for solving inhomogeneous recurrences (see Appendix B or any textbook on discrete mathematics), but for this particular recurrence, a special trick leads to a faster solution. We can reduce our inhomogeneous recurrence to a homogeneous one by rewriting it as

$$[A(n) + 1] - [A(n-1) + 1] - [A(n-2) + 1] = 0$$

and substituting $B(n) = A(n) + 1$:

$$B(n) - B(n-1) - B(n-2) = 0,$$
$$B(0) = 1, \qquad B(1) = 1.$$

This homogeneous recurrence can be solved exactly in the same manner as recurrence (2.6) was solved to find an explicit formula for $F(n)$. But it can actually be avoided by noting that $B(n)$ is, in fact, the same recurrence as $F(n)$ except that it starts with two 1's and thus runs one step ahead of $F(n)$. So $B(n) = F(n+1)$, and

$$A(n) = B(n) - 1 = F(n+1) - 1 = \frac{1}{\sqrt{5}}(\phi^{n+1} - \hat{\phi}^{n+1}) - 1.$$

Hence, $A(n) \in \Theta(\phi^n)$, and if we measure the size of n by the number of bits $b = \lfloor \log_2 n \rfloor + 1$ in its binary representation, the efficiency class will be even worse, namely, doubly exponential: $A(b) \in \Theta(\phi^{2^b})$.

The poor efficiency class of the algorithm could be anticipated by the nature of recurrence (2.11). Indeed, it contains two recursive calls with the sizes of smaller instances only slightly smaller than size n. (Have you encountered such a situation before?) We can also see the reason behind the algorithm's inefficiency by looking at a recursive tree of calls tracing the algorithm's execution. An example of such a tree for $n = 5$ is given in Figure 2.6. Note that the same values of the function are being evaluated here again and again, which is clearly extremely inefficient.

We can obtain a much faster algorithm by simply computing the successive elements of the Fibonacci sequence iteratively, as is done in the following algorithm.

ALGORITHM *Fib*(n)

//Computes the nth Fibonacci number iteratively by using its definition
//Input: A nonnegative integer n
//Output: The nth Fibonacci number
$F[0] \leftarrow 0$; $F[1] \leftarrow 1$
for $i \leftarrow 2$ **to** n **do**
  $F[i] \leftarrow F[i-1] + F[i-2]$
return $F[n]$

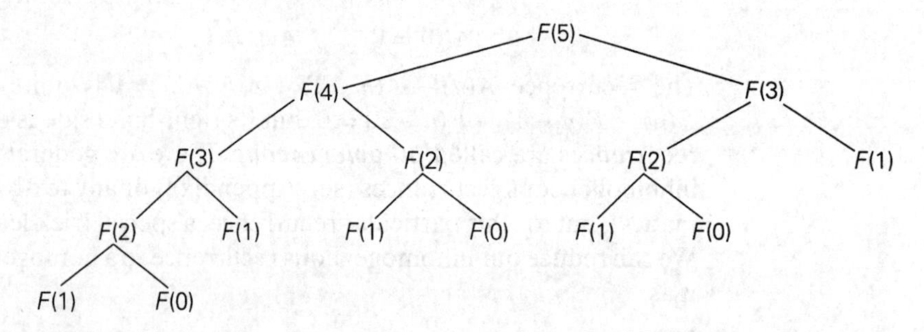

FIGURE 2.6 Tree of recursive calls for computing the 5th Fibonacci number by the definition-based algorithm.

This algorithm clearly makes $n - 1$ additions. Hence, it is linear as a function of n and "only" exponential as a function of the number of bits b in n's binary representation. Note that using an extra array for storing all the preceding elements of the Fibonacci sequence can be avoided: storing just two values is necessary to accomplish the task (see Problem 8 in this section's exercises).

The third alternative for computing the nth Fibonacci number lies in using formula (2.10). The efficiency of the algorithm will obviously be determined by the efficiency of an exponentiation algorithm used for computing ϕ^n. If it is done by simply multiplying ϕ by itself $n - 1$ times, the algorithm will be in $\Theta(n) = \Theta(2^b)$. There are faster algorithms for the exponentiation problem. For example, we will discuss $\Theta(\log n) = \Theta(b)$ algorithms for this problem in Chapters 4 and 6. Note also that special care should be exercised in implementing this approach to computing the nth Fibonacci number. Since all its intermediate results are irrational numbers, we would have to make sure that their approximations in the computer are accurate enough so that the final round-off yields a correct result.

Finally, there exists a $\Theta(\log n)$ algorithm for computing the nth Fibonacci number that manipulates only integers. It is based on the equality

$$\begin{bmatrix} F(n-1) & F(n) \\ F(n) & F(n+1) \end{bmatrix} = \begin{bmatrix} 0 & 1 \\ 1 & 1 \end{bmatrix}^n \quad \text{for } n \geq 1$$

and an efficient way of computing matrix powers.

Exercises 2.5

1. Find a Web site dedicated to applications of the Fibonacci numbers and study it.

2. *Fibonacci's rabbits problem* A man put a pair of rabbits in a place surrounded by a wall. How many pairs of rabbits will be there in a year if the initial pair of rabbits (male and female) are newborn and all rabbit pairs are not fertile during their first month of life but thereafter give birth to one new male/female pair at the end of every month?

3. *Climbing stairs* Find the number of different ways to climb an n-stair staircase if each step is either one or two stairs. For example, a 3-stair staircase can be climbed three ways: 1-1-1, 1-2, and 2-1.

4. How many even numbers are there among the first n Fibonacci numbers, i.e., among the numbers $F(0), F(1), \ldots, F(n-1)$? Give a closed-form formula valid for every $n > 0$.

5. Check by direct substitutions that the function $\frac{1}{\sqrt{5}}(\phi^n - \hat{\phi}^n)$ indeed satisfies recurrence (2.6) and initial conditions (2.7).

6. The maximum values of the Java primitive types `int` and `long` are $2^{31} - 1$ and $2^{63} - 1$, respectively. Find the smallest n for which the nth Fibonacci number is not going to fit in a memory allocated for

a. the type `int`. **b.** the type `long`.

7. Consider the recursive definition-based algorithm for computing the nth Fibonacci number $F(n)$. Let $C(n)$ and $Z(n)$ be the number of times $F(1)$ and $F(0)$ are computed, respectively. Prove that

a. $C(n) = F(n)$. **b.** $Z(n) = F(n-1)$.

8. Improve algorithm *Fib* of the text so that it requires only $\Theta(1)$ space.

9. Prove the equality

$$\begin{bmatrix} F(n-1) & F(n) \\ F(n) & F(n+1) \end{bmatrix} = \begin{bmatrix} 0 & 1 \\ 1 & 1 \end{bmatrix}^n \quad \text{for } n \geq 1.$$

10. How many modulo divisions are made by Euclid's algorithm on two consecutive Fibonacci numbers $F(n)$ and $F(n-1)$ as the algorithm's input?

11. *Dissecting a Fibonacci rectangle* Given a rectangle whose sides are two consecutive Fibonacci numbers, design an algorithm to dissect it into squares with no more than two squares being the same size. What is the time efficiency class of your algorithm?

12. In the language of your choice, implement two algorithms for computing the last five digits of the nth Fibonacci number that are based on (a) the recursive definition-based algorithm $F(n)$; (b) the iterative definition-based algorithm *Fib*(n). Perform an experiment to find the largest value of n for which your programs run under 1 minute on your computer.

2.6 Empirical Analysis of Algorithms

In Sections 2.3 and 2.4, we saw how algorithms, both nonrecursive and recursive, can be analyzed mathematically. Though these techniques can be applied successfully to many simple algorithms, the power of mathematics, even when enhanced with more advanced techniques (see [Sed96], [Pur04], [Gra94], and [Gre07]), is far from limitless. In fact, even some seemingly simple algorithms have proved to be very difficult to analyze with mathematical precision and certainty. As we pointed out in Section 2.1, this is especially true for the average-case analysis.

The principal alternative to the mathematical analysis of an algorithm's efficiency is its empirical analysis. This approach implies steps spelled out in the following plan.

General Plan for the Empirical Analysis of Algorithm Time Efficiency

1. Understand the experiment's purpose.
2. Decide on the efficiency metric M to be measured and the measurement unit (an operation count vs. a time unit).
3. Decide on characteristics of the input sample (its range, size, and so on).
4. Prepare a program implementing the algorithm (or algorithms) for the experimentation.

5. Generate a sample of inputs.
6. Run the algorithm (or algorithms) on the sample's inputs and record the data observed.
7. Analyze the data obtained.

Let us discuss these steps one at a time. There are several different goals one can pursue in analyzing algorithms empirically. They include checking the accuracy of a theoretical assertion about the algorithm's efficiency, comparing the efficiency of several algorithms for solving the same problem or different implementations of the same algorithm, developing a hypothesis about the algorithm's efficiency class, and ascertaining the efficiency of the program implementing the algorithm on a particular machine. Obviously, an experiment's design should depend on the question the experimenter seeks to answer.

In particular, the goal of the experiment should influence, if not dictate, how the algorithm's efficiency is to be measured. The first alternative is to insert a counter (or counters) into a program implementing the algorithm to count the number of times the algorithm's basic operation is executed. This is usually a straightforward operation; you should only be mindful of the possibility that the basic operation is executed in several places in the program and that all its executions need to be accounted for. As straightforward as this task usually is, you should always test the modified program to ensure that it works correctly, in terms of both the problem it solves and the counts it yields.

The second alternative is to time the program implementing the algorithm in question. The easiest way to do this is to use a system's command, such as the `time` command in UNIX. Alternatively, one can measure the running time of a code fragment by asking for the system time right before the fragment's start (t_{start}) and just after its completion (t_{finish}), and then computing the difference between the two ($t_{finish} - t_{start}$).[7] In C and C++, you can use the function `clock` for this purpose; in Java, the method `currentTimeMillis()` in the `System` class is available.

It is important to keep several facts in mind, however. First, a system's time is typically not very accurate, and you might get somewhat different results on repeated runs of the same program on the same inputs. An obvious remedy is to make several such measurements and then take their average (or the median) as the sample's observation point. Second, given the high speed of modern computers, the running time may fail to register at all and be reported as zero. The standard trick to overcome this obstacle is to run the program in an extra loop many times, measure the total running time, and then divide it by the number of the loop's repetitions. Third, on a computer running under a time-sharing system such as UNIX, the reported time may include the time spent by the CPU on other programs, which obviously defeats the purpose of the experiment. Therefore, you should take care to ask the system for the time devoted specifically to execution of

7. If the system time is given in units called "ticks," the difference should be divided by a constant indicating the number of ticks per time unit.

your program. (In UNIX, this time is called the "user time," and it is automatically provided by the `time` command.)

Thus, measuring the physical running time has several disadvantages, both principal (dependence on a particular machine being the most important of them) and technical, not shared by counting the executions of a basic operation. On the other hand, the physical running time provides very specific information about an algorithm's performance in a particular computing environment, which can be of more importance to the experimenter than, say, the algorithm's asymptotic efficiency class. In addition, measuring time spent on different segments of a program can pinpoint a bottleneck in the program's performance that can be missed by an abstract deliberation about the algorithm's basic operation. Getting such data—called ***profiling***—is an important resource in the empirical analysis of an algorithm's running time; the data in question can usually be obtained from the system tools available in most computing environments.

Whether you decide to measure the efficiency by basic operation counting or by time clocking, you will need to decide on a sample of inputs for the experiment. Often, the goal is to use a sample representing a "typical" input; so the challenge is to understand what a "typical" input is. For some classes of algorithms—e.g., for algorithms for the traveling salesman problem that we are going to discuss later in the book—researchers have developed a set of instances they use for benchmarking. But much more often than not, an input sample has to be developed by the experimenter. Typically, you will have to make decisions about the sample size (it is sensible to start with a relatively small sample and increase it later if necessary), the range of instance sizes (typically neither trivially small nor excessively large), and a procedure for generating instances in the range chosen. The instance sizes can either adhere to some pattern (e.g., 1000, 2000, 3000, . . . , 10,000 or 500, 1000, 2000, 4000, . . . , 128,000) or be generated randomly within the range chosen.

The principal advantage of size changing according to a pattern is that its impact is easier to analyze. For example, if a sample's sizes are generated by doubling, you can compute the ratios $M(2n)/M(n)$ of the observed metric M (the count or the time) to see whether the ratios exhibit a behavior typical of algorithms in one of the basic efficiency classes discussed in Section 2.2. The major disadvantage of nonrandom sizes is the possibility that the algorithm under investigation exhibits atypical behavior on the sample chosen. For example, if all the sizes in a sample are even and your algorithm runs much more slowly on odd-size inputs, the empirical results will be quite misleading.

Another important issue concerning sizes in an experiment's sample is whether several instances of the same size should be included. If you expect the observed metric to vary considerably on instances of the same size, it would be probably wise to include several instances for every size in the sample. (There are well-developed methods in statistics to help the experimenter make such decisions; you will find no shortage of books on this subject.) Of course, if several instances of the same size are included in the sample, the averages or medians of the observed values for each size should be computed and investigated instead of or in addition to individual sample points.

Much more often than not, an empirical analysis requires generating random numbers. Even if you decide to use a pattern for input sizes, you will typically want instances themselves generated randomly. Generating random numbers on a digital computer is known to present a difficult problem because, in principle, the problem can be solved only approximately. This is the reason computer scientists prefer to call such numbers ***pseudorandom***. As a practical matter, the easiest and most natural way of getting such numbers is to take advantage of a random number generator available in computer language libraries. Typically, its output will be a value of a (pseudo)random variable uniformly distributed in the interval between 0 and 1. If a different (pseudo)random variable is desired, an appropriate transformation needs to be made. For example, if x is a continuous random variable uniformly distributed on the interval $0 \le x < 1$, the variable $y = l + \lfloor x(r - l) \rfloor$ will be uniformly distributed among the integer values between integers l and $r - 1$ $(l < r)$.

Alternatively, you can implement one of several known algorithms for generating (pseudo)random numbers. The most widely used and thoroughly studied of such algorithms is the ***linear congruential method***.

ALGORITHM *Random*$(n, m, seed, a, b)$

//Generates a sequence of n pseudorandom numbers according to the linear
//   congruential method
//Input: A positive integer n and positive integer parameters m, $seed$, a, b
//Output: A sequence $r_1, \ldots, r_n$ of n pseudorandom integers uniformly
//   distri   among integer values between 0 and $m - 1$
//Note: Pseudorandom numbers between 0 and 1 can be obtained
//   by treating the integers generated as digits after the decimal point

$r_0 \leftarrow seed$
for $i \leftarrow 1$ **to** n **do**
    $r_i \leftarrow (a * r_{i-1} + b) \bmod m$

The simplicity of this pseudocode is misleading because the devil lies in the details of choosing the algorithm's parameters. Here is a partial list of recommendations based on the results of a sophisticated mathematical analysis (see [KnuII, pp. 184–185] for details): *seed* may be chosen arbitrarily and is often set to the current date and time; m should be large and may be conveniently taken as 2^w, where w is the computer's word size; a should be selected as an integer between $0.01m$ and $0.99m$ with no particular pattern in its digits but such that $a \bmod 8 = 5$; and the value of b can be chosen as 1.

The empirical data obtained as the result of an experiment need to be recorded and then presented for an analysis. Data can be presented numerically in a table or graphically in a ***scatterplot***, i.e., by points in a Cartesian coordinate system. It is a good idea to use both these options whenever it is feasible because both methods have their unique strengths and weaknesses.

The principal advantage of tabulated data lies in the opportunity to manipulate it easily. For example, one can compute the ratios $M(n)/g(n)$ where $g(n)$ is a candidate to represent the efficiency class of the algorithm in question. If the algorithm is indeed in $\Theta(g(n))$, most likely these ratios will converge to some positive constant as n gets large. (Note that careless novices sometimes assume that this constant must be 1, which is, of course, incorrect according to the definition of $\Theta(g(n))$.) Or one can compute the ratios $M(2n)/M(n)$ and see how the running time reacts to doubling of its input size. As we discussed in Section 2.2, such ratios should change only slightly for logarithmic algorithms and most likely converge to 2, 4, and 8 for linear, quadratic, and cubic algorithms, respectively—to name the most obvious and convenient cases.

On the other hand, the form of a scatterplot may also help in ascertaining the algorithm's probable efficiency class. For a logarithmic algorithm, the scatterplot will have a concave shape (Figure 2.7a); this fact distinguishes it from all the other basic efficiency classes. For a linear algorithm, the points will tend to aggregate around a straight line or, more generally, to be contained between two straight lines (Figure 2.7b). Scatterplots of functions in $\Theta(n \lg n)$ and $\Theta(n^2)$ will have a convex shape (Figure 2.7c), making them difficult to differentiate. A scatterplot of a cubic algorithm will also have a convex shape, but it will show a much more rapid increase in the metric's values. An exponential algorithm will most probably require a logarithmic scale for the vertical axis, in which the values of $\log_a M(n)$ rather than those of $M(n)$ are plotted. (The commonly used logarithm base is 2 or 10.) In such a coordinate system, a scatterplot of a truly exponential algorithm should resemble a linear function because $M(n) \approx ca^n$ implies $\log_b M(n) \approx \log_b c + n \log_b a$, and vice versa.

One of the possible applications of the empirical analysis is to predict the algorithm's performance on an instance not included in the experiment sample. For example, if you observe that the ratios $M(n)/g(n)$ are close to some constant c for the sample instances, it could be sensible to approximate $M(n)$ by the product $cg(n)$ for other instances, too. This approach should be used with caution, especially for values of n outside the sample range. (Mathematicians call such predictions ***extrapolation***, as opposed to ***interpolation***, which deals with values within the sample range.) Of course, you can try unleashing the standard techniques of statistical data analysis and prediction. Note, however, that the majority of such techniques are based on specific probabilistic assumptions that may or may not be valid for the experimental data in question.

It seems appropriate to end this section by pointing out the basic differences between mathematical and empirical analyses of algorithms. The principal strength of the mathematical analysis is its independence of specific inputs; its principal weakness is its limited applicability, especially for investigating the average-case efficiency. The principal strength of the empirical analysis lies in its applicability to any algorithm, but its results can depend on the particular sample of instances and the computer used in the experiment.

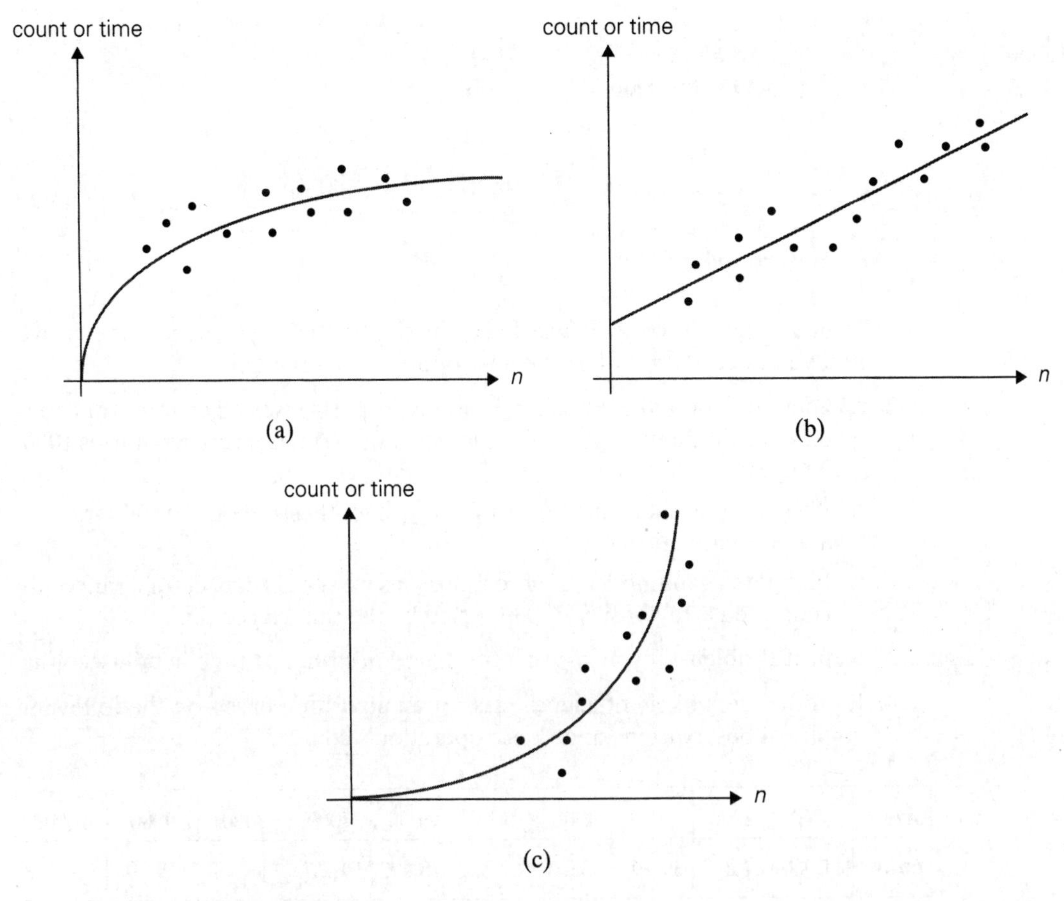

FIGURE 2.7 Typical scatter plots. (a) Logarithmic. (b) Linear. (c) One of the convex functions.

Exercises 2.6

1. Consider the following well-known sorting algorithm, which is studied later in the book, with a counter inserted to count the number of key comparisons.

ALGORITHM *SortAnalysis*$(A[0..n-1])$
//Input: An array $A[0..n-1]$ of n orderable elements
//Output: The total number of key comparisons made
count $\leftarrow 0$
for $i \leftarrow 1$ **to** $n-1$ **do**

```
v ← A[i]
j ← i − 1
while j ≥ 0 and A[j] > v do
    count ← count + 1
    A[j + 1] ← A[j]
    j ← j − 1
A[j + 1] ← v
return count
```

Is the comparison counter inserted in the right place? If you believe it is, prove it; if you believe it is not, make an appropriate correction.

2. **a.** Run the program of Problem 1, with a properly inserted counter (or counters) for the number of key comparisons, on 20 random arrays of sizes 1000, 2000, 3000, . . . , 20,000.

   b. Analyze the data obtained to form a hypothesis about the algorithm's average-case efficiency.

   c. Estimate the number of key comparisons we should expect for a randomly generated array of size 25,000 sorted by the same algorithm.

3. Repeat Problem 2 by measuring the program's running time in milliseconds.

4. Hypothesize a likely efficiency class of an algorithm based on the following empirical observations of its basic operation's count:

size	1000	2000	3000	4000	5000	6000	7000	8000	9000	10000
count	11,966	24,303	39,992	53,010	67,272	78,692	91,274	113,063	129,799	140,538

5. What scale transformation will make a logarithmic scatterplot look like a linear one?

6. How can one distinguish a scatterplot for an algorithm in $\Theta(\lg \lg n)$ from a scatterplot for an algorithm in $\Theta(\lg n)$?

7. **a.** Find empirically the largest number of divisions made by Euclid's algorithm for computing $\gcd(m, n)$ for $1 \le n \le m \le 100$.

   b. For each positive integer k, find empirically the smallest pair of integers $1 \le n \le m \le 100$ for which Euclid's algorithm needs to make k divisions in order to find $\gcd(m, n)$.

8. The average-case efficiency of Euclid's algorithm on inputs of size n can be measured by the average number of divisions $D_{avg}(n)$ made by the algorithm in computing $\gcd(n, 1)$, $\gcd(n, 2)$, . . . , $\gcd(n, n)$. For example,

$$D_{avg}(5) = \frac{1}{5}(1 + 2 + 3 + 2 + 1) = 1.8.$$

Produce a scatterplot of $D_{avg}(n)$ and indicate the algorithm's likely average-case efficiency class.

9. Run an experiment to ascertain the efficiency class of the sieve of Eratosthenes (see Section 1.1).

10. Run a timing experiment for the three algorithms for computing $\gcd(m, n)$ presented in Section 1.1.

2.7 Algorithm Visualization

In addition to the mathematical and empirical analyses of algorithms, there is yet a third way to study algorithms. It is called ***algorithm visualization*** and can be defined as the use of images to convey some useful information about algorithms. That information can be a visual illustration of an algorithm's operation, of its performance on different kinds of inputs, or of its execution speed versus that of other algorithms for the same problem. To accomplish this goal, an algorithm visualization uses graphic elements—points, line segments, two- or three-dimensional bars, and so on—to represent some "interesting events" in the algorithm's operation.

There are two principal variations of algorithm visualization:

- Static algorithm visualization
- Dynamic algorithm visualization, also called ***algorithm animation***

Static algorithm visualization shows an algorithm's progress through a series of still images. Algorithm animation, on the other hand, shows a continuous, movie-like presentation of an algorithm's operations. Animation is an arguably more sophisticated option, which, of course, is much more difficult to implement.

Early efforts in the area of algorithm visualization go back to the 1970s. The watershed event happened in 1981 with the appearance of a 30-minute color sound film titled *Sorting Out Sorting*. This algorithm visualization classic was produced at the University of Toronto by Ronald Baecker with the assistance of D. Sherman [Bae81, Bae98]. It contained visualizations of nine well-known sorting algorithms (more than half of them are discussed later in the book) and provided quite a convincing demonstration of their relative speeds.

The success of *Sorting Out Sorting* made sorting algorithms a perennial favorite for algorithm animation. Indeed, the sorting problem lends itself quite naturally to visual presentation via vertical or horizontal bars or sticks of different heights or lengths, which need to be rearranged according to their sizes (Figure 2.8). This presentation is convenient, however, only for illustrating actions of a typical sorting algorithm on small inputs. For larger files, *Sorting Out Sorting* used the ingenious idea of presenting data by a scatterplot of points on a coordinate plane, with the first coordinate representing an item's position in the file and the second one representing the item's value; with such a representation, the process of sorting looks like a transformation of a "random" scatterplot of points into the points along a frame's diagonal (Figure 2.9). In addition, most sorting algorithms

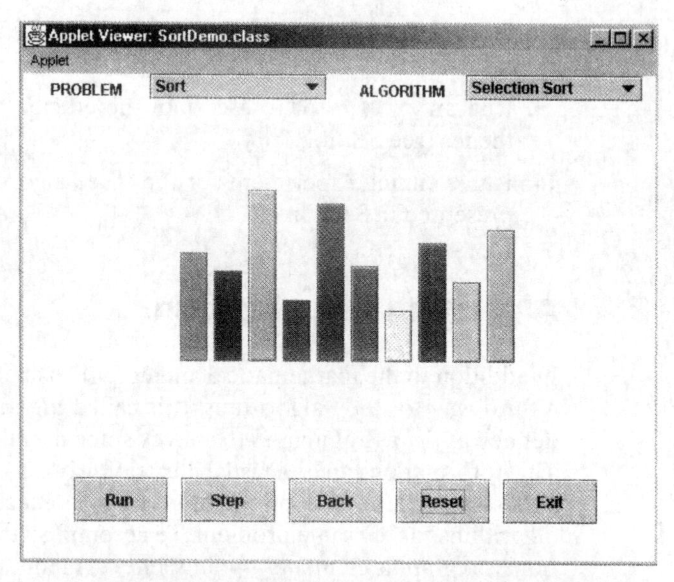

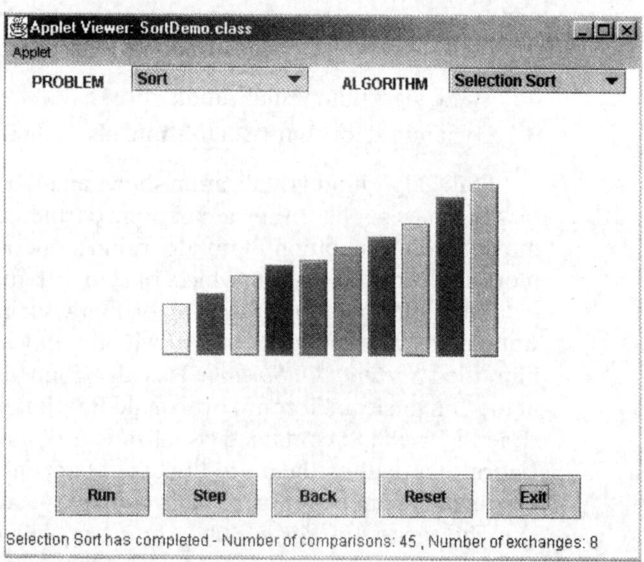

FIGURE 2.8 Initial and final screens of a typical visualization of a sorting algorithm using the bar representation.

work by comparing and exchanging two given items at a time—an event that can be animated relatively easily.

Since the appearance of *Sorting Out Sorting*, a great number of algorithm animations have been created, especially after the appearance of Java and the

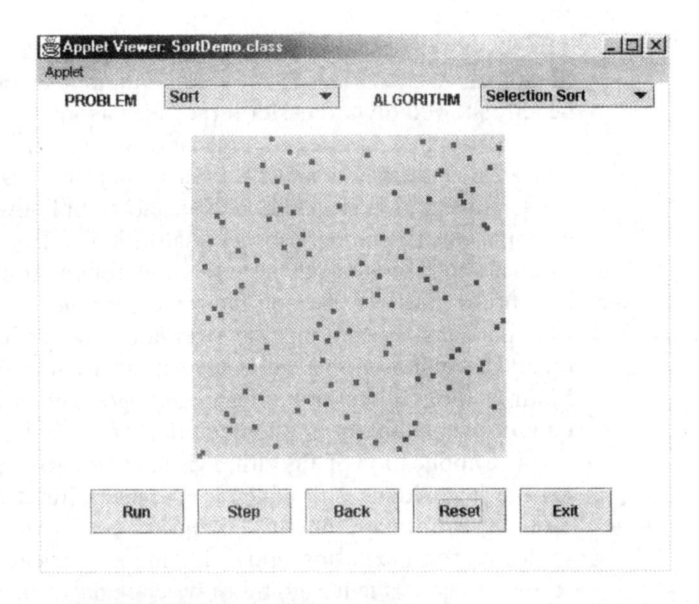

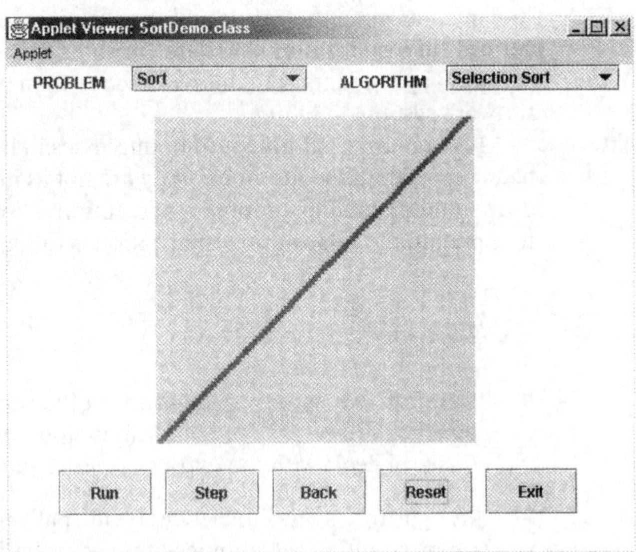

FIGURE 2.9 Initial and final screens of a typical visualization of a sorting algorithm using the scatterplot representation.

World Wide Web in the 1990s. They range in scope from one particular algorithm to a group of algorithms for the same problem (e.g., sorting) or the same application area (e.g., geometric algorithms) to general-purpose animation systems. At the end of 2010, a catalog of links to existing visualizations, maintained under the

NSF-supported AlgoVizProject, contained over 500 links. Unfortunately, a survey of existing visualizations found most of them to be of low quality, with the content heavily skewed toward easier topics such as sorting [Sha07].

There are two principal applications of algorithm visualization: research and education. Potential benefits for researchers are based on expectations that algorithm visualization may help uncover some unknown features of algorithms. For example, one researcher used a visualization of the recursive Tower of Hanoi algorithm in which odd- and even-numbered disks were colored in two different colors. He noticed that two disks of the same color never came in direct contact during the algorithm's execution. This observation helped him in developing a better nonrecursive version of the classic algorithm. To give another example, Bentley and McIlroy [Ben93] mentioned using an algorithm animation system in their work on improving a library implementation of a leading sorting algorithm.

The application of algorithm visualization to education seeks to help students learning algorithms. The available evidence of its effectiveness is decisively mixed. Although some experiments did register positive learning outcomes, others failed to do so. The increasing body of evidence indicates that creating sophisticated software systems is not going to be enough. In fact, it appears that the level of student involvement with visualization might be more important than specific features of visualization software. In some experiments, low-tech visualizations prepared by students were more effective than passive exposure to sophisticated software systems.

To summarize, although some successes in both research and education have been reported in the literature, they are not as impressive as one might expect. A deeper understanding of human perception of images will be required before the true potential of algorithm visualization is fulfilled.

SUMMARY

- There are two kinds of algorithm efficiency: time efficiency and space efficiency. *Time efficiency* indicates how fast the algorithm runs; *space efficiency* deals with the extra space it requires.

- An algorithm's time efficiency is principally measured as a function of its input size by counting the number of times its basic operation is executed. A *basic operation* is the operation that contributes the most to running time. Typically, it is the most time-consuming operation in the algorithm's innermost loop.

- For some algorithms, the running time can differ considerably for inputs of the same size, leading to *worst-case* efficiency, *average-case* efficiency, and *best-case* efficiency.

- The established framework for analyzing time efficiency is primarily grounded in the order of growth of the algorithm's running time as its input size goes to infinity.

- The notations O, Ω, and Θ are used to indicate and compare the asymptotic orders of growth of functions expressing algorithm efficiencies.
- The efficiencies of a large number of algorithms fall into the following few classes: *constant*, *logarithmic*, *linear*, *linearithmic*, *quadratic*, *cubic*, and *exponential*.
- The main tool for analyzing the time efficiency of a nonrecursive algorithm is to set up a sum expressing the number of executions of its basic operation and ascertain the sum's order of growth.
- The main tool for analyzing the time efficiency of a recursive algorithm is to set up a recurrence relation expressing the number of executions of its basic operation and ascertain the solution's order of growth.
- Succinctness of a recursive algorithm may mask its inefficiency.
- The *Fibonacci numbers* are an important sequence of integers in which every element is equal to the sum of its two immediate predecessors. There are several algorithms for computing the Fibonacci numbers, with drastically different efficiencies.
- Empirical analysis of an algorithm is performed by running a program implementing the algorithm on a sample of inputs and analyzing the data observed (the basic operation's count or physical running time). This often involves generating pseudorandom numbers. The applicability to any algorithm is the principal strength of this approach; the dependence of results on the particular computer and instance sample is its main weakness.
- *Algorithm visualization* is the use of images to convey useful information about algorithms. The two principal variations of algorithm visualization are static algorithm visualization and dynamic algorithm visualization (also called *algorithm animation*).

3

Brute Force and Exhaustive Search

Science is as far removed from brute force as this sword from a crowbar.
—Edward Lytton (1803–1873), *Leila*, Book II, Chapter I

Doing a thing well is often a waste of time.
—Robert Byrne, a master pool and billiards player and a writer

After introducing the framework and methods for algorithm analysis in the preceding chapter, we are ready to embark on a discussion of algorithm design strategies. Each of the next eight chapters is devoted to a particular design strategy. The subject of this chapter is brute force and its important special case, exhaustive search. Brute force can be described as follows:

> ***Brute force*** is a straightforward approach to solving a problem, usually directly based on the problem statement and definitions of the concepts involved.

The "force" implied by the strategy's definition is that of a computer and not that of one's intellect. "Just do it!" would be another way to describe the prescription of the brute-force approach. And often, the brute-force strategy is indeed the one that is easiest to apply.

As an example, consider the exponentiation problem: compute a^n for a nonzero number a and a nonnegative integer n. Although this problem might seem trivial, it provides a useful vehicle for illustrating several algorithm design strategies, including the brute force. (Also note that computing $a^n \bmod m$ for some large integers is a principal component of a leading encryption algorithm.) By the definition of exponentiation,

$$a^n = \underbrace{a * \cdots * a}_{n \text{ times}}.$$

This suggests simply computing a^n by multiplying 1 by a n times.

We have already encountered at least two brute-force algorithms in the book: the consecutive integer checking algorithm for computing $\gcd(m, n)$ in Section 1.1 and the definition-based algorithm for matrix multiplication in Section 2.3. Many other examples are given later in this chapter. (Can you identify a few algorithms you already know as being based on the brute-force approach?)

Though rarely a source of clever or efficient algorithms, the brute-force approach should not be overlooked as an important algorithm design strategy. First, unlike some of the other strategies, brute force is applicable to a very wide variety of problems. In fact, it seems to be the only general approach for which it is more difficult to point out problems it *cannot* tackle. Second, for some important problems—e.g., sorting, searching, matrix multiplication, string matching—the brute-force approach yields reasonable algorithms of at least some practical value with no limitation on instance size. Third, the expense of designing a more efficient algorithm may be unjustifiable if only a few instances of a problem need to be solved and a brute-force algorithm can solve those instances with acceptable speed. Fourth, even if too inefficient in general, a brute-force algorithm can still be useful for solving small-size instances of a problem. Finally, a brute-force algorithm can serve an important theoretical or educational purpose as a yardstick with which to judge more efficient alternatives for solving a problem.

3.1 Selection Sort and Bubble Sort

In this section, we consider the application of the brute-force approach to the problem of sorting: given a list of n orderable items (e.g., numbers, characters from some alphabet, character strings), rearrange them in nondecreasing order. As we mentioned in Section 1.3, dozens of algorithms have been developed for solving this very important problem. You might have learned several of them in the past. If you have, try to forget them for the time being and look at the problem afresh.

Now, after your mind is unburdened of previous knowledge of sorting algorithms, ask yourself a question: "What would be the most straightforward method for solving the sorting problem?" Reasonable people may disagree on the answer to this question. The two algorithms discussed here—selection sort and bubble sort—seem to be the two prime candidates.

Selection Sort

We start selection sort by scanning the entire given list to find its smallest element and exchange it with the first element, putting the smallest element in its final position in the sorted list. Then we scan the list, starting with the second element, to find the smallest among the last $n-1$ elements and exchange it with the second element, putting the second smallest element in its final position. Generally, on the

ith pass through the list, which we number from 0 to $n-2$, the algorithm searches for the smallest item among the last $n-i$ elements and swaps it with A_i:

$$\underbrace{A_0 \le A_1 \le \cdots \le A_{i-1}}_{\text{in their final positions}} \mid \underbrace{A_i, \ldots, A_{min}, \ldots, A_{n-1}}_{\text{the last } n-i \text{ elements}}$$

After $n-1$ passes, the list is sorted.

Here is pseudocode of this algorithm, which, for simplicity, assumes that the list is implemented as an array:

```
ALGORITHM  SelectionSort(A[0..n − 1])
    //Sorts a given array by selection sort
    //Input: An array A[0..n − 1] of orderable elements
    //Output: Array A[0..n − 1] sorted in nondecreasing order
    for i ← 0 to n − 2 do
        min ← i
        for j ← i + 1 to n − 1 do
            if A[j] < A[min]  min ← j
        swap A[i] and A[min]
```

As an example, the action of the algorithm on the list 89, 45, 68, 90, 29, 34, 17 is illustrated in Figure 3.1.

The analysis of selection sort is straightforward. The input size is given by the number of elements n; the basic operation is the key comparison $A[j] < A[min]$. The number of times it is executed depends only on the array size and is given by the following sum:

$$C(n) = \sum_{i=0}^{n-2} \sum_{j=i+1}^{n-1} 1 = \sum_{i=0}^{n-2} [(n-1) - (i+1) + 1] = \sum_{i=0}^{n-2} (n-1-i).$$

| 89 45 68 90 29 34 **17**
17 | 45 68 90 **29** 34 89
17 29 | 68 90 45 **34** 89
17 29 34 | 90 **45** 68 89
17 29 34 45 | 90 **68** 89
17 29 34 45 68 | 90 **89**
17 29 34 45 68 89 | 90

FIGURE 3.1 Example of sorting with selection sort. Each line corresponds to one iteration of the algorithm, i.e., a pass through the list's tail to the right of the vertical bar; an element in bold indicates the smallest element found. Elements to the left of the vertical bar are in their final positions and are not considered in this and subsequent iterations.

Since we have already encountered the last sum in analyzing the algorithm of Example 2 in Section 2.3, you should be able to compute it now on your own. Whether you compute this sum by distributing the summation symbol or by immediately getting the sum of decreasing integers, the answer, of course, must be the same:

$$C(n) = \sum_{i=0}^{n-2} \sum_{j=i+1}^{n-1} 1 = \sum_{i=0}^{n-2} (n-1-i) = \frac{(n-1)n}{2}.$$

Thus, selection sort is a $\Theta(n^2)$ algorithm on all inputs. Note, however, that the number of key swaps is only $\Theta(n)$, or, more precisely, $n-1$ (one for each repetition of the i loop). This property distinguishes selection sort positively from many other sorting algorithms.

Bubble Sort

Another brute-force application to the sorting problem is to compare adjacent elements of the list and exchange them if they are out of order. By doing it repeatedly, we end up "bubbling up" the largest element to the last position on the list. The next pass bubbles up the second largest element, and so on, until after $n-1$ passes the list is sorted. Pass i $(0 \le i \le n-2)$ of bubble sort can be represented by the following diagram:

$$A_0, \ldots, A_j \overset{?}{\leftrightarrow} A_{j+1}, \ldots, A_{n-i-1} \mid \underbrace{A_{n-i} \le \cdots \le A_{n-1}}_{\text{in their final positions}}$$

Here is pseudocode of this algorithm.

ALGORITHM *BubbleSort*($A[0..n-1]$)
    //Sorts a given array by bubble sort
    //Input: An array $A[0..n-1]$ of orderable elements
    //Output: Array $A[0..n-1]$ sorted in nondecreasing order
    for $i \leftarrow 0$ **to** $n-2$ **do**
        for $j \leftarrow 0$ **to** $n-2-i$ **do**
            if $A[j+1] < A[j]$ swap $A[j]$ and $A[j+1]$

The action of the algorithm on the list 89, 45, 68, 90, 29, 34, 17 is illustrated as an example in Figure 3.2.

The number of key comparisons for the bubble-sort version given above is the same for all arrays of size n; it is obtained by a sum that is almost identical to the sum for selection sort:

$$89 \overset{?}{\leftrightarrow} 45 \quad 68 \quad 90 \quad 29 \quad 34 \quad 17$$
$$45 \quad 89 \overset{?}{\leftrightarrow} 68 \quad 90 \quad 29 \quad 34 \quad 17$$
$$45 \quad 68 \quad 89 \overset{?}{\leftrightarrow} 90 \overset{?}{\leftrightarrow} 29 \quad 34 \quad 17$$
$$45 \quad 68 \quad 89 \quad 29 \quad 90 \overset{?}{\leftrightarrow} 34 \quad 17$$
$$45 \quad 68 \quad 89 \quad 29 \quad 34 \quad 90 \overset{?}{\leftrightarrow} 17$$
$$45 \quad 68 \quad 89 \quad 29 \quad 34 \quad 17 \mid 90$$

$$45 \overset{?}{\leftrightarrow} 68 \overset{?}{\leftrightarrow} 89 \overset{?}{\leftrightarrow} 29 \quad 34 \quad 17 \mid 90$$
$$45 \quad 68 \quad 29 \quad 89 \overset{?}{\leftrightarrow} 34 \quad 17 \mid 90$$
$$45 \quad 68 \quad 29 \quad 34 \quad 89 \overset{?}{\leftrightarrow} 17 \mid 90$$
$$45 \quad 68 \quad 29 \quad 34 \quad 17 \mid 89 \quad 90$$

etc.

FIGURE 3.2 First two passes of bubble sort on the list 89, 45, 68, 90, 29, 34, 17. A new line is shown after a swap of two elements is done. The elements to the right of the vertical bar are in their final positions and are not considered in subsequent iterations of the algorithm.

$$C(n) = \sum_{i=0}^{n-2} \sum_{j=0}^{n-2-i} 1 = \sum_{i=0}^{n-2} [(n-2-i) - 0 + 1]$$

$$= \sum_{i=0}^{n-2} (n-1-i) = \frac{(n-1)n}{2} \in \Theta(n^2).$$

The number of key swaps, however, depends on the input. In the worst case of decreasing arrays, it is the same as the number of key comparisons:

$$S_{worst}(n) = C(n) = \frac{(n-1)n}{2} \in \Theta(n^2).$$

As is often the case with an application of the brute-force strategy, the first version of an algorithm obtained can often be improved upon with a modest amount of effort. Specifically, we can improve the crude version of bubble sort given above by exploiting the following observation: if a pass through the list makes no exchanges, the list has been sorted and we can stop the algorithm (Problem 12a in this section's exercises). Though the new version runs faster on some inputs, it is still in $\Theta(n^2)$ in the worst and average cases. In fact, even among elementary sorting methods, bubble sort is an inferior choice, and if it were not for its catchy name, you would probably have never heard of it. However, the general lesson you just learned is important and worth repeating:

> A first application of the brute-force approach often results in an algorithm that can be improved with a modest amount of effort.

Exercises 3.1

1. **a.** Give an example of an algorithm that should not be considered an application of the brute-force approach.

   b. Give an example of a problem that cannot be solved by a brute-force algorithm.

2. **a.** What is the time efficiency of the brute-force algorithm for computing a^n as a function of n? As a function of the number of bits in the binary representation of n?

   b. If you are to compute $a^n \bmod m$ where $a > 1$ and n is a large positive integer, how would you circumvent the problem of a very large magnitude of a^n?

3. For each of the algorithms in Problems 4, 5, and 6 of Exercises 2.3, tell whether or not the algorithm is based on the brute-force approach.

4. **a.** Design a brute-force algorithm for computing the value of a polynomial

$$p(x) = a_n x^n + a_{n-1} x^{n-1} + \cdots + a_1 x + a_0$$

   at a given point x_0 and determine its worst-case efficiency class.

   b. If the algorithm you designed is in $\Theta(n^2)$, design a linear algorithm for this problem.

   c. Is it possible to design an algorithm with a better-than-linear efficiency for this problem?

5. A network topology specifies how computers, printers, and other devices are connected over a network. The figure below illustrates three common topologies of networks: the ring, the star, and the fully connected mesh.

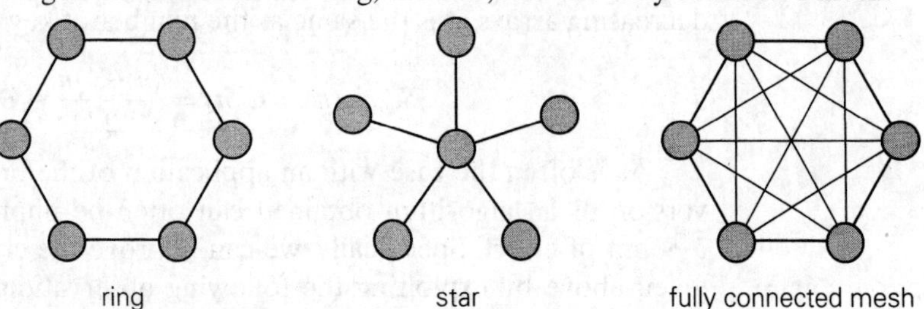

   You are given a boolean matrix $A[0..n-1, 0..n-1]$, where $n > 3$, which is supposed to be the adjacency matrix of a graph modeling a network with one of these topologies. Your task is to determine which of these three topologies, if any, the matrix represents. Design a brute-force algorithm for this task and indicate its time efficiency class.

6. *Tetromino tilings* Tetrominoes are tiles made of four 1×1 squares. There are five types of tetrominoes shown below:

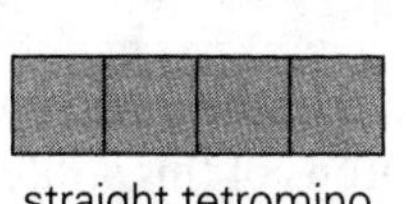
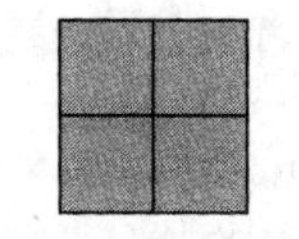
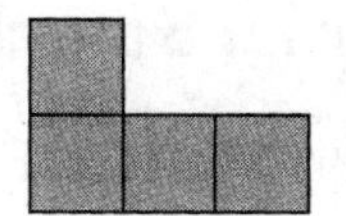
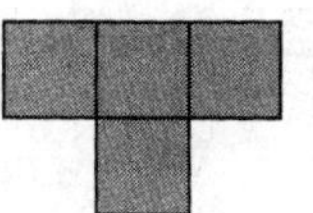
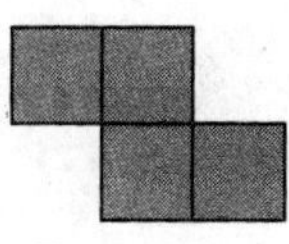

straight tetromino square tetromino L-tetromino T-tetromino Z-tetromino

Is it possible to tile—i.e., cover exactly without overlaps—an 8×8 chessboard with

a. straight tetrominoes? **b.** square tetrominoes?

c. L-tetrominoes? **d.** T-tetrominoes?

e. Z-tetrominoes?

7. *A stack of fake coins* There are n stacks of n identical-looking coins. All of the coins in one of these stacks are counterfeit, while all the coins in the other stacks are genuine. Every genuine coin weighs 10 grams; every fake weighs 11 grams. You have an analytical scale that can determine the exact weight of any number of coins.

a. Devise a brute-force algorithm to identify the stack with the fake coins and determine its worst-case efficiency class.

b. What is the minimum number of weighings needed to identify the stack with the fake coins?

8. Sort the list E, X, A, M, P, L, E in alphabetical order by selection sort.

9. Is selection sort stable? (The definition of a stable sorting algorithm was given in Section 1.3.)

10. Is it possible to implement selection sort for linked lists with the same $\Theta(n^2)$ efficiency as the array version?

11. Sort the list E, X, A, M, P, L, E in alphabetical order by bubble sort.

12. **a.** Prove that if bubble sort makes no exchanges on its pass through a list, the list is sorted and the algorithm can be stopped.

b. Write pseudocode of the method that incorporates this improvement.

c. Prove that the worst-case efficiency of the improved version is quadratic.

13. Is bubble sort stable?

14. *Alternating disks* You have a row of $2n$ disks of two colors, n dark and n light. They alternate: dark, light, dark, light, and so on. You want to get all the dark disks to the right-hand end, and all the light disks to the left-hand end. The only moves you are allowed to make are those that interchange the positions of two neighboring disks.

Design an algorithm for solving this puzzle and determine the number of moves it takes. [Gar99]

3.2 Sequential Search and Brute-Force String Matching

We saw in the previous section two applications of the brute-force approach to the sorting porblem. Here we discuss two applications of this strategy to the problem of searching. The first deals with the canonical problem of searching for an item of a given value in a given list. The second is different in that it deals with the string-matching problem.

Sequential Search

We have already encountered a brute-force algorithm for the general searching problem: it is called sequential search (see Section 2.1). To repeat, the algorithm simply compares successive elements of a given list with a given search key until either a match is encountered (successful search) or the list is exhausted without finding a match (unsuccessful search). A simple extra trick is often employed in implementing sequential search: if we append the search key to the end of the list, the search for the key will have to be successful, and therefore we can eliminate the end of list check altogether. Here is pseudocode of this enhanced version.

ALGORITHM *SequentialSearch2*($A[0..n]$, K)
    //Implements sequential search with a search key as a sentinel
    //Input: An array A of n elements and a search key K
    //Output: The index of the first element in $A[0..n-1]$ whose value is
    //   equal to K or -1 if no such element is found
    $A[n] \leftarrow K$
    $i \leftarrow 0$
    while $A[i] \neq K$ **do**
        $i \leftarrow i + 1$
    if $i < n$ **return** i
    else return -1

Another straightforward improvement can be incorporated in sequential search if a given list is known to be sorted: searching in such a list can be stopped as soon as an element greater than or equal to the search key is encountered.

Sequential search provides an excellent illustration of the brute-force approach, with its characteristic strength (simplicity) and weakness (inferior efficiency). The efficiency results obtained in Section 2.1 for the standard version of sequential search change for the enhanced version only very slightly, so that the algorithm remains linear in both the worst and average cases. We discuss later in the book several searching algorithms with a better time efficiency.

Brute-Force String Matching

Recall the string-matching problem introduced in Section 1.3: given a string of n characters called the ***text*** and a string of m characters ($m \le n$) called the ***pattern***, find a substring of the text that matches the pattern. To put it more precisely, we want to find i—the index of the leftmost character of the first matching substring in the text—such that $t_i = p_0, \ldots, t_{i+j} = p_j, \ldots, t_{i+m-1} = p_{m-1}$:

$$
\begin{array}{ccccccccccc}
t_0 & \ldots & t_i & \ldots & t_{i+j} & \ldots & t_{i+m-1} & & \ldots & t_{n-1} & \text{text } T \\
 & & \updownarrow & & \updownarrow & & \updownarrow & & & & \\
 & & p_0 & \ldots & p_j & \ldots & p_{m-1} & \text{pattern } P & & &
\end{array}
$$

If matches other than the first one need to be found, a string-matching algorithm can simply continue working until the entire text is exhausted.

A brute-force algorithm for the string-matching problem is quite obvious: align the pattern against the first m characters of the text and start matching the corresponding pairs of characters from left to right until either all the m pairs of the characters match (then the algorithm can stop) or a mismatching pair is encountered. In the latter case, shift the pattern one position to the right and resume the character comparisons, starting again with the first character of the pattern and its counterpart in the text. Note that the last position in the text that can still be a beginning of a matching substring is $n - m$ (provided the text positions are indexed from 0 to $n - 1$). Beyond that position, there are not enough characters to match the entire pattern; hence, the algorithm need not make any comparisons there.

```
ALGORITHM  BruteForceStringMatch(T[0..n − 1], P[0..m − 1])
    //Implements brute-force string matching
    //Input: An array T[0..n − 1] of n characters representing a text and
    //        an array P[0..m − 1] of m characters representing a pattern
    //Output: The index of the first character in the text that starts a
    //        matching substring or −1 if the search is unsuccessful
    for i ← 0 to n − m do
        j ← 0
        while j < m and P[j] = T[i + j] do
            j ← j + 1
        if j = m return i
    return −1
```

An operation of the algorithm is illustrated in Figure 3.3. Note that for this example, the algorithm shifts the pattern almost always after a single character comparison. The worst case is much worse: the algorithm may have to make all m comparisons before shifting the pattern, and this can happen for each of the $n - m + 1$ tries. (Problem 6 in this section's exercises asks you to give a specific example of such a situation.) Thus, in the worst case, the algorithm makes

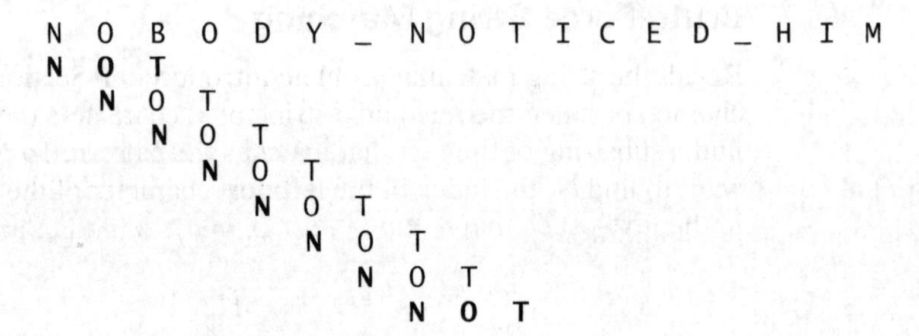

FIGURE 3.3 Example of brute-force string matching. The pattern's characters that are compared with their text counterparts are in bold type.

$m(n-m+1)$ character comparisons, which puts it in the $O(nm)$ class. For a typical word search in a natural language text, however, we should expect that most shifts would happen after very few comparisons (check the example again). Therefore, the average-case efficiency should be considerably better than the worst-case efficiency. Indeed it is: for searching in random texts, it has been shown to be linear, i.e., $\Theta(n)$. There are several more sophisticated and more efficient algorithms for string searching. The most widely known of them—by R. Boyer and J. Moore—is outlined in Section 7.2 along with its simplification suggested by R. Horspool.

Exercises 3.2

1. Find the number of comparisons made by the sentinel version of sequential search
   a. in the worst case.
   b. in the average case if the probability of a successful search is p $(0 \le p \le 1)$.
2. As shown in Section 2.1, the average number of key comparisons made by sequential search (without a sentinel, under standard assumptions about its inputs) is given by the formula
   $$C_{avg}(n) = \frac{p(n+1)}{2} + n(1-p),$$
   where p is the probability of a successful search. Determine, for a fixed n, the values of p $(0 \le p \le 1)$ for which this formula yields the maximum value of $C_{avg}(n)$ and the minimum value of $C_{avg}(n)$.

3. *Gadget testing* A firm wants to determine the highest floor of its n-story headquarters from which a gadget can fall without breaking. The firm has two identical gadgets to experiment with. If one of them gets broken, it cannot be repaired, and the experiment will have to be completed with the remaining gadget. Design an algorithm in the best efficiency class you can to solve this problem.

4. Determine the number of character comparisons made by the brute-force algorithm in searching for the pattern GANDHI in the text

THERE_IS_MORE_TO_LIFE_THAN_INCREASING_ITS_SPEED

Assume that the length of the text—it is 47 characters long—is known before the search starts.

5. How many comparisons (both successful and unsuccessful) will be made by the brute-force algorithm in searching for each of the following patterns in the binary text of one thousand zeros?

a. 00001 **b.** 10000 **c.** 01010

6. Give an example of a text of length n and a pattern of length m that constitutes a worst-case input for the brute-force string-matching algorithm. Exactly how many character comparisons will be made for such input?

7. In solving the string-matching problem, would there be any advantage in comparing pattern and text characters right-to-left instead of left-to-right?

8. Consider the problem of counting, in a given text, the number of substrings that start with an A and end with a B. For example, there are four such substrings in CABAAXBYA.

a. Design a brute-force algorithm for this problem and determine its efficiency class.

b. Design a more efficient algorithm for this problem. [Gin04]

9. Write a visualization program for the brute-force string-matching algorithm.

10. *Word Find* A popular diversion in the United States, "word find" (or "word search") puzzles ask the player to find each of a given set of words in a square table filled with single letters. A word can read horizontally (left or right), vertically (up or down), or along a 45 degree diagonal (in any of the four directions) formed by consecutively adjacent cells of the table; it may wrap around the table's boundaries, but it must read in the same direction with no zigzagging. The same cell of the table may be used in different words, but, in a given word, the same cell may be used no more than once. Write a computer program for solving this puzzle.

11. *Battleship game* Write a program based on a version of brute-force pattern matching for playing the game Battleship on the computer. The rules of the game are as follows. There are two opponents in the game (in this case, a human player and the computer). The game is played on two identical boards (10 × 10 tables of squares) on which each opponent places his or her ships, not seen by the opponent. Each player has five ships, each of which occupies a certain number of squares on the board: a destroyer (two squares), a submarine (three squares), a cruiser (three squares), a battleship (four squares), and an aircraft carrier (five squares). Each ship is placed either horizontally or vertically, with no two ships touching each other. The game is played by the opponents taking turns "shooting" at each other's ships. The

result of every shot is displayed as either a hit or a miss. In case of a hit, the player gets to go again and keeps playing until missing. The goal is to sink all the opponent's ships before the opponent succeeds in doing it first. To sink a ship, all squares occupied by the ship must be hit.

3.3 Closest-Pair and Convex-Hull Problems by Brute Force

In this section, we consider a straightforward approach to two well-known problems dealing with a finite set of points in the plane. These problems, aside from their theoretical interest, arise in two important applied areas: computational geometry and operations research.

Closest-Pair Problem

The closest-pair problem calls for finding the two closest points in a set of n points. It is the simplest of a variety of problems in computational geometry that deals with proximity of points in the plane or higher-dimensional spaces. Points in question can represent such physical objects as airplanes or post offices as well as database records, statistical samples, DNA sequences, and so on. An air-traffic controller might be interested in two closest planes as the most probable collision candidates. A regional postal service manager might need a solution to the closest-pair problem to find candidate post-office locations to be closed.

One of the important applications of the closest-pair problem is cluster analysis in statistics. Based on n data points, hierarchical cluster analysis seeks to organize them in a hierarchy of clusters based on some similarity metric. For numerical data, this metric is usually the Euclidean distance; for text and other nonnumerical data, metrics such as the Hamming distance (see Problem 5 in this section's exercises) are used. A bottom-up algorithm begins with each element as a separate cluster and merges them into successively larger clusters by combining the closest pair of clusters.

For simplicity, we consider the two-dimensional case of the closest-pair problem. We assume that the points in question are specified in a standard fashion by their (x, y) Cartesian coordinates and that the distance between two points $p_i(x_i, y_i)$ and $p_j(x_j, y_j)$ is the standard Euclidean distance

$$d(p_i, p_j) = \sqrt{(x_i - x_j)^2 + (y_i - y_j)^2}.$$

The brute-force approach to solving this problem leads to the following obvious algorithm: compute the distance between each pair of distinct points and find a pair with the smallest distance. Of course, we do not want to compute the distance between the same pair of points twice. To avoid doing so, we consider only the pairs of points (p_i, p_j) for which $i < j$.

Pseudocode below computes the distance between the two closest points; getting the closest points themselves requires just a trivial modification.

ALGORITHM *BruteForceClosestPair(P)*

```
//Finds distance between two closest points in the plane by brute force
//Input: A list P of n (n ≥ 2) points p_1(x_1, y_1), ..., p_n(x_n, y_n)
//Output: The distance between the closest pair of points
d ← ∞
for i ← 1 to n − 1 do
    for j ← i + 1 to n do
        d ← min(d, sqrt((x_i − x_j)^2 + (y_i − y_j)^2)) //sqrt is square root
return d
```

The basic operation of the algorithm is computing the square root. In the age of electronic calculators with a square-root button, one might be led to believe that computing the square root is as simple an operation as, say, addition or multiplication. Of course, it is not. For starters, even for most integers, square roots are irrational numbers that therefore can be found only approximately. Moreover, computing such approximations is not a trivial matter. But, in fact, computing square roots in the loop can be avoided! (Can you think how?) The trick is to realize that we can simply ignore the square-root function and compare the values $(x_i - x_j)^2 + (y_i - y_j)^2$ themselves. We can do this because the smaller a number of which we take the square root, the smaller its square root, or, as mathematicians say, the square-root function is strictly increasing.

Then the basic operation of the algorithm will be squaring a number. The number of times it will be executed can be computed as follows:

$$C(n) = \sum_{i=1}^{n-1} \sum_{j=i+1}^{n} 2 = 2 \sum_{i=1}^{n-1} (n-i)$$

$$= 2[(n-1) + (n-2) + \cdots + 1] = (n-1)n \in \Theta(n^2).$$

Of course, speeding up the innermost loop of the algorithm could only decrease the algorithm's running time by a constant factor (see Problem 1 in this section's exercises), but it cannot improve its asymptotic efficiency class. In Chapter 5, we discuss a linearithmic algorithm for this problem, which is based on a more sophisticated design technique.

Convex-Hull Problem

On to the other problem—that of computing the convex hull. Finding the convex hull for a given set of points in the plane or a higher dimensional space is one of the most important—some people believe the most important—problems in computational geometry. This prominence is due to a variety of applications in which this problem needs to be solved, either by itself or as a part of a larger task. Several such applications are based on the fact that convex hulls provide convenient approximations of object shapes and data sets given. For example, in computer animation, replacing objects by their convex hulls speeds up collision detection; the same idea is used in path planning for Mars mission rovers. Convex hulls are used in computing accessibility maps produced from satellite images by Geographic Information Systems. They are also used for detecting outliers by some statistical techniques. An efficient algorithm for computing a diameter of a set of points, which is the largest distance between two of the points, needs the set's convex hull to find the largest distance between two of its extreme points (see below). Finally, convex hulls are important for solving many optimization problems, because their extreme points provide a limited set of solution candidates.

We start with a definition of a convex set.

DEFINITION A set of points (finite or infinite) in the plane is called ***convex*** if for any two points p and q in the set, the entire line segment with the endpoints at p and q belongs to the set.

All the sets depicted in Figure 3.4a are convex, and so are a straight line, a triangle, a rectangle, and, more generally, any convex polygon,[1] a circle, and the entire plane. On the other hand, the sets depicted in Figure 3.4b, any finite set of two or more distinct points, the boundary of any convex polygon, and a circumference are examples of sets that are not convex.

Now we are ready for the notion of the convex hull. Intuitively, the convex hull of a set of n points in the plane is the smallest convex polygon that contains all of them either inside or on its boundary. If this formulation does not fire up your enthusiasm, consider the problem as one of barricading n sleeping tigers by a fence of the shortest length. This interpretation is due to D. Harel [Har92]; it is somewhat lively, however, because the fenceposts have to be erected right at the spots where some of the tigers sleep! There is another, much tamer interpretation of this notion. Imagine that the points in question are represented by nails driven into a large sheet of plywood representing the plane. Take a rubber band and stretch it to include all the nails, then let it snap into place. The convex hull is the area bounded by the snapped rubber band (Figure 3.5).

A formal definition of the convex hull that is applicable to arbitrary sets, including sets of points that happen to lie on the same line, follows.

DEFINITION The ***convex hull*** of a set S of points is the smallest convex set containing S. (The "smallest" requirement means that the convex hull of S must be a subset of any convex set containing S.)

If S is convex, its convex hull is obviously S itself. If S is a set of two points, its convex hull is the line segment connecting these points. If S is a set of three

---

1. By "a triangle, a rectangle, and, more generally, any convex polygon," we mean here a region, i.e., the set of points both inside and on the boundary of the shape in question.

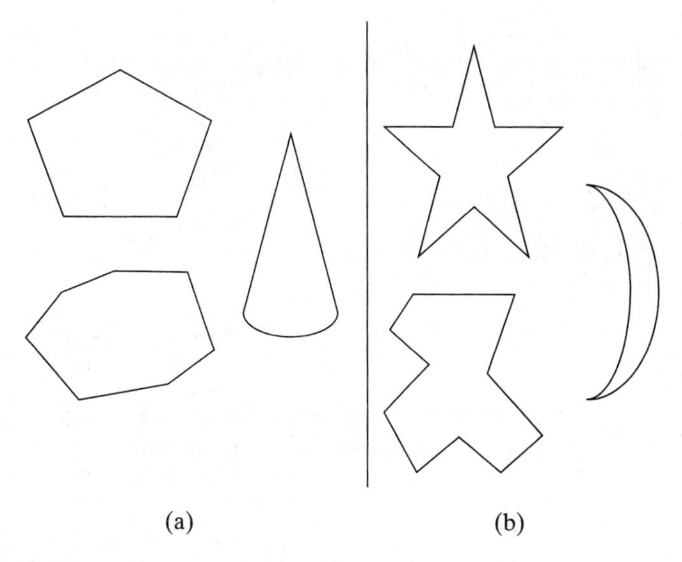

FIGURE 3.4 (a) Convex sets. (b) Sets that are not convex.

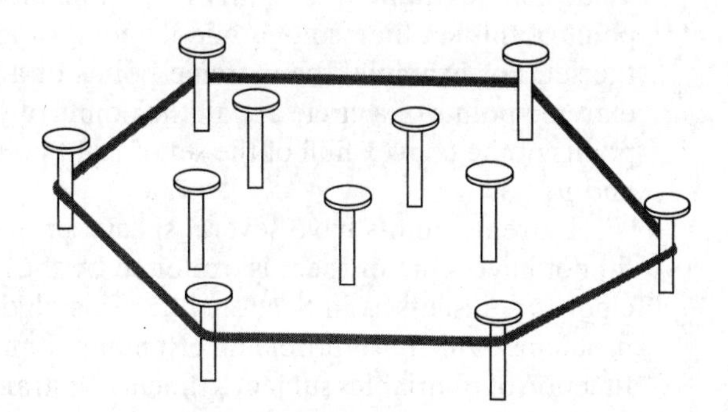

FIGURE 3.5 Rubber-band interpretation of the convex hull.

points not on the same line, its convex hull is the triangle with the vertices at the three points given; if the three points do lie on the same line, the convex hull is the line segment with its endpoints at the two points that are farthest apart. For an example of the convex hull for a larger set, see Figure 3.6.

A study of the examples makes the following theorem an expected result.

THEOREM The convex hull of any set S of $n > 2$ points not all on the same line is a convex polygon with the vertices at some of the points of S. (If all the points do lie on the same line, the polygon degenerates to a line segment but still with the endpoints at two points of S.)

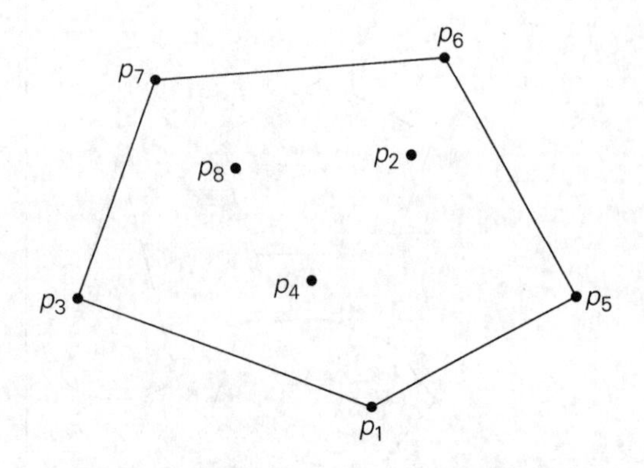

FIGURE 3.6 The convex hull for this set of eight points is the convex polygon with vertices at p_1, p_5, p_6, p_7, and p_3.

The ***convex-hull problem*** is the problem of constructing the convex hull for a given set S of n points. To solve it, we need to find the points that will serve as the vertices of the polygon in question. Mathematicians call the vertices of such a polygon "extreme points." By definition, an ***extreme point*** of a convex set is a point of this set that is not a middle point of any line segment with endpoints in the set. For example, the extreme points of a triangle are its three vertices, the extreme points of a circle are all the points of its circumference, and the extreme points of the convex hull of the set of eight points in Figure 3.6 are p_1, p_5, p_6, p_7, and p_3.

Extreme points have several special properties other points of a convex set do not have. One of them is exploited by the ***simplex method***, a very important algorithm discussed in Section 10.1. This algorithm solves ***linear programming*** problems, which are problems of finding a minimum or a maximum of a linear function of n variables subject to linear constraints (see Problem 12 in this section's exercises for an example and Sections 6.6 and 10.1 for a general discussion). Here, however, we are interested in extreme points because their identification solves the convex-hull problem. Actually, to solve this problem completely, we need to know a bit more than just which of n points of a given set are extreme points of the set's convex hull: we need to know which pairs of points need to be connected to form the boundary of the convex hull. Note that this issue can also be addressed by listing the extreme points in a clockwise or a counterclockwise order.

So how can we solve the convex-hull problem in a brute-force manner? If you do not see an immediate plan for a frontal attack, do not be dismayed: the convex-hull problem is one with no obvious algorithmic solution. Nevertheless, there is a simple but inefficient algorithm that is based on the following observation about line segments making up the boundary of a convex hull: a line segment connecting two points p_i and p_j of a set of n points is a part of the convex hull's boundary if and only if all the other points of the set lie on the same side of the straight line through these two points.[2] (Verify this property for the set in Figure 3.6.) Repeating this test for every pair of points yields a list of line segments that make up the convex hull's boundary.

A few elementary facts from analytical geometry are needed to implement this algorithm. First, the straight line through two points (x_1, y_1), (x_2, y_2) in the coordinate plane can be defined by the equation

$$ax + by = c,$$

where $a = y_2 - y_1$, $b = x_1 - x_2$, $c = x_1y_2 - y_1x_2$.

Second, such a line divides the plane into two half-planes: for all the points in one of them, $ax + by > c$, while for all the points in the other, $ax + by < c$. (For the points on the line itself, of course, $ax + by = c$.) Thus, to check whether certain points lie on the same side of the line, we can simply check whether the expression $ax + by - c$ has the same sign for each of these points. We leave the implementation details as an exercise.

What is the time efficiency of this algorithm? It is in $O(n^3)$: for each of $n(n-1)/2$ pairs of distinct points, we may need to find the sign of $ax + by - c$ for each of the other $n - 2$ points. There are much more efficient algorithms for this important problem, and we discuss one of them later in the book.

Exercises 3.3

1. Assuming that *sqrt* takes about 10 times longer than each of the other operations in the innermost loop of *BruteForceClosestPoints*, which are assumed to take the same amount of time, estimate how much faster the algorithm will run after the improvement discussed in Section 3.3.

2. Can you design a more efficient algorithm than the one based on the brute-force strategy to solve the closest-pair problem for n points $x_1, x_2, \ldots, x_n$ on the real line?

3. Let $x_1 < x_2 < \cdots < x_n$ be real numbers representing coordinates of n villages located along a straight road. A post office needs to be built in one of these villages.

a. Design an efficient algorithm to find the post-office location minimizing the average distance between the villages and the post office.

b. Design an efficient algorithm to find the post-office location minimizing the maximum distance from a village to the post office.

2. For the sake of simplicity, we assume here that no three points of a given set lie on the same line. A modification needed for the general case is left for the exercises.

4. a. There are several alternative ways to define a distance between two points $p_1(x_1, y_1)$ and $p_2(x_2, y_2)$ in the Cartesian plane. In particular, the ***Manhattan distance*** is defined as

$$d_M(p_1, p_2) = |x_1 - x_2| + |y_1 - y_2|.$$

Prove that d_M satisfies the following axioms, which every distance function must satisfy:

i. $d_M(p_1, p_2) \geq 0$ for any two points p_1 and p_2, and $d_M(p_1, p_2) = 0$ if and only if $p_1 = p_2$

ii. $d_M(p_1, p_2) = d_M(p_2, p_1)$

iii. $d_M(p_1, p_2) \leq d_M(p_1, p_3) + d_M(p_3, p_2)$ for any p_1, p_2, and p_3

b. Sketch all the points in the Cartesian plane whose Manhattan distance to the origin (0, 0) is equal to 1. Do the same for the Euclidean distance.

c. True or false: A solution to the closest-pair problem does not depend on which of the two metrics—d_E (Euclidean) or d_M (Manhattan)—is used?

5. The ***Hamming distance*** between two strings of equal length is defined as the number of positions at which the corresponding symbols are different. It is named after Richard Hamming (1915–1998), a prominent American scientist and engineer, who introduced it in his seminal paper on error-detecting and error-correcting codes.

a. Does the Hamming distance satisfy the three axioms of a distance metric listed in Problem 4?

b. What is the time efficiency class of the brute-force algorithm for the closest-pair problem if the points in question are strings of m symbols long and the distance between two of them is measured by the Hamming distance?

6. *Odd pie fight* There are $n \geq 3$ people positioned on a field (Euclidean plane) so that each has a unique nearest neighbor. Each person has a cream pie. At a signal, everybody hurls his or her pie at the nearest neighbor. Assuming that n is odd and that nobody can miss his or her target, true or false: There always remains at least one person not hit by a pie. [Car79]

7. The closest-pair problem can be posed in the k-dimensional space, in which the Euclidean distance between two points $p'(x'_1, \ldots, x'_k)$ and $p''(x''_1, \ldots, x''_k)$ is defined as

$$d(p', p'') = \sqrt{\sum_{s=1}^{k}(x'_s - x''_s)^2}.$$

What is the time-efficiency class of the brute-force algorithm for the k-dimensional closest-pair problem?

8. Find the convex hulls of the following sets and identify their extreme points (if they have any):

a. a line segment

b. a square

c. the boundary of a square

d. a straight line

9. Design a linear-time algorithm to determine two extreme points of the convex hull of a given set of $n > 1$ points in the plane.

10. What modification needs to be made in the brute-force algorithm for the convex-hull problem to handle more than two points on the same straight line?

11. Write a program implementing the brute-force algorithm for the convex-hull problem.

12. Consider the following small instance of the linear programming problem:

$$\begin{aligned} \text{maximize} \quad & 3x + 5y \\ \text{subject to} \quad & x + y \leq 4 \\ & x + 3y \leq 6 \\ & x \geq 0,\ y \geq 0. \end{aligned}$$

a. Sketch, in the Cartesian plane, the problem's ***feasible region***, defined as the set of points satisfying all the problem's constraints.

b. Identify the region's extreme points.

c. Solve this optimization problem by using the following theorem: A linear programming problem with a nonempty bounded feasible region always has a solution, which can be found at one of the extreme points of its feasible region.

3.4 Exhaustive Search

Many important problems require finding an element with a special property in a domain that grows exponentially (or faster) with an instance size. Typically, such problems arise in situations that involve—explicitly or implicitly—combinatorial objects such as permutations, combinations, and subsets of a given set. Many such problems are optimization problems: they ask to find an element that maximizes or minimizes some desired characteristic such as a path length or an assignment cost.

Exhaustive search is simply a brute-force approach to combinatorial problems. It suggests generating each and every element of the problem domain, selecting those of them that satisfy all the constraints, and then finding a desired element (e.g., the one that optimizes some objective function). Note that although the idea of exhaustive search is quite straightforward, its implementation typically requires an algorithm for generating certain combinatorial objects. We delay a discussion of such algorithms until the next chapter and assume here that they exist.

We illustrate exhaustive search by applying it to three important problems: the traveling salesman problem, the knapsack problem, and the assignment problem.

Traveling Salesman Problem

The ***traveling salesman problem (TSP)*** has been intriguing researchers for the last 150 years by its seemingly simple formulation, important applications, and interesting connections to other combinatorial problems. In layman's terms, the problem asks to find the shortest tour through a given set of n cities that visits each city exactly once before returning to the city where it started. The problem can be conveniently modeled by a weighted graph, with the graph's vertices representing the cities and the edge weights specifying the distances. Then the problem can be stated as the problem of finding the shortest ***Hamiltonian circuit*** of the graph. (A Hamiltonian circuit is defined as a cycle that passes through all the vertices of the graph exactly once. It is named after the Irish mathematician Sir William Rowan Hamilton (1805–1865), who became interested in such cycles as an application of his algebraic discoveries.)

It is easy to see that a Hamiltonian circuit can also be defined as a sequence of $n+1$ adjacent vertices $v_{i_0}, v_{i_1}, \ldots, v_{i_{n-1}}, v_{i_0}$, where the first vertex of the sequence is the same as the last one and all the other $n-1$ vertices are distinct. Further, we can assume, with no loss of generality, that all circuits start and end at one particular vertex (they are cycles after all, are they not?). Thus, we can get all the tours by generating all the permutations of $n-1$ intermediate cities, compute the tour lengths, and find the shortest among them. Figure 3.7 presents a small instance of the problem and its solution by this method.

An inspection of Figure 3.7 reveals three pairs of tours that differ only by their direction. Hence, we could cut the number of vertex permutations by half. We could, for example, choose any two intermediate vertices, say, b and c, and then consider only permutations in which b precedes c. (This trick implicitly defines a tour's direction.)

This improvement cannot brighten the efficiency picture much, however. The total number of permutations needed is still $\frac{1}{2}(n-1)!$, which makes the exhaustive-search approach impractical for all but very small values of n. On the other hand, if you always see your glass as half-full, you can claim that cutting the work by half is nothing to sneeze at, even if you solve a small instance of the problem, especially by hand. Also note that had we not limited our investigation to the circuits starting at the same vertex, the number of permutations would have been even larger, by a factor of n.

Knapsack Problem

Here is another well-known problem in algorithmics. Given n items of known weights $w_1, w_2, \ldots, w_n$ and values $v_1, v_2, \ldots, v_n$ and a knapsack of capacity W, find the most valuable subset of the items that fit into the knapsack. If you do not like the idea of putting yourself in the shoes of a thief who wants to steal the most

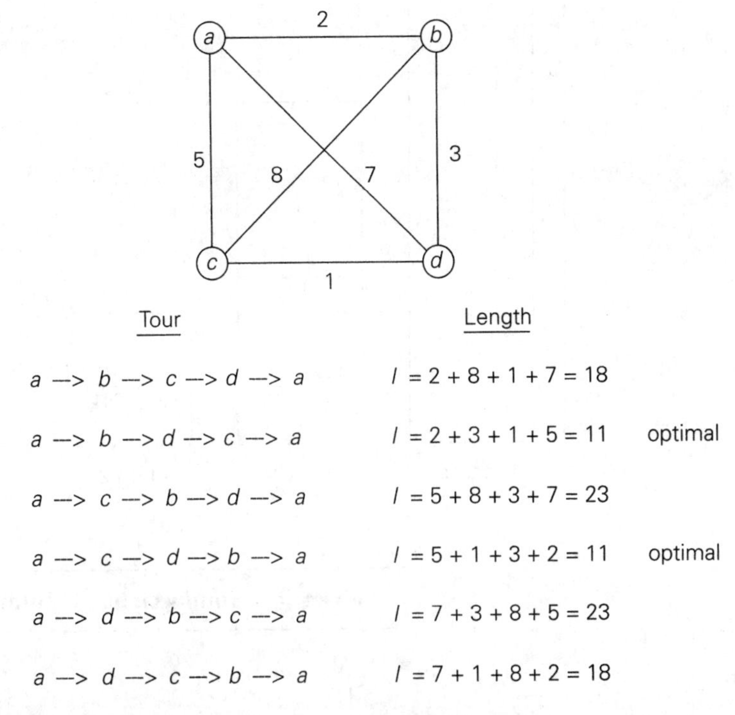

FIGURE 3.7 Solution to a small instance of the traveling salesman problem by exhaustive search.

valuable loot that fits into his knapsack, think about a transport plane that has to deliver the most valuable set of items to a remote location without exceeding the plane's capacity. Figure 3.8a presents a small instance of the knapsack problem.

The exhaustive-search approach to this problem leads to generating all the subsets of the set of n items given, computing the total weight of each subset in order to identify feasible subsets (i.e., the ones with the total weight not exceeding the knapsack capacity), and finding a subset of the largest value among them. As an example, the solution to the instance of Figure 3.8a is given in Figure 3.8b. Since the number of subsets of an n-element set is 2^n, the exhaustive search leads to a $\Omega(2^n)$ algorithm, no matter how efficiently individual subsets are generated.

Thus, for both the traveling salesman and knapsack problems considered above, exhaustive search leads to algorithms that are extremely inefficient on every input. In fact, these two problems are the best-known examples of so-called ***NP-hard problems***. No polynomial-time algorithm is known for any *NP*-hard problem. Moreover, most computer scientists believe that such algorithms do not exist, although this very important conjecture has never been proven. More-sophisticated approaches—backtracking and branch-and-bound (see Sections 12.1 and 12.2)—enable us to solve some but not all instances of these and

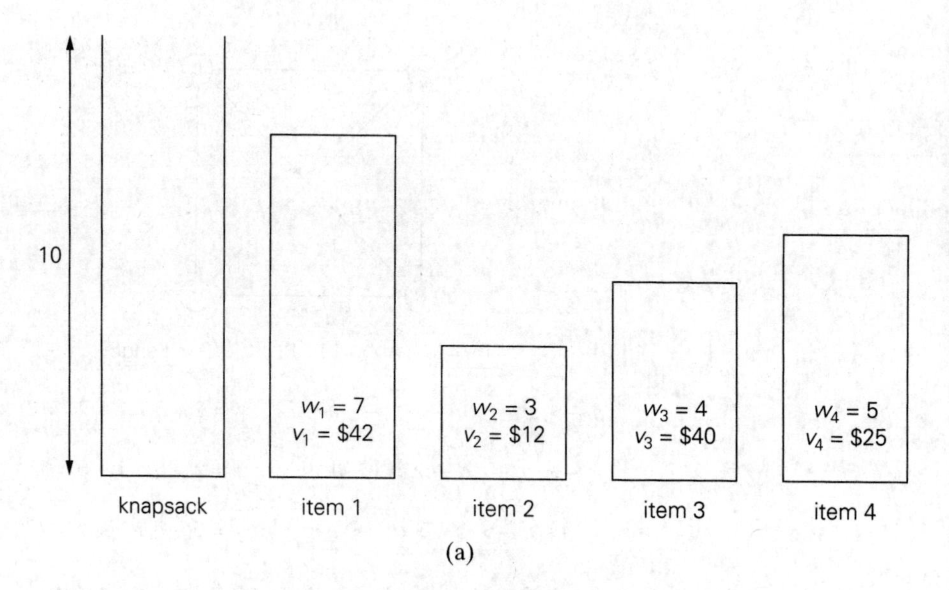

(a)

Subset	Total weight	Total value
∅	0	$ 0
{1}	7	$42
{2}	3	$12
{3}	4	$40
{4}	5	$25
{1, 2}	10	$54
{1, 3}	11	not feasible
{1, 4}	12	not feasible
{2, 3}	7	$52
{2, 4}	8	$37
{3, 4}	**9**	**$65**
{1, 2, 3}	14	not feasible
{1, 2, 4}	15	not feasible
{1, 3, 4}	16	not feasible
{2, 3, 4}	12	not feasible
{1, 2, 3, 4}	19	not feasible

(b)

FIGURE 3.8 (a) Instance of the knapsack problem. (b) Its solution by exhaustive search. The information about the optimal selection is in bold.

similar problems in less than exponential time. Alternatively, we can use one of many approximation algorithms, such as those described in Section 12.3.

Assignment Problem

In our third example of a problem that can be solved by exhaustive search, there are n people who need to be assigned to execute n jobs, one person per job. (That is, each person is assigned to exactly one job and each job is assigned to exactly one person.) The cost that would accrue if the ith person is assigned to the jth job is a known quantity $C[i, j]$ for each pair $i, j = 1, 2, \ldots, n$. The problem is to find an assignment with the minimum total cost.

A small instance of this problem follows, with the table entries representing the assignment costs $C[i, j]$:

	Job 1	**Job 2**	**Job 3**	**Job 4**
Person 1	9	2	7	8
Person 2	6	4	3	7
Person 3	5	8	1	8
Person 4	7	6	9	4

It is easy to see that an instance of the assignment problem is completely specified by its cost matrix C. In terms of this matrix, the problem is to select one element in each row of the matrix so that all selected elements are in different columns and the total sum of the selected elements is the smallest possible. Note that no obvious strategy for finding a solution works here. For example, we cannot select the smallest element in each row, because the smallest elements may happen to be in the same column. In fact, the smallest element in the entire matrix need not be a component of an optimal solution. Thus, opting for the exhaustive search may appear as an unavoidable evil.

We can describe feasible solutions to the assignment problem as n-tuples $\langle j_1, \ldots, j_n \rangle$ in which the ith component, $i = 1, \ldots, n$, indicates the column of the element selected in the ith row (i.e., the job number assigned to the ith person). For example, for the cost matrix above, $\langle 2, 3, 4, 1 \rangle$ indicates the assignment of Person 1 to Job 2, Person 2 to Job 3, Person 3 to Job 4, and Person 4 to Job 1. The requirements of the assignment problem imply that there is a one-to-one correspondence between feasible assignments and permutations of the first n integers. Therefore, the exhaustive-search approach to the assignment problem would require generating all the permutations of integers $1, 2, \ldots, n$, computing the total cost of each assignment by summing up the corresponding elements of the cost matrix, and finally selecting the one with the smallest sum. A few first iterations of applying this algorithm to the instance given above are shown in Figure 3.9; you are asked to complete it in the exercises.

$$C = \begin{bmatrix} 9 & 2 & 7 & 8 \\ 6 & 4 & 3 & 7 \\ 5 & 8 & 1 & 8 \\ 7 & 6 & 9 & 4 \end{bmatrix}$$

$\langle 1, 2, 3, 4 \rangle$ cost = 9 + 4 + 1 + 4 = 18
$\langle 1, 2, 4, 3 \rangle$ cost = 9 + 4 + 8 + 9 = 30
$\langle 1, 3, 2, 4 \rangle$ cost = 9 + 3 + 8 + 4 = 24
$\langle 1, 3, 4, 2 \rangle$ cost = 9 + 3 + 8 + 6 = 26
$\langle 1, 4, 2, 3 \rangle$ cost = 9 + 7 + 8 + 9 = 33
$\langle 1, 4, 3, 2 \rangle$ cost = 9 + 7 + 1 + 6 = 23

etc.

FIGURE 3.9 First few iterations of solving a small instance of the assignment problem by exhaustive search.

Since the number of permutations to be considered for the general case of the assignment problem is $n!$, exhaustive search is impractical for all but very small instances of the problem. Fortunately, there is a much more efficient algorithm for this problem called the ***Hungarian method*** after the Hungarian mathematicians König and Egerváry, whose work underlies the method (see, e.g., [Kol95]).

This is good news: the fact that a problem domain grows exponentially or faster does not necessarily imply that there can be no efficient algorithm for solving it. In fact, we present several other examples of such problems later in the book. However, such examples are more of an exception to the rule. More often than not, there are no known polynomial-time algorithms for problems whose domain grows exponentially with instance size, provided we want to solve them exactly. And, as we mentioned above, such algorithms quite possibly do not exist.

Exercises 3.4

1. **a.** Assuming that each tour can be generated in constant time, what will be the efficiency class of the exhaustive-search algorithm outlined in the text for the traveling salesman problem?

   b. If this algorithm is programmed on a computer that makes ten billion additions per second, estimate the maximum number of cities for which the problem can be solved in

   i. 1 hour. **ii.** 24 hours. **iii.** 1 year. **iv.** 1 century.

2. Outline an exhaustive-search algorithm for the Hamiltonian circuit problem.

3. Outline an algorithm to determine whether a connected graph represented by its adjacency matrix has an Eulerian circuit. What is the efficiency class of your algorithm?

4. Complete the application of exhaustive search to the instance of the assignment problem started in the text.

5. Give an example of the assignment problem whose optimal solution does not include the smallest element of its cost matrix.

6. Consider the ***partition problem***: given n positive integers, partition them into two disjoint subsets with the same sum of their elements. (Of course, the problem does not always have a solution.) Design an exhaustive-search algorithm for this problem. Try to minimize the number of subsets the algorithm needs to generate.

7. Consider the ***clique problem***: given a graph G and a positive integer k, determine whether the graph contains a ***clique*** of size k, i.e., a complete subgraph of k vertices. Design an exhaustive-search algorithm for this problem.

8. Explain how exhaustive search can be applied to the sorting problem and determine the efficiency class of such an algorithm.

9. *Eight-queens problem* Consider the classic puzzle of placing eight queens on an 8×8 chessboard so that no two queens are in the same row or in the same column or on the same diagonal. How many different positions are there so that

   a. no two queens are on the same square?

   b. no two queens are in the same row?

   c. no two queens are in the same row or in the same column?

   Also estimate how long it would take to find all the solutions to the problem by exhaustive search based on each of these approaches on a computer capable of checking 10 billion positions per second.

10. *Magic squares* A magic square of order n is an arrangement of the integers from 1 to n^2 in an $n \times n$ matrix, with each number occurring exactly once, so that each row, each column, and each main diagonal has the same sum.

    a. Prove that if a magic square of order n exists, the sum in question must be equal to $n(n^2 + 1)/2$.

    b. Design an exhaustive-search algorithm for generating all magic squares of order n.

    c. Go to the Internet or your library and find a better algorithm for generating magic squares.

    d. Implement the two algorithms—the exhaustive search and the one you have found—and run an experiment to determine the largest value of n for which each of the algorithms is able to find a magic square of order n in less than 1 minute on your computer.

11. *Famous alphametic* A puzzle in which the digits in a correct mathematical expression, such as a sum, are replaced by letters is called ***cryptarithm***; if, in addition, the puzzle's words make sense, it is said to be an ***alphametic***. The most well-known alphametic was published by the renowned British puzzlist Henry E. Dudeney (1857–1930):

$$\begin{array}{r} \text{S E N D} \\ +\ \text{M O R E} \\ \hline \text{M O N E Y} \end{array}$$

Two conditions are assumed: first, the correspondence between letters and decimal digits is one-to-one, i.e., each letter represents one digit only and different letters represent different digits. Second, the digit zero does not appear as the left-most digit in any of the numbers. To solve an alphametic means to find which digit each letter represents. Note that a solution's uniqueness cannot be assumed and has to be verified by the solver.

a. Write a program for solving cryptarithms by exhaustive search. Assume that a given cryptarithm is a sum of two words.

b. Solve Dudeney's puzzle the way it was expected to be solved when it was first published in 1924.

3.5 Depth-First Search and Breadth-First Search

The term "exhaustive search" can also be applied to two very important algorithms that systematically process all vertices and edges of a graph. These two traversal algorithms are ***depth-first search (DFS)*** and ***breadth-first search (BFS)***. These algorithms have proved to be very useful for many applications involving graphs in artificial intelligence and operations research. In addition, they are indispensable for efficient investigation of fundamental properties of graphs such as connectivity and cycle presence.

Depth-First Search

Depth-first search starts a graph's traversal at an arbitrary vertex by marking it as visited. On each iteration, the algorithm proceeds to an unvisited vertex that is adjacent to the one it is currently in. (If there are several such vertices, a tie can be resolved arbitrarily. As a practical matter, which of the adjacent unvisited candidates is chosen is dictated by the data structure representing the graph. In our examples, we always break ties by the alphabetical order of the vertices.) This process continues until a dead end—a vertex with no adjacent unvisited vertices—is encountered. At a dead end, the algorithm backs up one edge to the vertex it came from and tries to continue visiting unvisited vertices from there. The algorithm eventually halts after backing up to the starting vertex, with the latter being a dead end. By then, all the vertices in the same connected component as the starting vertex have been visited. If unvisited vertices still remain, the depth-first search must be restarted at any one of them.

It is convenient to use a stack to trace the operation of depth-first search. We push a vertex onto the stack when the vertex is reached for the first time (i.e., the

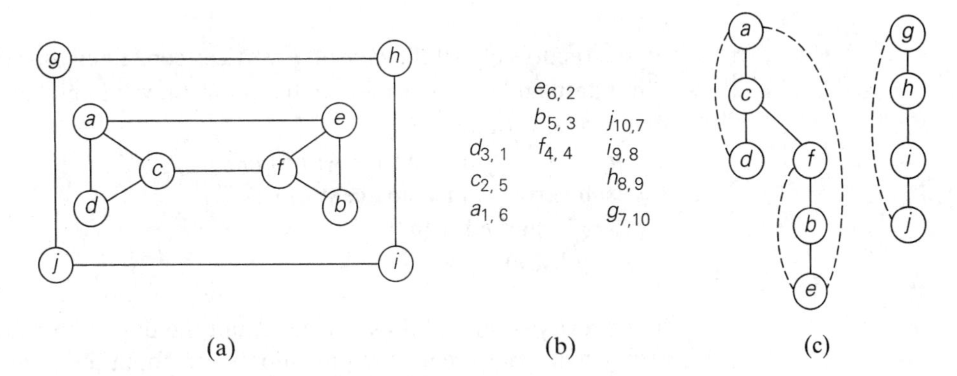

FIGURE 3.10 Example of a DFS traversal. (a) Graph. (b) Traversal's stack (the first subscript number indicates the order in which a vertex is visited, i.e., pushed onto the stack; the second one indicates the order in which it becomes a dead-end, i.e., popped off the stack). (c) DFS forest with the tree and back edges shown with solid and dashed lines, respectively.

visit of the vertex starts), and we pop a vertex off the stack when it becomes a dead end (i.e., the visit of the vertex ends).

It is also very useful to accompany a depth-first search traversal by constructing the so-called ***depth-first search forest***. The starting vertex of the traversal serves as the root of the first tree in such a forest. Whenever a new unvisited vertex is reached for the first time, it is attached as a child to the vertex from which it is being reached. Such an edge is called a ***tree edge*** because the set of all such edges forms a forest. The algorithm may also encounter an edge leading to a previously visited vertex other than its immediate predecessor (i.e., its parent in the tree). Such an edge is called a ***back edge*** because it connects a vertex to its ancestor, other than the parent, in the depth-first search forest. Figure 3.10 provides an example of a depth-first search traversal, with the traversal stack and corresponding depth-first search forest shown as well.

Here is pseudocode of the depth-first search.

ALGORITHM *DFS*(*G*)

```
//Implements a depth-first search traversal of a given graph
//Input: Graph G = ⟨V, E⟩
//Output: Graph G with its vertices marked with consecutive integers
//        in the order they are first encountered by the DFS traversal
mark each vertex in V with 0 as a mark of being "unvisited"
count ← 0
for each vertex v in V do
    if v is marked with 0
        dfs(v)
```

```
dfs(v)
//visits recursively all the unvisited vertices connected to vertex v
//by a path and numbers them in the order they are encountered
//via global variable count
count ← count + 1;   mark v with count
for each vertex w in V adjacent to v do
    if w is marked with 0
        dfs(w)
```

The brevity of the DFS pseudocode and the ease with which it can be performed by hand may create a wrong impression about the level of sophistication of this algorithm. To appreciate its true power and depth, you should trace the algorithm's action by looking not at a graph's diagram but at its adjacency matrix or adjacency lists. (Try it for the graph in Figure 3.10 or a smaller example.)

How efficient is depth-first search? It is not difficult to see that this algorithm is, in fact, quite efficient since it takes just the time proportional to the size of the data structure used for representing the graph in question. Thus, for the adjacency matrix representation, the traversal time is in $\Theta(|V|^2)$, and for the adjacency list representation, it is in $\Theta(|V| + |E|)$ where $|V|$ and $|E|$ are the number of the graph's vertices and edges, respectively.

A DFS forest, which is obtained as a by-product of a DFS traversal, deserves a few comments, too. To begin with, it is not actually a forest. Rather, we can look at it as the given graph with its edges classified by the DFS traversal into two disjoint classes: tree edges and back edges. (No other types are possible for a DFS forest of an undirected graph.) Again, tree edges are edges used by the DFS traversal to reach previously unvisited vertices. If we consider only the edges in this class, we will indeed get a forest. Back edges connect vertices to previously visited vertices other than their immediate predecessors in the traversal. They connect vertices to their ancestors in the forest other than their parents.

A DFS traversal itself and the forest-like representation of the graph it provides have proved to be extremely helpful for the development of efficient algorithms for checking many important properties of graphs.[3] Note that the DFS yields two orderings of vertices: the order in which the vertices are reached for the first time (pushed onto the stack) and the order in which the vertices become dead ends (popped off the stack). These orders are qualitatively different, and various applications can take advantage of either of them.

Important elementary applications of DFS include checking connectivity and checking acyclicity of a graph. Since *dfs* halts after visiting all the vertices con-

---

3. The discovery of several such applications was an important breakthrough achieved by the two American computer scientists John Hopcroft and Robert Tarjan in the 1970s. For this and other contributions, they were given the Turing Award—the most prestigious prize in the computing field [Hop87, Tar87].

nected by a path to the starting vertex, checking a graph's connectivity can be done as follows. Start a DFS traversal at an arbitrary vertex and check, after the algorithm halts, whether all the vertices of the graph will have been visited. If they have, the graph is connected; otherwise, it is not connected. More generally, we can use DFS for identifying connected components of a graph (how?).

As for checking for a cycle presence in a graph, we can take advantage of the graph's representation in the form of a DFS forest. If the latter does not have back edges, the graph is clearly acyclic. If there is a back edge from some vertex u to its ancestor v (e.g., the back edge from d to a in Figure 3.10c), the graph has a cycle that comprises the path from v to u via a sequence of tree edges in the DFS forest followed by the back edge from u to v.

You will find a few other applications of DFS later in the book, although more sophisticated applications, such as finding articulation points of a graph, are not included. (A vertex of a connected graph is said to be its ***articulation point*** if its removal with all edges incident to it breaks the graph into disjoint pieces.)

Breadth-First Search

If depth-first search is a traversal for the brave (the algorithm goes as far from "home" as it can), breadth-first search is a traversal for the cautious. It proceeds in a concentric manner by visiting first all the vertices that are adjacent to a starting vertex, then all unvisited vertices two edges apart from it, and so on, until all the vertices in the same connected component as the starting vertex are visited. If there still remain unvisited vertices, the algorithm has to be restarted at an arbitrary vertex of another connected component of the graph.

It is convenient to use a queue (note the difference from depth-first search!) to trace the operation of breadth-first search. The queue is initialized with the traversal's starting vertex, which is marked as visited. On each iteration, the algorithm identifies all unvisited vertices that are adjacent to the front vertex, marks them as visited, and adds them to the queue; after that, the front vertex is removed from the queue.

Similarly to a DFS traversal, it is useful to accompany a BFS traversal by constructing the so-called ***breadth-first search forest***. The traversal's starting vertex serves as the root of the first tree in such a forest. Whenever a new unvisited vertex is reached for the first time, the vertex is attached as a child to the vertex it is being reached from with an edge called a ***tree edge***. If an edge leading to a previously visited vertex other than its immediate predecessor (i.e., its parent in the tree) is encountered, the edge is noted as a ***cross edge***. Figure 3.11 provides an example of a breadth-first search traversal, with the traversal queue and corresponding breadth-first search forest shown.

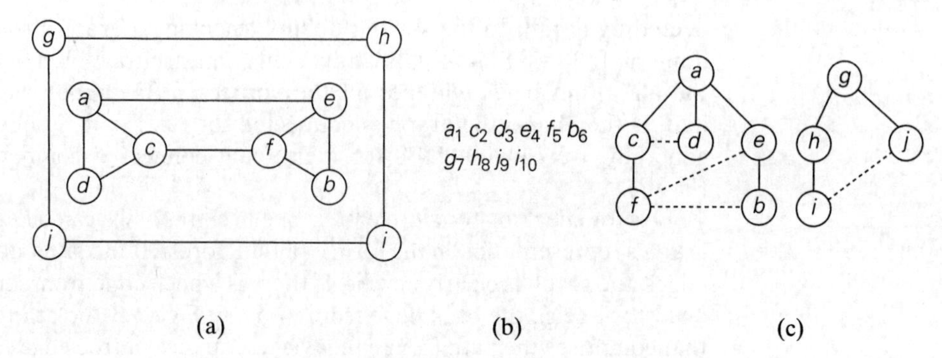

FIGURE 3.11 Example of a BFS traversal. (a) Graph. (b) Traversal queue, with the numbers indicating the order in which the vertices are visited, i.e., added to (and removed from) the queue. (c) BFS forest with the tree and cross edges shown with solid and dotted lines, respectively.

Here is pseudocode of the breadth-first search.

ALGORITHM *BFS(G)*

```
//Implements a breadth-first search traversal of a given graph
//Input: Graph G = ⟨V, E⟩
//Output: Graph G with its vertices marked with consecutive integers
//        in the order they are visited by the BFS traversal
mark each vertex in V with 0 as a mark of being "unvisited"
count ← 0
for each vertex v in V do
    if v is marked with 0
        bfs(v)

bfs(v)
//visits all the unvisited vertices connected to vertex v
//by a path and numbers them in the order they are visited
//via global variable count
count ← count + 1;  mark v with count and initialize a queue with v
while the queue is not empty do
    for each vertex w in V adjacent to the front vertex do
        if w is marked with 0
            count ← count + 1;  mark w with count
            add w to the queue
    remove the front vertex from the queue
```

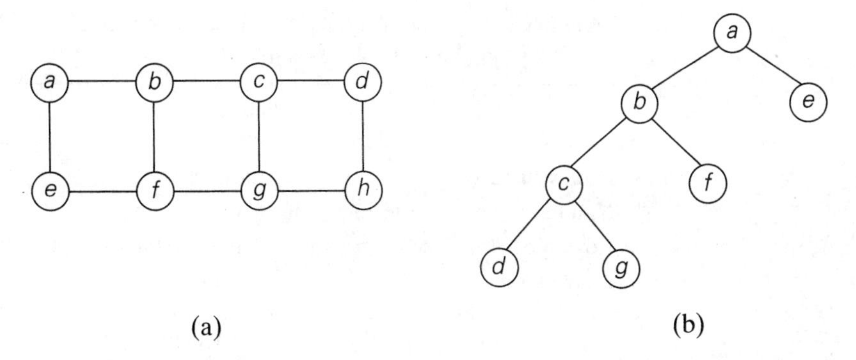

FIGURE 3.12 Illustration of the BFS-based algorithm for finding a minimum-edge path. (a) Graph. (b) Part of its BFS tree that identifies the minimum-edge path from a to g.

Breadth-first search has the same efficiency as depth-first search: it is in $\Theta(|V|^2)$ for the adjacency matrix representation and in $\Theta(|V| + |E|)$ for the adjacency list representation. Unlike depth-first search, it yields a single ordering of vertices because the queue is a FIFO (first-in first-out) structure and hence the order in which vertices are added to the queue is the same order in which they are removed from it. As to the structure of a BFS forest of an undirected graph, it can also have two kinds of edges: tree edges and cross edges. Tree edges are the ones used to reach previously unvisited vertices. Cross edges connect vertices to those visited before, but, unlike back edges in a DFS tree, they connect vertices either on the same or adjacent levels of a BFS tree.

BFS can be used to check connectivity and acyclicity of a graph, essentially in the same manner as DFS can. It is not applicable, however, for several less straightforward applications such as finding articulation points. On the other hand, it can be helpful in some situations where DFS cannot. For example, BFS can be used for finding a path with the fewest number of edges between two given vertices. To do this, we start a BFS traversal at one of the two vertices and stop it as soon as the other vertex is reached. The simple path from the root of the BFS tree to the second vertex is the path sought. For example, path $a - b - c - g$ in the graph in Figure 3.12 has the fewest number of edges among all the paths between vertices a and g. Although the correctness of this application appears to stem immediately from the way BFS operates, a mathematical proof of its validity is not quite elementary (see, e.g., [Cor09, Section 22.2]).

Table 3.1 summarizes the main facts about depth-first search and breadth-first search.

TABLE 3.1 Main facts about depth-first search (DFS) and breadth-first search (BFS)

	DFS	**BFS**
Data structure	a stack	a queue
Number of vertex orderings	two orderings	one ordering
Edge types (undirected graphs)	tree and back edges	tree and cross edges
Applications	connectivity, acyclicity, articulation points	connectivity, acyclicity, minimum-edge paths
Efficiency for adjacency matrix	$\Theta(\|V^2\|)$	$\Theta(\|V^2\|)$
Efficiency for adjacency lists	$\Theta(\|V\| + \|E\|)$	$\Theta(\|V\| + \|E\|)$

Exercises 3.5

1. Consider the following graph.

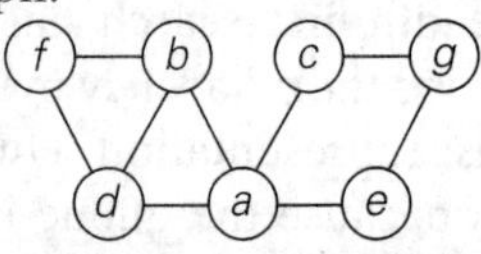

   a. Write down the adjacency matrix and adjacency lists specifying this graph. (Assume that the matrix rows and columns and vertices in the adjacency lists follow in the alphabetical order of the vertex labels.)

   b. Starting at vertex a and resolving ties by the vertex alphabetical order, traverse the graph by depth-first search and construct the corresponding depth-first search tree. Give the order in which the vertices were reached for the first time (pushed onto the traversal stack) and the order in which the vertices became dead ends (popped off the stack).

2. If we define sparse graphs as graphs for which $|E| \in O(|V|)$, which implementation of DFS will have a better time efficiency for such graphs, the one that uses the adjacency matrix or the one that uses the adjacency lists?

3. Let G be a graph with n vertices and m edges.

   a. True or false: All its DFS forests (for traversals starting at different vertices) will have the same number of trees?

   b. True or false: All its DFS forests will have the same number of tree edges and the same number of back edges?

4. Traverse the graph of Problem 1 by breadth-first search and construct the corresponding breadth-first search tree. Start the traversal at vertex a and resolve ties by the vertex alphabetical order.

5. Prove that a cross edge in a BFS tree of an undirected graph can connect vertices only on either the same level or on two adjacent levels of a BFS tree.

6. **a.** Explain how one can check a graph's acyclicity by using breadth-first search.

   b. Does either of the two traversals—DFS or BFS—always find a cycle faster than the other? If you answer yes, indicate which of them is better and explain why it is the case; if you answer no, give two examples supporting your answer.

7. Explain how one can identify connected components of a graph by using

   a. a depth-first search.

   b. a breadth-first search.

8. A graph is said to be ***bipartite*** if all its vertices can be partitioned into two disjoint subsets X and Y so that every edge connects a vertex in X with a vertex in Y. (One can also say that a graph is bipartite if its vertices can be colored in two colors so that every edge has its vertices colored in different colors; such graphs are also called ***2-colorable***.) For example, graph (i) is bipartite while graph (ii) is not.

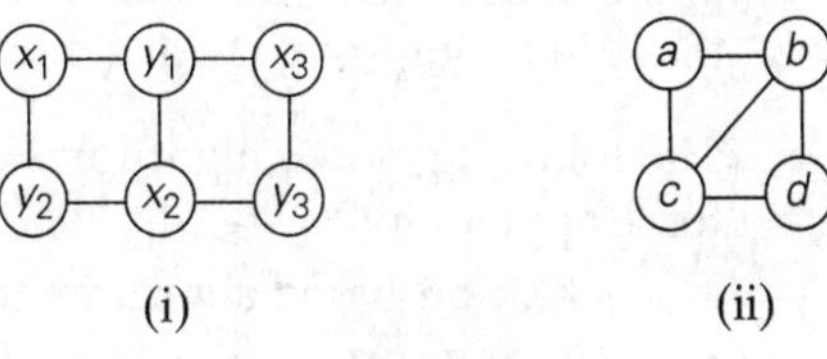

   a. Design a DFS-based algorithm for checking whether a graph is bipartite.

   b. Design a BFS-based algorithm for checking whether a graph is bipartite.

9. Write a program that, for a given graph, outputs:

   a. vertices of each connected component

   b. its cycle or a message that the graph is acyclic

10. One can model a maze by having a vertex for a starting point, a finishing point, dead ends, and all the points in the maze where more than one path can be taken, and then connecting the vertices according to the paths in the maze.

   a. Construct such a graph for the following maze.

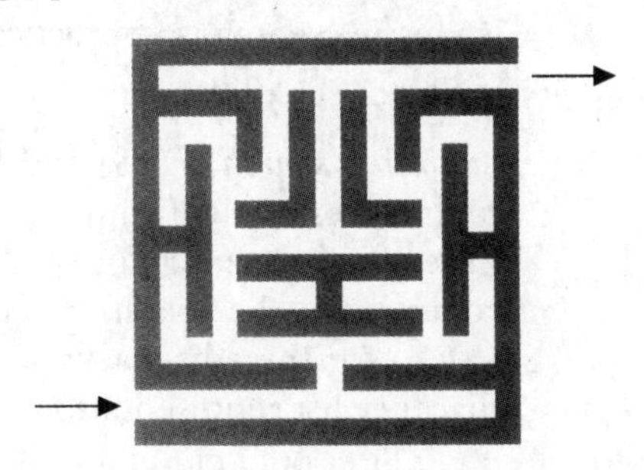

b. Which traversal—DFS or BFS—would you use if you found yourself in a maze and why?

11. *Three Jugs* Siméon Denis Poisson (1781–1840), a famous French mathematician and physicist, is said to have become interested in mathematics after encountering some version of the following old puzzle. Given an 8-pint jug full of water and two empty jugs of 5- and 3-pint capacity, get exactly 4 pints of water in one of the jugs by completely filling up and/or emptying jugs into others. Solve this puzzle by using breadth-first search.

SUMMARY

- *Brute force* is a straightforward approach to solving a problem, usually directly based on the problem statement and definitions of the concepts involved.
- The principal strengths of the brute-force approach are wide applicability and simplicity; its principal weakness is the subpar efficiency of most brute-force algorithms.
- A first application of the brute-force approach often results in an algorithm that can be improved with a modest amount of effort.
- The following noted algorithms can be considered as examples of the brute-force approach:
  - definition-based algorithm for matrix multiplication
  - *selection sort*
  - *sequential search*
  - straightforward string-matching algorithm
- *Exhaustive search* is a brute-force approach to combinatorial problems. It suggests generating each and every combinatorial object of the problem, selecting those of them that satisfy all the constraints, and then finding a desired object.
- The *traveling salesman problem,* the *knapsack problem,* and the *assignment problem* are typical examples of problems that can be solved, at least theoretically, by exhaustive-search algorithms.
- Exhaustive search is impractical for all but very small instances of problems it can be applied to.
- *Depth-first search* (*DFS*) and *breadth-first search* (*BFS*) are two principal graph-traversal algorithms. By representing a graph in a form of a depth-first or breadth-first search forest, they help in the investigation of many important properties of the graph. Both algorithms have the same time efficiency: $\Theta(|V|^2)$ for the adjacency matrix representation and $\Theta(|V| + |E|)$ for the adjacency list representation.

4

Decrease-and-Conquer

Plutarch says that Sertorius, in order to teach his soldiers that perseverance and wit are better than brute force, had two horses brought before them, and set two men to pull out their tails. One of the men was a burly Hercules, who tugged and tugged, but all to no purpose; the other was a sharp, weasel-faced tailor, who plucked one hair at a time, amidst roars of laughter, and soon left the tail quite bare.

—E. Cobham Brewer, *Dictionary of Phrase and Fable*, 1898

The ***decrease-and-conquer*** technique is based on exploiting the relationship between a solution to a given instance of a problem and a solution to its smaller instance. Once such a relationship is established, it can be exploited either top down or bottom up. The former leads naturally to a recursive implementation, although, as one can see from several examples in this chapter, an ultimate implementation may well be nonrecursive. The bottom-up variation is usually implemented iteratively, starting with a solution to the smallest instance of the problem; it is called sometimes the ***incremental approach***.

There are three major variations of decrease-and-conquer:

- decrease by a constant
- decrease by a constant factor
- variable size decrease

In the ***decrease-by-a-constant*** variation, the size of an instance is reduced by the same constant on each iteration of the algorithm. Typically, this constant is equal to one (Figure 4.1), although other constant size reductions do happen occasionally.

Consider, as an example, the exponentiation problem of computing a^n where $a \neq 0$ and n is a nonnegative integer. The relationship between a solution to an instance of size n and an instance of size $n-1$ is obtained by the obvious formula $a^n = a^{n-1} \cdot a$. So the function $f(n) = a^n$ can be computed either "top down" by using its recursive definition

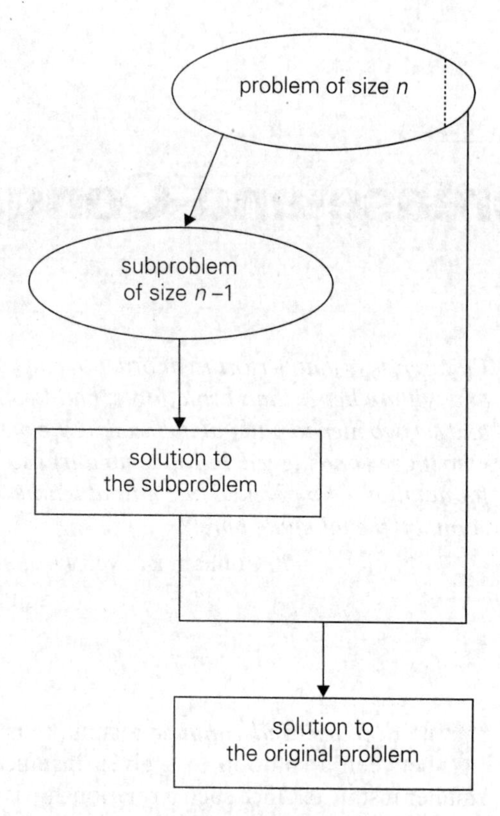

FIGURE 4.1 Decrease-(by one)-and-conquer technique.

$$f(n) = \begin{cases} f(n-1) \cdot a & \text{if } n > 0, \\ 1 & \text{if } n = 0, \end{cases} \tag{4.1}$$

or "bottom up" by multiplying 1 by a n times. (Yes, it is the same as the brute-force algorithm, but we have come to it by a different thought process.) More interesting examples of decrease-by-one algorithms appear in Sections 4.1–4.3.

The ***decrease-by-a-constant-factor*** technique suggests reducing a problem instance by the same constant factor on each iteration of the algorithm. In most applications, this constant factor is equal to two. (Can you give an example of such an algorithm?) The decrease-by-half idea is illustrated in Figure 4.2.

For an example, let us revisit the exponentiation problem. If the instance of size n is to compute a^n, the instance of half its size is to compute $a^{n/2}$, with the obvious relationship between the two: $a^n = (a^{n/2})^2$. But since we consider here instances with integer exponents only, the former does not work for odd n. If n is odd, we have to compute a^{n-1} by using the rule for even-valued exponents and then multiply the result by a. To summarize, we have the following formula:

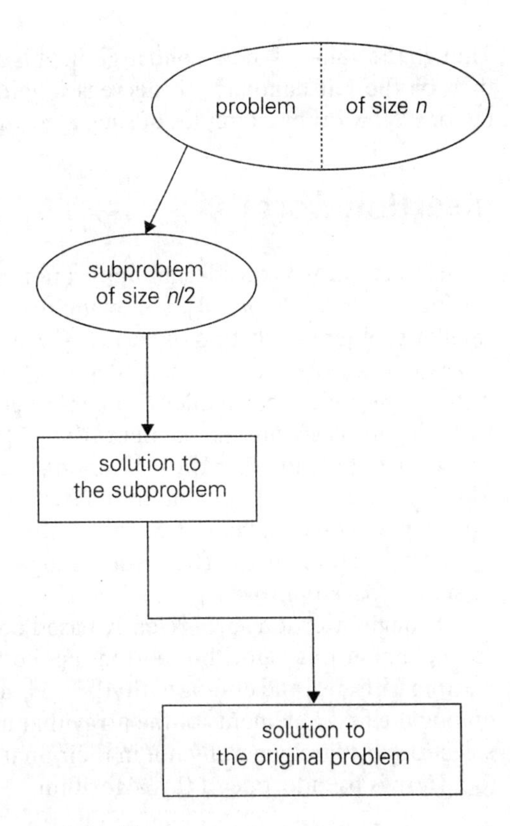

FIGURE 4.2 Decrease-(by half)-and-conquer technique.

$$a^n = \begin{cases} (a^{n/2})^2 & \text{if } n \text{ is even and positive,} \\ (a^{(n-1)/2})^2 \cdot a & \text{if } n \text{ is odd,} \\ 1 & \text{if } n = 0. \end{cases} \quad \textbf{(4.2)}$$

If we compute a^n recursively according to formula (4.2) and measure the algorithm's efficiency by the number of multiplications, we should expect the algorithm to be in $\Theta(\log n)$ because, on each iteration, the size is reduced by about a half at the expense of one or two multiplications.

A few other examples of decrease-by-a-constant-factor algorithms are given in Section 4.4 and its exercises. Such algorithms are so efficient, however, that there are few examples of this kind.

Finally, in the ***variable-size-decrease*** variety of decrease-and-conquer, the size-reduction pattern varies from one iteration of an algorithm to another. Euclid's algorithm for computing the greatest common divisor provides a good example of such a situation. Recall that this algorithm is based on the formula

$$\gcd(m, n) = \gcd(n, m \bmod n).$$

Though the value of the second argument is always smaller on the right-hand side than on the left-hand side, it decreases neither by a constant nor by a constant factor. A few other examples of such algorithms appear in Section 4.5.

4.1 Insertion Sort

In this section, we consider an application of the decrease-by-one technique to sorting an array $A[0..n-1]$. Following the technique's idea, we assume that the smaller problem of sorting the array $A[0..n-2]$ has already been solved to give us a sorted array of size $n-1$: $A[0] \leq \cdots \leq A[n-2]$. How can we take advantage of this solution to the smaller problem to get a solution to the original problem by taking into account the element $A[n-1]$? Obviously, all we need is to find an appropriate position for $A[n-1]$ among the sorted elements and insert it there. This is usually done by scanning the sorted subarray from right to left until the first element smaller than or equal to $A[n-1]$ is encountered to insert $A[n-1]$ right after that element. The resulting algorithm is called ***straight insertion sort*** or simply ***insertion sort***.

Though insertion sort is clearly based on a recursive idea, it is more efficient to implement this algorithm bottom up, i.e., iteratively. As shown in Figure 4.3, starting with $A[1]$ and ending with $A[n-1]$, $A[i]$ is inserted in its appropriate place among the first i elements of the array that have been already sorted (but, unlike selection sort, are generally not in their final positions).

Here is pseudocode of this algorithm.

```
ALGORITHM InsertionSort(A[0..n − 1])
    //Sorts a given array by insertion sort
    //Input: An array A[0..n − 1] of n orderable elements
    //Output: Array A[0..n − 1] sorted in nondecreasing order
    for i ← 1 to n − 1 do
        v ← A[i]
        j ← i − 1
        while j ≥ 0 and A[j] > v do
            A[j + 1] ← A[j]
            j ← j − 1
        A[j + 1] ← v
```

$$A[0] \leq \cdots \leq A[j] < A[j+1] \leq \cdots \leq A[i-1] \mid A[i] \cdots A[n-1]$$

smaller than or equal to $A[i]$ greater than $A[i]$

FIGURE 4.3 Iteration of insertion sort: $A[i]$ is inserted in its proper position among the preceding elements previously sorted.

89 I **45**	68	90	29	34	17	
45	89 I **68**	90	29	34	17	
45	68	89 I **90**	29	34	17	
45	68	89	90 I **29**	34	17	
29	45	68	89	90 I **34**	17	
29	34	45	68	89	90 I **17**	
17	29	34	45	68	89	90

FIGURE 4.4 Example of sorting with insertion sort. A vertical bar separates the sorted part of the array from the remaining elements; the element being inserted is in bold.

The operation of the algorithm is illustrated in Figure 4.4.

The basic operation of the algorithm is the key comparison $A[j] > v$. (Why not $j \geq 0$? Because it is almost certainly faster than the former in an actual computer implementation. Moreover, it is not germane to the algorithm: a better implementation with a sentinel—see Problem 8 in this section's exercises—eliminates it altogether.)

The number of key comparisons in this algorithm obviously depends on the nature of the input. In the worst case, $A[j] > v$ is executed the largest number of times, i.e., for every $j = i - 1, \ldots, 0$. Since $v = A[i]$, it happens if and only if $A[j] > A[i]$ for $j = i - 1, \ldots, 0$. (Note that we are using the fact that on the ith iteration of insertion sort all the elements preceding $A[i]$ are the first i elements in the input, albeit in the sorted order.) Thus, for the worst-case input, we get $A[0] > A[1]$ (for $i = 1$), $A[1] > A[2]$ (for $i = 2$), ..., $A[n-2] > A[n-1]$ (for $i = n - 1$). In other words, the worst-case input is an array of strictly decreasing values. The number of key comparisons for such an input is

$$C_{worst}(n) = \sum_{i=1}^{n-1} \sum_{j=0}^{i-1} 1 = \sum_{i=1}^{n-1} i = \frac{(n-1)n}{2} \in \Theta(n^2).$$

Thus, in the worst case, insertion sort makes exactly the same number of comparisons as selection sort (see Section 3.1).

In the best case, the comparison $A[j] > v$ is executed only once on every iteration of the outer loop. It happens if and only if $A[i-1] \leq A[i]$ for every $i = 1, \ldots, n-1$, i.e., if the input array is already sorted in nondecreasing order. (Though it "makes sense" that the best case of an algorithm happens when the problem is already solved, it is not always the case, as you are going to see in our discussion of quicksort in Chapter 5.) Thus, for sorted arrays, the number of key comparisons is

$$C_{best}(n) = \sum_{i=1}^{n-1} 1 = n - 1 \in \Theta(n).$$

This very good performance in the best case of sorted arrays is not very useful by itself, because we cannot expect such convenient inputs. However, almost-sorted files do arise in a variety of applications, and insertion sort preserves its excellent performance on such inputs.

A rigorous analysis of the algorithm's average-case efficiency is based on investigating the number of element pairs that are out of order (see Problem 11 in this section's exercises). It shows that on randomly ordered arrays, insertion sort makes on average half as many comparisons as on decreasing arrays, i.e.,

$$C_{avg}(n) \approx \frac{n^2}{4} \in \Theta(n^2).$$

This twice-as-fast average-case performance coupled with an excellent efficiency on almost-sorted arrays makes insertion sort stand out among its principal competitors among elementary sorting algorithms, selection sort and bubble sort. In addition, its extension named ***shellsort***, after its inventor D. L. Shell [She59], gives us an even better algorithm for sorting moderately large files (see Problem 12 in this section's exercises).

Exercises 4.1

1. *Ferrying soldiers* A detachment of n soldiers must cross a wide and deep river with no bridge in sight. They notice two 12-year-old boys playing in a rowboat by the shore. The boat is so tiny, however, that it can only hold two boys or one soldier. How can the soldiers get across the river and leave the boys in joint possession of the boat? How many times need the boat pass from shore to shore?

2. *Alternating glasses*

   a. There are $2n$ glasses standing next to each other in a row, the first n of them filled with a soda drink and the remaining n glasses empty. Make the glasses alternate in a filled-empty-filled-empty pattern in the minimum number of glass moves. [Gar78]

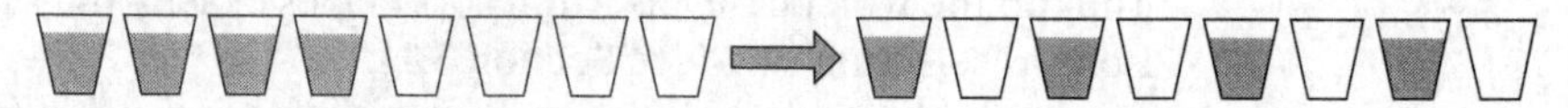

   b. Solve the same problem if $2n$ glasses—n with a drink and n empty—are initially in a random order.

3. *Marking cells* Design an algorithm for the following task. For any even n, mark n cells on an infinite sheet of graph paper so that each marked cell has an odd number of marked neighbors. Two cells are considered neighbors if they are next to each other either horizontally or vertically but not diagonally. The marked cells must form a contiguous region, i.e., a region in which there is a path between any pair of marked cells that goes through a sequence of marked neighbors. [Kor05]

4. Design a decrease-by-one algorithm for generating the power set of a set of n elements. (The power set of a set S is the set of all the subsets of S, including the empty set and S itself.)

5. Consider the following algorithm to check connectivity of a graph defined by its adjacency matrix.

```
ALGORITHM  Connected(A[0..n − 1, 0..n − 1])
    //Input: Adjacency matrix A[0..n − 1, 0..n − 1]) of an undirected graph G
    //Output: 1 (true) if G is connected and 0 (false) if it is not
    if n = 1 return 1  //one-vertex graph is connected by definition
    else
        if not Connected(A[0..n − 2, 0..n − 2]) return 0
        else for j ← 0 to n − 2 do
                if A[n − 1, j] return 1
            return 0
```

Does this algorithm work correctly for every undirected graph with $n > 0$ vertices? If you answer yes, indicate the algorithm's efficiency class in the worst case; if you answer no, explain why.

6. *Team ordering* You have the results of a completed round-robin tournament in which n teams played each other once. Each game ended either with a victory for one of the teams or with a tie. Design an algorithm that lists the teams in a sequence so that every team did not lose the game with the team listed immediately after it. What is the time efficiency class of your algorithm?

7. Apply insertion sort to sort the list E, X, A, M, P, L, E in alphabetical order.

8. **a.** What sentinel should be put before the first element of an array being sorted in order to avoid checking the in-bound condition $j \geq 0$ on each iteration of the inner loop of insertion sort?

   b. Is the sentinel version in the same efficiency class as the original version?

9. Is it possible to implement insertion sort for sorting linked lists? Will it have the same $O(n^2)$ time efficiency as the array version?

10. Compare the text's implementation of insertion sort with the following version.

```
ALGORITHM  InsertSort2(A[0..n − 1])
    for i ← 1 to n − 1 do
        j ← i − 1
        while j ≥ 0 and A[j] > A[j + 1] do
            swap(A[j], A[j + 1])
            j ← j − 1
```

What is the time efficiency of this algorithm? How is it compared to that of the version given in Section 4.1?

11. Let $A[0..n-1]$ be an array of n sortable elements. (For simplicity, you may assume that all the elements are distinct.) A pair $(A[i], A[j])$ is called an ***inversion*** if $i < j$ and $A[i] > A[j]$.

a. What arrays of size n have the largest number of inversions and what is this number? Answer the same questions for the smallest number of inversions.

b. Show that the average-case number of key comparisons in insertion sort is given by the formula

$$C_{avg}(n) \approx \frac{n^2}{4}.$$

12. Shellsort (more accurately Shell's sort) is an important sorting algorithm that works by applying insertion sort to each of several interleaving sublists of a given list. On each pass through the list, the sublists in question are formed by stepping through the list with an increment h_i taken from some predefined decreasing sequence of step sizes, $h_1 > \cdots > h_i > \cdots > 1$, which must end with 1. (The algorithm works for any such sequence, though some sequences are known to yield a better efficiency than others. For example, the sequence 1, 4, 13, 40, 121, . . . , used, of course, in reverse, is known to be among the best for this purpose.)

a. Apply shellsort to the list

$$S, H, E, L, L, S, O, R, T, I, S, U, S, E, F, U, L$$

b. Is shellsort a stable sorting algorithm?

c. Implement shellsort, straight insertion sort, selection sort, and bubble sort in the language of your choice and compare their performance on random arrays of sizes 10^n for $n = 2, 3, 4, 5,$ and 6 as well as on increasing and decreasing arrays of these sizes.

4.2 Topological Sorting

In this section, we discuss an important problem for directed graphs, with a variety of applications involving prerequisite-restricted tasks. Before we pose this problem, though, let us review a few basic facts about directed graphs themselves. A ***directed graph***, or ***digraph*** for short, is a graph with directions specified for all its edges (Figure 4.5a is an example). The adjacency matrix and adjacency lists are still two principal means of representing a digraph. There are only two notable differences between undirected and directed graphs in representing them: (1) the adjacency matrix of a directed graph does not have to be symmetric; (2) an edge in a directed graph has just one (not two) corresponding nodes in the digraph's adjacency lists.

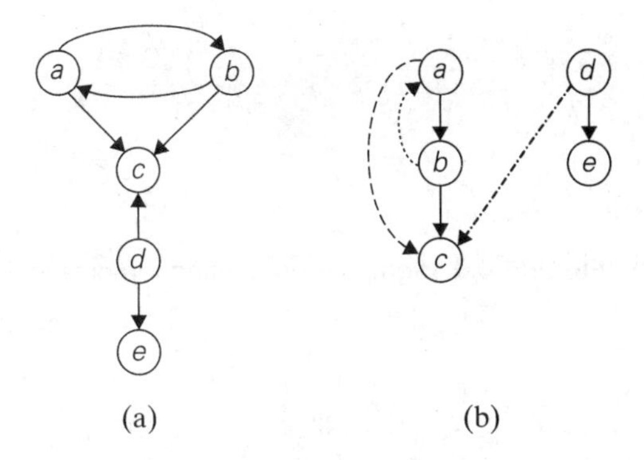

FIGURE 4.5 (a) Digraph. (b) DFS forest of the digraph for the DFS traversal started at a.

Depth-first search and breadth-first search are principal traversal algorithms for traversing digraphs as well, but the structure of corresponding forests can be more complex than for undirected graphs. Thus, even for the simple example of Figure 4.5a, the depth-first search forest (Figure 4.5b) exhibits all four types of edges possible in a DFS forest of a directed graph: ***tree edges*** (ab, bc, de), ***back edges*** (ba) from vertices to their ancestors, ***forward edges*** (ac) from vertices to their descendants in the tree other than their children, and ***cross edges*** (dc), which are none of the aforementioned types.

Note that a back edge in a DFS forest of a directed graph can connect a vertex to its parent. Whether or not it is the case, the presence of a back edge indicates that the digraph has a directed cycle. A ***directed cycle*** in a digraph is a sequence of three or more of its vertices that starts and ends with the same vertex and in which every vertex is connected to its immediate predecessor by an edge directed from the predecessor to the successor. For example, a, b, a is a directed cycle in the digraph in Figure 4.5a. Conversely, if a DFS forest of a digraph has no back edges, the digraph is a ***dag***, an acronym for ***directed acyclic graph***.

Edge directions lead to new questions about digraphs that are either meaningless or trivial for undirected graphs. In this section, we discuss one such question. As a motivating example, consider a set of five required courses {C1, C2, C3, C4, C5} a part-time student has to take in some degree program. The courses can be taken in any order as long as the following course prerequisites are met: C1 and C2 have no prerequisites, C3 requires C1 and C2, C4 requires C3, and C5 requires C3 and C4. The student can take only one course per term. In which order should the student take the courses?

The situation can be modeled by a digraph in which vertices represent courses and directed edges indicate prerequisite requirements (Figure 4.6). In terms of this digraph, the question is whether we can list its vertices in such an order that for every edge in the graph, the vertex where the edge starts is listed before the vertex where the edge ends. (Can you find such an ordering of this digraph's vertices?) This problem is called ***topological sorting***. It can be posed for an

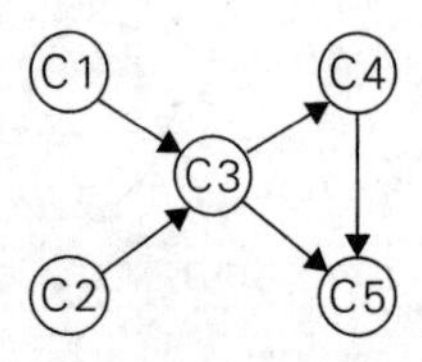

FIGURE 4.6 Digraph representing the prerequisite structure of five courses.

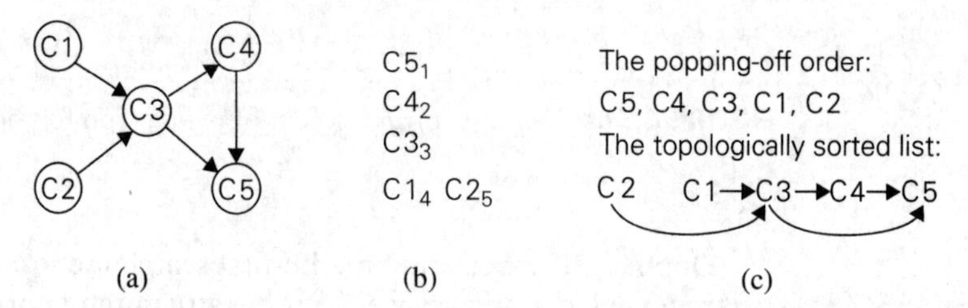

FIGURE 4.7 (a) Digraph for which the topological sorting problem needs to be solved. (b) DFS traversal stack with the subscript numbers indicating the popping-off order. (c) Solution to the problem.

arbitrary digraph, but it is easy to see that the problem cannot have a solution if a digraph has a directed cycle. Thus, for topological sorting to be possible, a digraph in question must be a dag. It turns out that being a dag is not only necessary but also sufficient for topological sorting to be possible; i.e., if a digraph has no directed cycles, the topological sorting problem for it has a solution. Moreover, there are two efficient algorithms that both verify whether a digraph is a dag and, if it is, produce an ordering of vertices that solves the topological sorting problem.

The first algorithm is a simple application of depth-first search: perform a DFS traversal and note the order in which vertices become dead-ends (i.e., popped off the traversal stack). Reversing this order yields a solution to the topological sorting problem, provided, of course, no back edge has been encountered during the traversal. If a back edge has been encountered, the digraph is not a dag, and topological sorting of its vertices is impossible.

Why does the algorithm work? When a vertex v is popped off a DFS stack, no vertex u with an edge from u to v can be among the vertices popped off before v. (Otherwise, (u, v) would have been a back edge.) Hence, any such vertex u will be listed after v in the popped-off order list, and before v in the reversed list.

Figure 4.7 illustrates an application of this algorithm to the digraph in Figure 4.6. Note that in Figure 4.7c, we have drawn the edges of the digraph, and they all point from left to right as the problem's statement requires. It is a convenient way to check visually the correctness of a solution to an instance of the topological sorting problem.

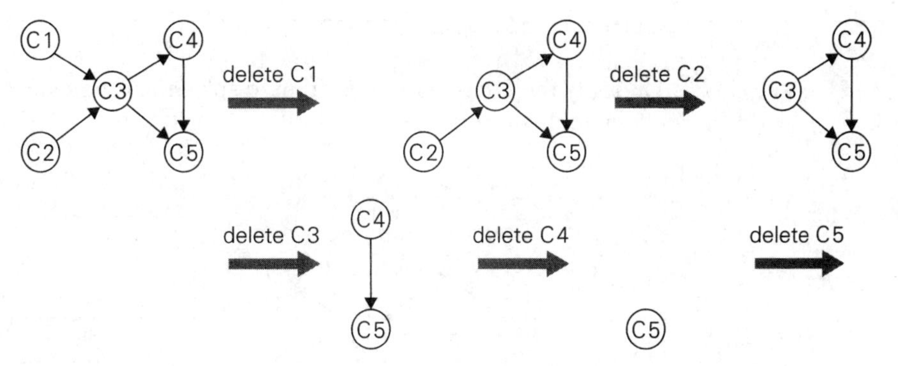

The solution obtained is C1, C2, C3, C4, C5

FIGURE 4.8 Illustration of the source-removal algorithm for the topological sorting problem. On each iteration, a vertex with no incoming edges is deleted from the digraph.

The second algorithm is based on a direct implementation of the decrease-(by one)-and-conquer technique: repeatedly, identify in a remaining digraph a ***source***, which is a vertex with no incoming edges, and delete it along with all the edges outgoing from it. (If there are several sources, break the tie arbitrarily. If there are none, stop because the problem cannot be solved—see Problem 6a in this section's exercises.) The order in which the vertices are deleted yields a solution to the topological sorting problem. The application of this algorithm to the same digraph representing the five courses is given in Figure 4.8.

Note that the solution obtained by the source-removal algorithm is different from the one obtained by the DFS-based algorithm. Both of them are correct, of course; the topological sorting problem may have several alternative solutions.

The tiny size of the example we used might create a wrong impression about the topological sorting problem. But imagine a large project—e.g., in construction, research, or software development—that involves a multitude of interrelated tasks with known prerequisites. The first thing to do in such a situation is to make sure that the set of given prerequisites is not contradictory. The convenient way of doing this is to solve the topological sorting problem for the project's digraph. Only then can one start thinking about scheduling tasks to, say, minimize the total completion time of the project. This would require, of course, other algorithms that you can find in general books on operations research or in special ones on CPM (Critical Path Method) and PERT (Program Evaluation and Review Technique) methodologies.

As to applications of topological sorting in computer science, they include instruction scheduling in program compilation, cell evaluation ordering in spreadsheet formulas, and resolving symbol dependencies in linkers.

Exercises 4.2

1. Apply the DFS-based algorithm to solve the topological sorting problem for the following digraphs:

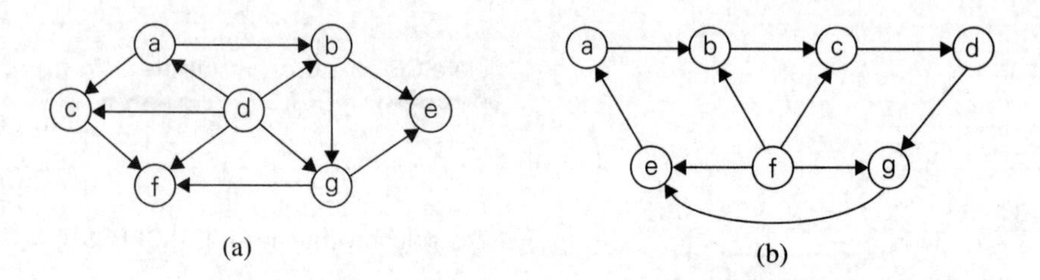

2. **a.** Prove that the topological sorting problem has a solution if and only if it is a dag.

   b. For a digraph with n vertices, what is the largest number of distinct solutions the topological sorting problem can have?

3. **a.** What is the time efficiency of the DFS-based algorithm for topological sorting?

   b. How can one modify the DFS-based algorithm to avoid reversing the vertex ordering generated by DFS?

4. Can one use the order in which vertices are pushed onto the DFS stack (instead of the order they are popped off it) to solve the topological sorting problem?

5. Apply the source-removal algorithm to the digraphs of Problem 1 above.

6. **a.** Prove that a nonempty dag must have at least one source.

   b. How would you find a source (or determine that such a vertex does not exist) in a digraph represented by its adjacency matrix? What is the time efficiency of this operation?

   c. How would you find a source (or determine that such a vertex does not exist) in a digraph represented by its adjacency lists? What is the time efficiency of this operation?

7. Can you implement the source-removal algorithm for a digraph represented by its adjacency lists so that its running time is in $O(|V| + |E|)$?

8. Implement the two topological sorting algorithms in the language of your choice. Run an experiment to compare their running times.

9. A digraph is called ***strongly connected*** if for any pair of two distinct vertices u and v there exists a directed path from u to v and a directed path from v to u. In general, a digraph's vertices can be partitioned into disjoint maximal subsets of vertices that are mutually accessible via directed paths; these subsets are called ***strongly connected components*** of the digraph. There are two DFS-

based algorithms for identifying strongly connected components. Here is the simpler (but somewhat less efficient) one of the two:

Step 1 Perform a DFS traversal of the digraph given and number its vertices in the order they become dead ends.

Step 2 Reverse the directions of all the edges of the digraph.

Step 3 Perform a DFS traversal of the new digraph by starting (and, if necessary, restarting) the traversal at the highest numbered vertex among still unvisited vertices.

The strongly connected components are exactly the vertices of the DFS trees obtained during the last traversal.

a. Apply this algorithm to the following digraph to determine its strongly connected components:

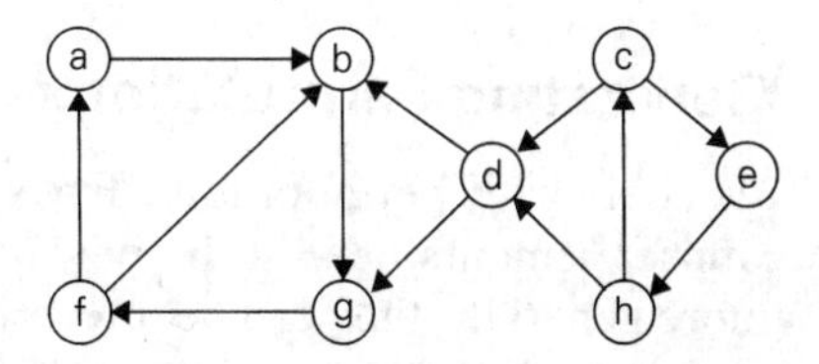

b. What is the time efficiency class of this algorithm? Give separate answers for the adjacency matrix representation and adjacency list representation of an input digraph.

c. How many strongly connected components does a dag have?

10. *Spider's web* A spider sits at the bottom (point S) of its web, and a fly sits at the top (F). How many different ways can the spider reach the fly by moving along the web's lines in the directions indicated by the arrows? [Kor05]

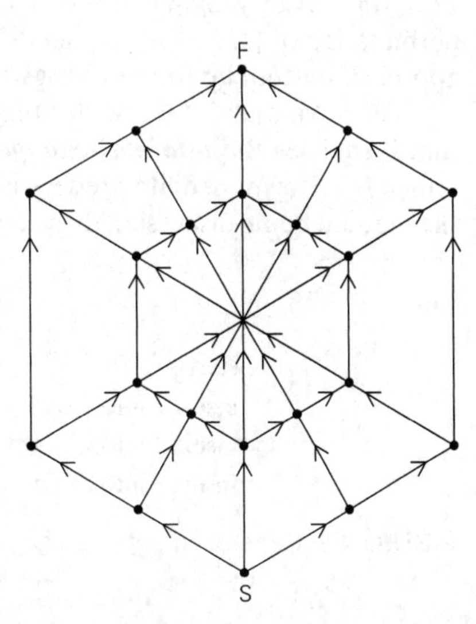

4.3 Algorithms for Generating Combinatorial Objects

In this section, we keep our promise to discuss algorithms for generating combinatorial objects. The most important types of combinatorial objects are permutations, combinations, and subsets of a given set. They typically arise in problems that require a consideration of different choices. We already encountered them in Chapter 3 when we discussed exhaustive search. Combinatorial objects are studied in a branch of discrete mathematics called combinatorics. Mathematicians, of course, are primarily interested in different counting formulas; we should be grateful for such formulas because they tell us how many items need to be generated. In particular, they warn us that the number of combinatorial objects typically grows exponentially or even faster as a function of the problem size. But our primary interest here lies in algorithms for generating combinatorial objects, not just in counting them.

Generating Permutations

We start with permutations. For simplicity, we assume that the underlying set whose elements need to be permuted is simply the set of integers from 1 to n; more generally, they can be interpreted as indices of elements in an n-element set $\{a_1, \ldots, a_n\}$. What would the decrease-by-one technique suggest for the problem of generating all $n!$ permutations of $\{1, \ldots, n\}$? The smaller-by-one problem is to generate all $(n-1)!$ permutations. Assuming that the smaller problem is solved, we can get a solution to the larger one by inserting n in each of the n possible positions among elements of every permutation of $n-1$ elements. All the permutations obtained in this fashion will be distinct (why?), and their total number will be $n(n-1)! = n!$. Hence, we will obtain all the permutations of $\{1, \ldots, n\}$.

We can insert n in the previously generated permutations either left to right or right to left. It turns out that it is beneficial to start with inserting n into $12 \ldots (n-1)$ by moving right to left and then switch direction every time a new permutation of $\{1, \ldots, n-1\}$ needs to be processed. An example of applying this approach bottom up for $n = 3$ is given in Figure 4.9.

The advantage of this order of generating permutations stems from the fact that it satisfies the ***minimal-change requirement***: each permutation can be obtained from its immediate predecessor by exchanging just two elements in it. (For the method being discussed, these two elements are always adjacent to each other.

start	1		
insert 2 into 1 right to left	12	21	
insert 3 into 12 right to left	123	132	312
insert 3 into 21 left to right	321	231	213

FIGURE 4.9 Generating permutations bottom up.

Check this for the permutations generated in Figure 4.9.) The minimal-change requirement is beneficial both for the algorithm's speed and for applications using the permutations. For example, in Section 3.4, we needed permutations of cities to solve the traveling salesman problem by exhaustive search. If such permutations are generated by a minimal-change algorithm, we can compute the length of a new tour from the length of its predecessor in constant rather than linear time (how?).

It is possible to get the same ordering of permutations of n elements without explicitly generating permutations for smaller values of n. It can be done by associating a direction with each element k in a permutation. We indicate such a direction by a small arrow written above the element in question, e.g.,

$$\overrightarrow{3}\ \overleftarrow{2}\ \overrightarrow{4}\ \overleftarrow{1}.$$

The element k is said to be ***mobile*** in such an arrow-marked permutation if its arrow points to a smaller number adjacent to it. For example, for the permutation $\overrightarrow{3}\overleftarrow{2}\overrightarrow{4}\overleftarrow{1}$, 3 and 4 are mobile while 2 and 1 are not. Using the notion of a mobile element, we can give the following description of the ***Johnson-Trotter algorithm*** for generating permutations.

ALGORITHM *JohnsonTrotter*(n)

//Implements Johnson-Trotter algorithm for generating permutations
//Input: A positive integer n
//Output: A list of all permutations of $\{1, \ldots, n\}$
initialize the first permutation with $\overleftarrow{1}\ \overleftarrow{2} \ldots \overleftarrow{n}$
while the last permutation has a mobile element **do**
    find its largest mobile element k
    swap k with the adjacent element k's arrow points to
    reverse the direction of all the elements that are larger than k
    add the new permutation to the list

Here is an application of this algorithm for $n = 3$, with the largest mobile element shown in bold:

$$\overleftarrow{1}\,\overleftarrow{2}\,\overleftarrow{\mathbf{3}}\quad \overleftarrow{1}\,\overleftarrow{\mathbf{3}}\,\overleftarrow{2}\quad \overleftarrow{3}\,\overleftarrow{1}\,\overleftarrow{\mathbf{2}}\quad \overrightarrow{\mathbf{3}}\,\overleftarrow{2}\,\overleftarrow{1}\quad \overleftarrow{2}\,\overrightarrow{\mathbf{3}}\,\overleftarrow{1}\quad \overleftarrow{2}\,\overleftarrow{1}\,\overrightarrow{3}.$$

This algorithm is one of the most efficient for generating permutations; it can be implemented to run in time proportional to the number of permutations, i.e., in $\Theta(n!)$. Of course, it is horribly slow for all but very small values of n; however, this is not the algorithm's "fault" but rather the fault of the problem: it simply asks to generate too many items.

One can argue that the permutation ordering generated by the Johnson-Trotter algorithm is not quite natural; for example, the natural place for permutation $n(n-1)\ldots 1$ seems to be the last one on the list. This would be the case if permutations were listed in increasing order—also called the ***lexicographic or-***

der—which is the order in which they would be listed in a dictionary if the numbers were interpreted as letters of an alphabet. For example, for $n = 3$,

$$123 \quad 132 \quad 213 \quad 231 \quad 312 \quad 321.$$

So how can we generate the permutation following $a_1a_2 \ldots a_{n-1}a_n$ in lexicographic order? If $a_{n-1} < a_n$, which is the case for exactly one half of all the permutations, we can simply transpose these last two elements. For example, 123 is followed by 132. If $a_{n-1} > a_n$, we find the permutation's longest decreasing suffix $a_{i+1} > a_{i+2} > \cdots > a_n$ (but $a_i < a_{i+1}$); increase a_i by exchanging it with the smallest element of the suffix that is greater than a_i; and reverse the new suffix to put it in increasing order. For example, 362541 is followed by 364125. Here is pseudocode of this simple algorithm whose origins go as far back as 14th-century India.

ALGORITHM *LexicographicPermute*(n)
 //Generates permutations in lexicographic order
 //Input: A positive integer n
 //Output: A list of all permutations of $\{1, \ldots, n\}$ in lexicographic order
 initialize the first permutation with $12 \ldots n$
 while last permutation has two consecutive elements in increasing order **do**
  let i be its largest index such that $a_i < a_{i+1}$ //$a_{i+1} > a_{i+2} > \cdots > a_n$
  find the largest index j such that $a_i < a_j$ //$j \geq i + 1$ since $a_i < a_{i+1}$
  swap a_i with a_j //$a_{i+1}a_{i+2} \ldots a_n$ will remain in decreasing order
  reverse the order of the elements from a_{i+1} to a_n inclusive
  add the new permutation to the list

Generating Subsets

Recall that in Section 3.4 we examined the knapsack problem, which asks to find the most valuable subset of items that fits a knapsack of a given capacity. The exhaustive-search approach to solving this problem discussed there was based on generating all subsets of a given set of items. In this section, we discuss algorithms for generating all 2^n subsets of an abstract set $A = \{a_1, \ldots, a_n\}$. (Mathematicians call the set of all subsets of a set its ***power set***.)

The decrease-by-one idea is immediately applicable to this problem, too. All subsets of $A = \{a_1, \ldots, a_n\}$ can be divided into two groups: those that do not contain a_n and those that do. The former group is nothing but all the subsets of $\{a_1, \ldots, a_{n-1}\}$, while each and every element of the latter can be obtained by adding a_n to a subset of $\{a_1, \ldots, a_{n-1}\}$. Thus, once we have a list of all subsets of $\{a_1, \ldots, a_{n-1}\}$, we can get all the subsets of $\{a_1, \ldots, a_n\}$ by adding to the list all its elements with a_n put into each of them. An application of this algorithm to generate all subsets of $\{a_1, a_2, a_3\}$ is illustrated in Figure 4.10.

Similarly to generating permutations, we do not have to generate power sets of smaller sets. A convenient way of solving the problem directly is based on a one-to-one correspondence between all 2^n subsets of an n element set $A = \{a_1, \ldots, a_n\}$

n	subsets							
0	$\varnothing$							
1	$\varnothing$	$\{a_1\}$						
2	$\varnothing$	$\{a_1\}$	$\{a_2\}$	$\{a_1, a_2\}$				
3	$\varnothing$	$\{a_1\}$	$\{a_2\}$	$\{a_1, a_2\}$	$\{a_3\}$	$\{a_1, a_3\}$	$\{a_2, a_3\}$	$\{a_1, a_2, a_3\}$

FIGURE 4.10 Generating subsets bottom up.

and all 2^n bit strings $b_1, \ldots, b_n$ of length n. The easiest way to establish such a correspondence is to assign to a subset the bit string in which $b_i = 1$ if a_i belongs to the subset and $b_i = 0$ if a_i does not belong to it. (We mentioned this idea of bit vectors in Section 1.4.) For example, the bit string 000 will correspond to the empty subset of a three-element set, 111 will correspond to the set itself, i.e., $\{a_1, a_2, a_3\}$, and 110 will represent $\{a_1, a_2\}$. With this correspondence in place, we can generate all the bit strings of length n by generating successive binary numbers from 0 to $2^n - 1$, padded, when necessary, with an appropriate number of leading 0's. For example, for the case of $n = 3$, we obtain

bit strings	000	001	010	011	100	101	110	111
subsets	$\varnothing$	$\{a_3\}$	$\{a_2\}$	$\{a_2, a_3\}$	$\{a_1\}$	$\{a_1, a_3\}$	$\{a_1, a_2\}$	$\{a_1, a_2, a_3\}$

Note that although the bit strings are generated by this algorithm in lexicographic order (in the two-symbol alphabet of 0 and 1), the order of the subsets looks anything but natural. For example, we might want to have the so-called ***squashed order***, in which any subset involving a_j can be listed only after all the subsets involving $a_1, \ldots, a_{j-1}$, as was the case for the list of the three-element set in Figure 4.10. It is easy to adjust the bit string–based algorithm above to yield a squashed ordering of the subsets involved (see Problem 6 in this section's exercises).

A more challenging question is whether there exists a minimal-change algorithm for generating bit strings so that every one of them differs from its immediate predecessor by only a single bit. (In the language of subsets, we want every subset to differ from its immediate predecessor by either an addition or a deletion, but not both, of a single element.) The answer to this question is yes. For example, for $n = 3$, we can get

000 001 011 010 110 111 101 100.

Such a sequence of bit strings is called the ***binary reflected Gray code***. Frank Gray, a researcher at AT&T Bell Laboratories, reinvented it in the 1940s to minimize the effect of errors in transmitting digital signals (see, e.g., [Ros07], pp. 642–643). Seventy years earlier, the French engineer Émile Baudot used such codes in telegraphy. Here is pseudocode that generates the binary reflected Gray code recursively.

```
ALGORITHM  BRGC(n)
    //Generates recursively the binary reflected Gray code of order n
    //Input: A positive integer n
    //Output: A list of all bit strings of length n composing the Gray code
    if n = 1 make list L containing bit strings 0 and 1 in this order
    else generate list L1 of bit strings of size n − 1 by calling BRGC(n − 1)
         copy list L1 to list L2 in reversed order
         add 0 in front of each bit string in list L1
         add 1 in front of each bit string in list L2
         append L2 to L1 to get list L
    return L
```

The correctness of the algorithm stems from the fact that it generates 2^n bit strings and all of them are distinct. Both these assertions are easy to check by mathematical induction. Note that the binary reflected Gray code is cyclic: its last bit string differs from the first one by a single bit. For a nonrecursive algorithm for generating the binary reflected Gray code see Problem 9 in this section's exercises.

Exercises 4.3

1. Is it realistic to implement an algorithm that requires generating all permutations of a 25-element set on your computer? What about all the subsets of such a set?
2. Generate all permutations of {1, 2, 3, 4} by
   a. the bottom-up minimal-change algorithm.
   b. the Johnson-Trotter algorithm.
   c. the lexicographic-order algorithm.
3. Apply *LexicographicPermute* to multiset {1, 2, 2, 3}. Does it generate correctly all the permutations in lexicographic order?
4. Consider the following implementation of the algorithm for generating permutations discovered by B. Heap [Hea63].

```
ALGORITHM  HeapPermute(n)
    //Implements Heap's algorithm for generating permutations
    //Input: A positive integer n and a global array A[1..n]
    //Output: All permutations of elements of A
    if n = 1
        write A
```

```
else
    for i ← 1 to n do
        HeapPermute(n − 1)
        if n is odd
            swap A[1] and A[n]
        else swap A[i] and A[n]
```

a. Trace the algorithm by hand for $n = 2$, 3, and 4.

b. Prove the correctness of Heap's algorithm.

c. What is the time efficiency of *HeapPermute*?

5. Generate all the subsets of a four-element set $A = \{a_1, a_2, a_3, a_4\}$ by each of the two algorithms outlined in this section.

6. What simple trick would make the bit string–based algorithm generate subsets in squashed order?

7. Write pseudocode for a recursive algorithm for generating all 2^n bit strings of length n.

8. Write a nonrecursive algorithm for generating 2^n bit strings of length n that implements bit strings as arrays and does not use binary additions.

9. a. Generate the binary reflexive Gray code of order 4.

b. Trace the following nonrecursive algorithm to generate the binary reflexive Gray code of order 4. Start with the n-bit string of all 0's. For $i = 1, 2, \ldots, 2^{n-1}$, generate the ith bit string by flipping bit b in the previous bit string, where b is the position of the least significant 1 in the binary representation of i.

10. Design a decrease-and-conquer algorithm for generating all combinations of k items chosen from n, i.e., all k-element subsets of a given n-element set. Is your algorithm a minimal-change algorithm?

11. *Gray code and the Tower of Hanoi*

a. Show that the disk moves made in the classic recursive algorithm for the Tower of Hanoi puzzle can be used for generating the binary reflected Gray code.

b. Show how the binary reflected Gray code can be used for solving the Tower of Hanoi puzzle.

12. *Fair attraction* In olden days, one could encounter the following attraction at a fair. A light bulb was connected to several switches in such a way that it lighted up only when all the switches were closed. Each switch was controlled by a push button; pressing the button toggled the switch, but there was no way to know the state of the switch. The object was to turn the light bulb on. Design an algorithm to turn on the light bulb with the minimum number of button pushes needed in the worst case for n switches.

4.4 Decrease-by-a-Constant-Factor Algorithms

You may recall from the introduction to this chapter that decrease-by-a-constant-factor is the second major variety of decrease-and-conquer. As an example of an algorithm based on this technique, we mentioned there exponentiation by squaring defined by formula (4.2). In this section, you will find a few other examples of such algorithms.. The most important and well-known of them is binary search. Decrease-by-a-constant-factor algorithms usually run in logarithmic time, and, being very efficient, do not happen often; a reduction by a factor other than two is especially rare.

Binary Search

Binary search is a remarkably efficient algorithm for searching in a sorted array. It works by comparing a search key K with the array's middle element $A[m]$. If they match, the algorithm stops; otherwise, the same operation is repeated recursively for the first half of the array if $K < A[m]$, and for the second half if $K > A[m]$:

$$\underbrace{A[0] \ldots A[m-1]}_{\substack{\text{search here if}\\ K<A[m]}} \quad \overset{\displaystyle K}{\overset{\updownarrow}{A[m]}} \quad \underbrace{A[m+1] \ldots A[n-1]}_{\substack{\text{search here if}\\ K>A[m]}}.$$

As an example, let us apply binary search to searching for $K = 70$ in the array

3	14	27	31	39	42	55	70	74	81	85	93	98

The iterations of the algorithm are given in the following table:

index	0	1	2	3	4	5	6	7	8	9	10	11	12
value	3	14	27	31	39	42	55	70	74	81	85	93	98
iteration 1	*l*						*m*						*r*
iteration 2								*l*			*m*		*r*
iteration 3								*l,m*	*r*				

Though binary search is clearly based on a recursive idea, it can be easily implemented as a nonrecursive algorithm, too. Here is pseudocode of this nonrecursive version.

ALGORITHM *BinarySearch*$(A[0..n-1], K)$

```
//Implements nonrecursive binary search
//Input: An array A[0..n − 1] sorted in ascending order and
//       a search key K
//Output: An index of the array's element that is equal to K
//       or −1 if there is no such element
l ← 0;   r ← n − 1
while l ≤ r do
    m ← ⌊(l + r)/2⌋
    if K = A[m] return m
    else if K < A[m] r ← m − 1
    else l ← m + 1
return −1
```

The standard way to analyze the efficiency of binary search is to count the number of times the search key is compared with an element of the array. Moreover, for the sake of simplicity, we will count the so-called three-way comparisons. This assumes that after one comparison of K with $A[m]$, the algorithm can determine whether K is smaller, equal to, or larger than $A[m]$.

How many such comparisons does the algorithm make on an array of n elements? The answer obviously depends not only on n but also on the specifics of a particular instance of the problem. Let us find the number of key comparisons in the worst case $C_{worst}(n)$. The worst-case inputs include all arrays that do not contain a given search key, as well as some successful searches. Since after one comparison the algorithm faces the same situation but for an array half the size, we get the following recurrence relation for $C_{worst}(n)$:

$$C_{worst}(n) = C_{worst}(\lfloor n/2 \rfloor) + 1 \quad \text{for } n > 1, \quad C_{worst}(1) = 1. \qquad \textbf{(4.3)}$$

(Stop and convince yourself that $n/2$ must be, indeed, rounded down and that the initial condition must be written as specified.)

We already encountered recurrence (4.3), with a different initial condition, in Section 2.4 (see recurrence (2.4) and its solution there for $n = 2^k$). For the initial condition $C_{worst}(1) = 1$, we obtain

$$C_{worst}(2^k) = k + 1 = \log_2 n + 1. \qquad \textbf{(4.4)}$$

Further, similarly to the case of recurrence (2.4) (Problem 7 in Exercises 2.4), the solution given by formula (4.4) for $n = 2^k$ can be tweaked to get a solution valid for an arbitrary positive integer n:

$$C_{worst}(n) = \lfloor \log_2 n \rfloor + 1 = \lceil \log_2(n+1) \rceil. \qquad \textbf{(4.5)}$$

Formula (4.5) deserves attention. First, it implies that the worst-case time efficiency of binary search is in $\Theta(\log n)$. Second, it is the answer we should have fully expected: since the algorithm simply reduces the size of the remaining array by about half on each iteration, the number of such iterations needed to reduce the initial size n to the final size 1 has to be about $\log_2 n$. Third, to reiterate the point made in Section 2.1, the logarithmic function grows so slowly that its values remain small even for very large values of n. In particular, according to formula (4.5), it will take no more than $\lceil \log_2(10^3+1) \rceil = 10$ three-way comparisons to find an element of a given value (or establish that there is no such element) in any sorted array of one thousand elements, and it will take no more than $\lceil \log_2(10^6+1) \rceil = 20$ comparisons to do it for any sorted array of size one million!

What can we say about the average-case efficiency of binary search? A sophisticated analysis shows that the average number of key comparisons made by binary search is only slightly smaller than that in the worst case:

$$C_{avg}(n) \approx \log_2 n.$$

(More accurate formulas for the average number of comparisons in a successful and an unsuccessful search are $C_{avg}^{yes}(n) \approx \log_2 n - 1$ and $C_{avg}^{no}(n) \approx \log_2(n+1)$, respectively.)

Though binary search is an optimal searching algorithm if we restrict our operations only to comparisons between keys (see Section 11.2), there are searching algorithms (see interpolation search in Section 4.5 and hashing in Section 7.3) with a better average-case time efficiency, and one of them (hashing) does not even require the array to be sorted! These algorithms do require some special calculations in addition to key comparisons, however. Finally, the idea behind binary search has several applications beyond searching (see, e.g., [Ben00]). In addition, it can be applied to solving nonlinear equations in one unknown; we discuss this continuous analogue of binary search, called the method of bisection, in Section 12.4.

Fake-Coin Problem

Of several versions of the fake-coin identification problem, we consider here the one that best illustrates the decrease-by-a-constant-factor strategy. Among n identical-looking coins, one is fake. With a balance scale, we can compare any two sets of coins. That is, by tipping to the left, to the right, or staying even, the balance scale will tell whether the sets weigh the same or which of the sets is heavier than the other but not by how much. The problem is to design an efficient algorithm for detecting the fake coin. An easier version of the problem—the one we discuss here—assumes that the fake coin is known to be, say, lighter than the genuine one.[1]

The most natural idea for solving this problem is to divide n coins into two piles of $\lfloor n/2 \rfloor$ coins each, leaving one extra coin aside if n is odd, and put the two

1. A much more challenging version assumes no additional information about the relative weights of the fake and genuine coins or even the presence of the fake coin among n given coins. We pursue this more difficult version in the exercises for Section 11.2.

piles on the scale. If the piles weigh the same, the coin put aside must be fake; otherwise, we can proceed in the same manner with the lighter pile, which must be the one with the fake coin.

We can easily set up a recurrence relation for the number of weighings $W(n)$ needed by this algorithm in the worst case:

$$W(n) = W(\lfloor n/2 \rfloor) + 1 \quad \text{for } n > 1, \quad W(1) = 0.$$

This recurrence should look familiar to you. Indeed, it is almost identical to the one for the worst-case number of comparisons in binary search. (The difference is in the initial condition.) This similarity is not really surprising, since both algorithms are based on the same technique of halving an instance size. The solution to the recurrence for the number of weighings is also very similar to the one we had for binary search: $W(n) = \lfloor \log_2 n \rfloor$.

This stuff should look elementary by now, if not outright boring. But wait: the interesting point here is the fact that the above algorithm is not the most efficient solution. It would be more efficient to divide the coins not into two but into *three* piles of about $n/3$ coins each. (Details of a precise formulation are developed in this section's exercises. Do not miss it! If your instructor forgets, demand the instructor to assign Problem 10.) After weighing two of the piles, we can reduce the instance size by a factor of three. Accordingly, we should expect the number of weighings to be about $\log_3 n$, which is smaller than $\log_2 n$.

Russian Peasant Multiplication

Now we consider a nonorthodox algorithm for multiplying two positive integers called ***multiplication à la russe*** or the ***Russian peasant method***. Let n and m be positive integers whose product we want to compute, and let us measure the instance size by the value of n. Now, if n is even, an instance of half the size has to deal with $n/2$, and we have an obvious formula relating the solution to the problem's larger instance to the solution to the smaller one:

$$n \cdot m = \frac{n}{2} \cdot 2m.$$

If n is odd, we need only a slight adjustment of this formula:

$$n \cdot m = \frac{n-1}{2} \cdot 2m + m.$$

Using these formulas and the trivial case of $1 \cdot m = m$ to stop, we can compute product $n \cdot m$ either recursively or iteratively. An example of computing $50 \cdot 65$ with this algorithm is given in Figure 4.11. Note that all the extra addends shown in parentheses in Figure 4.11a are in the rows that have odd values in the first column. Therefore, we can find the product by simply adding all the elements in the m column that have an odd number in the n column (Figure 4.11b).

Also note that the algorithm involves just the simple operations of halving, doubling, and adding—a feature that might be attractive, for example, to those

n	m	
50	65	
25	130	
12	260	(+130)
6	520	
3	1040	
1	2080	(+1040)
	2080	+(130 + 1040) = 3250

(a)

n	m	
50	65	
25	130	130
12	260	
6	520	
3	1040	1040
1	2080	2080
		3250

(b)

FIGURE 4.11 Computing 50 · 65 by the Russian peasant method.

who do not want to memorize the table of multiplications. It is this feature of the algorithm that most probably made it attractive to Russian peasants who, according to Western visitors, used it widely in the nineteenth century and for whom the method is named. (In fact, the method was known to Egyptian mathematicians as early as 1650 B.C. [Cha98, p. 16].) It also leads to very fast hardware implementation since doubling and halving of binary numbers can be performed using shifts, which are among the most basic operations at the machine level.

Josephus Problem

Our last example is the ***Josephus problem***, named for Flavius Josephus, a famous Jewish historian who participated in and chronicled the Jewish revolt of 66–70 C.E. against the Romans. Josephus, as a general, managed to hold the fortress of Jotapata for 47 days, but after the fall of the city he took refuge with 40 diehards in a nearby cave. There, the rebels voted to perish rather than surrender. Josephus proposed that each man in turn should dispatch his neighbor, the order to be determined by casting lots. Josephus contrived to draw the last lot, and, as one of the two surviving men in the cave, he prevailed upon his intended victim to surrender to the Romans.

So let n people numbered 1 to n stand in a circle. Starting the grim count with person number 1, we eliminate every second person until only one survivor is left. The problem is to determine the survivor's number $J(n)$. For example (Figure 4.12), if n is 6, people in positions 2, 4, and 6 will be eliminated on the first pass through the circle, and people in initial positions 3 and 1 will be eliminated on the second pass, leaving a sole survivor in initial position 5—thus, $J(6) = 5$. To give another example, if n is 7, people in positions 2, 4, 6, and 1 will be eliminated on the first pass (it is more convenient to include 1 in the first pass) and people in positions 5 and, for convenience, 3 on the second—thus, $J(7) = 7$.

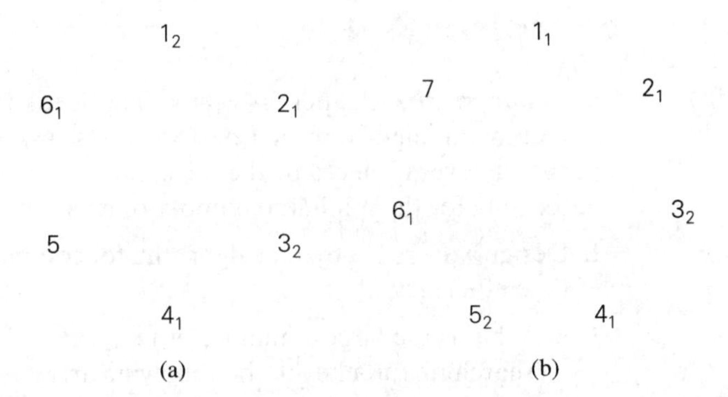

FIGURE 4.12 Instances of the Josephus problem for (a) $n = 6$ and (b) $n = 7$. Subscript numbers indicate the pass on which the person in that position is eliminated. The solutions are $J(6) = 5$ and $J(7) = 7$, respectively.

It is convenient to consider the cases of even and odd n's separately. If n is even, i.e., $n = 2k$, the first pass through the circle yields an instance of exactly the same problem but half its initial size. The only difference is in position numbering; for example, a person in initial position 3 will be in position 2 for the second pass, a person in initial position 5 will be in position 3, and so on (check Figure 4.12a). It is easy to see that to get the initial position of a person, we simply need to multiply his new position by 2 and subtract 1. This relationship will hold, in particular, for the survivor, i.e.,

$$J(2k) = 2J(k) - 1.$$

Let us now consider the case of an odd n ($n > 1$), i.e., $n = 2k + 1$. The first pass eliminates people in all even positions. If we add to this the elimination of the person in position 1 right after that, we are left with an instance of size k. Here, to get the initial position that corresponds to the new position numbering, we have to multiply the new position number by 2 and add 1 (check Figure 4.12b). Thus, for odd values of n, we get

$$J(2k + 1) = 2J(k) + 1.$$

Can we get a closed-form solution to the two-case recurrence subject to the initial condition $J(1) = 1$? The answer is yes, though getting it requires more ingenuity than just applying backward substitutions. In fact, one way to find a solution is to apply forward substitutions to get, say, the first 15 values of $J(n)$, discern a pattern, and then prove its general validity by mathematical induction. We leave the execution of this plan to the exercises; alternatively, you can look it up in [GKP94], whose exposition of the Josephus problem we have been following. Interestingly, the most elegant form of the closed-form answer involves the binary representation of size n: $J(n)$ can be obtained by a 1-bit cyclic shift left of n itself! For example, $J(6) = J(110_2) = 101_2 = 5$ and $J(7) = J(111_2) = 111_2 = 7$.

Exercises 4.4

1. *Cutting a stick* A stick n inches long needs to be cut into n 1-inch pieces. Outline an algorithm that performs this task with the minimum number of cuts if several pieces of the stick can be cut at the same time. Also give a formula for the minimum number of cuts.

2. Design a decrease-by-half algorithm for computing $\lfloor \log_2 n \rfloor$ and determine its time efficiency.

3. **a.** What is the largest number of key comparisons made by binary search in searching for a key in the following array?

3	14	27	31	39	42	55	70	74	81	85	93	98

   b. List all the keys of this array that will require the largest number of key comparisons when searched for by binary search.

   c. Find the average number of key comparisons made by binary search in a successful search in this array. Assume that each key is searched for with the same probability.

   d. Find the average number of key comparisons made by binary search in an unsuccessful search in this array. Assume that searches for keys in each of the 14 intervals formed by the array's elements are equally likely.

4. Estimate how many times faster an average successful search will be in a sorted array of one million elements if it is done by binary search versus sequential search.

5. The time efficiency of sequential search does not depend on whether a list is implemented as an array or as a linked list. Is it also true for searching a sorted list by binary search?

6. **a.** Design a version of binary search that uses only two-way comparisons such as $\le$ and $=$. Implement your algorithm in the language of your choice and carefully debug it: such programs are notorious for being prone to bugs.

   b. Analyze the time efficiency of the two-way comparison version designed in part a.

7. *Picture guessing* A version of the popular problem-solving task involves presenting people with an array of 42 pictures—seven rows of six pictures each—and asking them to identify the target picture by asking questions that can be answered yes or no. Further, people are then required to identify the picture with as few questions as possible. Suggest the most efficient algorithm for this problem and indicate the largest number of questions that may be necessary.

8. Consider ***ternary search***—the following algorithm for searching in a sorted array $A[0..n-1]$. If $n = 1$, simply compare the search key K with the single

element of the array; otherwise, search recursively by comparing K with $A[\lfloor n/3 \rfloor]$, and if K is larger, compare it with $A[\lfloor 2n/3 \rfloor]$ to determine in which third of the array to continue the search.

a. What design technique is this algorithm based on?

b. Set up a recurrence for the number of key comparisons in the worst case. You may assume that $n = 3^k$.

c. Solve the recurrence for $n = 3^k$.

d. Compare this algorithm's efficiency with that of binary search.

9. An array $A[0..n-2]$ contains $n-1$ integers from 1 to n in increasing order. (Thus one integer in this range is missing.) Design the most efficient algorithm you can to find the missing integer and indicate its time efficiency.

10. a. Write pseudocode for the divide-into-three algorithm for the fake-coin problem. Make sure that your algorithm handles properly all values of n, not only those that are multiples of 3.

b. Set up a recurrence relation for the number of weighings in the divide-into-three algorithm for the fake-coin problem and solve it for $n = 3^k$.

c. For large values of n, about how many times faster is this algorithm than the one based on dividing coins into two piles? Your answer should not depend on n.

11. a. Apply the Russian peasant algorithm to compute $26 \cdot 47$.

b. From the standpoint of time efficiency, does it matter whether we multiply n by m or m by n by the Russian peasant algorithm?

12. a. Write pseudocode for the Russian peasant multiplication algorithm.

b. What is the time efficiency class of Russian peasant multiplication?

13. Find $J(40)$—the solution to the Josephus problem for $n = 40$.

14. Prove that the solution to the Josephus problem is 1 for every n that is a power of 2.

15. For the Josephus problem,

a. compute $J(n)$ for $n = 1, 2, \ldots, 15$.

b. discern a pattern in the solutions for the first fifteen values of n and prove its general validity.

c. prove the validity of getting $J(n)$ by a 1-bit cyclic shift left of the binary representation of n.

4.5 Variable-Size-Decrease Algorithms

In the third principal variety of decrease-and-conquer, the size reduction pattern varies from one iteration of the algorithm to another. Euclid's algorithm for computing the greatest common divisor (Section 1.1) provides a good example of this kind of algorithm. In this section, we encounter a few more examples of this variety.

Computing a Median and the Selection Problem

The ***selection problem*** is the problem of finding the kth smallest element in a list of n numbers. This number is called the kth ***order statistic***. Of course, for $k = 1$ or $k = n$, we can simply scan the list in question to find the smallest or largest element, respectively. A more interesting case of this problem is for $k = \lceil n/2 \rceil$, which asks to find an element that is not larger than one half of the list's elements and not smaller than the other half. This middle value is called the ***median***, and it is one of the most important notions in mathematical statistics. Obviously, we can find the kth smallest element in a list by sorting the list first and then selecting the kth element in the output of a sorting algorithm. The time of such an algorithm is determined by the efficiency of the sorting algorithm used. Thus, with a fast sorting algorithm such as mergesort (discussed in the next chapter), the algorithm's efficiency is in $O(n \log n)$.

You should immediately suspect, however, that sorting the entire list is most likely overkill since the problem asks not to order the entire list but just to find its kth smallest element. Indeed, we can take advantage of the idea of ***partitioning*** a given list around some value p of, say, its first element. In general, this is a rearrangement of the list's elements so that the left part contains all the elements smaller than or equal to p, followed by the ***pivot*** p itself, followed by all the elements greater than or equal to p.

Of the two principal algorithmic alternatives to partition an array, here we discuss the ***Lomuto partitioning*** [Ben00, p. 117]; we introduce the better known Hoare's algorithm in the next chapter. To get the idea behind the Lomuto partitioning, it is helpful to think of an array—or, more generally, a subarray $A[l..r]$ $(0 \le l \le r \le n-1)$—under consideration as composed of three contiguous segments. Listed in the order they follow pivot p, they are as follows: a segment with elements known to be smaller than p, the segment of elements known to be greater than or equal to p, and the segment of elements yet to be compared to p (see Figure 4.13a). Note that the segments can be empty; for example, it is always the case for the first two segments before the algorithm starts.

Starting with $i = l + 1$, the algorithm scans the subarray $A[l..r]$ left to right, maintaining this structure until a partition is achieved. On each iteration, it compares the first element in the unknown segment (pointed to by the scanning index i in Figure 4.13a) with the pivot p. If $A[i] \ge p$, i is simply incremented to expand the segment of the elements greater than or equal to p while shrinking the unprocessed segment. If $A[i] < p$, it is the segment of the elements smaller than p that needs to be expanded. This is done by incrementing s, the index of the last

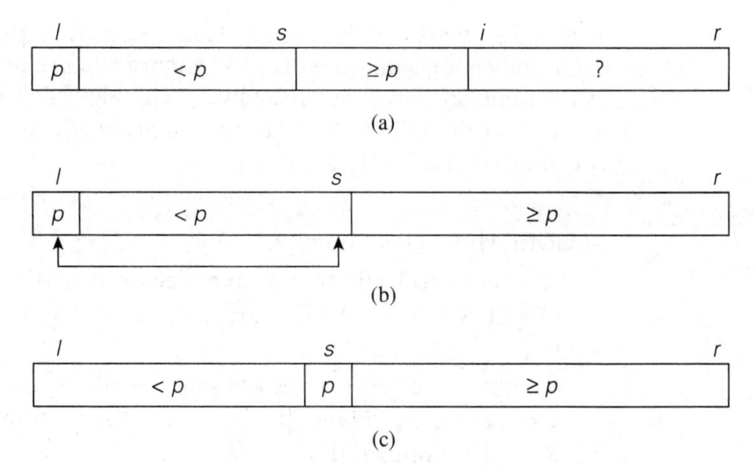

FIGURE 4.13 Illustration of the Lomuto partitioning.

element in the first segment, swapping $A[i]$ and $A[s]$, and then incrementing i to point to the new first element of the shrunk unprocessed segment. After no unprocessed elements remain (Figure 4.13b), the algorithm swaps the pivot with $A[s]$ to achieve a partition being sought (Figure 4.13c).

Here is pseudocode implementing this partitioning procedure.

ALGORITHM *LomutoPartition*($A[l..r]$)

//Partitions subarray by Lomuto's algorithm using first element as pivot
//Input: A subarray $A[l..r]$ of array $A[0..n-1]$, defined by its left and right
// indices l and r $(l \le r)$
//Output: Partition of $A[l..r]$ and the new position of the pivot
$p \leftarrow A[l]$
$s \leftarrow l$
for $i \leftarrow l+1$ **to** r **do**
  if $A[i] < p$
    $s \leftarrow s+1$; swap($A[s]$, $A[i]$)
swap($A[l]$, $A[s]$)
return s

How can we take advantage of a list partition to find the kth smallest element in it? Let us assume that the list is implemented as an array whose elements are indexed starting with a 0, and let s be the partition's split position, i.e., the index of the array's element occupied by the pivot after partitioning. If $s = k-1$, pivot p itself is obviously the kth smallest element, which solves the problem. If $s > k-1$, the kth smallest element in the entire array can be found as the kth smallest element in the left part of the partitioned array. And if $s < k-1$, it can be found as the $(k - s)$th smallest element in its right part. Thus, if we do not solve the problem outright, we reduce its instance to a smaller one, which can be solved by the same approach, i.e., recursively. This algorithm is called ***quickselect***.

To find the kth smallest element in array $A[0..n-1]$ by this algorithm, call *Quickselect*$(A[0..n-1], k)$ where

ALGORITHM *Quickselect*$(A[l..r], k)$

//Solves the selection problem by recursive partition-based algorithm
//Input: Subarray $A[l..r]$ of array $A[0..n-1]$ of orderable elements and
// integer k $(1 \le k \le r - l + 1)$
//Output: The value of the kth smallest element in $A[l..r]$
$s \leftarrow$ *LomutoPartition*$(A[l..r])$ //or another partition algorithm
if $s = k - 1$ **return** $A[s]$
else if $s > l + k - 1$ *Quickselect*$(A[l..s-1], k)$
else *Quickselect*$(A[s+1..r], k - 1 - s)$

In fact, the same idea can be implemented without recursion as well. For the nonrecursive version, we need not even adjust the value of k but just continue until $s = k - 1$.

EXAMPLE Apply the partition-based algorithm to find the median of the following list of nine numbers: 4, 1, 10, 8, 7, 12, 9, 2, 15. Here, $k = \lceil 9/2 \rceil = 5$ and our task is to find the 5th smallest element in the array.

We use the above version of array partitioning, showing the pivots in bold.

0	1	2	3	4	5	6	7	8
s	i							
4	1	10	8	7	12	9	2	15
	s	i						
4	1	10	8	7	12	9	2	15
	s						i	
4	1	10	8	7	12	9	2	15
		s					i	
4	1	2	8	7	12	9	10	15
		s						i
4	1	2	8	7	12	9	10	15
2	1	**4**	8	7	12	9	10	15

Since $s = 2$ is smaller than $k - 1 = 4$, we proceed with the right part of the array:

0	1	2	3	4	5	6	7	8
			s	*i*				
			8	7	12	9	10	15
				s	*i*			
			8	7	12	9	10	15
				s				*i*
			8	7	12	9	10	15
			7	**8**	12	9	10	15

Now $s = k - 1 = 4$, and hence we can stop: the found median is 8, which is greater than 2, 1, 4, and 7 but smaller than 12, 9, 10, and 15. ■

How efficient is quickselect? Partitioning an n-element array always requires $n - 1$ key comparisons. If it produces the split that solves the selection problem without requiring more iterations, then for this best case we obtain $C_{best}(n) = n - 1 \in \Theta(n)$. Unfortunately, the algorithm can produce an extremely unbalanced partition of a given array, with one part being empty and the other containing $n - 1$ elements. In the worst case, this can happen on each of the $n - 1$ iterations. (For a specific example of the worst-case input, consider, say, the case of $k = n$ and a strictly increasing array.) This implies that

$$C_{worst}(n) = (n - 1) + (n - 2) + \cdots + 1 = (n - 1)n/2 \in \Theta(n^2),$$

which compares poorly with the straightforward sorting-based approach mentioned in the beginning of our selection problem discussion. Thus, the usefulness of the partition-based algorithm depends on the algorithm's efficiency in the average case. Fortunately, a careful mathematical analysis has shown that the average-case efficiency is linear. In fact, computer scientists have discovered a more sophisticated way of choosing a pivot in quickselect that guarantees linear time even in the worst case [Blo73], but it is too complicated to be recommended for practical applications.

It is also worth noting that the partition-based algorithm solves a somewhat more general problem of identifying the k smallest and $n - k$ largest elements of a given list, not just the value of its kth smallest element.

Interpolation Search

As the next example of a variable-size-decrease algorithm, we consider an algorithm for searching in a sorted array called ***interpolation search***. Unlike binary search, which always compares a search key with the middle value of a given sorted array (and hence reduces the problem's instance size by half), interpolation search takes into account the value of the search key in order to find the array's element to be compared with the search key. In a sense, the algorithm mimics the way we

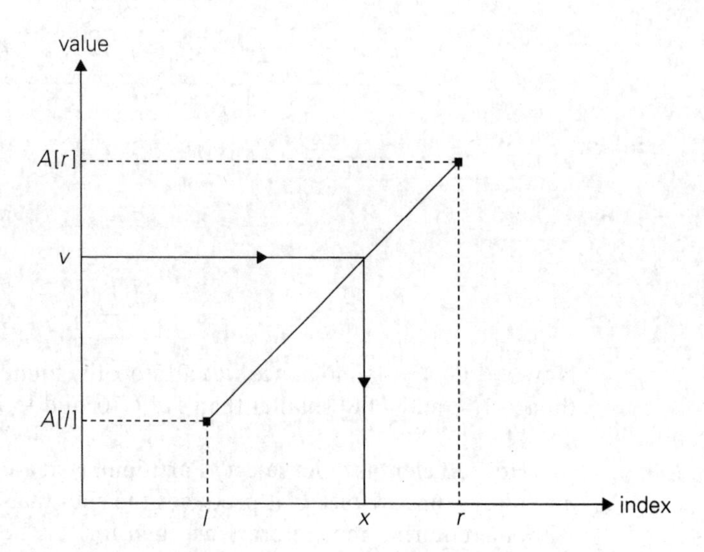

FIGURE 4.14 Index computation in interpolation search.

search for a name in a telephone book: if we are searching for someone named Brown, we open the book not in the middle but very close to the beginning, unlike our action when searching for someone named, say, Smith.

More precisely, on the iteration dealing with the array's portion between the leftmost element $A[l]$ and the rightmost element $A[r]$, the algorithm assumes that the array values increase linearly, i.e., along the straight line through the points $(l, A[l])$ and $(r, A[r])$. (The accuracy of this assumption can influence the algorithm's efficiency but not its correctness.) Accordingly, the search key's value v is compared with the element whose index is computed as (the round-off of) the x coordinate of the point on the straight line through the points $(l, A[l])$ and $(r, A[r])$ whose y coordinate is equal to the search value v (Figure 4.14).

Writing down a standard equation for the straight line passing through the points $(l, A[l])$ and $(r, A[r])$, substituting v for y, and solving it for x leads to the following formula:

$$x = l + \left\lfloor \frac{(v - A[l])(r - l)}{A[r] - A[l]} \right\rfloor. \tag{4.6}$$

The logic behind this approach is quite straightforward. We know that the array values are increasing (more accurately, not decreasing) from $A[l]$ to $A[r]$, but we do not know how they do it. Had these values increased linearly, which is the simplest manner possible, the index computed by formula (4.4) would be the expected location of the array's element with the value equal to v. Of course, if v is not between $A[l]$ and $A[r]$, formula (4.4) need not be applied (why?).

After comparing v with $A[x]$, the algorithm either stops (if they are equal) or proceeds by searching in the same manner among the elements indexed either between l and $x - 1$ or between $x + 1$ and r, depending on whether $A[x]$ is smaller or larger than v. Thus, the size of the problem's instance is reduced, but we cannot tell a priori by how much.

The analysis of the algorithm's efficiency shows that interpolation search uses fewer than $\log_2 \log_2 n + 1$ key comparisons on the average when searching in a list of n random keys. This function grows so slowly that the number of comparisons is a very small constant for all practically feasible inputs (see Problem 6 in this section's exercises). But in the worst case, interpolation search is only linear, which must be considered a bad performance (why?).

Assessing the worthiness of interpolation search versus that of binary search, Robert Sedgewick wrote in the second edition of his *Algorithms* that binary search is probably better for smaller files but interpolation search is worth considering for large files and for applications where comparisons are particularly expensive or access costs are very high. Note that in Section 12.4 we discuss a continuous counterpart of interpolation search, which can be seen as one more example of a variable-size-decrease algorithm.

Searching and Insertion in a Binary Search Tree

Let us revisit the binary search tree. Recall that this is a binary tree whose nodes contain elements of a set of orderable items, one element per node, so that for every node all elements in the left subtree are smaller and all the elements in the right subtree are greater than the element in the subtree's root. When we need to search for an element of a given value v in such a tree, we do it recursively in the following manner. If the tree is empty, the search ends in failure. If the tree is not empty, we compare v with the tree's root $K(r)$. If they match, a desired element is found and the search can be stopped; if they do not match, we continue with the search in the left subtree of the root if $v < K(r)$ and in the right subtree if $v > K(r)$. Thus, on each iteration of the algorithm, the problem of searching in a binary search tree is reduced to searching in a smaller binary search tree. The most sensible measure of the size of a search tree is its height; obviously, the decrease in a tree's height normally changes from one iteration to another of the binary tree search—thus giving us an excellent example of a variable-size-decrease algorithm.

In the worst case of the binary tree search, the tree is severely skewed. This happens, in particular, if a tree is constructed by successive insertions of an increasing or decreasing sequence of keys (Figure 4.15).

Obviously, the search for a_{n-1} in such a tree requires n comparisons, making the worst-case efficiency of the search operation fall into $\Theta(n)$. Fortunately, the average-case efficiency turns out to be in $\Theta(\log n)$. More precisely, the number of key comparisons needed for a search in a binary search tree built from n random keys is about $2\ln n \approx 1.39 \log_2 n$. Since insertion of a new key into a binary search tree is almost identical to that of searching there, it also exemplifies the variable-size-decrease technique and has the same efficiency characteristics as the search operation.

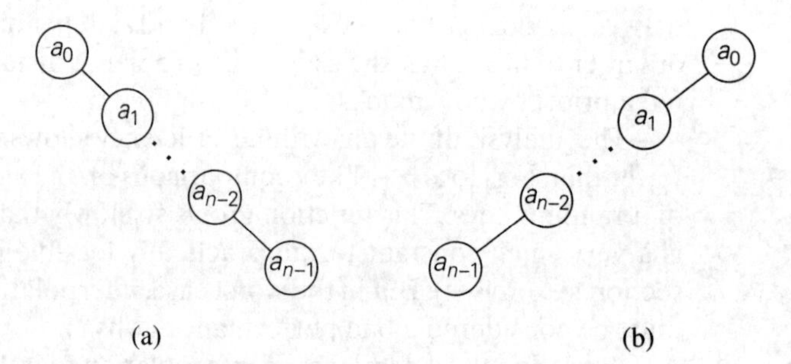

FIGURE 4.15 Binary search trees for (a) an increasing sequence of keys and (b) a decreasing sequence of keys.

The Game of Nim

There are several well-known games that share the following features. There are two players, who move in turn. No randomness or hidden information is permitted: all players know all information about gameplay. A game is impartial: each player has the same moves available from the same game position. Each of a finite number of available moves leads to a smaller instance of the same game. The game ends with a win by one of the players (there are no ties). The winner is the last player who is able to move.

A prototypical example of such games is ***Nim***. Generally, the game is played with several piles of chips, but we consider the one-pile version first. Thus, there is a single pile of n chips. Two players take turns by removing from the pile at least one and at most m chips; the number of chips taken may vary from one move to another, but both the lower and upper limits stay the same. Who wins the game by taking the last chip, the player moving first or second, if both players make the best moves possible?

Let us call an instance of the game a winning position for the player to move next if that player has a winning strategy, i.e., a sequence of moves that results in a victory no matter what moves the opponent makes. Let us call an instance of the game a losing position for the player to move next if every move available for that player leads to a winning position for the opponent. The standard approach to determining which positions are winning and which are losing is to investigate small values of n first. It is logical to consider the instance of $n = 0$ as a losing one for the player to move next because this player is the first one who cannot make a move. Any instance with $1 \le n \le m$ chips is obviously a winning position for the player to move next (why?). The instance with $n = m + 1$ chips is a losing one because taking any allowed number of chips puts the opponent in a winning position. (See an illustration for $m = 4$ in Figure 4.16.) Any instance with $m + 2 \le n \le 2m + 1$ chips is a winning position for the player to move next because there is a move that leaves the opponent with $m + 1$ chips, which is a losing

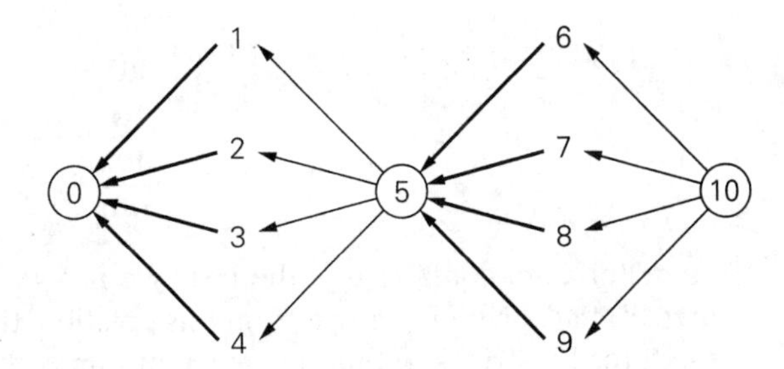

FIGURE 4.16 Illustration of one-pile Nim with the maximum number of chips that may be taken on each move $m = 4$. The numbers indicate n, the number of chips in the pile. The losing positions for the player to move are circled. Only winning moves from the winning positions are shown (in bold).

position. $2m + 2 = 2(m + 1)$ chips is the next losing position, and so on. It is not difficult to see the pattern that can be formally proved by mathematical induction: an instance with n chips is a winning position for the player to move next if and only if n is not a multiple of $m + 1$. The winning strategy is to take $n \bmod (m + 1)$ chips on every move; any deviation from this strategy puts the opponent in a winning position.

One-pile Nim has been known for a very long time. It appeared, in particular, as the ***summation game*** in the first published book on recreational mathematics, authored by Claude-Gaspar Bachet, a French aristocrat and mathematician, in 1612: a player picks a positive integer less than, say, 10, and then his opponent and he take turns adding any integer less than 10; the first player to reach 100 exactly is the winner [Dud70].

In general, Nim is played with $I > 1$ piles of chips of sizes $n_1, n_2, \ldots, n_I$. On each move, a player can take any available number of chips, including all of them, from any single pile. The goal is the same—to be the last player able to make a move. Note that for $I = 2$, it is easy to figure out who wins this game and how. Here is a hint: the answer for the game's instances with $n_1 = n_2$ differs from the answer for those with $n_1 \neq n_2$.

A solution to the general case of Nim is quite unexpected because it is based on the binary representation of the pile sizes. Let $b_1, b_2, \ldots, b_I$ be the pile sizes in binary. Compute their ***binary digital sum***, also known as the ***nim sum***, defined as the sum of binary digits discarding any carry. (In other words, a binary digit s_i in the sum is 0 if the number of 1's in the ith position in the addends is even, and it is 1 if the number of 1's is odd.) It turns out that an instance of Nim is a winning one for the player to move next if and only if its nim sum contains at least one 1; consequently, Nim's instance is a losing instance if and only if its nim sum contains only zeros. For example, for the commonly played instance with $n_1 = 3$, $n_2 = 4$, $n_3 = 5$, the nim sum is

$$\begin{array}{r} 011 \\ 100 \\ 101 \\ \hline 010 \end{array}$$

Since this sum contains a 1, the instance is a winning one for the player moving first. To find a winning move from this position, the player needs to change one of the three bit strings so that the new nim sum contains only 0's. It is not difficult to see that the only way to accomplish this is to remove two chips from the first pile.

This ingenious solution to the game of Nim was discovered by Harvard mathematics professor C. L. Bouton more than 100 years ago. Since then, mathematicians have developed a much more general theory of such games. An excellent account of this theory, with applications to many specific games, is given in the monograph by E. R. Berlekamp, J. H. Conway, and R. K. Guy [Ber03].

Exercises 4.5

1. **a.** If we measure an instance size of computing the greatest common divisor of m and n by the size of the second number n, by how much can the size decrease after one iteration of Euclid's algorithm?

   b. Prove that an instance size will always decrease at least by a factor of two after two successive iterations of Euclid's algorithm.

2. Apply quickselect to find the median of the list of numbers 9, 12, 5, 17, 20, 30, 8.

3. Write pseudocode for a nonrecursive implementation of quickselect.

4. Derive the formula underlying interpolation search.

5. Give an example of the worst-case input for interpolation search and show that the algorithm is linear in the worst case.

6. **a.** Find the smallest value of n for which $\log_2 \log_2 n + 1$ is greater than 6.

   b. Determine which, if any, of the following assertions are true:
   i. $\log \log n \in o(\log n)$ **ii.** $\log \log n \in \Theta(\log n)$ **iii.** $\log \log n \in \Omega(\log n)$

7. **a.** Outline an algorithm for finding the largest key in a binary search tree. Would you classify your algorithm as a variable-size-decrease algorithm?

   b. What is the time efficiency class of your algorithm in the worst case?

8. **a.** Outline an algorithm for deleting a key from a binary search tree. Would you classify this algorithm as a variable-size-decrease algorithm?

   b. What is the time efficiency class of your algorithm in the worst case?

9. Outline a variable-size-decrease algorithm for constructing an Eulerian circuit in a connected graph with all vertices of even degrees.

10. *Misère one-pile Nim* Consider the so-called ***misère version*** of the one-pile Nim, in which the player taking the last chip loses the game. All the other conditions of the game remain the same, i.e., the pile contains n chips and on each move a player takes at least one but no more than m chips. Identify the winning and losing positions (for the player to move next) in this game.

11. a. *Moldy chocolate* Two players take turns by breaking an $m \times n$ chocolate bar, which has one spoiled 1×1 square. Each break must be a single straight line cutting all the way across the bar along the boundaries between the squares. After each break, the player who broke the bar last eats the piece that does not contain the spoiled square. The player left with the spoiled square loses the game. Is it better to go first or second in this game?

b. Write an interactive program to play this game with the computer. Your program should make a winning move in a winning position and a random legitimate move in a losing position.

12. *Flipping pancakes* There are n pancakes all of different sizes that are stacked on top of each other. You are allowed to slip a flipper under one of the pancakes and flip over the whole stack above the flipper. The purpose is to arrange pancakes according to their size with the biggest at the bottom. (You can see a visualization of this puzzle on the *Interactive Mathematics Miscellany and Puzzles* site [Bog].) Design an algorithm for solving this puzzle.

13. You need to search for a given number in an $n \times n$ matrix in which every row and every column is sorted in increasing order. Can you design a $O(n)$ algorithm for this problem? [Laa10]

SUMMARY

- *Decrease-and-conquer* is a general algorithm design technique, based on exploiting a relationship between a solution to a given instance of a problem and a solution to a smaller instance of the same problem. Once such a relationship is established, it can be exploited either top down (usually recursively) or bottom up.

- There are three major variations of decrease-and-conquer:
  - *decrease-by-a-constant*, most often by one (e.g., insertion sort)
  - *decrease-by-a-constant-factor*, most often by the factor of two (e.g., binary search)
  - *variable-size-decrease* (e.g., Euclid's algorithm)

- *Insertion sort* is a direct application of the decrease-(by one)-and-conquer technique to the sorting problem. It is a $\Theta(n^2)$ algorithm both in the worst and average cases, but it is about twice as fast on average than in the worst case. The algorithm's notable advantage is a good performance on almost-sorted arrays.

- A *digraph* is a graph with directions on its edges. The *topological sorting problem* asks to list vertices of a digraph in an order such that for every edge of the digraph, the vertex it starts at is listed before the vertex it points to. This problem has a solution if and only if a digraph is a *dag* (*directed acyclic graph*), i.e., it has no directed cycles.

- There are two algorithms for solving the topological sorting problem. The first one is based on depth-first search; the second is based on a direct application of the decrease-by-one technique.

- The decrease-by-one technique is a natural approach to developing algorithms for generating elementary combinatorial objects. The most efficient class of such algorithms are minimal-change algorithms. However, the number of combinatorial objects grows so fast that even the best algorithms are of practical interest only for very small instances of such problems.

- *Binary search* is a very efficient algorithm for searching in a sorted array. It is a principal example of a decrease-by-a-constant-factor algorithm. Other examples include *exponentiation by squaring*, identifying a fake coin with a balance scale, *Russian peasant multiplication*, and the *Josephus problem*.

- For some decrease-and-conquer algorithms, the size reduction varies from one iteration of the algorithm to another. Examples of such *variable-size-decrease* algorithms include Euclid's algorithm, the partition-based algorithm for the *selection problem*, *interpolation search*, and searching and insertion in a binary search tree. *Nim* exemplifies games that proceed through a series of diminishing instances of the same game.

5

Divide-and-Conquer

Whatever man prays for, he prays for a miracle. Every prayer reduces itself to this—Great God, grant that twice two be not four.

—Ivan Turgenev (1818–1883), Russian novelist and short-story writer

Divide-and-conquer is probably the best-known general algorithm design technique. Though its fame may have something to do with its catchy name, it is well deserved: quite a few very efficient algorithms are specific implementations of this general strategy. Divide-and-conquer algorithms work according to the following general plan:

1. A problem is divided into several subproblems of the same type, ideally of about equal size.
2. The subproblems are solved (typically recursively, though sometimes a different algorithm is employed, especially when subproblems become small enough).
3. If necessary, the solutions to the subproblems are combined to get a solution to the original problem.

The divide-and-conquer technique is diagrammed in Figure 5.1, which depicts the case of dividing a problem into two smaller subproblems, by far the most widely occurring case (at least for divide-and-conquer algorithms designed to be executed on a single-processor computer).

As an example, let us consider the problem of computing the sum of n numbers $a_0, \ldots, a_{n-1}$. If $n > 1$, we can divide the problem into two instances of the same problem: to compute the sum of the first $\lfloor n/2 \rfloor$ numbers and to compute the sum of the remaining $\lceil n/2 \rceil$ numbers. (Of course, if $n = 1$, we simply return a_0 as the answer.) Once each of these two sums is computed by applying the same method recursively, we can add their values to get the sum in question:

$$a_0 + \cdots + a_{n-1} = (a_0 + \cdots + a_{\lfloor n/2 \rfloor - 1}) + (a_{\lfloor n/2 \rfloor} + \cdots + a_{n-1}).$$

Is this an efficient way to compute the sum of n numbers? A moment of reflection (why could it be more efficient than the brute-force summation?), a

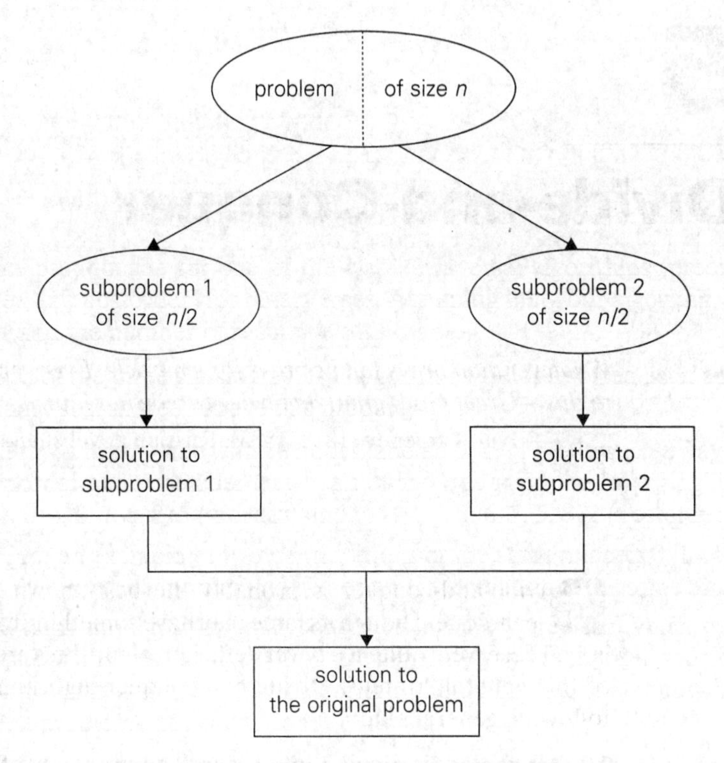

FIGURE 5.1 Divide-and-conquer technique (typical case).

small example of summing, say, four numbers by this algorithm, a formal analysis (which follows), and common sense (we do not normally compute sums this way, do we?) all lead to a negative answer to this question.[1]

Thus, not every divide-and-conquer algorithm is necessarily more efficient than even a brute-force solution. But often our prayers to the Goddess of Algorithmics—see the chapter's epigraph—*are* answered, and the time spent on executing the divide-and-conquer plan turns out to be significantly smaller than solving a problem by a different method. In fact, the divide-and-conquer approach yields some of the most important and efficient algorithms in computer science. We discuss a few classic examples of such algorithms in this chapter. Though we consider only sequential algorithms here, it is worth keeping in mind that the divide-and-conquer technique is ideally suited for parallel computations, in which each subproblem can be solved simultaneously by its own processor.

1. Actually, the divide-and-conquer algorithm, called the ***pairwise summation***, may substantially reduce the accumulated round-off error of the sum of numbers that can be represented only approximately in a digital computer [Hig93].

As mentioned above, in the most typical case of divide-and-conquer a problem's instance of size n is divided into two instances of size $n/2$. More generally, an instance of size n can be divided into b instances of size n/b, with a of them needing to be solved. (Here, a and b are constants; $a \geq 1$ and $b > 1$.) Assuming that size n is a power of b to simplify our analysis, we get the following recurrence for the running time $T(n)$:

$$T(n) = aT(n/b) + f(n), \qquad \textbf{(5.1)}$$

where $f(n)$ is a function that accounts for the time spent on dividing an instance of size n into instances of size n/b and combining their solutions. (For the sum example above, $a = b = 2$ and $f(n) = 1$.) Recurrence (5.1) is called the ***general divide-and-conquer recurrence***. Obviously, the order of growth of its solution $T(n)$ depends on the values of the constants a and b and the order of growth of the function $f(n)$. The efficiency analysis of many divide-and-conquer algorithms is greatly simplified by the following theorem (see Appendix B).

Master Theorem If $f(n) \in \Theta(n^d)$ where $d \geq 0$ in recurrence (5.1), then

$$T(n) \in \begin{cases} \Theta(n^d) & \text{if } a < b^d, \\ \Theta(n^d \log n) & \text{if } a = b^d, \\ \Theta(n^{\log_b a}) & \text{if } a > b^d. \end{cases}$$

Analogous results hold for the O and Ω notations, too.

For example, the recurrence for the number of additions $A(n)$ made by the divide-and-conquer sum-computation algorithm (see above) on inputs of size $n = 2^k$ is

$$A(n) = 2A(n/2) + 1.$$

Thus, for this example, $a = 2$, $b = 2$, and $d = 0$; hence, since $a > b^d$,

$$A(n) \in \Theta(n^{\log_b a}) = \Theta(n^{\log_2 2}) = \Theta(n).$$

Note that we were able to find the solution's efficiency class without going through the drudgery of solving the recurrence. But, of course, this approach can only establish a solution's order of growth to within an unknown multiplicative constant, whereas solving a recurrence equation with a specific initial condition yields an exact answer (at least for n's that are powers of b).

It is also worth pointing out that if $a = 1$, recurrence (5.1) covers decrease-by-a-constant-factor algorithms discussed in the previous chapter. In fact, some people consider such algorithms as binary search degenerate cases of divide-and-conquer, where just one of two subproblems of half the size needs to be solved. It is better not to do this and consider decrease-by-a-constant-factor and divide-and-conquer as different design paradigms.

5.1 Mergesort

Mergesort is a perfect example of a successful application of the divide-and-conquer technique. It sorts a given array $A[0..n-1]$ by dividing it into two halves $A[0..\lfloor n/2\rfloor-1]$ and $A[\lfloor n/2\rfloor..n-1]$, sorting each of them recursively, and then merging the two smaller sorted arrays into a single sorted one.

```
ALGORITHM  Mergesort(A[0..n − 1])
    //Sorts array A[0..n − 1] by recursive mergesort
    //Input: An array A[0..n − 1] of orderable elements
    //Output: Array A[0..n − 1] sorted in nondecreasing order
    if n > 1
        copy A[0..⌊n/2⌋ − 1] to B[0..⌊n/2⌋ − 1]
        copy A[⌊n/2⌋..n − 1] to C[0..⌈n/2⌉ − 1]
        Mergesort(B[0..⌊n/2⌋ − 1])
        Mergesort(C[0..⌈n/2⌉ − 1])
        Merge(B, C, A)   //see below
```

The ***merging*** of two sorted arrays can be done as follows. Two pointers (array indices) are initialized to point to the first elements of the arrays being merged. The elements pointed to are compared, and the smaller of them is added to a new array being constructed; after that, the index of the smaller element is incremented to point to its immediate successor in the array it was copied from. This operation is repeated until one of the two given arrays is exhausted, and then the remaining elements of the other array are copied to the end of the new array.

```
ALGORITHM  Merge(B[0..p − 1], C[0..q − 1], A[0..p + q − 1])
    //Merges two sorted arrays into one sorted array
    //Input: Arrays B[0..p − 1] and C[0..q − 1] both sorted
    //Output: Sorted array A[0..p + q − 1] of the elements of B and C
    i ← 0; j ← 0; k ← 0
    while i < p and j < q do
        if B[i] ≤ C[j]
            A[k] ← B[i]; i ← i + 1
        else A[k] ← C[j]; j ← j + 1
        k ← k + 1
    if i = p
        copy C[j..q − 1] to A[k..p + q − 1]
    else copy B[i..p − 1] to A[k..p + q − 1]
```

The operation of the algorithm on the list 8, 3, 2, 9, 7, 1, 5, 4 is illustrated in Figure 5.2.

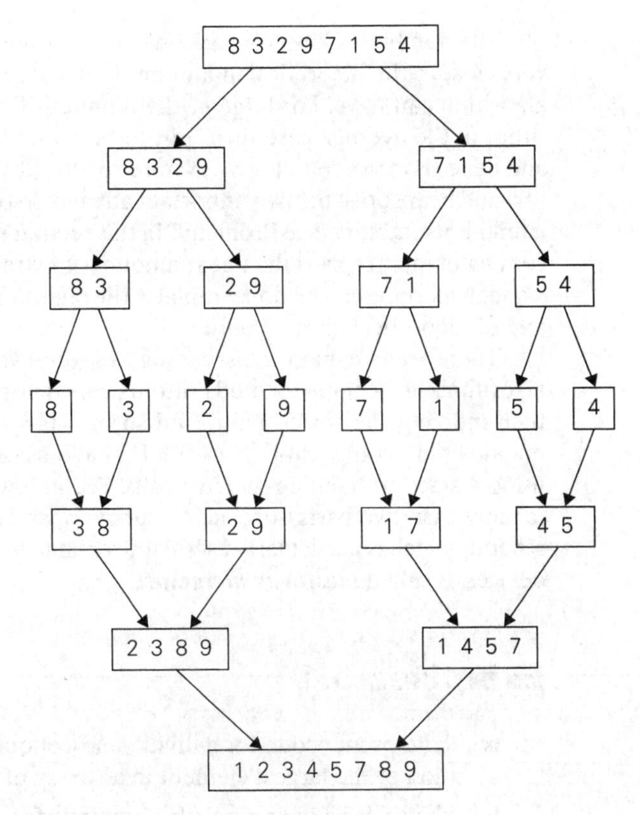

FIGURE 5.2 Example of mergesort operation.

How efficient is mergesort? Assuming for simplicity that n is a power of 2, the recurrence relation for the number of key comparisons $C(n)$ is

$$C(n) = 2C(n/2) + C_{merge}(n) \quad \text{for } n > 1, \quad C(1) = 0.$$

Let us analyze $C_{merge}(n)$, the number of key comparisons performed during the merging stage. At each step, exactly one comparison is made, after which the total number of elements in the two arrays still needing to be processed is reduced by 1. In the worst case, neither of the two arrays becomes empty before the other one contains just one element (e.g., smaller elements may come from the alternating arrays). Therefore, for the worst case, $C_{merge}(n) = n - 1$, and we have the recurrence

$$C_{worst}(n) = 2C_{worst}(n/2) + n - 1 \quad \text{for } n > 1, \quad C_{worst}(1) = 0.$$

Hence, according to the Master Theorem, $C_{worst}(n) \in \Theta(n \log n)$ (why?). In fact, it is easy to find the exact solution to the worst-case recurrence for $n = 2^k$:

$$C_{worst}(n) = n \log_2 n - n + 1.$$

The number of key comparisons made by mergesort in the worst case comes very close to the theoretical minimum[2] that any general comparison-based sorting algorithm can have. For large n, the number of comparisons made by this algorithm in the average case turns out to be about $0.25n$ less (see [Gon91, p. 173]) and hence is also in $\Theta(n \log n)$. A noteworthy advantage of mergesort over quicksort and heapsort—the two important advanced sorting algorithms to be discussed later—is its stability (see Problem 7 in this section's exercises). The principal shortcoming of mergesort is the linear amount of extra storage the algorithm requires. Though merging can be done in-place, the resulting algorithm is quite complicated and of theoretical interest only.

There are two main ideas leading to several variations of mergesort. First, the algorithm can be implemented bottom up by merging pairs of the array's elements, then merging the sorted pairs, and so on. (If n is not a power of 2, only slight bookkeeping complications arise.) This avoids the time and space overhead of using a stack to handle recursive calls. Second, we can divide a list to be sorted in more than two parts, sort each recursively, and then merge them together. This scheme, which is particularly useful for sorting files residing on secondary memory devices, is called ***multiway mergesort***.

Exercises 5.1

1. **a.** Write pseudocode for a divide-and-conquer algorithm for finding the position of the largest element in an array of n numbers.

   b. What will be your algorithm's output for arrays with several elements of the largest value?

   c. Set up and solve a recurrence relation for the number of key comparisons made by your algorithm.

   d. How does this algorithm compare with the brute-force algorithm for this problem?

2. **a.** Write pseudocode for a divide-and-conquer algorithm for finding values of both the largest and smallest elements in an array of n numbers.

   b. Set up and solve (for $n = 2^k$) a recurrence relation for the number of key comparisons made by your algorithm.

   c. How does this algorithm compare with the brute-force algorithm for this problem?

3. **a.** Write pseudocode for a divide-and-conquer algorithm for the exponentiation problem of computing a^n where n is a positive integer.

   b. Set up and solve a recurrence relation for the number of multiplications made by this algorithm.

2. As we shall see in Section 11.2, this theoretical minimum is $\lceil \log_2 n! \rceil \approx \lceil n \log_2 n - 1.44n \rceil$.

c. How does this algorithm compare with the brute-force algorithm for this problem?

4. As mentioned in Chapter 2, logarithm bases are irrelevant in most contexts arising in analyzing an algorithm's efficiency class. Is this true for both assertions of the Master Theorem that include logarithms?

5. Find the order of growth for solutions of the following recurrences.

a. $T(n) = 4T(n/2) + n$, $T(1) = 1$

b. $T(n) = 4T(n/2) + n^2$, $T(1) = 1$

c. $T(n) = 4T(n/2) + n^3$, $T(1) = 1$

6. Apply mergesort to sort the list *E, X, A, M, P, L, E* in alphabetical order.

7. Is mergesort a stable sorting algorithm?

8. a. Solve the recurrence relation for the number of key comparisons made by mergesort in the worst case. You may assume that $n = 2^k$.

b. Set up a recurrence relation for the number of key comparisons made by mergesort on best-case inputs and solve it for $n = 2^k$.

c. Set up a recurrence relation for the number of key moves made by the version of mergesort given in Section 5.1. Does taking the number of key moves into account change the algorithm's efficiency class?

9. Let $A[0..n-1]$ be an array of n real numbers. A pair $(A[i], A[j])$ is said to be an ***inversion*** if these numbers are out of order, i.e., $i < j$ but $A[i] > A[j]$. Design an $O(n \log n)$ algorithm for counting the number of inversions.

10. Implement the bottom-up version of mergesort in the language of your choice.

11. *Tromino puzzle* A tromino (more accurately, a right tromino) is an L-shaped tile formed by three 1×1 squares. The problem is to cover any $2^n \times 2^n$ chessboard with a missing square with trominoes. Trominoes can be oriented in an arbitrary way, but they should cover all the squares of the board except the missing one exactly and with no overlaps. [Gol94]

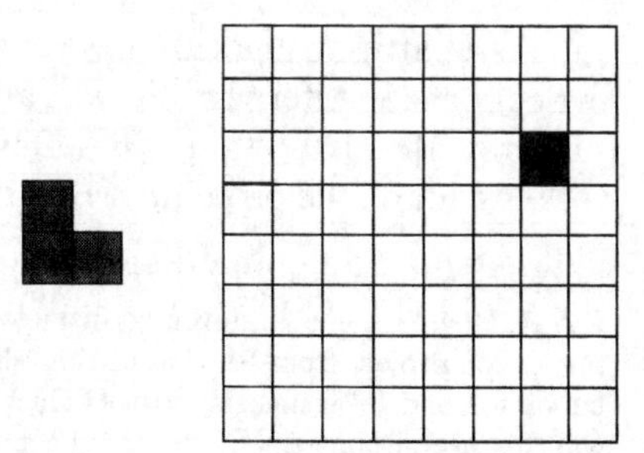

Design a divide-and-conquer algorithm for this problem.

5.2 Quicksort

Quicksort is the other important sorting algorithm that is based on the divide-and-conquer approach. Unlike mergesort, which divides its input elements according to their position in the array, quicksort divides them according to their value. We already encountered this idea of an array partition in Section 4.5, where we discussed the selection problem. A partition is an arrangement of the array's elements so that all the elements to the left of some element $A[s]$ are less than or equal to $A[s]$, and all the elements to the right of $A[s]$ are greater than or equal to it:

$$\underbrace{A[0] \ldots A[s-1]}_{\text{all are } \leq A[s]} \quad A[s] \quad \underbrace{A[s+1] \ldots A[n-1]}_{\text{all are } \geq A[s]}$$

Obviously, after a partition is achieved, $A[s]$ will be in its final position in the sorted array, and we can continue sorting the two subarrays to the left and to the right of $A[s]$ independently (e.g., by the same method). Note the difference with mergesort: there, the division of the problem into two subproblems is immediate and the entire work happens in combining their solutions; here, the entire work happens in the division stage, with no work required to combine the solutions to the subproblems.

Here is pseudocode of quicksort: call *Quicksort*($A[0..n-1]$) where

ALGORITHM *Quicksort*($A[l..r]$)
//Sorts a subarray by quicksort
//Input: Subarray of array $A[0..n-1]$, defined by its left and right
// indices l and r
//Output: Subarray $A[l..r]$ sorted in nondecreasing order
if $l < r$
  $s \leftarrow$ *Partition*($A[l..r]$) //s is a split position
  Quicksort($A[l..s-1]$)
  Quicksort($A[s+1..r]$)

As a partition algorithm, we can certainly use the Lomuto partition discussed in Section 4.5. Alternatively, we can partition $A[0..n-1]$ and, more generally, its subarray $A[l..r]$ $(0 \leq l < r \leq n-1)$ by the more sophisticated method suggested by C.A.R. Hoare, the prominent British computer scientist who invented quicksort.[3]

---

3. C.A.R. Hoare, at age 26, invented his algorithm in 1960 while trying to sort words for a machine translation project from Russian to English. Says Hoare, "My first thought on how to do this was bubblesort and, by an amazing stroke of luck, my second thought was Quicksort." It is hard to disagree with his overall assessment: "I have been very lucky. What a wonderful way to start a career in Computing, by discovering a new sorting algorithm!" [Hoa96]. Twenty years later, he received the Turing Award for "fundamental contributions to the definition and design of programming languages"; in 1980, he was also knighted for services to education and computer science.

As before, we start by selecting a pivot—an element with respect to whose value we are going to divide the subarray. There are several different strategies for selecting a pivot; we will return to this issue when we analyze the algorithm's efficiency. For now, we use the simplest strategy of selecting the subarray's first element: $p = A[l]$.

Unlike the Lomuto algorithm, we will now scan the subarray from both ends, comparing the subarray's elements to the pivot. The left-to-right scan, denoted below by index pointer i, starts with the second element. Since we want elements smaller than the pivot to be in the left part of the subarray, this scan skips over elements that are smaller than the pivot and stops upon encountering the first element greater than or equal to the pivot. The right-to-left scan, denoted below by index pointer j, starts with the last element of the subarray. Since we want elements larger than the pivot to be in the right part of the subarray, this scan skips over elements that are larger than the pivot and stops on encountering the first element smaller than or equal to the pivot. (Why is it worth stopping the scans after encountering an element equal to the pivot? Because doing this tends to yield more even splits for arrays with a lot of duplicates, which makes the algorithm run faster. For example, if we did otherwise for an array of n equal elements, we would have gotten a split into subarrays of sizes $n - 1$ and 0, reducing the problem size just by 1 after scanning the entire array.)

After both scans stop, three situations may arise, depending on whether or not the scanning indices have crossed. If scanning indices i and j have not crossed, i.e., $i < j$, we simply exchange $A[i]$ and $A[j]$ and resume the scans by incrementing i and decrementing j, respectively:

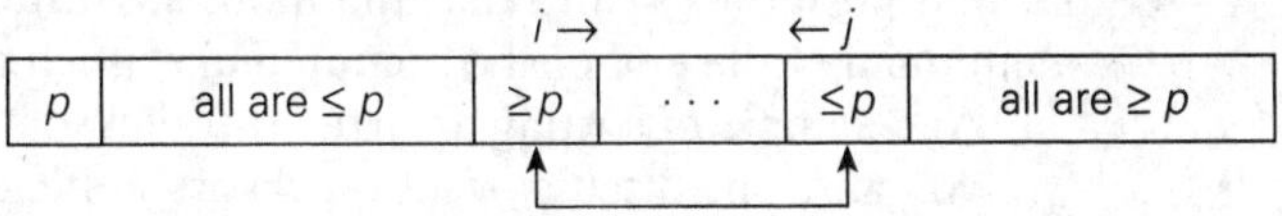

		$i \rightarrow$		$\leftarrow j$	
p	all are $\leq p$	$\geq p$	. . .	$\leq p$	all are $\geq p$

If the scanning indices have crossed over, i.e., $i > j$, we will have partitioned the subarray after exchanging the pivot with $A[j]$:

		$\leftarrow j$	$i \rightarrow$	
p	all are $\leq p$	$\leq p$	$\geq p$	all are $\geq p$

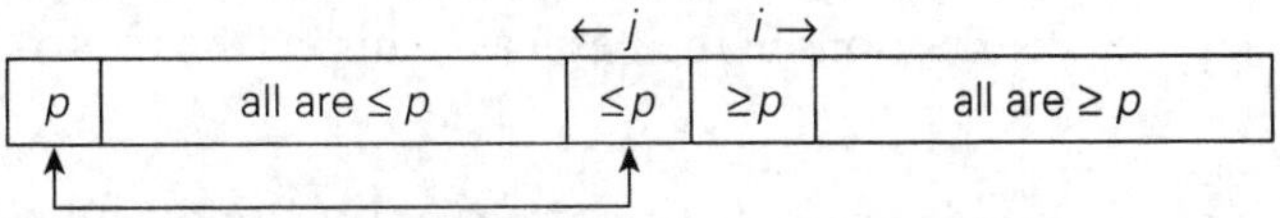

Finally, if the scanning indices stop while pointing to the same element, i.e., $i = j$, the value they are pointing to must be equal to p (why?). Thus, we have the subarray partitioned, with the split position $s = i = j$:

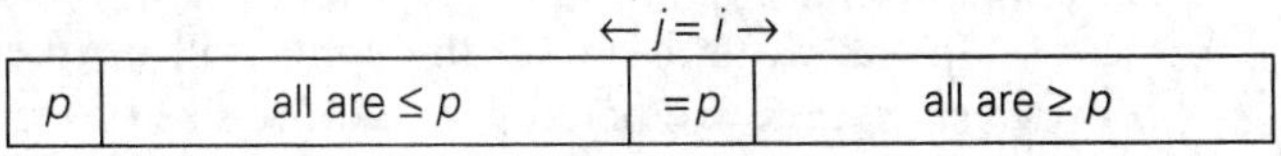

		$\leftarrow j = i \rightarrow$	
p	all are $\leq p$	$= p$	all are $\geq p$

We can combine the last case with the case of crossed-over indices ($i > j$) by exchanging the pivot with $A[j]$ whenever $i \geq j$.

Here is pseudocode implementing this partitioning procedure.

ALGORITHM *HoarePartition*$(A[l..r])$

```
//Partitions a subarray by Hoare's algorithm, using the first element
//        as a pivot
//Input: Subarray of array A[0..n − 1], defined by its left and right
//        indices l and r (l < r)
//Output: Partition of A[l..r], with the split position returned as
//        this function's value
p ← A[l]
i ← l; j ← r + 1
repeat
    repeat i ← i + 1 until A[i] ≥ p
    repeat j ← j − 1 until A[j] ≤ p
    swap(A[i], A[j])
until i ≥ j
swap(A[i], A[j])   //undo last swap when i ≥ j
swap(A[l], A[j])
return j
```

Note that index i can go out of the subarray's bounds in this pseudocode. Rather than checking for this possibility every time index i is incremented, we can append to array $A[0..n-1]$ a "sentinel" that would prevent index i from advancing beyond position n. Note that the more sophisticated method of pivot selection mentioned at the end of the section makes such a sentinel unnecessary.

An example of sorting an array by quicksort is given in Figure 5.3.

We start our discussion of quicksort's efficiency by noting that the number of key comparisons made before a partition is achieved is $n+1$ if the scanning indices cross over and n if they coincide (why?). If all the splits happen in the middle of corresponding subarrays, we will have the best case. The number of key comparisons in the best case satisfies the recurrence

$$C_{best}(n) = 2C_{best}(n/2) + n \quad \text{for } n > 1, \quad C_{best}(1) = 0.$$

According to the Master Theorem, $C_{best}(n) \in \Theta(n \log_2 n)$; solving it exactly for $n = 2^k$ yields $C_{best}(n) = n \log_2 n$.

In the worst case, all the splits will be skewed to the extreme: one of the two subarrays will be empty, and the size of the other will be just 1 less than the size of the subarray being partitioned. This unfortunate situation will happen, in particular, for increasing arrays, i.e., for inputs for which the problem is already solved! Indeed, if $A[0..n-1]$ is a strictly increasing array and we use $A[0]$ as the pivot, the left-to-right scan will stop on $A[1]$ while the right-to-left scan will go all the way to reach $A[0]$, indicating the split at position 0:

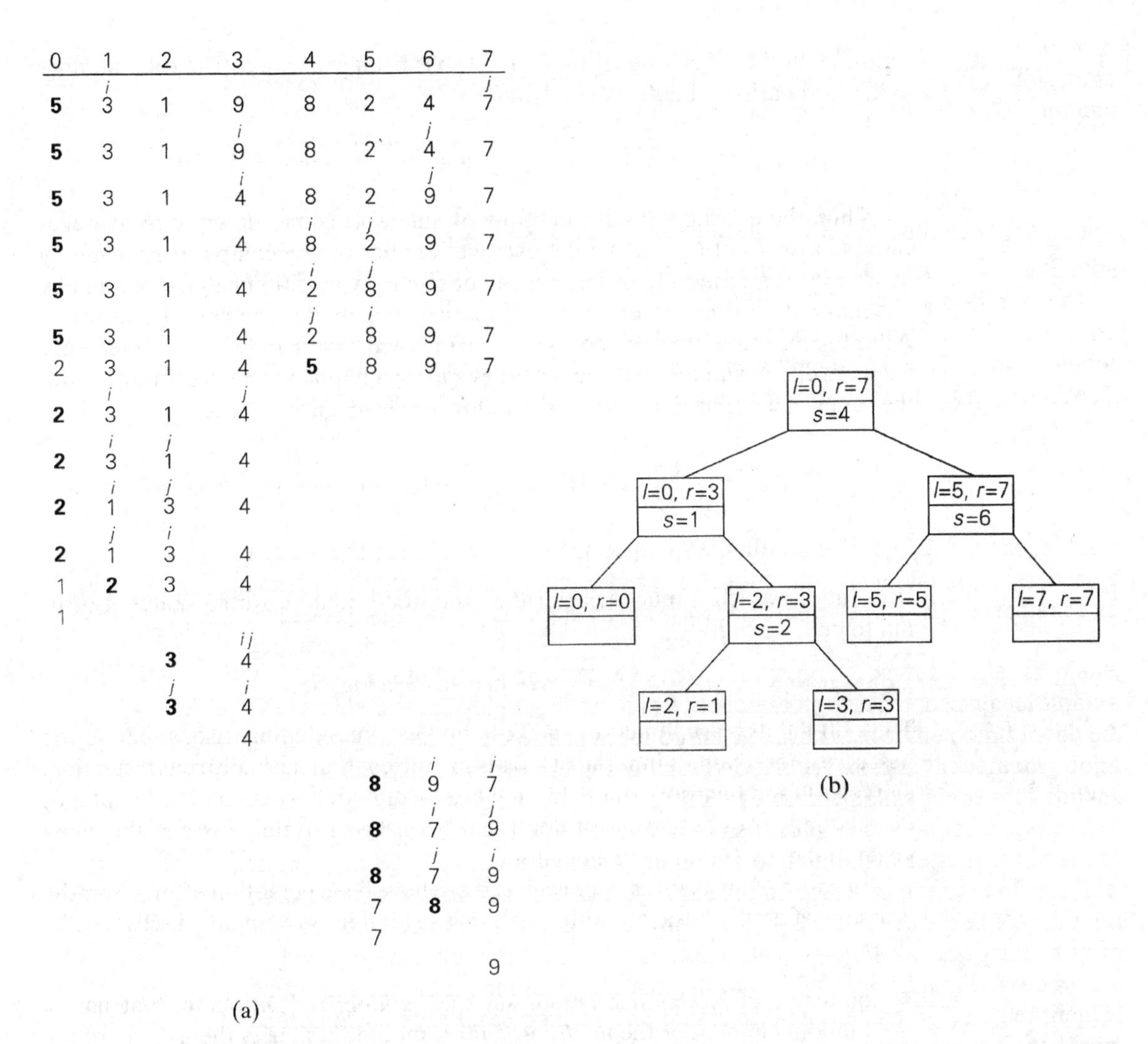

FIGURE 5.3 Example of quicksort operation. (a) Array's transformations with pivots shown in bold. (b) Tree of recursive calls to *Quicksort* with input values l and r of subarray bounds and split position s of a partition obtained.

←j	i→		
A[0]	A[1]	. . .	A[n–1]

So, after making $n + 1$ comparisons to get to this partition and exchanging the pivot $A[0]$ with itself, the algorithm will be left with the strictly increasing array $A[1..n-1]$ to sort. This sorting of strictly increasing arrays of diminishing sizes will continue until the last one $A[n-2..n-1]$ has been processed. The total number of key comparisons made will be equal to

$$C_{worst}(n) = (n+1) + n + \cdots + 3 = \frac{(n+1)(n+2)}{2} - 3 \in \Theta(n^2).$$

Thus, the question about the utility of quicksort comes down to its average-case behavior. Let $C_{avg}(n)$ be the average number of key comparisons made by quicksort on a randomly ordered array of size n. A partition can happen in any position s $(0 \le s \le n-1)$ after $n+1$ comparisons are made to achieve the partition. After the partition, the left and right subarrays will have s and $n-1-s$ elements, respectively. Assuming that the partition split can happen in each position s with the same probability $1/n$, we get the following recurrence relation:

$$C_{avg}(n) = \frac{1}{n}\sum_{s=0}^{n-1}[(n+1) + C_{avg}(s) + C_{avg}(n-1-s)] \quad \text{for } n > 1,$$

$$C_{avg}(0) = 0, \quad C_{avg}(1) = 0.$$

Its solution, which is much trickier than the worst- and best-case analyses, turns out to be

$$C_{avg}(n) \approx 2n \ln n \approx 1.39n \log_2 n.$$

Thus, on the average, quicksort makes only 39% more comparisons than in the best case. Moreover, its innermost loop is so efficient that it usually runs faster than mergesort (and heapsort, another $n \log n$ algorithm that we discuss in Chapter 6) on randomly ordered arrays of nontrivial sizes. This certainly justifies the name given to the algorithm by its inventor.

Because of quicksort's importance, there have been persistent efforts over the years to refine the basic algorithm. Among several improvements discovered by researchers are:

- better pivot selection methods such as ***randomized quicksort*** that uses a random element or the ***median-of-three*** method that uses the median of the leftmost, rightmost, and the middle element of the array
- switching to insertion sort on very small subarrays (between 5 and 15 elements for most computer systems) or not sorting small subarrays at all and finishing the algorithm with insertion sort applied to the entire nearly sorted array
- modifications of the partitioning algorithm such as the three-way partition into segments smaller than, equal to, and larger than the pivot (see Problem 9 in this section's exercises)

According to Robert Sedgewick [Sed11, p. 296], the world's leading expert on quicksort, such improvements in combination can cut the running time of the algorithm by 20%–30%.

Like any sorting algorithm, quicksort has weaknesses. It is not stable. It requires a stack to store parameters of subarrays that are yet to be sorted. While the size of this stack can be made to be in $O(\log n)$ by always sorting first the smaller of two subarrays obtained by partitioning, it is worse than the $O(1)$ space efficiency of heapsort. Although more sophisticated ways of choosing a pivot make the quadratic running time of the worst case very unlikely, they do not eliminate it completely. And even the performance on randomly ordered arrays is known to be sensitive not only to implementation details of the algorithm but also to both computer architecture and data type. Still, the January/February 2000 issue of *Computing in Science & Engineering*, a joint publication of the American Institute of Physics and the IEEE Computer Society, selected quicksort as one of the 10 algorithms "with the greatest influence on the development and practice of science and engineering in the 20th century."

Exercises 5.2

1. Apply quicksort to sort the list E, X, A, M, P, L, E in alphabetical order. Draw the tree of the recursive calls made.

2. For the partitioning procedure outlined in this section:

a. Prove that if the scanning indices stop while pointing to the same element, i.e., $i = j$, the value they are pointing to must be equal to p.

b. Prove that when the scanning indices stop, j cannot point to an element more than one position to the left of the one pointed to by i.

3. Give an example showing that quicksort is not a stable sorting algorithm.

4. Give an example of an array of n elements for which the sentinel mentioned in the text is actually needed. What should be its value? Also explain why a single sentinel suffices for any input.

5. For the version of quicksort given in this section:

a. Are arrays made up of all equal elements the worst-case input, the best-case input, or neither?

b. Are strictly decreasing arrays the worst-case input, the best-case input, or neither?

6. a. For quicksort with the median-of-three pivot selection, are strictly increasing arrays the worst-case input, the best-case input, or neither?

b. Answer the same question for strictly decreasing arrays.

7. a. Estimate how many times faster quicksort will sort an array of one million random numbers than insertion sort.

b. True or false: For every $n > 1$, there are n-element arrays that are sorted faster by insertion sort than by quicksort?

8. Design an algorithm to rearrange elements of a given array of n real numbers so that all its negative elements precede all its positive elements. Your algorithm should be both time efficient and space efficient.

9. **a.** The ***Dutch national flag problem*** is to rearrange an array of characters R, W, and B (red, white, and blue are the colors of the Dutch national flag) so that all the R's come first, the W's come next, and the B's come last. [Dij76] Design a linear in-place algorithm for this problem.

b. Explain how a solution to the Dutch national flag problem can be used in quicksort.

10. Implement quicksort in the language of your choice. Run your program on a sample of inputs to verify the theoretical assertions about the algorithm's efficiency.

11. *Nuts and bolts* You are given a collection of n bolts of different widths and n corresponding nuts. You are allowed to try a nut and bolt together, from which you can determine whether the nut is larger than the bolt, smaller than the bolt, or matches the bolt exactly. However, there is no way to compare two nuts together or two bolts together. The problem is to match each bolt to its nut. Design an algorithm for this problem with average-case efficiency in $\Theta(n \log n)$. [Raw91]

5.3 Binary Tree Traversals and Related Properties

In this section, we see how the divide-and-conquer technique can be applied to binary trees. A ***binary tree*** T is defined as a finite set of nodes that is either empty or consists of a root and two disjoint binary trees T_L and T_R called, respectively, the left and right subtree of the root. We usually think of a binary tree as a special case of an ordered tree (Figure 5.4). (This standard interpretation was an alternative definition of a binary tree in Section 1.4.)

Since the definition itself divides a binary tree into two smaller structures of the same type, the left subtree and the right subtree, many problems about binary trees can be solved by applying the divide-and-conquer technique. As an example, let us consider a recursive algorithm for computing the height of a binary tree. Recall that the height is defined as the length of the longest path from the root to a leaf. Hence, it can be computed as the maximum of the heights of the root's left

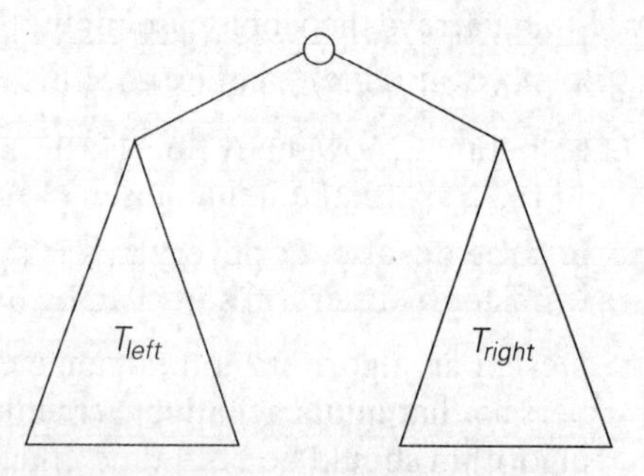

FIGURE 5.4 Standard representation of a binary tree.

and right subtrees plus 1. (We have to add 1 to account for the extra level of the root.) Also note that it is convenient to define the height of the empty tree as −1. Thus, we have the following recursive algorithm.

ALGORITHM *Height(T)*

//Computes recursively the height of a binary tree
//Input: A binary tree T
//Output: The height of T
if $T = \varnothing$ **return** -1
else return $\max\{Height(T_{left}), Height(T_{right})\} + 1$

We measure the problem's instance size by the number of nodes $n(T)$ in a given binary tree T. Obviously, the number of comparisons made to compute the maximum of two numbers and the number of additions $A(n(T))$ made by the algorithm are the same. We have the following recurrence relation for $A(n(T))$:

$$A(n(T)) = A(n(T_{left})) + A(n(T_{right})) + 1 \quad \text{for } n(T) > 0,$$
$$A(0) = 0.$$

Before we solve this recurrence (can you tell what its solution is?), let us note that addition is not the most frequently executed operation of this algorithm. What is? Checking—and this is very typical for binary tree algorithms—that the tree is not empty. For example, for the empty tree, the comparison $T = \varnothing$ is executed once but there are no additions, and for a single-node tree, the comparison and addition numbers are 3 and 1, respectively.

It helps in the analysis of tree algorithms to draw the tree's extension by replacing the empty subtrees by special nodes. The extra nodes (shown by little squares in Figure 5.5) are called ***external***; the original nodes (shown by little circles) are called ***internal***. By definition, the extension of the empty binary tree is a single external node.

It is easy to see that the *Height* algorithm makes exactly one addition for every internal node of the extended tree, and it makes one comparison to check whether

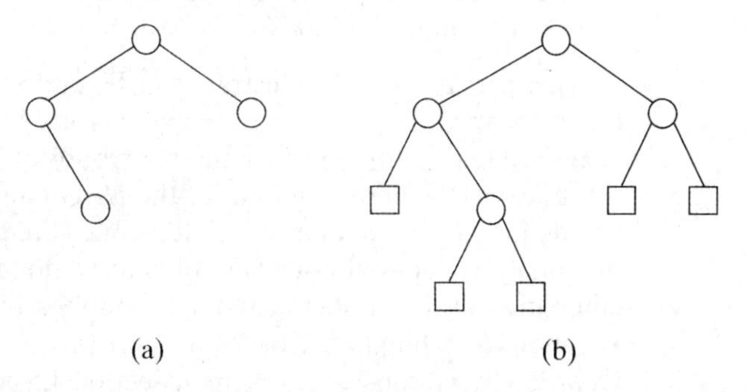

FIGURE 5.5 Binary tree (on the left) and its extension (on the right). Internal nodes are shown as circles; external nodes are shown as squares.

the tree is empty for every internal and external node. Therefore, to ascertain the algorithm's efficiency, we need to know how many external nodes an extended binary tree with n internal nodes can have. After checking Figure 5.5 and a few similar examples, it is easy to hypothesize that the number of external nodes x is always 1 more than the number of internal nodes n:

$$x = n + 1. \tag{5.2}$$

To prove this equality, consider the total number of nodes, both internal and external. Since every node, except the root, is one of the two children of an internal node, we have the equation

$$2n + 1 = x + n,$$

which immediately implies equality (5.2).

Note that equality (5.2) also applies to any nonempty ***full binary tree***, in which, by definition, every node has either zero or two children: for a full binary tree, n and x denote the numbers of parental nodes and leaves, respectively.

Returning to algorithm *Height*, the number of comparisons to check whether the tree is empty is

$$C(n) = n + x = 2n + 1,$$

and the number of additions is

$$A(n) = n.$$

The most important divide-and-conquer algorithms for binary trees are the three classic traversals: preorder, inorder, and postorder. All three traversals visit nodes of a binary tree recursively, i.e., by visiting the tree's root and its left and right subtrees. They differ only by the timing of the root's visit:

In the ***preorder traversal***, the root is visited before the left and right subtrees are visited (in that order).
In the ***inorder traversal***, the root is visited after visiting its left subtree but before visiting the right subtree.
In the ***postorder traversal***, the root is visited after visiting the left and right subtrees (in that order).

These traversals are illustrated in Figure 5.6. Their pseudocodes are quite straightforward, repeating the descriptions given above. (These traversals are also a standard feature of data structures textbooks.) As to their efficiency analysis, it is identical to the above analysis of the *Height* algorithm because a recursive call is made for each node of an extended binary tree.

Finally, we should note that, obviously, not all questions about binary trees require traversals of both left and right subtrees. For example, the search and insert operations for a binary search tree require processing only one of the two subtrees. Accordingly, we considered them in Section 4.5 not as applications of divide-and-conquer but rather as examples of the variable-size-decrease technique.

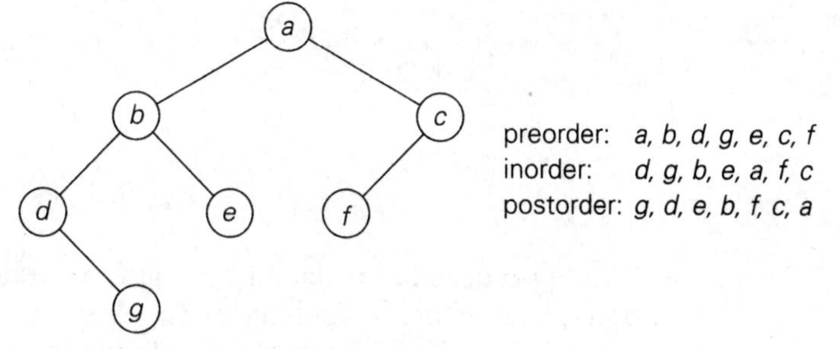

FIGURE 5.6 Binary tree and its traversals.

Exercises 5.3

1. Design a divide-and-conquer algorithm for computing the number of levels in a binary tree. (In particular, the algorithm must return 0 and 1 for the empty and single-node trees, respectively.) What is the time efficiency class of your algorithm?

2. The following algorithm seeks to compute the number of leaves in a binary tree.

   ALGORITHM *LeafCounter(T)*
   //Computes recursively the number of leaves in a binary tree
   //Input: A binary tree T
   //Output: The number of leaves in T
   if $T = \varnothing$ **return** 0
   else return $LeafCounter(T_{left}) + LeafCounter(T_{right})$

   Is this algorithm correct? If it is, prove it; if it is not, make an appropriate correction.

3. Can you compute the height of a binary tree with the same asymptotic efficiency as the section's divide-and-conquer algorithm but without using a stack explicitly or implicitly? Of course, you may use a different algorithm altogether.

4. Prove equality (5.2) by mathematical induction.

5. Traverse the following binary tree
   a. in preorder.
   b. in inorder.
   c. in postorder.

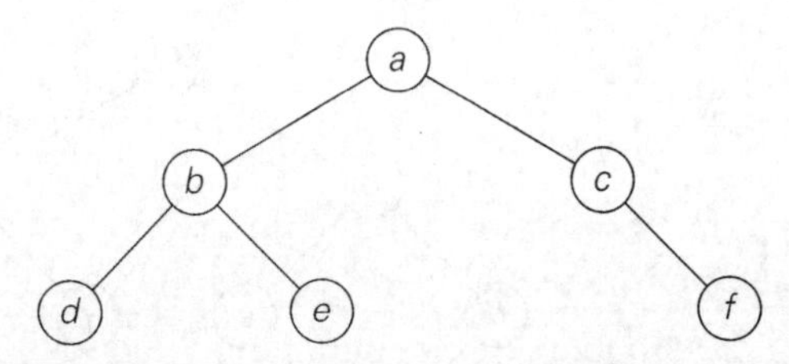

6. Write pseudocode for one of the classic traversal algorithms (preorder, inorder, and postorder) for binary trees. Assuming that your algorithm is recursive, find the number of recursive calls made.

7. Which of the three classic traversal algorithms yields a sorted list if applied to a binary search tree? Prove this property.

8. **a.** Draw a binary tree with 10 nodes labeled 0, 1, . . . , 9 in such a way that the inorder and postorder traversals of the tree yield the following lists: 9, 3, 1, 0, 4, 2, 7, 6, 8, 5 (inorder) and 9, 1, 4, 0, 3, 6, 7, 5, 8, 2 (postorder).

   b. Give an example of two permutations of the same n labels $0, 1, \ldots, n-1$ that cannot be inorder and postorder traversal lists of the same binary tree.

   c. Design an algorithm that constructs a binary tree for which two given lists of n labels $0, 1, \ldots, n-1$ are generated by the inorder and postorder traversals of the tree. Your algorithm should also identify inputs for which the problem has no solution.

9. The ***internal path length*** I of an extended binary tree is defined as the sum of the lengths of the paths—taken over all internal nodes—from the root to each internal node. Similarly, the ***external path length*** E of an extended binary tree is defined as the sum of the lengths of the paths—taken over all external nodes—from the root to each external node. Prove that $E = I + 2n$ where n is the number of internal nodes in the tree.

10. Write a program for computing the internal path length of an extended binary tree. Use it to investigate empirically the average number of key comparisons for searching in a randomly generated binary search tree.

11. *Chocolate bar puzzle* Given an $n \times m$ chocolate bar, you need to break it into nm 1×1 pieces. You can break a bar only in a straight line, and only one bar can be broken at a time. Design an algorithm that solves the problem with the minimum number of bar breaks. What is this minimum number? Justify your answer by using properties of a binary tree.

5.4 Multiplication of Large Integers and Strassen's Matrix Multiplication

In this section, we examine two surprising algorithms for seemingly straightforward tasks: multiplying two integers and multiplying two square matrices. Both achieve a better asymptotic efficiency by ingenious application of the divide-and-conquer technique.

Multiplication of Large Integers

Some applications, notably modern cryptography, require manipulation of integers that are over 100 decimal digits long. Since such integers are too long to fit in a single word of a modern computer, they require special treatment. This practical need supports investigations of algorithms for efficient manipulation of large integers. In this section, we outline an interesting algorithm for multiplying such numbers. Obviously, if we use the conventional pen-and-pencil algorithm for multiplying two n-digit integers, each of the n digits of the first number is multiplied by each of the n digits of the second number for the total of n^2 digit multiplications. (If one of the numbers has fewer digits than the other, we can pad the shorter number with leading zeros to equalize their lengths.) Though it might appear that it would be impossible to design an algorithm with fewer than n^2 digit multiplications, this turns out not to be the case. The miracle of divide-and-conquer comes to the rescue to accomplish this feat.

To demonstrate the basic idea of the algorithm, let us start with a case of two-digit integers, say, 23 and 14. These numbers can be represented as follows:

$$23 = 2 \cdot 10^1 + 3 \cdot 10^0 \quad \text{and} \quad 14 = 1 \cdot 10^1 + 4 \cdot 10^0.$$

Now let us multiply them:

$$\begin{aligned} 23 * 14 &= (2 \cdot 10^1 + 3 \cdot 10^0) * (1 \cdot 10^1 + 4 \cdot 10^0) \\ &= (2 * 1)10^2 + (2 * 4 + 3 * 1)10^1 + (3 * 4)10^0. \end{aligned}$$

The last formula yields the correct answer of 322, of course, but it uses the same four digit multiplications as the pen-and-pencil algorithm. Fortunately, we can compute the middle term with just one digit multiplication by taking advantage of the products $2 * 1$ and $3 * 4$ that need to be computed anyway:

$$2 * 4 + 3 * 1 = (2 + 3) * (1 + 4) - 2 * 1 - 3 * 4.$$

Of course, there is nothing special about the numbers we just multiplied. For any pair of two-digit numbers $a = a_1a_0$ and $b = b_1b_0$, their product c can be computed by the formula

$$c = a * b = c_2 10^2 + c_1 10^1 + c_0,$$

where

$c_2 = a_1 * b_1$ is the product of their first digits,

$c_0 = a_0 * b_0$ is the product of their second digits,

$c_1 = (a_1 + a_0) * (b_1 + b_0) - (c_2 + c_0)$ is the product of the sum of the a's digits and the sum of the b's digits minus the sum of c_2 and c_0.

Now we apply this trick to multiplying two n-digit integers a and b where n is a positive even number. Let us divide both numbers in the middle—after all, we promised to take advantage of the divide-and-conquer technique. We denote the first half of the a's digits by a_1 and the second half by a_0; for b, the notations are b_1 and b_0, respectively. In these notations, $a = a_1a_0$ implies that $a = a_1 10^{n/2} + a_0$ and $b = b_1b_0$ implies that $b = b_1 10^{n/2} + b_0$. Therefore, taking advantage of the same trick we used for two-digit numbers, we get

$$\begin{aligned} c = a * b &= (a_1 10^{n/2} + a_0) * (b_1 10^{n/2} + b_0) \\ &= (a_1 * b_1)10^n + (a_1 * b_0 + a_0 * b_1)10^{n/2} + (a_0 * b_0) \\ &= c_2 10^n + c_1 10^{n/2} + c_0, \end{aligned}$$

where

$c_2 = a_1 * b_1$ is the product of their first halves,

$c_0 = a_0 * b_0$ is the product of their second halves,

$c_1 = (a_1 + a_0) * (b_1 + b_0) - (c_2 + c_0)$ is the product of the sum of the a's halves and the sum of the b's halves minus the sum of c_2 and c_0.

If $n/2$ is even, we can apply the same method for computing the products c_2, c_0, and c_1. Thus, if n is a power of 2, we have a recursive algorithm for computing the product of two n-digit integers. In its pure form, the recursion is stopped when n becomes 1. It can also be stopped when we deem n small enough to multiply the numbers of that size directly.

How many digit multiplications does this algorithm make? Since multiplication of n-digit numbers requires three multiplications of $n/2$-digit numbers, the recurrence for the number of multiplications $M(n)$ is

$$M(n) = 3M(n/2) \quad \text{for } n > 1, \quad M(1) = 1.$$

Solving it by backward substitutions for $n = 2^k$ yields

$$\begin{aligned} M(2^k) &= 3M(2^{k-1}) = 3[3M(2^{k-2})] = 3^2 M(2^{k-2}) \\ &= \cdots = 3^i M(2^{k-i}) = \cdots = 3^k M(2^{k-k}) = 3^k. \end{aligned}$$

Since $k = \log_2 n$,

$$M(n) = 3^{\log_2 n} = n^{\log_2 3} \approx n^{1.585}.$$

(On the last step, we took advantage of the following property of logarithms: $a^{\log_b c} = c^{\log_b a}$.)

But what about additions and subtractions? Have we not decreased the number of multiplications by requiring more of those operations? Let $A(n)$ be the number of digit additions and subtractions executed by the above algorithm in multiplying two n-digit decimal integers. Besides $3A(n/2)$ of these operations needed to compute the three products of $n/2$-digit numbers, the above formulas require five additions and one subtraction. Hence, we have the recurrence

$$A(n) = 3A(n/2) + cn \quad \text{for } n > 1, \ A(1) = 1.$$

Applying the Master Theorem, which was stated in the beginning of the chapter, we obtain $A(n) \in \Theta(n^{\log_2 3})$, which means that the total number of additions and subtractions have the same asymptotic order of growth as the number of multiplications.

The asymptotic advantage of this algorithm notwithstanding, how practical is it? The answer depends, of course, on the computer system and program quality implementing the algorithm, which might explain the rather wide disparity of reported results. On some machines, the divide-and-conquer algorithm has been reported to outperform the conventional method on numbers only 8 decimal digits long and to run more than twice faster with numbers over 300 decimal digits long—the area of particular importance for modern cryptography. Whatever this outperformance "crossover point" happens to be on a particular machine, it is worth switching to the conventional algorithm after the multiplicands become smaller than the crossover point. Finally, if you program in an object-oriented language such as Java, C++, or Smalltalk, you should also be aware that these languages have special classes for dealing with large integers.

Discovered by 23-year-old Russian mathematician Anatoly Karatsuba in 1960, the divide-and-conquer algorithm proved wrong the then-prevailing opinion that the time efficiency of any integer multiplication algorithm must be in $\Omega(n^2)$. The discovery encouraged researchers to look for even (asymptotically) faster algorithms for this and other algebraic problems. We will see such an algorithm in the next section.

Strassen's Matrix Multiplication

Now that we have seen that the divide-and-conquer approach can reduce the number of one-digit multiplications in multiplying two integers, we should not be surprised that a similar feat can be accomplished for multiplying matrices. Such an algorithm was published by V. Strassen in 1969 [Str69]. The principal insight of the algorithm lies in the discovery that we can find the product C of two 2×2 matrices A and B with just seven multiplications as opposed to the eight required by the brute-force algorithm (see Example 3 in Section 2.3). This is accomplished by using the following formulas:

$$\begin{bmatrix} c_{00} & c_{01} \\ c_{10} & c_{11} \end{bmatrix} = \begin{bmatrix} a_{00} & a_{01} \\ a_{10} & a_{11} \end{bmatrix} * \begin{bmatrix} b_{00} & b_{01} \\ b_{10} & b_{11} \end{bmatrix}$$

$$= \begin{bmatrix} m_1 + m_4 - m_5 + m_7 & m_3 + m_5 \\ m_2 + m_4 & m_1 + m_3 - m_2 + m_6 \end{bmatrix},$$

where

$$\begin{aligned}
m_1 &= (a_{00} + a_{11}) * (b_{00} + b_{11}),\\
m_2 &= (a_{10} + a_{11}) * b_{00},\\
m_3 &= a_{00} * (b_{01} - b_{11}),\\
m_4 &= a_{11} * (b_{10} - b_{00}),\\
m_5 &= (a_{00} + a_{01}) * b_{11},\\
m_6 &= (a_{10} - a_{00}) * (b_{00} + b_{01}),\\
m_7 &= (a_{01} - a_{11}) * (b_{10} + b_{11}).
\end{aligned}$$

Thus, to multiply two 2×2 matrices, Strassen's algorithm makes seven multiplications and 18 additions/subtractions, whereas the brute-force algorithm requires eight multiplications and four additions. These numbers should not lead us to multiplying 2×2 matrices by Strassen's algorithm. Its importance stems from its *asymptotic* superiority as matrix order n goes to infinity.

Let A and B be two $n \times n$ matrices where n is a power of 2. (If n is not a power of 2, matrices can be padded with rows and columns of zeros.) We can divide A, B, and their product C into four $n/2 \times n/2$ submatrices each as follows:

$$\left[\begin{array}{c|c} C_{00} & C_{01} \\ \hline C_{10} & C_{11} \end{array}\right] = \left[\begin{array}{c|c} A_{00} & A_{01} \\ \hline A_{10} & A_{11} \end{array}\right] * \left[\begin{array}{c|c} B_{00} & B_{01} \\ \hline B_{10} & B_{11} \end{array}\right].$$

It is not difficult to verify that one can treat these submatrices as numbers to get the correct product. For example, C_{00} can be computed either as $A_{00} * B_{00} + A_{01} * B_{10}$ or as $M_1 + M_4 - M_5 + M_7$ where M_1, M_4, M_5, and M_7 are found by Strassen's formulas, with the numbers replaced by the corresponding submatrices. If the seven products of $n/2 \times n/2$ matrices are computed recursively by the same method, we have Strassen's algorithm for matrix multiplication.

Let us evaluate the asymptotic efficiency of this algorithm. If $M(n)$ is the number of multiplications made by Strassen's algorithm in multiplying two $n \times n$ matrices (where n is a power of 2), we get the following recurrence relation for it:

$$M(n) = 7M(n/2) \quad \text{for } n > 1, \quad M(1) = 1.$$

Since $n = 2^k$,

$$\begin{aligned}
M(2^k) &= 7M(2^{k-1}) = 7[7M(2^{k-2})] = 7^2 M(2^{k-2}) = \cdots\\
&= 7^i M(2^{k-i}) \cdots = 7^k M(2^{k-k}) = 7^k.
\end{aligned}$$

Since $k = \log_2 n$,

$$M(n) = 7^{\log_2 n} = n^{\log_2 7} \approx n^{2.807},$$

which is smaller than n^3 required by the brute-force algorithm.

Since this savings in the number of multiplications was achieved at the expense of making extra additions, we must check the number of additions $A(n)$ made by Strassen's algorithm. To multiply two matrices of order $n > 1$, the algorithm needs to multiply seven matrices of order $n/2$ and make 18 additions/subtractions of matrices of size $n/2$; when $n = 1$, no additions are made since two numbers are simply multiplied. These observations yield the following recurrence relation:

$$A(n) = 7A(n/2) + 18(n/2)^2 \quad \text{for } n > 1, \ \ A(1) = 0.$$

Though one can obtain a closed-form solution to this recurrence (see Problem 8 in this section's exercises), here we simply establish the solution's order of growth. According to the Master Theorem, $A(n) \in \Theta(n^{\log_2 7})$. In other words, the number of additions has the same order of growth as the number of multiplications. This puts Strassen's algorithm in $\Theta(n^{\log_2 7})$, which is a better efficiency class than $\Theta(n^3)$ of the brute-force method.

Since the time of Strassen's discovery, several other algorithms for multiplying two $n \times n$ matrices of real numbers in $O(n^\alpha)$ time with progressively smaller constants α have been invented. The fastest algorithm so far is that of Coopersmith and Winograd [Coo87] with its efficiency in $O(n^{2.376})$. The decreasing values of the exponents have been obtained at the expense of the increasing complexity of these algorithms. Because of large multiplicative constants, none of them is of practical value. However, they are interesting from a theoretical point of view. On one hand, they get closer and closer to the best theoretical lower bound known for matrix multiplication, which is n^2 multiplications, though the gap between this bound and the best available algorithm remains unresolved. On the other hand, matrix multiplication is known to be computationally equivalent to some other important problems, such as solving systems of linear equations (discussed in the next chapter).

Exercises 5.4

1. What are the smallest and largest numbers of digits the product of two decimal n-digit integers can have?
2. Compute 2101 * 1130 by applying the divide-and-conquer algorithm outlined in the text.
3. **a.** Prove the equality $a^{\log_b c} = c^{\log_b a}$, which was used in Section 5.4.

   b. Why is $n^{\log_2 3}$ better than $3^{\log_2 n}$ as a closed-form formula for $M(n)$?
4. **a.** Why did we not include multiplications by 10^n in the multiplication count $M(n)$ of the large-integer multiplication algorithm?

   b. In addition to assuming that n is a power of 2, we made, for the sake of simplicity, another, more subtle, assumption in setting up the recurrences for $M(n)$ and $A(n)$, which is not always true (it does not change the final answers, however). What is this assumption?
5. How many one-digit additions are made by the pen-and-pencil algorithm in multiplying two n-digit integers? You may disregard potential carries.
6. Verify the formulas underlying Strassen's algorithm for multiplying 2×2 matrices.

7. Apply Strassen's algorithm to compute

$$\begin{bmatrix} 1 & 0 & 2 & 1 \\ 4 & 1 & 1 & 0 \\ 0 & 1 & 3 & 0 \\ 5 & 0 & 2 & 1 \end{bmatrix} * \begin{bmatrix} 0 & 1 & 0 & 1 \\ 2 & 1 & 0 & 4 \\ 2 & 0 & 1 & 1 \\ 1 & 3 & 5 & 0 \end{bmatrix}$$

exiting the recursion when $n = 2$, i.e., computing the products of 2×2 matrices by the brute-force algorithm.

8. Solve the recurrence for the number of additions required by Strassen's algorithm. Assume that n is a power of 2.

9. V. Pan [Pan78] has discovered a divide-and-conquer matrix multiplication algorithm that is based on multiplying two 70×70 matrices using 143,640 multiplications. Find the asymptotic efficiency of Pan's algorithm (you may ignore additions) and compare it with that of Strassen's algorithm.

10. Practical implementations of Strassen's algorithm usually switch to the brute-force method after matrix sizes become smaller than some crossover point. Run an experiment to determine such a crossover point on your computer system.

5.5 The Closest-Pair and Convex-Hull Problems by Divide-and-Conquer

In Section 3.3, we discussed the brute-force approach to solving two classic problems of computational geometry: the closest-pair problem and the convex-hull problem. We saw that the two-dimensional versions of these problems can be solved by brute-force algorithms in $\Theta(n^2)$ and $O(n^3)$ time, respectively. In this section, we discuss more sophisticated and asymptotically more efficient algorithms for these problems, which are based on the divide-and-conquer technique.

The Closest-Pair Problem

Let P be a set of $n > 1$ points in the Cartesian plane. For the sake of simplicity, we assume that the points are distinct. We can also assume that the points are ordered in nondecreasing order of their x coordinate. (If they were not, we could sort them first by an efficeint sorting algorithm such as mergesort.) It will also be convenient to have the points sorted in a separate list in nondecreasing order of the y coordinate; we will denote such a list Q.

If $2 \le n \le 3$, the problem can be solved by the obvious brute-force algorithm. If $n > 3$, we can divide the points into two subsets P_l and P_r of $\lceil n/2 \rceil$ and $\lfloor n/2 \rfloor$ points, respectively, by drawing a vertical line through the median m of their x coordinates so that $\lceil n/2 \rceil$ points lie to the left of or on the line itself, and $\lfloor n/2 \rfloor$ points lie to the right of or on the line. Then we can solve the closest-pair problem

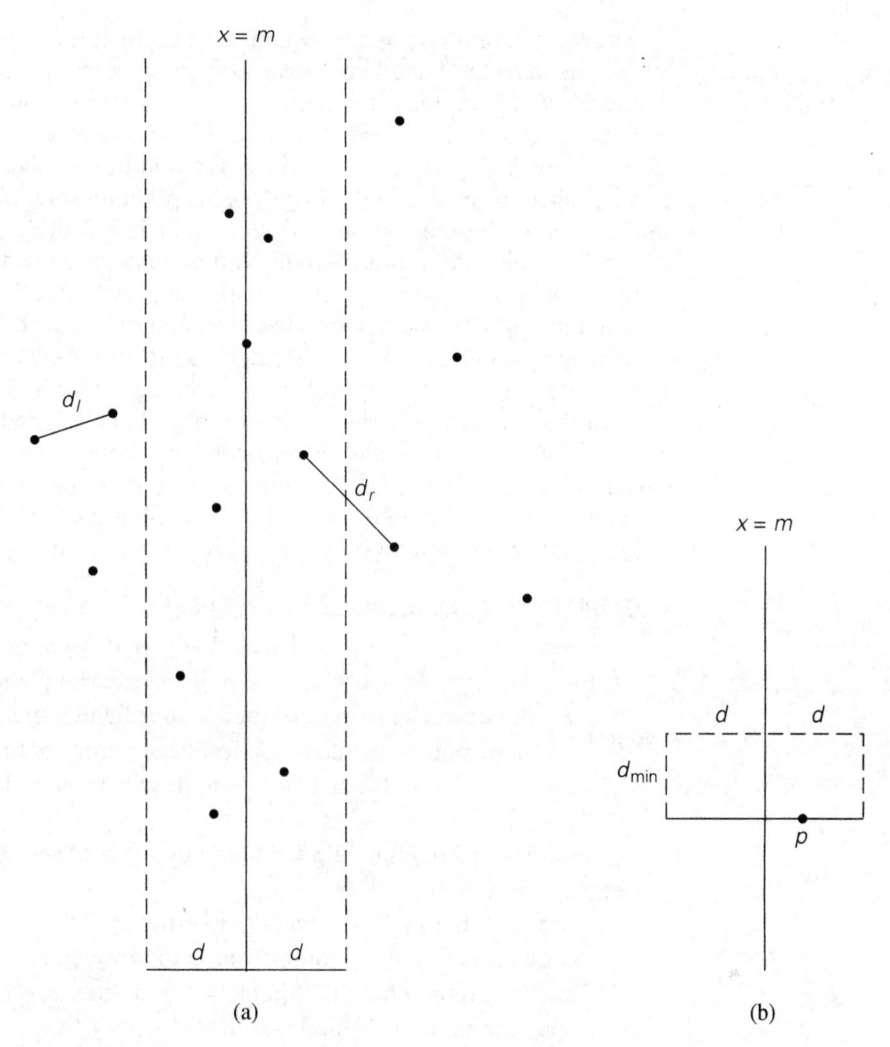

FIGURE 5.7 (a) Idea of the divide-and-conquer algorithm for the closest-pair problem. (b) Rectangle that may contain points closer than $d_{\min}$ to point p.

recursively for subsets P_l and P_r. Let d_l and d_r be the smallest distances between pairs of points in P_l and P_r, respectively, and let $d = \min\{d_l, d_r\}$.

Note that d is not necessarily the smallest distance between all the point pairs because points of a closer pair can lie on the opposite sides of the separating line. Therefore, as a step combining the solutions to the smaller subproblems, we need to examine such points. Obviously, we can limit our attention to the points inside the symmetric vertical strip of width $2d$ around the separating line, since the distance between any other pair of points is at least d (Figure 5.7a).

Let S be the list of points inside the strip of width $2d$ around the separating line, obtained from Q and hence ordered in nondecreasing order of their y coordinate. We will scan this list, updating the information about $d_{\min}$, the minimum distance seen so far, if we encounter a closer pair of points. Initially, $d_{\min} = d$, and subsequently $d_{\min} \le d$. Let $p(x, y)$ be a point on this list. For a point $p'(x', y')$ to have a chance to be closer to p than $d_{\min}$, the point must follow p on list S and the difference between their y coordinates must be less than $d_{\min}$ (why?). Geometrically, this means that p' must belong to the rectangle shown in Figure 5.7b. The principal insight exploited by the algorithm is the observation that the rectangle can contain just a few such points, because the points in each half (left and right) of the rectangle must be at least distance d apart. It is easy to prove that the total number of such points in the rectangle, including p, does not exceed eight (Problem 2 in this section's exercises); a more careful analysis reduces this number to six (see [Joh04, p. 695]). Thus, the algorithm can consider no more than five next points following p on the list S, before moving up to the next point.

Here is pseudocode of the algorithm. We follow the advice given in Section 3.3 to avoid computing square roots inside the innermost loop of the algorithm.

```
ALGORITHM EfficientClosestPair(P, Q)
    //Solves the closest-pair problem by divide-and-conquer
    //Input: An array P of n ≥ 2 points in the Cartesian plane sorted in
    //       nondecreasing order of their x coordinates and an array Q of the
    //       same points sorted in nondecreasing order of the y coordinates
    //Output: Euclidean distance between the closest pair of points
    if n ≤ 3
        return the minimal distance found by the brute-force algorithm
    else
        copy the first ⌈n/2⌉ points of P to array P_l
        copy the same ⌈n/2⌉ points from Q to array Q_l
        copy the remaining ⌊n/2⌋ points of P to array P_r
        copy the same ⌊n/2⌋ points from Q to array Q_r
        d_l ← EfficientClosestPair(P_l, Q_l)
        d_r ← EfficientClosestPair(P_r, Q_r)
        d ← min{d_l, d_r}
        m ← P[⌈n/2⌉ − 1].x
        copy all the points of Q for which |x − m| < d into array S[0..num − 1]
        dminsq ← d^2
        for i ← 0 to num − 2 do
            k ← i + 1
            while k ≤ num − 1 and (S[k].y − S[i].y)^2 < dminsq
                dminsq ← min((S[k].x − S[i].x)^2 + (S[k].y − S[i].y)^2, dminsq)
                k ← k + 1
    return sqrt(dminsq)
```

The algorithm spends linear time both for dividing the problem into two problems half the size and combining the obtained solutions. Therefore, assuming as usual that n is a power of 2, we have the following recurrence for the running time of the algorithm:

$$T(n) = 2T(n/2) + f(n),$$

where $f(n) \in \Theta(n)$. Applying the Master Theorem (with $a = 2$, $b = 2$, and $d = 1$), we get $T(n) \in \Theta(n \log n)$. The necessity to presort input points does not change the overall efficiency class if sorting is done by a $O(n \log n)$ algorithm such as mergesort. In fact, this is the best efficiency class one can achieve, because it has been proved that any algorithm for this problem must be in $\Omega(n \log n)$ under some natural assumptions about operations an algorithm can perform (see [Pre85, p. 188]).

Convex-Hull Problem

Let us revisit the convex-hull problem, introduced in Section 3.3: find the smallest convex polygon that contains n given points in the plane. We consider here a divide-and-conquer algorithm called ***quickhull*** because of its resemblance to quicksort.

Let S be a set of $n > 1$ points $p_1(x_1, y_1), \ldots, p_n(x_n, y_n)$ in the Cartesian plane. We assume that the points are sorted in nondecreasing order of their x coordinates, with ties resolved by increasing order of the y coordinates of the points involved. It is not difficult to prove the geometrically obvious fact that the leftmost point p_1 and the rightmost point p_n are two distinct extreme points of the set's convex hull (Figure 5.8). Let $\overrightarrow{p_1 p_n}$ be the straight line through points p_1 and p_n directed from p_1 to p_n. This line separates the points of S into two sets: S_1 is the set of points to the left of this line, and S_2 is the set of points to the right of this line. (We say that point q_3 is to the left of the line $\overrightarrow{q_1 q_2}$ directed from point q_1 to point q_2 if $q_1 q_2 q_3$ forms a counterclockwise cycle. Later, we cite an analytical way to check this condition, based on checking the sign of a determinant formed by the coordinates of the three points.) The points of S on the line $\overrightarrow{p_1 p_n}$, other than p_1 and p_n, cannot be extreme points of the convex hull and hence are excluded from further consideration.

The boundary of the convex hull of S is made up of two polygonal chains: an "upper" boundary and a "lower" boundary. The "upper" boundary, called the ***upper hull***, is a sequence of line segments with vertices at p_1, some of the points in S_1 (if S_1 is not empty) and p_n. The "lower" boundary, called the ***lower hull***, is a sequence of line segments with vertices at p_1, some of the points in S_2 (if S_2 is not empty) and p_n. The fact that the convex hull of the entire set S is composed of the upper and lower hulls, which can be constructed independently and in a similar fashion, is a very useful observation exploited by several algorithms for this problem.

For concreteness, let us discuss how quickhull proceeds to construct the upper hull; the lower hull can be constructed in the same manner. If S_1 is empty, the

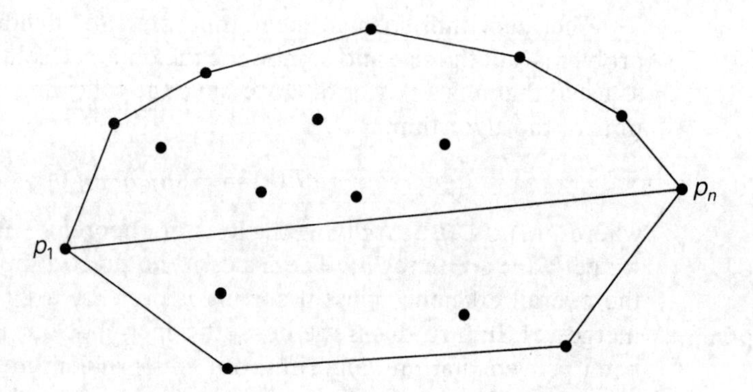

FIGURE 5.8 Upper and lower hulls of a set of points.

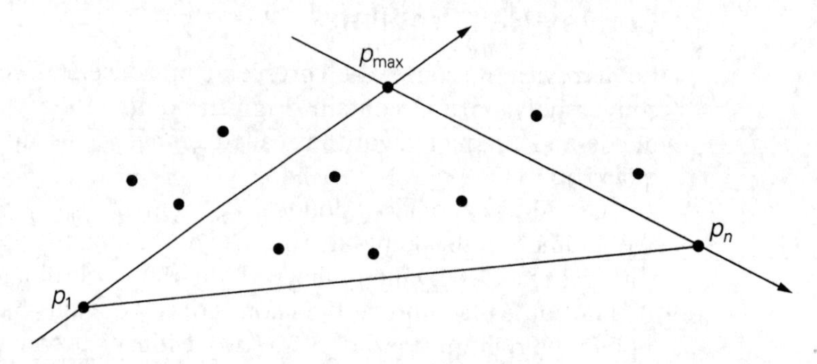

FIGURE 5.9 The idea of quickhull.

upper hull is simply the line segment with the endpoints at p_1 and p_n. If S_1 is not empty, the algorithm identifies point p_{max} in S_1, which is the farthest from the line $\overrightarrow{p_1 p_n}$ (Figure 5.9). If there is a tie, the point that maximizes the angle $\angle p_{max} p p_n$ can be selected. (Note that point p_{max} maximizes the area of the triangle with two vertices at p_1 and p_n and the third one at some other point of S_1.) Then the algorithm identifies all the points of set S_1 that are to the left of the line $\overrightarrow{p_1 p_{max}}$; these are the points that will make up the set $S_{1,1}$. The points of S_1 to the left of the line $\overrightarrow{p_{max} p_n}$ will make up the set $S_{1,2}$. It is not difficult to prove the following:

- p_{max} is a vertex of the upper hull.
- The points inside $\triangle p_1 p_{max} p_n$ cannot be vertices of the upper hull (and hence can be eliminated from further consideration).
- There are no points to the left of both lines $\overrightarrow{p_1 p_{max}}$ and $\overrightarrow{p_{max} p_n}$.

Therefore, the algorithm can continue constructing the upper hulls of $p_1 \cup S_{1,1} \cup p_{max}$ and $p_{max} \cup S_{1,2} \cup p_n$ recursively and then simply concatenate them to get the upper hull of the entire set $p_1 \cup S_1 \cup p_n$.

Now we have to figure out how the algorithm's geometric operations can be actually implemented. Fortunately, we can take advantage of the following very useful fact from analytical geometry: if $q_1(x_1, y_1)$, $q_2(x_2, y_2)$, and $q_3(x_3, y_3)$ are three arbitrary points in the Cartesian plane, then the area of the triangle $\Delta q_1q_2q_3$ is equal to one-half of the magnitude of the determinant

$$\begin{vmatrix} x_1 & y_1 & 1 \\ x_2 & y_2 & 1 \\ x_3 & y_3 & 1 \end{vmatrix} = x_1y_2 + x_3y_1 + x_2y_3 - x_3y_2 - x_2y_1 - x_1y_3,$$

while the sign of this expression is positive if and only if the point $q_3 = (x_3, y_3)$ is to the left of the line $\overrightarrow{q_1q_2}$. Using this formula, we can check in constant time whether a point lies to the left of the line determined by two other points as well as find the distance from the point to the line.

Quickhull has the same $\Theta(n^2)$ worst-case efficiency as quicksort (Problem 9 in this section's exercises). In the average case, however, we should expect a much better performance. First, the algorithm should benefit from the quicksort-like savings from the on-average balanced split of the problem into two smaller subproblems. Second, a significant fraction of the points—namely, those inside $\Delta p_1p_{\max}p_n$ (see Figure 5.9)—are eliminated from further processing. Under a natural assumption that points given are chosen randomly from a uniform distribution over some convex region (e.g., a circle or a rectangle), the average-case efficiency of quickhull turns out to be linear [Ove80].

Exercises 5.5

1. a. For the one-dimensional version of the closest-pair problem, i.e., for the problem of finding two closest numbers among a given set of n real numbers, design an algorithm that is directly based on the divide-and-conquer technique and determine its efficiency class.

b. Is it a good algorithm for this problem?

2. Prove that the divide-and-conquer algorithm for the closest-pair problem examines, for every point p in the vertical strip (see Figures 5.7a and 5.7b), no more than seven other points that can be closer to p than $d_{\min}$, the minimum distance between two points encountered by the algorithm up to that point.

3. Consider the version of the divide-and-conquer two-dimensional closest-pair algorithm in which, instead of presorting input set P, we simply sort each of the two sets P_l and P_r in nondecreasing order of their y coordinates on each recursive call. Assuming that sorting is done by mergesort, set up a recurrence relation for the running time in the worst case and solve it for $n = 2^k$.

4. Implement the divide-and-conquer closest-pair algorithm, outlined in this section, in the language of your choice.

5. Find on the Web a visualization of an algorithm for the closest-pair problem. What algorithm does this visualization represent?

6. The **Voronoi polygon** for a point p of a set S of points in the plane is defined to be the perimeter of the set of all points in the plane closer to p than to any other point in S. The union of all the Voronoi polygons of the points in S is called the ***Voronoi diagram*** of S.

   a. What is the Voronoi diagram for a set of three points?

   b. Find a visualization of an algorithm for generating the Voronoi diagram on the Web and study a few examples of such diagrams. Based on your observations, can you tell how the solution to the previous question is generalized to the general case?

7. Explain how one can find point p_{max} in the quickhull algorithm analytically.

8. What is the best-case efficiency of quickhull?

9. Give a specific example of inputs that make quickhull run in quadratic time.

10. Implement quickhull in the language of your choice.

11. *Creating decagons* There are 1000 points in the plane, no three of them on the same line. Devise an algorithm to construct 100 decagons with their vertices at these points. The decagons need not be convex, but each of them has to be simple, i.e., its boundary should not cross itself, and no two decagons may have a common point.

12. *Shortest path around* There is a fenced area in the two-dimensional Euclidean plane in the shape of a convex polygon with vertices at points $p_1(x_1, y_1), p_2(x_2, y_2), \ldots, p_n(x_n, y_n)$ (not necessarily in this order). There are two more points, $a(x_a, y_a)$ and $b(x_b, y_b)$ such that $x_a < \min\{x_1, x_2, \ldots, x_n\}$ and $x_b > \max\{x_1, x_2, \ldots, x_n\}$. Design a reasonably efficient algorithm for computing the length of the shortest path between a and b. [ORo98]

SUMMARY

- *Divide-and-conquer* is a general algorithm design technique that solves a problem by dividing it into several smaller subproblems of the same type (ideally, of about equal size), solving each of them recursively, and then combining their solutions to get a solution to the original problem. Many efficient algorithms are based on this technique, although it can be both inapplicable and inferior to simpler algorithmic solutions.

- Running time $T(n)$ of many divide-and-conquer algorithms satisfies the recurrence $T(n) = aT(n/b) + f(n)$. The *Master Theorem* establishes the order of growth of its solutions.

- *Mergesort* is a divide-and-conquer sorting algorithm. It works by dividing an input array into two halves, sorting them recursively, and then *merging* the two sorted halves to get the original array sorted. The algorithm's time efficiency is in $\Theta(n \log n)$ in all cases, with the number of key comparisons being very close to the theoretical minimum. Its principal drawback is a significant extra storage requirement.

- *Quicksort* is a divide-and-conquer sorting algorithm that works by partitioning its input elements according to their value relative to some preselected element. Quicksort is noted for its superior efficiency among $n \log n$ algorithms for sorting randomly ordered arrays but also for the quadratic worst-case efficiency.

- The classic traversals of a binary tree—*preorder*, *inorder*, and *postorder*—and similar algorithms that require recursive processing of both left and right subtrees can be considered examples of the divide-and-conquer technique. Their analysis is helped by replacing all the empty subtrees of a given tree by special *external nodes*.

- There is a divide-and-conquer algorithm for multiplying two n-digit integers that requires about $n^{1.585}$ one-digit multiplications.

- *Strassen's algorithm* needs only seven multiplications to multiply two 2×2 matrices. By exploiting the divide-and-conquer technique, this algorithm can multiply two $n \times n$ matrices with about $n^{2.807}$ multiplications.

- The divide-and-conquer technique can be successfully applied to two important problems of computational geometry: the closest-pair problem and the convex-hull problem.

6

Transform-and-Conquer

That's the secret to life . . . replace one worry with another.

—Charles M. Schulz (1922–2000), American cartoonist, the creator of *Peanuts*

This chapter deals with a group of design methods that are based on the idea of transformation. We call this general technique ***transform-and-conquer*** because these methods work as two-stage procedures. First, in the transformation stage, the problem's instance is modified to be, for one reason or another, more amenable to solution. Then, in the second or conquering stage, it is solved.

There are three major variations of this idea that differ by what we transform a given instance to (Figure 6.1):

- Transformation to a simpler or more convenient instance of the same problem—we call it ***instance simplification***.
- Transformation to a different representation of the same instance—we call it ***representation change***.
- Transformation to an instance of a different problem for which an algorithm is already available—we call it ***problem reduction***.

In the first three sections of this chapter, we encounter examples of the instance-simplification variety. Section 6.1 deals with the simple but fruitful idea of presorting. Many algorithmic problems are easier to solve if their input is sorted. Of course, the benefits of sorting should more than compensate for the

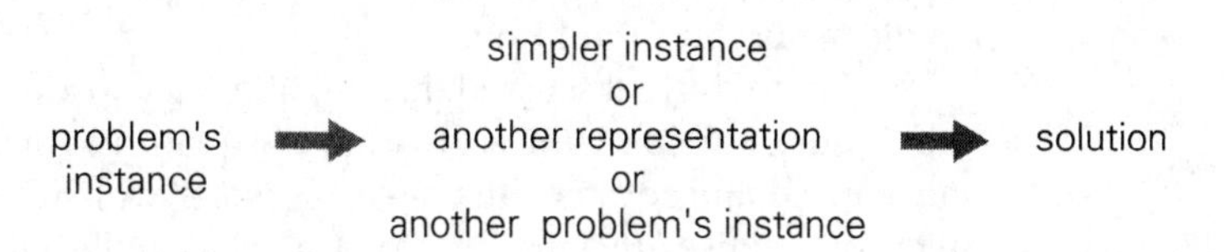

FIGURE 6.1 Transform-and-conquer strategy.

time spent on it; otherwise, we would be better off dealing with an unsorted input directly. Section 6.2 introduces one of the most important algorithms in applied mathematics: Gaussian elimination. This algorithm solves a system of linear equations by first transforming it to another system with a special property that makes finding a solution quite easy. In Section 6.3, the ideas of instance simplification and representation change are applied to search trees. The results are AVL trees and multiway balanced search trees; of the latter we consider the simplest case, 2-3 trees.

Section 6.4 presents heaps and heapsort. Even if you are already familiar with this important data structure and its application to sorting, you can still benefit from looking at them in this new light of transform-and-conquer design. In Section 6.5, we discuss Horner's rule, a remarkable algorithm for evaluating polynomials. If there were an Algorithm Hall of Fame, Horner's rule would be a serious candidate for induction based on the algorithm's elegance and efficiency. We also consider there two interesting algorithms for the exponentiation problem, both based on the representation-change idea.

The chapter concludes with a review of several applications of the third variety of transform-and-conquer: problem reduction. This variety should be considered the most radical of the three: one problem is reduced to another, i.e., transformed into an entirely different problem. This is a very powerful idea, and it is extensively used in the complexity theory (Chapter 11). Its application to designing practical algorithms is not trivial, however. First, we need to identify a new problem into which the given problem should be transformed. Then we must make sure that the transformation algorithm followed by the algorithm for solving the new problem is time efficient compared to other algorithmic alternatives. Among several examples, we discuss an important special case of ***mathematical modeling***, or expressing a problem in terms of purely mathematical objects such as variables, functions, and equations.

6.1 Presorting

Presorting is an old idea in computer science. In fact, interest in sorting algorithms is due, to a significant degree, to the fact that many questions about a list are easier to answer if the list is sorted. Obviously, the time efficiency of algorithms that involve sorting may depend on the efficiency of the sorting algorithm being used. For the sake of simplicity, we assume throughout this section that lists are implemented as arrays, because some sorting algorithms are easier to implement for the array representation.

So far, we have discussed three elementary sorting algorithms—selection sort, bubble sort, and insertion sort—that are quadratic in the worst and average cases, and two advanced algorithms—mergesort, which is always in $\Theta(n \log n)$, and quicksort, whose efficiency is also $\Theta(n \log n)$ in the average case but is quadratic in the worst case. Are there faster sorting algorithms? As we have already stated in Section 1.3 (see also Section 11.2), no general comparison-based sorting algorithm can have a better efficiency than $n \log n$ in the worst case, and the same result holds for the average-case efficiency.[1]

Following are three examples that illustrate the idea of presorting. More examples can be found in this section's exercises.

EXAMPLE 1 *Checking element uniqueness in an array* If this element uniqueness problem looks familiar to you, it should; we considered a brute-force algorithm for the problem in Section 2.3 (see Example 2). The brute-force algorithm compared pairs of the array's elements until either two equal elements were found or no more pairs were left. Its worst-case efficiency was in $\Theta(n^2)$.

Alternatively, we can sort the array first and then check only its consecutive elements: if the array has equal elements, a pair of them must be next to each other, and vice versa.

```
ALGORITHM  PresortElementUniqueness(A[0..n − 1])
    //Solves the element uniqueness problem by sorting the array first
    //Input: An array A[0..n − 1] of orderable elements
    //Output: Returns "true" if A has no equal elements, "false" otherwise
    sort the array A
    for i ← 0 to n − 2 do
        if A[i] = A[i + 1] return false
    return true
```

The running time of this algorithm is the sum of the time spent on sorting and the time spent on checking consecutive elements. Since the former requires at least $n \log n$ comparisons and the latter needs no more than $n - 1$ comparisons, it is the sorting part that will determine the overall efficiency of the algorithm. So, if we use a quadratic sorting algorithm here, the entire algorithm will not be more efficient than the brute-force one. But if we use a good sorting algorithm, such as mergesort, with worst-case efficiency in $\Theta(n \log n)$, the worst-case efficiency of the entire presorting-based algorithm will be also in $\Theta(n \log n)$:

$$T(n) = T_{sort}(n) + T_{scan}(n) \in \Theta(n \log n) + \Theta(n) = \Theta(n \log n). \quad \blacksquare$$

EXAMPLE 2 *Computing a mode* A ***mode*** is a value that occurs most often in a given list of numbers. For example, for 5, 1, 5, 7, 6, 5, 7, the mode is 5. (If several different values occur most often, any of them can be considered a mode.) The brute-force approach to computing a mode would scan the list and compute the frequencies of all its distinct values, then find the value with the largest frequency.

---

1. Sorting algorithms called ***radix sorts*** are linear but in terms of the total number of input bits. These algorithms work by comparing individual bits or pieces of keys rather than keys in their entirety. Although the running time of these algorithms is proportional to the number of input bits, they are still essentially $n \log n$ algorithms because the number of bits per key must be at least $\log_2 n$ in order to accommodate n distinct keys of input.

In order to implement this idea, we can store the values already encountered, along with their frequencies, in a separate list. On each iteration, the ith element of the original list is compared with the values already encountered by traversing this auxiliary list. If a matching value is found, its frequency is incremented; otherwise, the current element is added to the list of distinct values seen so far with a frequency of 1.

It is not difficult to see that the worst-case input for this algorithm is a list with no equal elements. For such a list, its ith element is compared with $i-1$ elements of the auxiliary list of distinct values seen so far before being added to the list with a frequency of 1. As a result, the worst-case number of comparisons made by this algorithm in creating the frequency list is

$$C(n) = \sum_{i=1}^{n}(i-1) = 0 + 1 + \cdots + (n-1) = \frac{(n-1)n}{2} \in \Theta(n^2).$$

The additional $n-1$ comparisons needed to find the largest frequency in the auxiliary list do not change the quadratic worst-case efficiency class of the algorithm.

As an alternative, let us first sort the input. Then all equal values will be adjacent to each other. To compute the mode, all we need to do is to find the longest run of adjacent equal values in the sorted array.

```
ALGORITHM PresortMode(A[0..n − 1])
    //Computes the mode of an array by sorting it first
    //Input: An array A[0..n − 1] of orderable elements
    //Output: The array's mode
    sort the array A
    i ← 0                  //current run begins at position i
    modefrequency ← 0      //highest frequency seen so far
    while i ≤ n − 1 do
        runlength ← 1;  runvalue ← A[i]
        while i + runlength ≤ n − 1 and A[i + runlength] = runvalue
            runlength ← runlength + 1
        if runlength > modefrequency
            modefrequency ← runlength;  modevalue ← runvalue
        i ← i + runlength
    return modevalue
```

The analysis here is similar to the analysis of Example 1: the running time of the algorithm will be dominated by the time spent on sorting since the remainder of the algorithm takes linear time (why?). Consequently, with an $n \log n$ sort, this method's worst-case efficiency will be in a better asymptotic class than the worst-case efficiency of the brute-force algorithm. ■

EXAMPLE 3 *Searching problem* Consider the problem of searching for a given value v in a given array of n sortable items. The brute-force solution here is sequential search (Section 3.1), which needs n comparisons in the worst case. If the array is sorted first, we can then apply binary search, which requires only $\lfloor \log_2 n \rfloor + 1$ comparisons in the worst case. Assuming the most efficient $n \log n$ sort, the total running time of such a searching algorithm in the worst case will be

$$T(n) = T_{sort}(n) + T_{search}(n) = \Theta(n \log n) + \Theta(\log n) = \Theta(n \log n),$$

which is inferior to sequential search. The same will also be true for the average-case efficiency. Of course, if we are to search in the same list more than once, the time spent on sorting might well be justified. (Problem 4 in this section's exercises asks to estimate the minimum number of searches needed to justify presorting.) ■

Before we finish our discussion of presorting, we should mention that many, if not most, geometric algorithms dealing with sets of points use presorting in one way or another. Points can be sorted by one of their coordinates, or by their distance from a particular line, or by some angle, and so on. For example, presorting was used in the divide-and-conquer algorithms for the closest-pair problem and for the convex-hull problem, which were discussed in Section 5.5.

Further, some problems for directed acyclic graphs can be solved more easily after topologically sorting the digraph in question. The problems of finding the longest and shortest paths in such digraphs (see the exercises for Sections 8.1 and 9.3) illustrate this point.

Finally, most algorithms based on the greedy technique, which is the subject of Chapter 9, require presorting of their inputs as an intrinsic part of their operations.

Exercises 6.1

1. Consider the problem of finding the distance between the two closest numbers in an array of n numbers. (The distance between two numbers x and y is computed as $|x - y|$.)
   a. Design a presorting-based algorithm for solving this problem and determine its efficiency class.
   b. Compare the efficiency of this algorithm with that of the brute-force algorithm (see Problem 9 in Exercises 1.2).
2. Let $A = \{a_1, \ldots, a_n\}$ and $B = \{b_1, \ldots, b_m\}$ be two sets of numbers. Consider the problem of finding their intersection, i.e., the set C of all the numbers that are in both A and B.
   a. Design a brute-force algorithm for solving this problem and determine its efficiency class.
   b. Design a presorting-based algorithm for solving this problem and determine its efficiency class.

3. Consider the problem of finding the smallest and largest elements in an array of n numbers.

a. Design a presorting-based algorithm for solving this problem and determine its efficiency class.

b. Compare the efficiency of the three algorithms: (i) the brute-force algorithm, (ii) this presorting-based algorithm, and (iii) the divide-and-conquer algorithm (see Problem 2 in Exercises 5.1).

4. Estimate how many searches will be needed to justify time spent on presorting an array of 10^3 elements if sorting is done by mergesort and searching is done by binary search. (You may assume that all searches are for elements known to be in the array.) What about an array of 10^6 elements?

5. To sort or not to sort? Design a reasonably efficient algorithm for solving each of the following problems and determine its efficiency class.

a. You are given n telephone bills and m checks sent to pay the bills ($n \geq m$). Assuming that telephone numbers are written on the checks, find out who failed to pay. (For simplicity, you may also assume that only one check is written for a particular bill and that it covers the bill in full.)

b. You have a file of n student records indicating each student's number, name, home address, and date of birth. Find out the number of students from each of the 50 U.S. states.

6. Given a set of $n \geq 3$ points in the Cartesian plane, connect them in a simple polygon, i.e., a closed path through all the points so that its line segments (the polygon's edges) do not intersect (except for neighboring edges at their common vertex). For example,

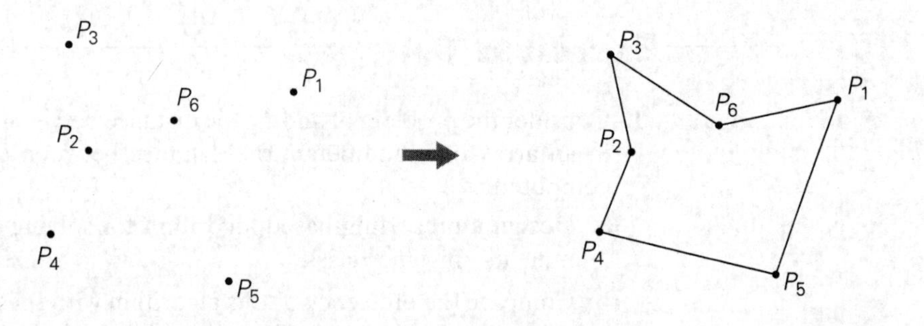

a. Does the problem always have a solution? Does it always have a unique solution?

b. Design a reasonably efficient algorithm for solving this problem and indicate its efficiency class.

7. You have an array of n real numbers and another integer s. Find out whether the array contains two elements whose sum is s. (For example, for the array 5, 9, 1, 3 and $s = 6$, the answer is yes, but for the same array and $s = 7$, the answer is no.) Design an algorithm for this problem with a better than quadratic time efficiency.

8. You have a list of n open intervals $(a_1, b_1), (a_2, b_2), \ldots, (a_n, b_n)$ on the real line. (An open interval (a, b) comprises all the points strictly between its endpoints a and b, i.e., $(a, b) = \{x \mid a < x < b\}$.) Find the maximum number of these intervals that have a common point. For example, for the intervals (1, 4), (0, 3), (−1.5, 2), (3.6, 5), this maximum number is 3. Design an algorithm for this problem with a better than quadratic time efficiency.

9. *Number placement* Given a list of n distinct integers and a sequence of n boxes with pre-set inequality signs inserted between them, design an algorithm that places the numbers into the boxes to satisfy those inequalities. For example, the numbers 4, 6, 3, 1, 8 can be placed in the five boxes as shown below:

$$\boxed{1} < \boxed{8} > \boxed{3} < \boxed{4} < \boxed{6}$$

10. *Maxima search*

a. A point (x_i, y_i) in the Cartesian plane is said to be dominated by point (x_j, y_j) if $x_i \le x_j$ and $y_i \le y_j$ with at least one of the two inequalities being strict. Given a set of n points, one of them is said to be a **maximum** of the set if it is not dominated by any other point in the set. For example, in the figure below, all the maximum points of the set of 10 points are circled.

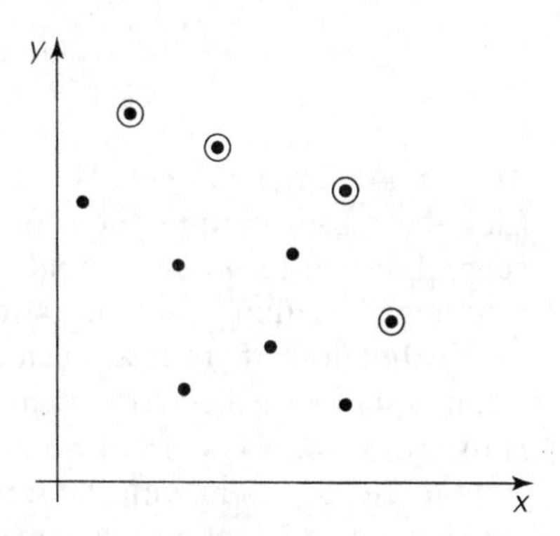

Design an efficient algorithm for finding all the maximum points of a given set of n points in the Cartesian plane. What is the time efficiency class of your algorithm?

b. Give a few real-world applications of this algorithm.

11. *Anagram detection*

a. Design an efficient algorithm for finding all sets of anagrams in a large file such as a dictionary of English words [Ben00]. For example, *eat*, *ate*, and *tea* belong to one such set.

b. Write a program implementing the algorithm.

6.2 Gaussian Elimination

You are certainly familiar with systems of two linear equations in two unknowns:

$$a_{11}x + a_{12}y = b_1$$
$$a_{21}x + a_{22}y = b_2.$$

Recall that unless the coefficients of one equation are proportional to the coefficients of the other, the system has a unique solution. The standard method for finding this solution is to use either equation to express one of the variables as a function of the other and then substitute the result into the other equation, yielding a linear equation whose solution is then used to find the value of the second variable.

In many applications, we need to solve a system of n equations in n unknowns:

$$\begin{aligned} a_{11}x_1 + a_{12}x_2 + \cdots + a_{1n}x_n &= b_1 \\ a_{21}x_1 + a_{22}x_2 + \cdots + a_{2n}x_n &= b_2 \\ &\vdots \\ a_{n1}x_1 + a_{n2}x_2 + \cdots + a_{nn}x_n &= b_n \end{aligned}$$

where n is a large number. Theoretically, we can solve such a system by generalizing the substitution method for solving systems of two linear equations (what general design technique would such a method be based upon?); however, the resulting algorithm would be extremely cumbersome.

Fortunately, there is a much more elegant algorithm for solving systems of linear equations called ***Gaussian elimination***.[2] The idea of Gaussian elimination is to transform a system of n linear equations in n unknowns to an equivalent system (i.e., a system with the same solution as the original one) with an upper-triangular coefficient matrix, a matrix with all zeros below its main diagonal:

2. The method is named after Carl Friedrich Gauss (1777–1855), who—like other giants in the history of mathematics such as Isaac Newton and Leonhard Euler—made numerous fundamental contributions to both theoretical and computational mathematics. The method was known to the Chinese 1800 years before the Europeans rediscovered it.

$$\begin{aligned} a_{11}x_1 + a_{12}x_2 + \cdots + a_{1n}x_n &= b_1 \\ a_{21}x_1 + a_{22}x_2 + \cdots + a_{2n}x_n &= b_2 \\ &\vdots \\ a_{n1}x_1 + a_{n2}x_2 + \cdots + a_{nn}x_n &= b_n \end{aligned} \quad \Longrightarrow \quad \begin{aligned} a'_{11}x_1 + a'_{12}x_2 + \cdots + a'_{1n}x_n &= b'_1 \\ a'_{22}x_2 + \cdots + a'_{2n}x_n &= b'_2 \\ &\vdots \\ a'_{nn}x_n &= b'_n. \end{aligned}$$

In matrix notations, we can write this as

$$Ax = b \quad \Longrightarrow \quad A'x = b',$$

where

$$A = \begin{bmatrix} a_{11} & a_{12} & \dots & a_{1n} \\ a_{21} & a_{22} & \dots & a_{2n} \\ \vdots & & & \\ a_{n1} & a_{n2} & \dots & a_{nn} \end{bmatrix}, \; b = \begin{bmatrix} b_1 \\ b_2 \\ \vdots \\ b_n \end{bmatrix}, \; A' = \begin{bmatrix} a'_{11} & a'_{12} & \dots & a'_{1n} \\ 0 & a'_{22} & \dots & a'_{2n} \\ \vdots & & & \\ 0 & 0 & \dots & a'_{nn} \end{bmatrix}, \; b' = \begin{bmatrix} b'_1 \\ b'_2 \\ \vdots \\ b'_n \end{bmatrix}.$$

(We added primes to the matrix elements and right-hand sides of the new system to stress the point that their values differ from their counterparts in the original system.)

Why is the system with the upper-triangular coefficient matrix better than a system with an arbitrary coefficient matrix? Because we can easily solve the system with an upper-triangular coefficient matrix by back substitutions as follows. First, we can immediately find the value of x_n from the last equation; then we can substitute this value into the next to last equation to get x_{n-1}, and so on, until we substitute the known values of the last $n-1$ variables into the first equation, from which we find the value of x_1.

So how can we get from a system with an arbitrary coefficient matrix A to an equivalent system with an upper-triangular coefficient matrix A'? We can do that through a series of the so-called ***elementary operations***:

- exchanging two equations of the system
- replacing an equation with its nonzero multiple
- replacing an equation with a sum or difference of this equation and some multiple of another equation

Since no elementary operation can change a solution to a system, any system that is obtained through a series of such operations will have the same solution as the original one.

Let us see how we can get to a system with an upper-triangular matrix. First, we use a_{11} as a ***pivot*** to make all x_1 coefficients zeros in the equations below the first one. Specifically, we replace the second equation with the difference between it and the first equation multiplied by a_{21}/a_{11} to get an equation with a zero coefficient for x_1. Doing the same for the third, fourth, and finally nth equation—with the multiples $a_{31}/a_{11}, a_{41}/a_{11}, \dots, a_{n1}/a_{11}$ of the first equation, respectively—makes all the coefficients of x_1 below the first equation zero. Then we get rid of all the coefficients of x_2 by subtracting an appropriate multiple of the second equation from each of the equations below the second one. Repeating this elimination for each of the first $n-1$ variables ultimately yields a system with an upper-triangular coefficient matrix.

Before we look at an example of Gaussian elimination, let us note that we can operate with just a system's coefficient matrix augmented, as its $(n+1)$st column, with the equations' right-hand side values. In other words, we need to write explicitly neither the variable names nor the plus and equality signs.

EXAMPLE 1 Solve the system by Gaussian elimination.

$$\begin{aligned} 2x_1 - x_2 + x_3 &= 1 \\ 4x_1 + x_2 - x_3 &= 5 \\ x_1 + x_2 + x_3 &= 0. \end{aligned}$$

$$\begin{bmatrix} 2 & -1 & 1 & 1 \\ 4 & 1 & -1 & 5 \\ 1 & 1 & 1 & 0 \end{bmatrix} \begin{matrix} \\ \text{row 2} - \frac{4}{2}\text{ row 1} \\ \text{row 3} - \frac{1}{2}\text{ row 1} \end{matrix}$$

$$\begin{bmatrix} 2 & -1 & 1 & 1 \\ 0 & 3 & -3 & 3 \\ 0 & \frac{3}{2} & \frac{1}{2} & -\frac{1}{2} \end{bmatrix} \begin{matrix} \\ \\ \text{row 3} - \frac{1}{2}\text{ row 2} \end{matrix}$$

$$\begin{bmatrix} 2 & -1 & 1 & 1 \\ 0 & 3 & -3 & 3 \\ 0 & 0 & 2 & -2 \end{bmatrix}$$

Now we can obtain the solution by back substitutions:

$x_3 = (-2)/2 = -1,\ \ x_2 = (3 - (-3)x_3)/3 = 0,\ \ \text{and}\ \ x_1 = (1 - x_3 - (-1)x_2)/2 = 1.$

■

Here is pseudocode of the first stage, called ***forward elimination***, of the algorithm.

```
Algorithm ForwardElimination(A[1..n; 1::n]; b[1::n])
//Applies Gaussian elimination to matrix A of a system's coefficients,
//augmented with vector b of the system's right-hand side values
//Input: Matrix A[1::n; 1; ::n] and column-vector b[1::n]
//Output: An equivalent upper-triangular matrix in place of A with the
//corresponding right-hand side values in the (n + 1)st column
for i ← 1 to n do A[i; n + 1] ← b[i]      //augments the matrix
for i ← 1 to n − 1 do
    for j ← i + 1 to n do
        for k ← n + 1 downto i do
            A[j; k] ← A[j; k] ← A[i; k] * A[j; i] / A[i; i]
```

There are two important observations to make about this pseudocode. First, it is not always correct: if $A[i, i] = 0$, we cannot divide by it and hence cannot use the ith row as a pivot for the ith iteration of the algorithm. In such a case, we should take advantage of the first elementary operation and exchange the ith row with some row below it that has a nonzero coefficient in the ith column. (If the system has a unique solution, which is the normal case for systems under consideration, such a row must exist.)

Since we have to be prepared for the possibility of row exchanges anyway, we can take care of another potential difficulty: the possibility that $A[i, i]$ is so small and consequently the scaling factor $A[j, i]/A[i, i]$ so large that the new value of $A[j, k]$ might become distorted by a round-off error caused by a subtraction of two numbers of greatly different magnitudes.[3] To avoid this problem, we can always look for a row with the largest absolute value of the coefficient in the ith column, exchange it with the ith row, and then use the new $A[i, i]$ as the ith iteration's pivot. This modification, called ***partial pivoting***, guarantees that the magnitude of the scaling factor will never exceed 1.

The second observation is the fact that the innermost loop is written with a glaring inefficiency. Can you find it before checking the following pseudocode, which both incorporates partial pivoting and eliminates this inefficiency?

```
ALGORITHM BetterForwardElimination(A[1..n, 1..n], b[1..n])
    //Implements Gaussian elimination with partial pivoting
    //Input: Matrix A[1..n, 1..n] and column-vector b[1..n]
    //Output: An equivalent upper-triangular matrix in place of A and the
    //corresponding right-hand side values in place of the (n + 1)st column
    for i ← 1 to n do A[i, n + 1] ← b[i] //appends b to A as the last column
    for i ← 1 to n − 1 do
        pivotrow ← i
        for j ← i + 1 to n do
            if |A[j, i]| > |A[pivotrow, i]| pivotrow ← j
        for k ← i to n + 1 do
            swap(A[i, k], A[pivotrow, k])
        for j ← i + 1 to n do
            temp ← A[j, i] / A[i, i]
            for k ← i to n + 1 do
                A[j, k] ← A[j, k] − A[i, k] ∗ temp
```

Let us find the time efficiency of this algorithm. Its innermost loop consists of a single line,

$$A[j, k] \leftarrow A[j, k] - A[i, k] * temp,$$

3. We discuss round-off errors in more detail in Section 11.4.

which contains one multiplication and one subtraction. On most computers, multiplication is unquestionably more expensive than addition/subtraction, and hence it is multiplication that is usually quoted as the algorithm's basic operation.[4] The standard summation formulas and rules reviewed in Section 2.3 (see also Appendix A) are very helpful in the following derivation:

$$
\begin{aligned}
C(n) &= \sum_{i=1}^{n-1}\sum_{j=i+1}^{n}\sum_{k=i}^{n+1} 1 = \sum_{i=1}^{n-1}\sum_{j=i+1}^{n}(n+1-i+1) = \sum_{i=1}^{n-1}\sum_{j=i+1}^{n}(n+2-i)\\
&= \sum_{i=1}^{n-1}(n+2-i)(n-(i+1)+1) = \sum_{i=1}^{n-1}(n+2-i)(n-i)\\
&= (n+1)(n-1)+n(n-2)+\cdots+3\cdot 1\\
&= \sum_{j=1}^{n-1}(j+2)j = \sum_{j=1}^{n-1}j^2 + \sum_{j=1}^{n-1}2j = \frac{(n-1)n(2n-1)}{6} + 2\frac{(n-1)n}{2}\\
&= \frac{n(n-1)(2n+5)}{6} \approx \frac{1}{3}n^3 \in \Theta(n^3).
\end{aligned}
$$

Since the second (***back substitution***) stage of Gaussian elimination is in $\Theta(n^2)$, as you are asked to show in the exercises, the running time is dominated by the cubic elimination stage, making the entire algorithm cubic as well.

Theoretically, Gaussian elimination always either yields an exact solution to a system of linear equations when the system has a unique solution or discovers that no such solution exists. In the latter case, the system will have either no solutions or infinitely many of them. In practice, solving systems of significant size on a computer by this method is not nearly so straightforward as the method would lead us to believe. The principal difficulty lies in preventing an accumulation of round-off errors (see Section 11.4). Consult textbooks on numerical analysis that analyze this and other implementation issues in great detail.

LU Decomposition

Gaussian elimination has an interesting and very useful byproduct called ***LU decomposition*** of the coefficient matrix. In fact, modern commercial implementations of Gaussian elimination are based on such a decomposition rather than on the basic algorithm outlined above.

EXAMPLE 2 Let us return to the example in the beginning of this section, where we applied Gaussian elimination to the matrix

4. As we mentioned in Section 2.1, on some computers multiplication is not necessarily more expensive than addition/subtraction. For this algorithm, this point is moot since we can simply count the number of times the innermost loop is executed, which is, of course, exactly the same number as the number of multiplications and the number of subtractions there.

$$A = \begin{bmatrix} 2 & -1 & 1 \\ 4 & 1 & -1 \\ 1 & 1 & 1 \end{bmatrix}.$$

Consider the lower-triangular matrix L made up of 1's on its main diagonal and the row multiples used in the forward elimination process

$$L = \begin{bmatrix} 1 & 0 & 0 \\ 2 & 1 & 0 \\ \frac{1}{2} & \frac{1}{2} & 1 \end{bmatrix}$$

and the upper-triangular matrix U that was the result of this elimination

$$U = \begin{bmatrix} 2 & -1 & 1 \\ 0 & 3 & -3 \\ 0 & 0 & 2 \end{bmatrix}.$$

It turns out that the product LU of these matrices is equal to matrix A. (For this particular pair of L and U, you can verify this fact by direct multiplication, but as a general proposition, it needs, of course, a proof, which we omit here.)

Therefore, solving the system $Ax = b$ is equivalent to solving the system $LUx = b$. The latter system can be solved as follows. Denote $y = Ux$, then $Ly = b$. Solve the system $Ly = b$ first, which is easy to do because L is a lower-triangular matrix; then solve the system $Ux = y$, with the upper-triangular matrix U, to find x. Thus, for the system at the beginning of this section, we first solve $Ly = b$:

$$\begin{bmatrix} 1 & 0 & 0 \\ 2 & 1 & 0 \\ \frac{1}{2} & \frac{1}{2} & 1 \end{bmatrix} \begin{bmatrix} y_1 \\ y_2 \\ y_3 \end{bmatrix} = \begin{bmatrix} 1 \\ 5 \\ 0 \end{bmatrix}.$$

Its solution is

$$y_1 = 1, \qquad y_2 = 5 - 2y_1 = 3, \qquad y_3 = 0 - \frac{1}{2}y_1 - \frac{1}{2}y_2 = -2.$$

Solving $Ux = y$ means solving

$$\begin{bmatrix} 2 & -1 & 1 \\ 0 & 3 & -3 \\ 0 & 0 & 2 \end{bmatrix} \begin{bmatrix} x_1 \\ x_2 \\ x_3 \end{bmatrix} = \begin{bmatrix} 1 \\ 3 \\ -2 \end{bmatrix},$$

and the solution is

$$x_3 = (-2)/2 = -1, \quad x_2 = (3 - (-3)x_3)/3 = 0, \quad x_1 = (1 - x_3 - (-1)x_2)/2 = 1. \qquad \blacksquare$$

Note that once we have the LU decomposition of matrix A, we can solve systems $Ax = b$ with as many right-hand side vectors b as we want to, one at a time. This is a distinct advantage over the classic Gaussian elimination discussed earlier. Also note that the LU decomposition does not actually require extra memory, because we can store the nonzero part of U in the upper-triangular part of A

(including the main diagonal) and store the nontrivial part of L below the main diagonal of A.

Computing a Matrix Inverse

Gaussian elimination is a very useful algorithm that tackles one of the most important problems of applied mathematics: solving systems of linear equations. In fact, Gaussian elimination can also be applied to several other problems of linear algebra, such as computing a matrix ***inverse***. The inverse of an $n \times n$ matrix A is an $n \times n$ matrix, denoted A^{-1}, such that

$$AA^{-1} = I,$$

where I is the $n \times n$ identity matrix (the matrix with all zero elements except the main diagonal elements, which are all ones). Not every square matrix has an inverse, but when it exists, the inverse is unique. If a matrix A does not have an inverse, it is called ***singular***. One can prove that a matrix is singular if and only if one of its rows is a linear combination (a sum of some multiples) of the other rows. A convenient way to check whether a matrix is nonsingular is to apply Gaussian elimination: if it yields an upper-triangular matrix with no zeros on the main diagonal, the matrix is nonsingular; otherwise, it is singular. So being singular is a very special situation, and most square matrices do have their inverses.

Theoretically, inverse matrices are very important because they play the role of reciprocals in matrix algebra, overcoming the absence of the explicit division operation for matrices. For example, in a complete analogy with a linear equation in one unknown $ax = b$ whose solution can be written as $x = a^{-1}b$ (if a is not zero), we can express a solution to a system of n equations in n unknowns $Ax = b$ as $x = A^{-1}b$ (if A is nonsingular) where b is, of course, a vector, not a number.

According to the definition of the inverse matrix for a nonsingular $n \times n$ matrix A, to compute it we need to find n^2 numbers x_{ij}, $1 \le i, j \le n$, such that

$$\begin{bmatrix} a_{11} & a_{12} & \dots & a_{1n} \\ a_{21} & a_{22} & \dots & a_{2n} \\ \vdots & & & \\ a_{n1} & a_{n2} & \dots & a_{nn} \end{bmatrix} \begin{bmatrix} x_{11} & x_{12} & \dots & x_{1n} \\ x_{21} & x_{22} & \dots & x_{2n} \\ \vdots & & & \\ x_{n1} & x_{n2} & \dots & x_{nn} \end{bmatrix} = \begin{bmatrix} 1 & 0 & \dots & 0 \\ 0 & 1 & \dots & 0 \\ \vdots & & & \\ 0 & 0 & \dots & 1 \end{bmatrix}.$$

We can find the unknowns by solving n systems of linear equations that have the same coefficient matrix A, the vector of unknowns x^j is the jth column of the inverse, and the right-hand side vector e^j is the jth column of the identity matrix $(1 \le j \le n)$:

$$Ax^j = e^j.$$

We can solve these systems by applying Gaussian elimination to matrix A augmented by the $n \times n$ identity matrix. Better yet, we can use forward elimination to find the LU decomposition of A and then solve the systems $LUx^j = e^j$, $j = 1, \dots, n$, as explained earlier.

Computing a Determinant

Another problem that can be solved by Gaussian elimination is computing a determinant. The ***determinant*** of an $n \times n$ matrix A, denoted $\det A$ or $|A|$, is a number whose value can be defined recursively as follows. If $n = 1$, i.e., if A consists of a single element a_{11}, $\det A$ is equal to a_{11}; for $n > 1$, $\det A$ is computed by the recursive formula

$$\det A = \sum_{j=1}^{n} s_j a_{1j} \det A_j,$$

where s_j is $+1$ if j is odd and -1 if j is even, a_{1j} is the element in row 1 and column j, and A_j is the $(n-1) \times (n-1)$ matrix obtained from matrix A by deleting its row 1 and column j.

In particular, for a 2×2 matrix, the definition implies a formula that is easy to remember:

$$\det \begin{bmatrix} a_{11} & a_{12} \\ a_{21} & a_{22} \end{bmatrix} = a_{11} \det [a_{22}] - a_{12} \det [a_{21}] = a_{11}a_{22} - a_{12}a_{21}.$$

In other words, the determinant of a 2×2 matrix is simply equal to the difference between the products of its diagonal elements.

For a 3×3 matrix, we get

$$\det \begin{bmatrix} a_{11} & a_{12} & a_{13} \\ a_{21} & a_{22} & a_{23} \\ a_{31} & a_{32} & a_{33} \end{bmatrix}$$

$$= a_{11} \det \begin{bmatrix} a_{22} & a_{23} \\ a_{32} & a_{33} \end{bmatrix} - a_{12} \det \begin{bmatrix} a_{21} & a_{23} \\ a_{31} & a_{33} \end{bmatrix} + a_{13} \det \begin{bmatrix} a_{21} & a_{22} \\ a_{31} & a_{32} \end{bmatrix}$$

$$= a_{11}a_{22}a_{33} + a_{12}a_{23}a_{31} + a_{13}a_{21}a_{32} - a_{11}a_{23}a_{32} - a_{12}a_{21}a_{33} - a_{13}a_{22}a_{31}.$$

Incidentally, this formula is very handy in a variety of applications. In particular, we used it twice already in Section 5.5 as a part of the quickhull algorithm.

But what if we need to compute a determinant of a large matrix? Although this is a task that is rarely needed in practice, it is worth discussing nevertheless. Using the recursive definition can be of little help because it implies computing the sum of $n!$ terms. Here, Gaussian elimination comes to the rescue again. The central point is the fact that the determinant of an upper-triangular matrix is equal to the product of elements on its main diagonal, and it is easy to see how elementary operations employed by the elimination algorithm influence the determinant's value. (Basically, it either remains unchanged or changes a sign or is multiplied by the constant used by the elimination algorithm.) As a result, we can compute the determinant of an $n \times n$ matrix in cubic time.

Determinants play an important role in the theory of systems of linear equations. Specifically, a system of n linear equations in n unknowns $Ax = b$ has a unique solution if and only if the determinant of its coefficient matrix $\det A$ is not equal to zero. Moreover, this solution can be found by the formulas called ***Cramer's rule***,

$$x_1 = \frac{\det A_1}{\det A}, \ldots, x_j = \frac{\det A_j}{\det A}, \ldots, x_n = \frac{\det A_n}{\det A},$$

where $\det A_j$ is the determinant of the matrix obtained by replacing the jth column of A by the column b. You are asked to investigate in the exercises whether using Cramer's rule is a good algorithm for solving systems of linear equations.

Exercises 6.2

1. Solve the following system by Gaussian elimination:

$$\begin{aligned} x_1 + x_2 + x_3 &= 2 \\ 2x_1 + x_2 + x_3 &= 3 \\ x_1 - x_2 + 3x_3 &= 8. \end{aligned}$$

2. **a.** Solve the system of the previous question by the LU decomposition method.

   b. From the standpoint of general algorithm design techniques, how would you classify the LU decomposition method?

3. Solve the system of Problem 1 by computing the inverse of its coefficient matrix and then multiplying it by the vector on the right-hand side.

4. Would it be correct to get the efficiency class of the forward elimination stage of Gaussian elimination as follows?

$$\begin{aligned} C(n) &= \sum_{i=1}^{n-1} \sum_{j=i+1}^{n} \sum_{k=i}^{n+1} 1 = \sum_{i=1}^{n-1} (n+2-i)(n-i) \\ &= \sum_{i=1}^{n-1} [(n+2)n - i(2n+2) + i^2] \\ &= \sum_{i=1}^{n-1} (n+2)n - \sum_{i=1}^{n-1} (2n+2)i + \sum_{i=1}^{n-1} i^2. \end{aligned}$$

   Since $s_1(n) = \sum_{i=1}^{n-1} (n+2)n \in \Theta(n^3)$, $s_2(n) = \sum_{i=1}^{n-1} (2n+2)i \in \Theta(n^3)$, and $s_3(n) = \sum_{i=1}^{n-1} i^2 \in \Theta(n^3)$, $s_1(n) - s_2(n) + s_3(n) \in \Theta(n^3)$.

5. Write pseudocode for the back-substitution stage of Gaussian elimination and show that its running time is in $\Theta(n^2)$.

6. Assuming that division of two numbers takes three times longer than their multiplication, estimate how much faster *BetterForwardElimination* is than *ForwardElimination*. (Of course, you should also assume that a compiler is not going to eliminate the inefficiency in *ForwardElimination*.)

7. a. Give an example of a system of two linear equations in two unknowns that has a unique solution and solve it by Gaussian elimination.

b. Give an example of a system of two linear equations in two unknowns that has no solution and apply Gaussian elimination to it.

c. Give an example of a system of two linear equations in two unknowns that has infinitely many solutions and apply Gaussian elimination to it.

8. The ***Gauss-Jordan elimination*** method differs from Gaussian elimination in that the elements above the main diagonal of the coefficient matrix are made zero at the same time and by the same use of a pivot row as the elements below the main diagonal.

a. Apply the Gauss-Jordan method to the system of Problem 1 of these exercises.

b. What general design strategy is this algorithm based on?

c. In general, how many multiplications are made by this method in solving a system of n equations in n unknowns? How does this compare with the number of multiplications made by the Gaussian elimination method in both its elimination and back-substitution stages?

9. A system $Ax = b$ of n linear equations in n unknowns has a unique solution if and only if $\det A \neq 0$. Is it a good idea to check this condition before applying Gaussian elimination to the system?

10. a. Apply Cramer's rule to solve the system of Problem 1 of these exercises.

b. Estimate how many times longer it will take to solve a system of n linear equations in n unknowns by Cramer's rule than by Gaussian elimination. Assume that all the determinants in Cramer's rule formulas are computed independently by Gaussian elimination.

11. *Lights out* This one-person game is played on an $n \times n$ board composed of 1×1 light panels. Each panel has a switch that can be turned on and off, thereby toggling the on/off state of this and four vertically and horizontally adjacent panels. (Of course, toggling a corner square affects a total of three panels, and toggling a noncorner panel on the board's border affects a total of four squares.) Given an initial subset of lighted squares, the goal is to turn all the lights off.

a. Show that an answer can be found by solving a system of linear equations with 0/1 coefficients and right-hand sides using the modulo 2 arithmetic.

b. Use Gaussian elimination to solve the 2×2 "all-ones" instance of this problem, where all the panels of the 2×2 board are initially lit.

c. Use Gaussian elimination to solve the 3×3 "all-ones" instance of this problem, where all the panels of the 3×3 board are initially lit.

6.3 Balanced Search Trees

In Sections 1.4, 4.5, and 5.3, we discussed the binary search tree—one of the principal data structures for implementing dictionaries. It is a binary tree whose nodes contain elements of a set of orderable items, one element per node, so that all elements in the left subtree are smaller than the element in the subtree's root, and all the elements in the right subtree are greater than it. Note that this transformation from a set to a binary search tree is an example of the representation-change technique. What do we gain by such transformation compared to the straightforward implementation of a dictionary by, say, an array? We gain in the time efficiency of searching, insertion, and deletion, which are all in $\Theta(\log n)$, but only in the average case. In the worst case, these operations are in $\Theta(n)$ because the tree can degenerate into a severely unbalanced one with its height equal to $n - 1$.

Computer scientists have expended a lot of effort in trying to find a structure that preserves the good properties of the classical binary search tree—principally, the logarithmic efficiency of the dictionary operations and having the set's elements sorted—but avoids its worst-case degeneracy. They have come up with two approaches.

- The first approach is of the instance-simplification variety: an unbalanced binary search tree is transformed into a balanced one. Because of this, such trees are called ***self-balancing***. Specific implementations of this idea differ by their definition of balance. An ***AVL tree*** requires the difference between the heights of the left and right subtrees of every node never exceed 1. A ***red-black tree*** tolerates the height of one subtree being twice as large as the other subtree of the same node. If an insertion or deletion of a new node creates a tree with a violated balance requirement, the tree is restructured by one of a family of special transformations called ***rotations*** that restore the balance required. In this section, we will discuss only AVL trees. Information about other types of binary search trees that utilize the idea of rebalancing via rotations, including red-black trees and ***splay trees***, can be found in the references [Cor09], [Sed02], and [Tar83].
- The second approach is of the representation-change variety: allow more than one element in a node of a search tree. Specific cases of such trees are ***2-3 trees***, ***2-3-4 trees***, and more general and important ***B-trees***. They differ in the number of elements admissible in a single node of a search tree, but all are perfectly balanced. We discuss the simplest case of such trees, the 2-3 tree, in this section, leaving the discussion of *B*-trees for Chapter 7.

AVL Trees

AVL trees were invented in 1962 by two Russian scientists, G. M. Adelson-Velsky and E. M. Landis [Ade62], after whom this data structure is named.

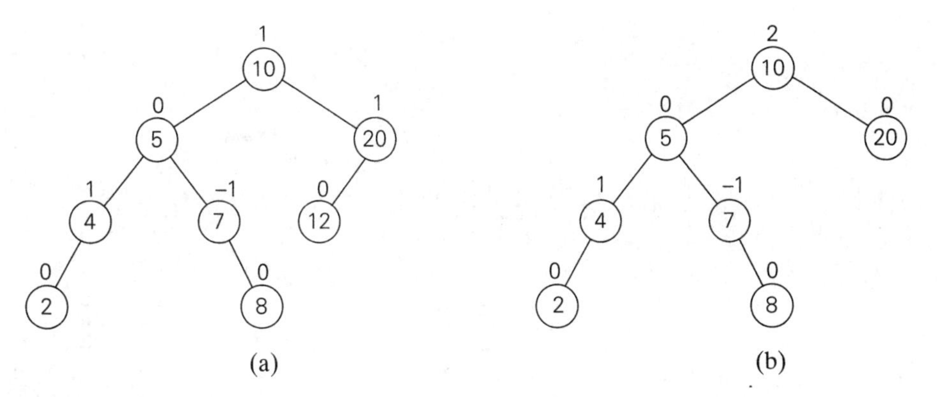

FIGURE 6.2 (a) AVL tree. (b) Binary search tree that is not an AVL tree. The numbers above the nodes indicate the nodes' balance factors.

DEFINITION An ***AVL tree*** is a binary search tree in which the ***balance factor*** of every node, which is defined as the difference between the heights of the node's left and right subtrees, is either 0 or +1 or −1. (The height of the empty tree is defined as −1. Of course, the balance factor can also be computed as the difference between the numbers of levels rather than the height difference of the node's left and right subtrees.)

For example, the binary search tree in Figure 6.2a is an AVL tree but the one in Figure 6.2b is not.

If an insertion of a new node makes an AVL tree unbalanced, we transform the tree by a rotation. A ***rotation*** in an AVL tree is a local transformation of its subtree rooted at a node whose balance has become either +2 or −2. If there are several such nodes, we rotate the tree rooted at the unbalanced node that is the closest to the newly inserted leaf. There are only four types of rotations; in fact, two of them are mirror images of the other two. In their simplest form, the four rotations are shown in Figure 6.3.

The first rotation type is called the ***single right rotation***, or ***R-rotation***. (Imagine rotating the edge connecting the root and its left child in the binary tree in Figure 6.3a to the right.) Figure 6.4 presents the single R-rotation in its most general form. Note that this rotation is performed after a new key is inserted into the left subtree of the left child of a tree whose root had the balance of +1 before the insertion.

The symmetric ***single left rotation***, or ***L-rotation***, is the mirror image of the single R-rotation. It is performed after a new key is inserted into the right subtree of the right child of a tree whose root had the balance of −1 before the insertion. (You are asked to draw a diagram of the general case of the single L-rotation in the exercises.)

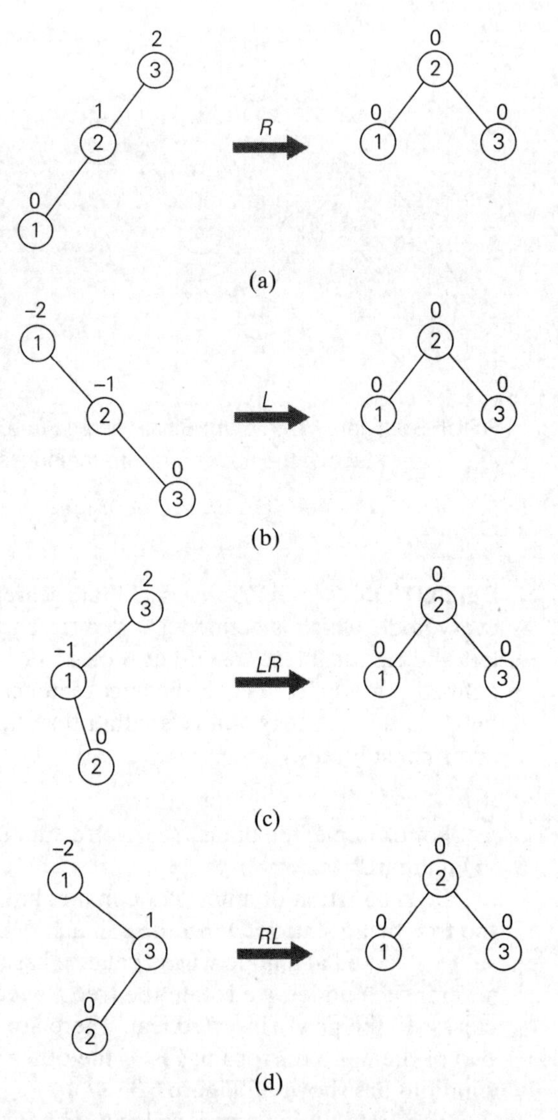

FIGURE 6.3 Four rotation types for AVL trees with three nodes. (a) Single *R*-rotation. (b) Single *L*-rotation. (c) Double *LR*-rotation. (d) Double *RL*-rotation.

The second rotation type is called the ***double left-right rotation*** (***LR-rotation***). It is, in fact, a combination of two rotations: we perform the *L*-rotation of the left subtree of root *r* followed by the *R*-rotation of the new tree rooted at *r* (Figure 6.5). It is performed after a new key is inserted into the right subtree of the left child of a tree whose root had the balance of +1 before the insertion.

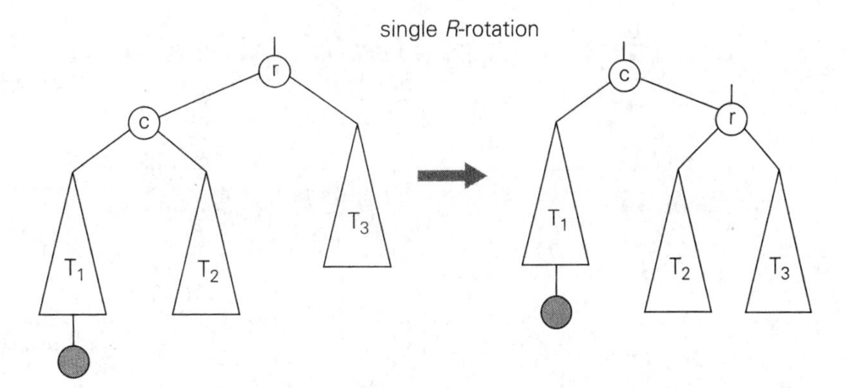

FIGURE 6.4 General form of the *R*-rotation in the AVL tree. A shaded node is the last one inserted.

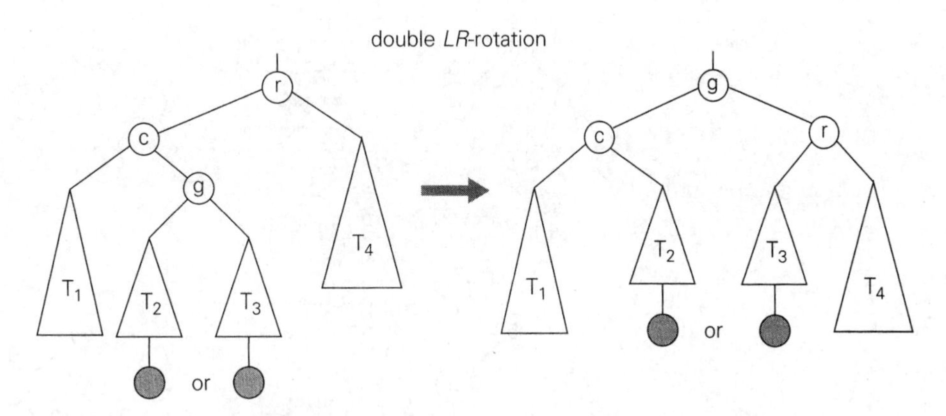

FIGURE 6.5 General form of the double *LR*-rotation in the AVL tree. A shaded node is the last one inserted. It can be either in the left subtree or in the right subtree of the root's grandchild.

The ***double right-left rotation*** (***RL-rotation***) is the mirror image of the double *LR*-rotation and is left for the exercises.

Note that the rotations are not trivial transformations, though fortunately they can be done in constant time. Not only should they guarantee that a resulting tree is balanced, but they should also preserve the basic requirements of a binary search tree. For example, in the initial tree of Figure 6.4, all the keys of subtree T_1 are smaller than c, which is smaller than all the keys of subtree T_2, which are smaller than r, which is smaller than all the keys of subtree T_3. And the same relationships among the key values hold, as they must, for the balanced tree after the rotation.

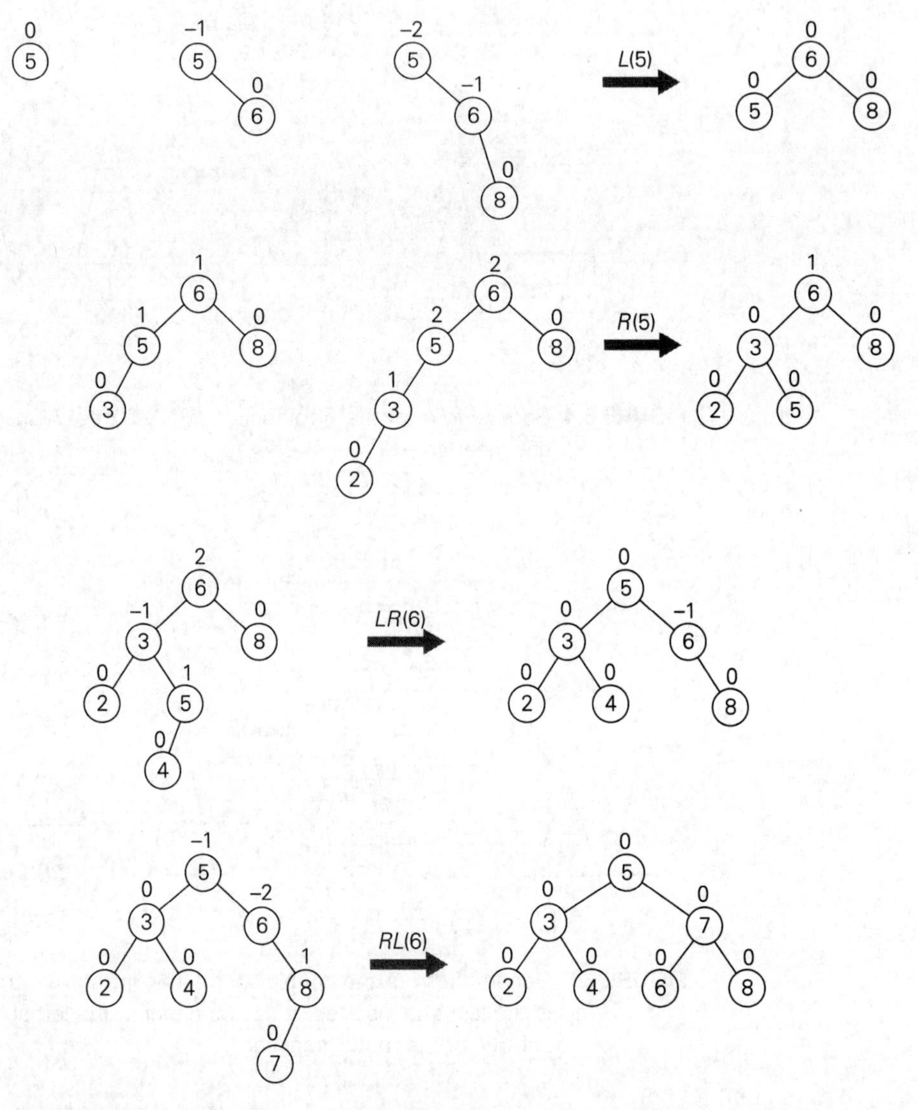

FIGURE 6.6 Construction of an AVL tree for the list 5, 6, 8, 3, 2, 4, 7 by successive insertions. The parenthesized number of a rotation's abbreviation indicates the root of the tree being reorganized.

An example of constructing an AVL tree for a given list of numbers is shown in Figure 6.6. As you trace the algorithm's operations, keep in mind that if there are several nodes with the ± 2 balance, the rotation is done for the tree rooted at the unbalanced node that is the closest to the newly inserted leaf.

How efficient are AVL trees? As with any search tree, the critical characteristic is the tree's height. It turns out that it is bounded both above and below by logarithmic functions. Specifically, the height h of any AVL tree with n nodes satisfies the inequalities

$$\lfloor \log_2 n \rfloor \leq h < 1.4405 \log_2(n + 2) - 1.3277.$$

(These weird-looking constants are round-offs of some irrational numbers related to Fibonacci numbers and the golden ratio—see Section 2.5.)

The inequalities immediately imply that the operations of searching and insertion are $\Theta(\log n)$ in the worst case. Getting an exact formula for the average height of an AVL tree constructed for random lists of keys has proved to be difficult, but it is known from extensive experiments that it is about $1.01\log_2 n + 0.1$ except when n is small [KnuIII, p. 468]. Thus, searching in an AVL tree requires, on average, almost the same number of comparisons as searching in a sorted array by binary search.

The operation of key deletion in an AVL tree is considerably more difficult than insertion, but fortunately it turns out to be in the same efficiency class as insertion, i.e., logarithmic.

These impressive efficiency characteristics come at a price, however. The drawbacks of AVL trees are frequent rotations and the need to maintain balances for its nodes. These drawbacks have prevented AVL trees from becoming the standard structure for implementing dictionaries. At the same time, their underlying idea—that of rebalancing a binary search tree via rotations—has proved to be very fruitful and has led to discoveries of other interesting variations of the classical binary search tree.

2-3 Trees

As mentioned at the beginning of this section, the second idea of balancing a search tree is to allow more than one key in the same node of such a tree. The simplest implementation of this idea is 2-3 trees, introduced by the U.S. computer scientist John Hopcroft in 1970. A ***2-3 tree*** is a tree that can have nodes of two kinds: 2-nodes and 3-nodes. A ***2-node*** contains a single key K and has two children: the left child serves as the root of a subtree whose keys are less than K, and the right child serves as the root of a subtree whose keys are greater than K. (In other words, a 2-node is the same kind of node we have in the classical binary search tree.) A ***3-node*** contains two ordered keys K_1 and K_2 ($K_1 < K_2$) and has three children. The leftmost child serves as the root of a subtree with keys less than K_1, the middle child serves as the root of a subtree with keys between K_1 and K_2, and the rightmost child serves as the root of a subtree with keys greater than K_2 (Figure 6.7).

The last requirement of the 2-3 tree is that all its leaves must be on the same level. In other words, a 2-3 tree is always perfectly height-balanced: the length of a path from the root to a leaf is the same for every leaf. It is this property that we "buy" by allowing more than one key in the same node of a search tree.

Searching for a given key K in a 2-3 tree is quite straightforward. We start at the root. If the root is a 2-node, we act as if it were a binary search tree: we either stop if K is equal to the root's key or continue the search in the left or right

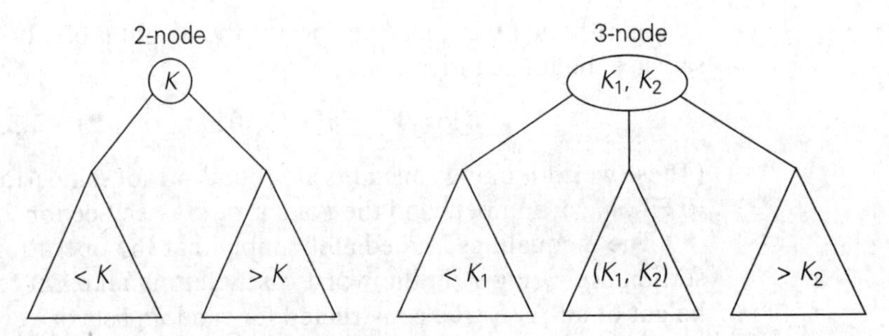

FIGURE 6.7 Two kinds of nodes of a 2-3 tree.

subtree if K is, respectively, smaller or larger than the root's key. If the root is a 3-node, we know after no more than two key comparisons whether the search can be stopped (if K is equal to one of the root's keys) or in which of the root's three subtrees it needs to be continued.

Inserting a new key in a 2-3 tree is done as follows. First of all, we always insert a new key K in a leaf, except for the empty tree. The appropriate leaf is found by performing a search for K. If the leaf in question is a 2-node, we insert K there as either the first or the second key, depending on whether K is smaller or larger than the node's old key. If the leaf is a 3-node, we split the leaf in two: the smallest of the three keys (two old ones and the new key) is put in the first leaf, the largest key is put in the second leaf, and the middle key is promoted to the old leaf's parent. (If the leaf happens to be the tree's root, a new root is created to accept the middle key.) Note that promotion of a middle key to its parent can cause the parent's overflow (if it was a 3-node) and hence can lead to several node splits along the chain of the leaf's ancestors.

An example of a 2-3 tree construction is given in Figure 6.8.

As for any search tree, the efficiency of the dictionary operations depends on the tree's height. So let us first find an upper bound for it. A 2-3 tree of height h with the smallest number of keys is a full tree of 2-nodes (such as the final tree in Figure 6.8 for $h = 2$). Therefore, for any 2-3 tree of height h with n nodes, we get the inequality

$$n \geq 1 + 2 + \cdots + 2^h = 2^{h+1} - 1,$$

and hence

$$h \leq \log_2(n + 1) - 1.$$

On the other hand, a 2-3 tree of height h with the largest number of keys is a full tree of 3-nodes, each with two keys and three children. Therefore, for any 2-3 tree with n nodes,

$$n \leq 2 \cdot 1 + 2 \cdot 3 + \cdots + 2 \cdot 3^h = 2(1 + 3 + \cdots + 3^h) = 3^{h+1} - 1$$

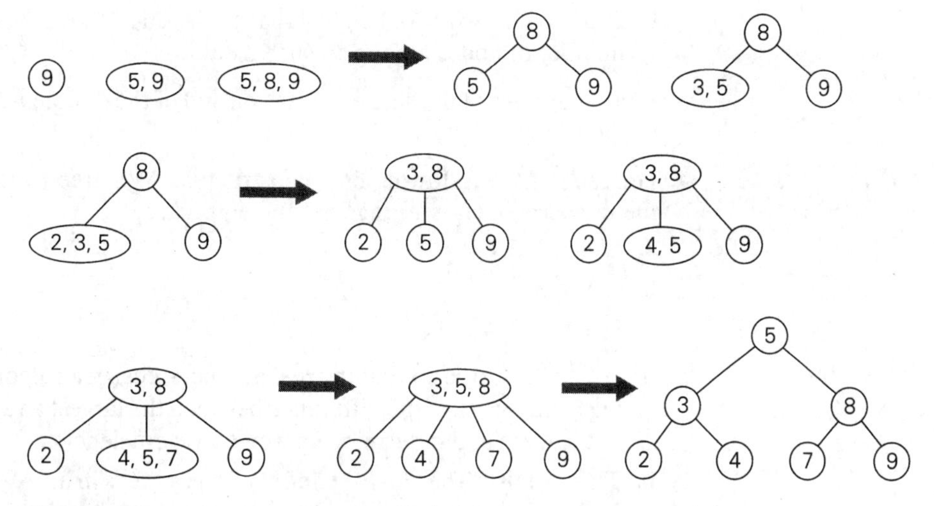

FIGURE 6.8 Construction of a 2-3 tree for the list 9, 5, 8, 3, 2, 4, 7.

and hence

$$h \geq \log_3(n+1) - 1.$$

These lower and upper bounds on height h,

$$\log_3(n+1) - 1 \leq h \leq \log_2(n+1) - 1,$$

imply that the time efficiencies of searching, insertion, and deletion are all in $\Theta(\log n)$ in both the worst and average case. We consider a very important generalization of 2-3 trees, called *B*-trees, in Section 7.4.

Exercises 6.3

1. Which of the following binary trees are AVL trees?

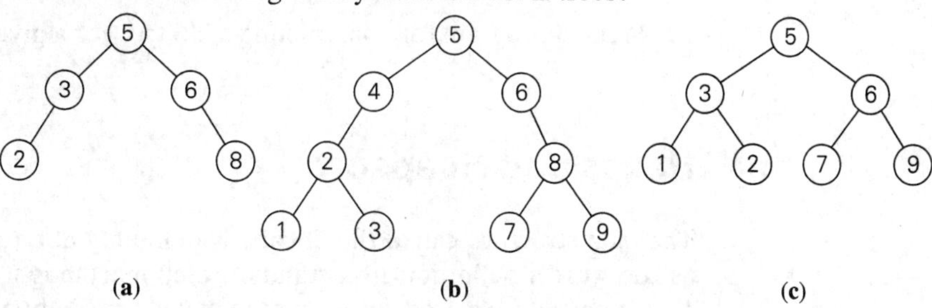

2. a. For $n = 1, 2, 3, 4$, and 5, draw all the binary trees with n nodes that satisfy the balance requirement of AVL trees.

b. Draw a binary tree of height 4 that can be an AVL tree and has the smallest number of nodes among all such trees.

3. Draw diagrams of the single L-rotation and of the double RL-rotation in their general form.

4. For each of the following lists, construct an AVL tree by inserting their elements successively, starting with the empty tree.

a. 1, 2, 3, 4, 5, 6

b. 6, 5, 4, 3, 2, 1

c. 3, 6, 5, 1, 2, 4

5. a. For an AVL tree containing real numbers, design an algorithm for computing the range (i.e., the difference between the largest and smallest numbers in the tree) and determine its worst-case efficiency.

b. True or false: The smallest and the largest keys in an AVL tree can always be found on either the last level or the next-to-last level?

6. Write a program for constructing an AVL tree for a given list of n distinct integers.

7. a. Construct a 2-3 tree for the list C, O, M, P, U, T, I, N, G. Use the alphabetical order of the letters and insert them successively starting with the empty tree.

b. Assuming that the probabilities of searching for each of the keys (i.e., the letters) are the same, find the largest number and the average number of key comparisons for successful searches in this tree.

8. Let T_B and $T_{2\text{-}3}$ be, respectively, a classical binary search tree and a 2-3 tree constructed for the same list of keys inserted in the corresponding trees in the same order. True or false: Searching for the same key in $T_{2\text{-}3}$ always takes fewer or the same number of key comparisons as searching in T_B?

9. For a 2-3 tree containing real numbers, design an algorithm for computing the range (i.e., the difference between the largest and smallest numbers in the tree) and determine its worst-case efficiency.

10. Write a program for constructing a 2-3 tree for a given list of n integers.

6.4 Heaps and Heapsort

The data structure called the "heap" is definitely not a disordered pile of items as the word's definition in a standard dictionary might suggest. Rather, it is a clever, partially ordered data structure that is especially suitable for implementing priority queues. Recall that a ***priority queue*** is a multiset of items with an orderable characteristic called an item's ***priority***, with the following operations:

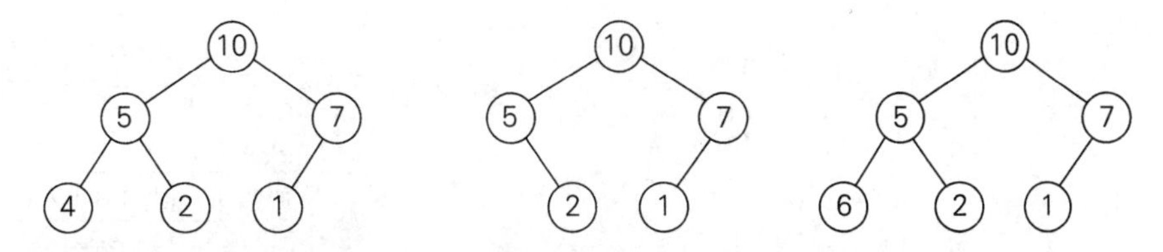

FIGURE 6.9 Illustration of the definition of heap: only the leftmost tree is a heap.

- finding an item with the highest (i.e., largest) priority
- deleting an item with the highest priority
- adding a new item to the multiset

It is primarily an efficient implementation of these operations that makes the heap both interesting and useful. Priority queues arise naturally in such applications as scheduling job executions by computer operating systems and traffic management by communication networks. They also arise in several important algorithms, e.g., Prim's algorithm (Section 9.1), Dijkstra's algorithm (Section 9.3), Huffman encoding (Section 9.4), and branch-and-bound applications (Section 12.2). The heap is also the data structure that serves as a cornerstone of a theoretically important sorting algorithm called heapsort. We discuss this algorithm after we define the heap and investigate its basic properties.

Notion of the Heap

DEFINITION A ***heap*** can be defined as a binary tree with keys assigned to its nodes, one key per node, provided the following two conditions are met:

1. The ***shape property***—the binary tree is ***essentially complete*** (or simply ***complete***), i.e., all its levels are full except possibly the last level, where only some rightmost leaves may be missing.
2. The ***parental dominance*** or ***heap property***—the key in each node is greater than or equal to the keys in its children. (This condition is considered automatically satisfied for all leaves.)[5]

For example, consider the trees of Figure 6.9. The first tree is a heap. The second one is not a heap, because the tree's shape property is violated. And the third one is not a heap, because the parental dominance fails for the node with key 5.

Note that key values in a heap are ordered top down; i.e., a sequence of values on any path from the root to a leaf is decreasing (nonincreasing, if equal keys are allowed). However, there is no left-to-right order in key values; i.e., there is no

5. Some authors require the key at each node to be *less* than or equal to the keys at its children. We call this variation a ***min-heap***.

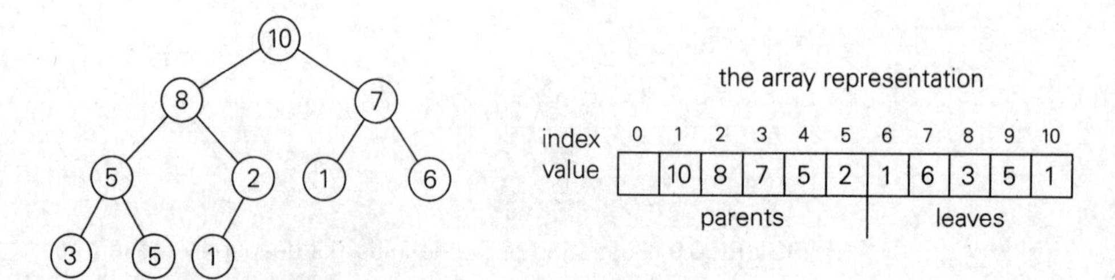

FIGURE 6.10 Heap and its array representation.

relationship among key values for nodes either on the same level of the tree or, more generally, in the left and right subtrees of the same node.

Here is a list of important properties of heaps, which are not difficult to prove (check these properties for the heap of Figure 6.10, as an example).

1. There exists exactly one essentially complete binary tree with n nodes. Its height is equal to $\lfloor \log_2 n \rfloor$.
2. The root of a heap always contains its largest element.
3. A node of a heap considered with all its descendants is also a heap.
4. A heap can be implemented as an array by recording its elements in the top-down, left-to-right fashion. It is convenient to store the heap's elements in positions 1 through n of such an array, leaving $H[0]$ either unused or putting there a sentinel whose value is greater than every element in the heap. In such a representation,
   a. the parental node keys will be in the first $\lfloor n/2 \rfloor$ positions of the array, while the leaf keys will occupy the last $\lceil n/2 \rceil$ positions;
   b. the children of a key in the array's parental position i $(1 \le i \le \lfloor n/2 \rfloor)$ will be in positions $2i$ and $2i + 1$, and, correspondingly, the parent of a key in position i $(2 \le i \le n)$ will be in position $\lfloor i/2 \rfloor$.

Thus, we could also define a heap as an array $H[1..n]$ in which every element in position i in the first half of the array is greater than or equal to the elements in positions $2i$ and $2i + 1$, i.e.,

$$H[i] \ge \max\{H[2i], H[2i+1]\} \quad \text{for } i = 1, \ldots, \lfloor n/2 \rfloor.$$

(Of course, if $2i + 1 > n$, just $H[i] \ge H[2i]$ needs to be satisfied.) While the ideas behind the majority of algorithms dealing with heaps are easier to understand if we think of heaps as binary trees, their actual implementations are usually much simpler and more efficient with arrays.

How can we construct a heap for a given list of keys? There are two principal alternatives for doing this. The first is the ***bottom-up heap construction*** algorithm illustrated in Figure 6.11. It initializes the essentially complete binary tree with n nodes by placing keys in the order given and then "heapifies" the tree as follows. Starting with the last parental node, the algorithm checks whether the parental

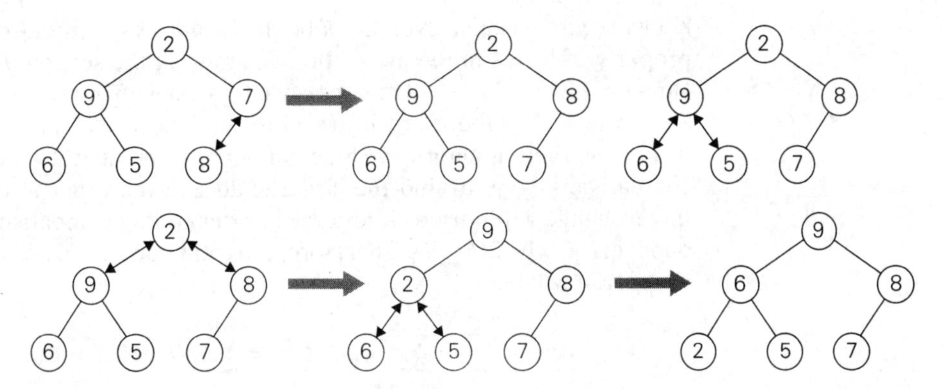

FIGURE 6.11 Bottom-up construction of a heap for the list 2, 9, 7, 6, 5, 8. The double-headed arrows show key comparisons verifying the parental dominance.

dominance holds for the key in this node. If it does not, the algorithm exchanges the node's key K with the larger key of its children and checks whether the parental dominance holds for K in its new position. This process continues until the parental dominance for K is satisfied. (Eventually, it has to because it holds automatically for any key in a leaf.) After completing the "heapification" of the subtree rooted at the current parental node, the algorithm proceeds to do the same for the node's immediate predecessor. The algorithm stops after this is done for the root of the tree.

ALGORITHM *HeapBottomUp*$(H[1..n])$

```
//Constructs a heap from elements of a given array
// by the bottom-up algorithm
//Input: An array H[1..n] of orderable items
//Output: A heap H[1..n]
for i ← ⌊n/2⌋ downto 1 do
    k ← i;   v ← H[k]
    heap ← false
    while not heap and 2 ∗ k ≤ n do
        j ← 2 ∗ k
        if j < n   //there are two children
            if H[j] < H[j + 1] j ← j + 1
        if v ≥ H[j]
            heap ← true
        else H[k] ← H[j];   k ← j
    H[k] ← v
```

How efficient is this algorithm in the worst case? Assume, for simplicity, that $n = 2^k - 1$ so that a heap's tree is full, i.e., the largest possible number of nodes occurs on each level. Let h be the height of the tree. According to the first property of heaps in the list at the beginning of the section, $h = \lfloor \log_2 n \rfloor$ or just $\lceil \log_2 (n+1) \rceil - 1 = k - 1$ for the specific values of n we are considering. Each key on level i of the tree will travel to the leaf level h in the worst case of the heap construction algorithm. Since moving to the next level down requires two comparisons—one to find the larger child and the other to determine whether the exchange is required—the total number of key comparisons involving a key on level i will be $2(h-i)$. Therefore, the total number of key comparisons in the worst case will be

$$C_{worst}(n) = \sum_{i=0}^{h-1} \sum_{\text{level } i \text{ keys}} 2(h-i) = \sum_{i=0}^{h-1} 2(h-i)2^i = 2(n - \log_2(n+1)),$$

where the validity of the last equality can be proved either by using the closed-form formula for the sum $\sum_{i=1}^{h} i2^i$ (see Appendix A) or by mathematical induction on h. Thus, with this bottom-up algorithm, a heap of size n can be constructed with fewer than $2n$ comparisons.

The alternative (and less efficient) algorithm constructs a heap by successive insertions of a new key into a previously constructed heap; some people call it the ***top-down heap construction*** algorithm. So how can we insert a new key K into a heap? First, attach a new node with key K in it after the last leaf of the existing heap. Then sift K up to its appropriate place in the new heap as follows. Compare K with its parent's key: if the latter is greater than or equal to K, stop (the structure is a heap); otherwise, swap these two keys and compare K with its new parent. This swapping continues until K is not greater than its last parent or it reaches the root (illustrated in Figure 6.12).

Obviously, this insertion operation cannot require more key comparisons than the heap's height. Since the height of a heap with n nodes is about $\log_2 n$, the time efficiency of insertion is in $O(\log n)$.

How can we delete an item from a heap? We consider here only the most important case of deleting the root's key, leaving the question about deleting an arbitrary key in a heap for the exercises. (Authors of textbooks like to do such things to their readers, do they not?) Deleting the root's key from a heap can be done with the following algorithm, illustrated in Figure 6.13.

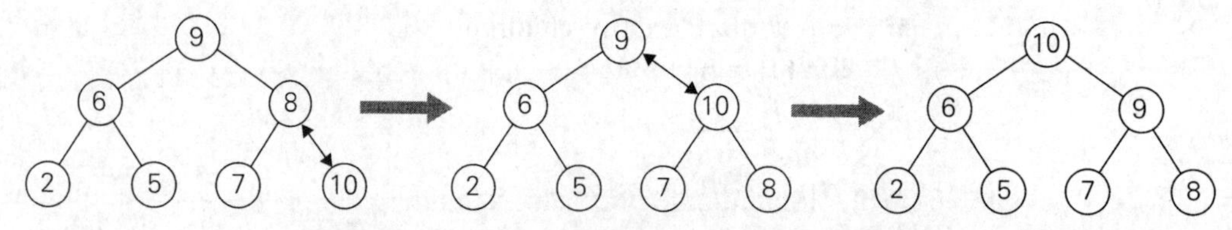

FIGURE 6.12 Inserting a key (10) into the heap constructed in Figure 6.11. The new key is sifted up via a swap with its parent until it is not larger than its parent (or is in the root).

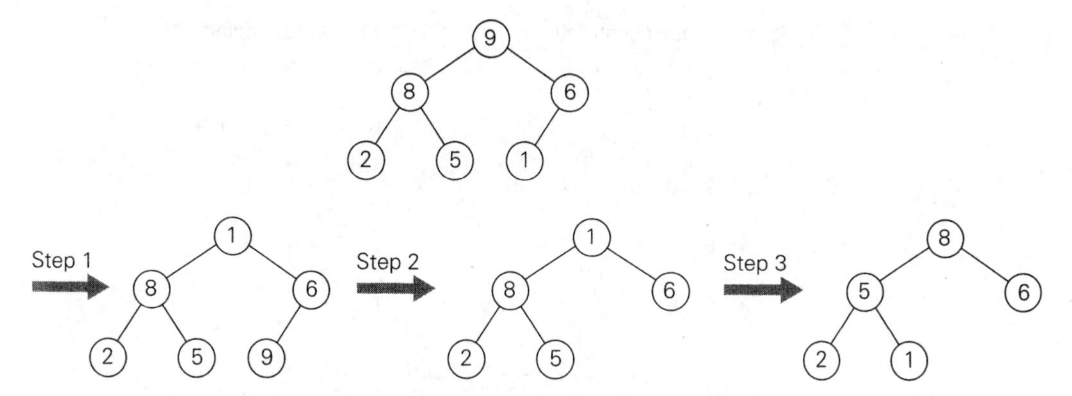

FIGURE 6.13 Deleting the root's key from a heap. The key to be deleted is swapped with the last key after which the smaller tree is "heapified" by exchanging the new key in its root with the larger key in its children until the parental dominance requirement is satisfied.

Maximum Key Deletion from a heap

Step 1 Exchange the root's key with the last key K of the heap.

Step 2 Decrease the heap's size by 1.

Step 3 "Heapify" the smaller tree by sifting K down the tree exactly in the same way we did it in the bottom-up heap construction algorithm. That is, verify the parental dominance for K: if it holds, we are done; if not, swap K with the larger of its children and repeat this operation until the parental dominance condition holds for K in its new position.

The efficiency of deletion is determined by the number of key comparisons needed to "heapify" the tree after the swap has been made and the size of the tree is decreased by 1. Since this cannot require more key comparisons than twice the heap's height, the time efficiency of deletion is in $O(\log n)$ as well.

Heapsort

Now we can describe ***heapsort***—an interesting sorting algorithm discovered by J. W. J. Williams [Wil64]. This is a two-stage algorithm that works as follows.

Stage 1 (heap construction): Construct a heap for a given array.

Stage 2 (maximum deletions): Apply the root-deletion operation $n - 1$ times to the remaining heap.

As a result, the array elements are eliminated in decreasing order. But since under the array implementation of heaps an element being deleted is placed last, the resulting array will be exactly the original array sorted in increasing order. Heapsort is traced on a specific input in Figure 6.14. (The same input as the one

Stage 1 (heap construction)	Stage 2 (maximum deletions)
2 9 **7** 6 5 8	**9** 6 8 2 5 7
2 **9** 8 6 5 7	7 6 8 2 5 | **9**
2 9 8 6 5 7	**8** 6 7 2 5
9 **2** 8 6 5 7	5 6 7 2 | **8**
9 6 8 2 5 7	**7** 6 5 2
	2 6 5 | **7**
	6 2 5
	5 2 | **6**
	5 2
	2 | **5**
	2

FIGURE 6.14 Sorting the array 2, 9, 7, 6, 5, 8 by heapsort.

in Figure 6.11 is intentionally used so that you can compare the tree and array implementations of the bottom-up heap construction algorithm.)

Since we already know that the heap construction stage of the algorithm is in $O(n)$, we have to investigate just the time efficiency of the second stage. For the number of key comparisons, $C(n)$, needed for eliminating the root keys from the heaps of diminishing sizes from n to 2, we get the following inequality:

$$C(n) \le 2\lfloor \log_2(n-1) \rfloor + 2\lfloor \log_2(n-2) \rfloor + \cdots + 2\lfloor \log_2 1 \rfloor \le 2\sum_{i=1}^{n-1} \log_2 i$$

$$\le 2\sum_{i=1}^{n-1} \log_2(n-1) = 2(n-1)\log_2(n-1) \le 2n \log_2 n.$$

This means that $C(n) \in O(n \log n)$ for the second stage of heapsort. For both stages, we get $O(n) + O(n \log n) = O(n \log n)$. A more detailed analysis shows that the time efficiency of heapsort is, in fact, in $\Theta(n \log n)$ in both the worst and average cases. Thus, heapsort's time efficiency falls in the same class as that of mergesort. Unlike the latter, heapsort is in-place, i.e., it does not require any extra storage. Timing experiments on random files show that heapsort runs more slowly than quicksort but can be competitive with mergesort.

Exercises 6.4

1. **a.** Construct a heap for the list 1, 8, 6, 5, 3, 7, 4 by the bottom-up algorithm.
   b. Construct a heap for the list 1, 8, 6, 5, 3, 7, 4 by successive key insertions (top-down algorithm).
   c. Is it always true that the bottom-up and top-down algorithms yield the same heap for the same input?
2. Outline an algorithm for checking whether an array $H[1..n]$ is a heap and determine its time efficiency.
3. **a.** Find the smallest and the largest number of keys that a heap of height h can contain.
   b. Prove that the height of a heap with n nodes is equal to $\lfloor \log_2 n \rfloor$.
4. Prove the following equality used in Section 6.4:

$$\sum_{i=0}^{h-1} 2(h-i)2^i = 2(n - \log_2(n+1)), \quad \text{where } n = 2^{h+1} - 1.$$

5. **a.** Design an efficient algorithm for finding and deleting an element of the smallest value in a heap and determine its time efficiency.
   b. Design an efficient algorithm for finding and deleting an element of a given value v in a heap H and determine its time efficiency.
6. Indicate the time efficiency classes of the three main operations of the priority queue implemented as
   a. an unsorted array.
   b. a sorted array.
   c. a binary search tree.
   d. an AVL tree.
   e. a heap.
7. Sort the following lists by heapsort by using the array representation of heaps.
   a. 1, 2, 3, 4, 5 (in increasing order)
   b. 5, 4, 3, 2, 1 (in increasing order)
   c. S, O, R, T, I, N, G (in alphabetical order)
8. Is heapsort a stable sorting algorithm?
9. What variety of the transform-and-conquer technique does heapsort represent?
10. Which sorting algorithm other than heapsort uses a priority queue?
11. Implement three advanced sorting algorithms—mergesort, quicksort, and heapsort—in the language of your choice and investigate their performance on arrays of sizes $n = 10^3$, 10^4, 10^5, and 10^6. For each of these sizes consider

a. randomly generated files of integers in the range [1..n].
b. increasing files of integers 1, 2, . . . , n.
c. decreasing files of integers n, $n-1$, . . . , 1.

12. *Spaghetti sort* Imagine a handful of uncooked spaghetti, individual rods whose lengths represent numbers that need to be sorted.
a. Outline a "spaghetti sort"—a sorting algorithm that takes advantage of this unorthodox representation.
b. What does this example of computer science folklore (see [Dew93]) have to do with the topic of this chapter in general and heapsort in particular?

6.5 Horner's Rule and Binary Exponentiation

In this section, we discuss the problem of computing the value of a polynomial

$$p(x) = a_n x^n + a_{n-1}x^{n-1} + \cdots + a_1 x + a_0 \quad \textbf{(6.1)}$$

at a given point x and its important special case of computing x^n. Polynomials constitute the most important class of functions because they possess a wealth of good properties on the one hand and can be used for approximating other types of functions on the other. The problem of manipulating polynomials efficiently has been important for several centuries; new discoveries were still being made the last 50 years. By far the most important of them was the ***fast Fourier transform (FFT)***. The practical importance of this remarkable algorithm, which is based on representing a polynomial by its values at specially chosen points, was such that some people consider it one of the most important algorithmic discoveries of all time. Because of its relative complexity, we do not discuss the FFT algorithm in this book. An interested reader will find a wealth of literature on the subject, including reasonably accessible treatments in such textbooks as [Kle06] and [Cor09].

Horner's Rule

Horner's rule is an old but very elegant and efficient algorithm for evaluating a polynomial. It is named after the British mathematician W. G. Horner, who published it in the early 19th century. But according to Knuth [KnuII, p. 486], the method was used by Isaac Newton 150 years before Horner. You will appreciate this method much more if you first design an algorithm for the polynomial evaluation problem by yourself and investigate its efficiency (see Problems 1 and 2 in this section's exercises).

Horner's rule is a good example of the representation-change technique since it is based on representing $p(x)$ by a formula different from (6.1). This new formula is obtained from (6.1) by successively taking x as a common factor in the remaining polynomials of diminishing degrees:

$$p(x) = (\cdots(a_n x + a_{n-1})x + \cdots)x + a_0. \quad \textbf{(6.2)}$$

For example, for the polynomial $p(x) = 2x^4 - x^3 + 3x^2 + x - 5$, we get

$$\begin{aligned} p(x) &= 2x^4 - x^3 + 3x^2 + x - 5 \\ &= x(2x^3 - x^2 + 3x + 1) - 5 \\ &= x(x(2x^2 - x + 3) + 1) - 5 \\ &= x(x(x(2x - 1) + 3) + 1) - 5. \end{aligned} \quad \textbf{(6.3)}$$

It is in formula (6.2) that we will substitute a value of x at which the polynomial needs to be evaluated. It is hard to believe that this is a way to an efficient algorithm, but the unpleasant appearance of formula (6.2) is just that, an appearance. As we shall see, there is no need to go explicitly through the transformation leading to it: all we need is an original list of the polynomial's coefficients.

The pen-and-pencil calculation can be conveniently organized with a two-row table. The first row contains the polynomial's coefficients (including all the coefficients equal to zero, if any) listed from the highest a_n to the lowest a_0. Except for its first entry, which is a_n, the second row is filled left to right as follows: the next entry is computed as the x's value times the last entry in the second row plus the next coefficient from the first row. The final entry computed in this fashion is the value being sought.

EXAMPLE 1 Evaluate $p(x) = 2x^4 - x^3 + 3x^2 + x - 5$ at $x = 3$.

coefficients	2	−1	3	1	−5
$x = 3$	2	$3 \cdot 2 + (-1) = 5$	$3 \cdot 5 + 3 = 18$	$3 \cdot 18 + 1 = 55$	$3 \cdot 55 + (-5) = 160$

Thus, $p(3) = 160$. (On comparing the table's entries with formula (6.3), you will see that $3 \cdot 2 + (-1) = 5$ is the value of $2x - 1$ at $x = 3$, $3 \cdot 5 + 3 = 18$ is the value of $x(2x - 1) + 3$ at $x = 3$, $3 \cdot 18 + 1 = 55$ is the value of $x(x(2x - 1) + 3) + 1$ at $x = 3$, and, finally, $3 \cdot 55 + (-5) = 160$ is the value of $x(x(x(2x - 1) + 3) + 1) - 5 = p(x)$ at $x = 3$.) ■

Pseudocode of this algorithm is the shortest one imaginable for a nontrivial algorithm:

```
ALGORITHM  Horner(P[0..n], x)
    //Evaluates a polynomial at a given point by Horner's rule
    //Input: An array P[0..n] of coefficients of a polynomial of degree n,
    //        stored from the lowest to the highest and a number x
    //Output: The value of the polynomial at x
    p ← P[n]
    for i ← n − 1 downto 0 do
        p ← x ∗ p + P[i]
    return p
```

The number of multiplications and the number of additions are given by the same sum:

$$M(n) = A(n) = \sum_{i=0}^{n-1} 1 = n.$$

To appreciate how efficient Horner's rule is, consider only the first term of a polynomial of degree n: $a_n x^n$. Just computing this single term by the brute-force algorithm would require n multiplications, whereas Horner's rule computes, in addition to this term, $n-1$ other terms, and it still uses the same number of multiplications! It is not surprising that Horner's rule is an optimal algorithm for polynomial evaluation without preprocessing the polynomial's coefficients. But it took scientists 150 years after Horner's publication to come to the realization that such a question was worth investigating.

Horner's rule also has some useful byproducts. The intermediate numbers generated by the algorithm in the process of evaluating $p(x)$ at some point x_0 turn out to be the coefficients of the quotient of the division of $p(x)$ by $x - x_0$, and the final result, in addition to being $p(x_0)$, is equal to the remainder of this division. Thus, according to Example 1, the quotient and the remainder of the division of $2x^4 - x^3 + 3x^2 + x - 5$ by $x - 3$ are $2x^3 + 5x^2 + 18x + 55$ and 160, respectively. This division algorithm, known as ***synthetic division***, is more convenient than so-called long division.

Binary Exponentiation

The amazing efficiency of Horner's rule fades if the method is applied to computing a^n, which is the value of x^n at $x = a$. In fact, it degenerates to the brute-force multiplication of a by itself, with wasteful additions of zeros in between. Since computing a^n (actually, $a^n \bmod m$) is an essential operation in several important primality-testing and encryption methods, we consider now two algorithms for computing a^n that are based on the representation-change idea. They both exploit the binary representation of exponent n, but one of them processes this binary string left to right, whereas the second does it right to left.

Let

$$n = b_I \dots b_i \dots b_0$$

be the bit string representing a positive integer n in the binary number system. This means that the value of n can be computed as the value of the polynomial

$$p(x) = b_I x^I + \cdots + b_i x^i + \cdots + b_0 \qquad \textbf{(6.4)}$$

at $x = 2$. For example, if $n = 13$, its binary representation is 1101 and

$$13 = 1 \cdot 2^3 + 1 \cdot 2^2 + 0 \cdot 2^1 + 1 \cdot 2^0.$$

Let us now compute the value of this polynomial by applying Horner's rule and see what the method's operations imply for computing the power

$$a^n = a^{p(2)} = a^{b_I 2^I + \cdots + b_i 2^i + \cdots + b_0}.$$

Horner's rule for the binary polynomial $p(2)$	**Implications for** $a^n = a^{p(2)}$
$p \leftarrow 1$ //the leading digit is always 1 for $n \geq 1$	$a^p \leftarrow a^1$
for $i \leftarrow I - 1$ **downto** 0 **do**	**for** $i \leftarrow I - 1$ **downto** 0 **do**
$p \leftarrow 2p + b_i$	$a^p \leftarrow a^{2p+b_i}$

But

$$a^{2p+b_i} = a^{2p} \cdot a^{b_i} = (a^p)^2 \cdot a^{b_i} = \begin{cases} (a^p)^2 & \text{if } b_i = 0, \\ (a^p)^2 \cdot a & \text{if } b_i = 1. \end{cases}$$

Thus, after initializing the accumulator's value to a, we can scan the bit string representing the exponent n to always square the last value of the accumulator and, if the current binary digit is 1, also to multiply it by a. These observations lead to the following ***left-to-right binary exponentiation*** method of computing a^n.

ALGORITHM *LeftRightBinaryExponentiation*$(a, b(n))$

```
//Computes a^n by the left-to-right binary exponentiation algorithm
//Input: A number a and a list b(n) of binary digits b_I, ..., b_0
//       in the binary expansion of a positive integer n
//Output: The value of a^n
product ← a
for i ← I − 1 downto 0 do
    product ← product ∗ product
    if b_i = 1 product ← product ∗ a
return product
```

EXAMPLE 2 Compute a^{13} by the left-to-right binary exponentiation algorithm. Here, $n = 13 = 1101_2$. So we have

binary digits of n	1	1	0	1
product accumulator	a	$a^2 \cdot a = a^3$	$(a^3)^2 = a^6$	$(a^6)^2 \cdot a = a^{13}$

■

Since the algorithm makes one or two multiplications on each repetition of its only loop, the total number of multiplications $M(n)$ made by it in computing a^n is

$$(b - 1) \leq M(n) \leq 2(b - 1),$$

where b is the length of the bit string representing the exponent n. Taking into account that $b - 1 = \lfloor \log_2 n \rfloor$, we can conclude that the efficiency of the left-to-right binary exponentiation is logarithmic. Thus, this algorithm is in a better efficiency class than the brute-force exponentiation, which always requires $n - 1$ multiplications.

The ***right-to-left binary exponentiation*** uses the same binary polynomial $p(2)$ (see (6.4)) yielding the value of n. But rather than applying Horner's rule to it as the previous method did, this one exploits it differently:

$$a^n = a^{b_I 2^I + \cdots + b_i 2^i + \cdots + b_0} = a^{b_I 2^I} \cdot \cdots \cdot a^{b_i 2^i} \cdot \cdots \cdot a^{b_0}.$$

Thus, a^n can be computed as the product of the terms

$$a^{b_i 2^i} = \begin{cases} a^{2^i} & \text{if } b_i = 1, \\ 1 & \text{if } b_i = 0, \end{cases}$$

i.e., the product of consecutive terms a^{2^i}, skipping those for which the binary digit b_i is zero. In addition, we can compute a^{2^i} by simply squaring the same term we computed for the previous value of i since $a^{2^i} = (a^{2^{i-1}})^2$. So we compute all such powers of a from the smallest to the largest (from right to left), but we include in the product accumulator only those whose corresponding binary digit is 1. Here is pseudocode of this algorithm.

ALGORITHM *RightLeftBinaryExponentiation*$(a, b(n))$

```
//Computes a^n by the right-to-left binary exponentiation algorithm
//Input: A number a and a list b(n) of binary digits b_I, ..., b_0
//       in the binary expansion of a nonnegative integer n
//Output: The value of a^n
term ← a   //initializes a^(2^i)
if b_0 = 1 product ← a
else product ← 1
for i ← 1 to I do
    term ← term * term
    if b_i = 1 product ← product * term
return product
```

EXAMPLE 3 Compute a^{13} by the right-to-left binary exponentiation method.

Here, $n = 13 = 1101_2$. So we have the following table filled in from right to left:

1	1	0	1	binary digits of n
a^8	a^4	a^2	a	terms a^{2^i}
$a^5 \cdot a^8 = a^{13}$	$a \cdot a^4 = a^5$		a	product accumulator

■

Obviously, the algorithm's efficiency is also logarithmic for the same reason the left-to-right binary multiplication is. The usefulness of both binary exponentiation algorithms is reduced somewhat by their reliance on availability of the explicit binary expansion of exponent n. Problem 9 in this section's exercises asks you to design an algorithm that does not have this shortcoming.

Exercises 6.5

1. Consider the following brute-force algorithm for evaluating a polynomial.

```
ALGORITHM BruteForcePolynomialEvaluation(P[0..n], x)
    //Computes the value of polynomial P at a given point x
    //by the "highest to lowest term" brute-force algorithm
    //Input: An array P[0..n] of the coefficients of a polynomial of degree n,
    //        stored from the lowest to the highest and a number x
    //Output: The value of the polynomial at the point x
    p ← 0.0
    for i ← n downto 0 do
        power ← 1
        for j ← 1 to i do
            power ← power * x
        p ← p + P[i] * power
    return p
```

Find the total number of multiplications and the total number of additions made by this algorithm.

2. Write pseudocode for the brute-force polynomial evaluation that stems from substituting a given value of the variable into the polynomial's formula and evaluating it from the lowest term to the highest one. Determine the number of multiplications and the number of additions made by this algorithm.

3. **a.** Estimate how much faster Horner's rule is compared to the "lowest-to-highest term" brute-force algorithm of Problem 2 if (i) the time of one multiplication is significantly larger than the time of one addition; (ii) the time of one multiplication is about the same as the time of one addition.

   b. Is Horner's rule more time efficient at the expense of being less space efficient than the brute-force algorithm?

4. **a.** Apply Horner's rule to evaluate the polynomial

$$p(x) = 3x^4 - x^3 + 2x + 5 \quad \text{at } x = -2.$$

   b. Use the results of the above application of Horner's rule to find the quotient and remainder of the division of $p(x)$ by $x + 2$.

5. Apply Horner's rule to convert 110100101 from binary to decimal.

6. Compare the number of multiplications and additions/subtractions needed by the "long division" of a polynomial $p(x) = a_n x^n + a_{n-1}x^{n-1} + \cdots + a_0$ by

$x - c$, where c is some constant, with the number of these operations in the "synthetic division."

7. a. Apply the left-to-right binary exponentiation algorithm to compute a^{17}.

b. Is it possible to extend the left-to-right binary exponentiation algorithm to work for every nonnegative integer exponent?

8. Apply the right-to-left binary exponentiation algorithm to compute a^{17}.

9. Design a nonrecursive algorithm for computing a^n that mimics the right-to-left binary exponentiation but does not explicitly use the binary representation of n.

10. Is it a good idea to use a general-purpose polynomial-evaluation algorithm such as Horner's rule to evaluate the polynomial $p(x) = x^n + x^{n-1} + \cdots + x + 1$?

11. According to the corollary of the Fundamental Theorem of Algebra, every polynomial

$$p(x) = a_n x^n + a_{n-1} x^{n-1} + \cdots + a_0$$

can be represented in the form

$$p(x) = a_n (x - x_1)(x - x_2) \dots (x - x_n)$$

where $x_1, x_2, \dots, x_n$ are the roots of the polynomial (generally, complex and not necessarily distinct). Discuss which of the two representations is more convenient for each of the following operations:

a. polynomial evaluation at a given point

b. addition of two polynomials

c. multiplication of two polynomials

12. *Polynomial interpolation* Given a set of n data points (x_i, y_i) where no two x_i are the same, find a polynomial $p(x)$ of degree at most $n - 1$ such that $p(x_i) = y_i$ for every $i = 1, 2, \dots, n$.

6.6 Problem Reduction

Here is my version of a well-known joke about mathematicians. Professor X, a noted mathematician, noticed that when his wife wanted to boil water for their tea, she took their kettle from their cupboard, filled it with water, and put it on the stove. Once, when his wife was away (if you have to know, she was signing her best-seller in a local bookstore), the professor had to boil water by himself. He saw that the kettle was sitting on the kitchen counter. What did Professor X do? He put the kettle in the cupboard first and then proceeded to follow his wife's routine.

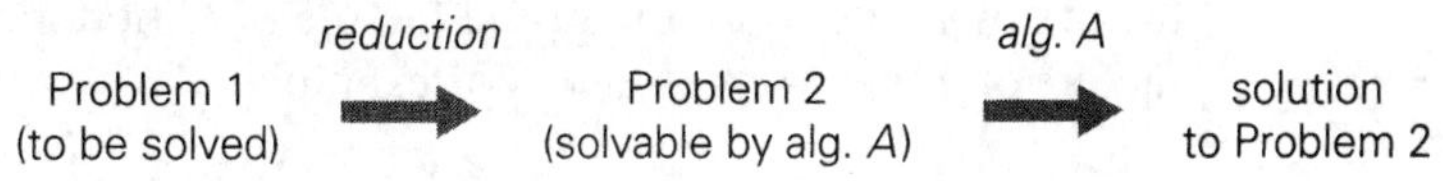

reduction *alg. A*

Problem 1 (to be solved) → Problem 2 (solvable by alg. *A*) → solution to Problem 2

FIGURE 6.15 Problem reduction strategy.

The way Professor X approached his task is an example of an important problem-solving strategy called ***problem reduction***. If you need to solve a problem, reduce it to another problem that you know how to solve (Figure 6.15).

The joke about the professor notwithstanding, the idea of problem reduction plays a central role in theoretical computer science, where it is used to classify problems according to their complexity. We will touch on this classification in Chapter 11. But the strategy can be used for actual problem solving, too. The practical difficulty in applying it lies, of course, in finding a problem to which the problem at hand should be reduced. Moreover, if we want our efforts to be of practical value, we need our reduction-based algorithm to be more efficient than solving the original problem directly.

Note that we have already encountered this technique earlier in the book. In Section 6.5, for example, we mentioned the so-called synthetic division done by applying Horner's rule for polynomial evaluation. In Section 5.5, we used the following fact from analytical geometry: if $p_1(x_1, y_1)$, $p_2(x_2, y_2)$, and $p_3(x_3, y_3)$ are three arbitrary points in the plane, then the determinant

$$\begin{vmatrix} x_1 & y_1 & 1 \\ x_2 & y_2 & 1 \\ x_3 & y_3 & 1 \end{vmatrix} = x_1y_2 + x_3y_1 + x_2y_3 - x_3y_2 - x_1y_3 - x_2y_1$$

is positive if and only if the point p_3 is to the left of the directed line $\overrightarrow{p_1p_2}$ through points p_1 and p_2. In other words, we reduced a geometric question about the relative locations of three points to a question about the sign of a determinant. In fact, the entire idea of analytical geometry is based on reducing geometric problems to algebraic ones. And the vast majority of geometric algorithms take advantage of this historic insight by René Descartes (1596–1650). In this section, we give a few more examples of algorithms based on the strategy of problem reduction.

Computing the Least Common Multiple

Recall that the ***least common multiple*** of two positive integers m and n, denoted $\text{lcm}(m, n)$, is defined as the smallest integer that is divisible by both m and n. For example, $\text{lcm}(24, 60) = 120$, and $\text{lcm}(11, 5) = 55$. The least common multiple is one of the most important notions in elementary arithmetic and algebra. Perhaps you remember the following middle-school method for computing it: Given the prime factorizations of m and n, compute the product of all the common prime factors of m and n, all the prime factors of m that are not in n, and all the prime factors of n that are not in m. For example,

$$24 = 2 \cdot 2 \cdot 2 \cdot 3,$$
$$60 = 2 \cdot 2 \cdot 3 \cdot 5,$$
$$\text{lcm}(24, 60) = (2 \cdot 2 \cdot 3) \cdot 2 \cdot 5 = 120.$$

As a computational procedure, this algorithm has the same drawbacks as the middle-school algorithm for computing the greatest common divisor discussed in Section 1.1: it is inefficient and requires a list of consecutive primes.

A much more efficient algorithm for computing the least common multiple can be devised by using problem reduction. After all, there is a very efficient algorithm (Euclid's algorithm) for finding the greatest common divisor, which is a product of all the common prime factors of m and n. Can we find a formula relating $\text{lcm}(m, n)$ and $\text{gcd}(m, n)$? It is not difficult to see that the product of $\text{lcm}(m, n)$ and $\text{gcd}(m, n)$ includes every factor of m and n exactly once and hence is simply equal to the product of m and n. This observation leads to the formula

$$\text{lcm}(m, n) = \frac{m \cdot n}{\text{gcd}(m, n)},$$

where $\text{gcd}(m, n)$ can be computed very efficiently by Euclid's algorithm.

Counting Paths in a Graph

As our next example, we consider the problem of counting paths between two vertices in a graph. It is not difficult to prove by mathematical induction that the number of different paths of length $k > 0$ from the ith vertex to the jth vertex of a graph (undirected or directed) equals the (i, j)th element of A^k where A is the adjacency matrix of the graph. Therefore, the problem of counting a graph's paths can be solved with an algorithm for computing an appropriate power of its adjacency matrix. Note that the exponentiation algorithms we discussed before for computing powers of numbers are applicable to matrices as well.

As a specific example, consider the graph of Figure 6.16. Its adjacency matrix A and its square A^2 indicate the numbers of paths of length 1 and 2, respectively, between the corresponding vertices of the graph. In particular, there are three

$$A = \begin{array}{c} \\ a \\ b \\ c \\ d \end{array} \begin{array}{c} \begin{array}{cccc} a & b & c & d \end{array} \\ \begin{bmatrix} 0 & 1 & 1 & 1 \\ 1 & 0 & 0 & 0 \\ 1 & 0 & 0 & 1 \\ 1 & 0 & 1 & 0 \end{bmatrix} \end{array} \qquad A^2 = \begin{array}{c} \\ a \\ b \\ c \\ d \end{array} \begin{array}{c} \begin{array}{cccc} a & b & c & d \end{array} \\ \begin{bmatrix} 3 & 0 & 1 & 1 \\ 0 & 1 & 1 & 1 \\ 1 & 1 & 2 & 1 \\ 1 & 1 & 1 & 2 \end{bmatrix} \end{array}$$

FIGURE 6.16 A graph, its adjacency matrix A, and its square A^2. The elements of A and A^2 indicate the numbers of paths of lengths 1 and 2, respectively.

paths of length 2 that start and end at vertex a ($a-b-a$, $a-c-a$, and $a-d-a$); but there is only one path of length 2 from a to c ($a-d-c$).

Reduction of Optimization Problems

Our next example deals with solving optimization problems. If a problem asks to find a maximum of some function, it is said to be a ***maximization problem***; if it asks to find a function's minimum, it is called a ***minimization problem***. Suppose now that you need to find a minimum of some function $f(x)$ and you have an algorithm for function maximization. How can you take advantage of the latter? The answer lies in the simple formula

$$\min f(x) = -\max[-f(x)].$$

In other words, to minimize a function, we can maximize its negative instead and, to get a correct minimal value of the function itself, change the sign of the answer. This property is illustrated for a function of one real variable in Figure 6.17.

Of course, the formula

$$\max f(x) = -\min[-f(x)]$$

is valid as well; it shows how a maximization problem can be reduced to an equivalent minimization problem.

This relationship between minimization and maximization problems is very general: it holds for functions defined on any domain D. In particular, we can

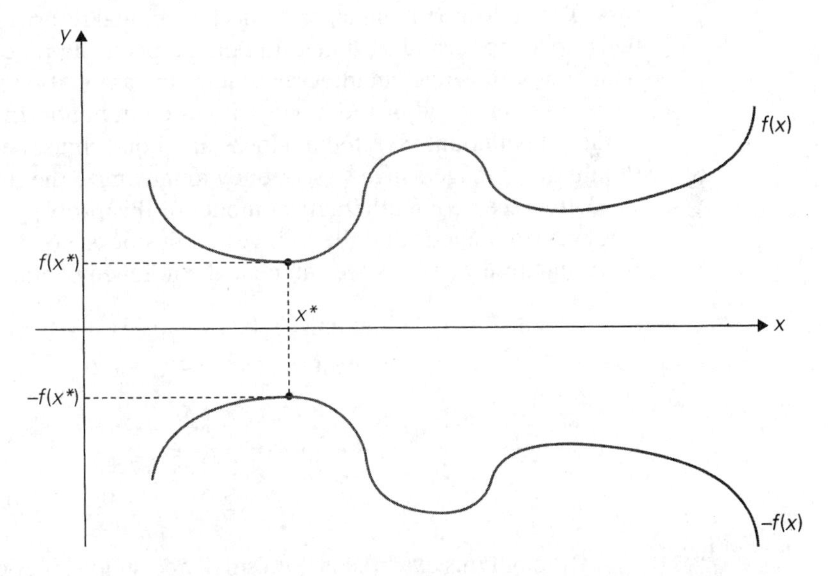

FIGURE 6.17 Relationship between minimization and maximization problems: $\min f(x) = -\max[-f(x)]$.

apply it to functions of several variables subject to additional constraints. A very important class of such problems is introduced below in this section.

Now that we are on the topic of function optimization, it is worth pointing out that the standard calculus procedure for finding extremum points of a function is, in fact, also based on problem reduction. Indeed, it suggests finding the function's derivative $f'(x)$ and then solving the equation $f'(x) = 0$ to find the function's critical points. In other words, the optimization problem is reduced to the problem of solving an equation as the principal part of finding extremum points. Note that we are not calling the calculus procedure an algorithm, since it is not clearly defined. In fact, there is no general method for solving equations. A little secret of calculus textbooks is that problems are carefully selected so that critical points can always be found without difficulty. This makes the lives of both students and instructors easier but, in the process, may unintentionally create a wrong impression in students' minds.

Linear Programming

Many problems of optimal decision making can be reduced to an instance of the ***linear programming*** problem—a problem of optimizing a linear function of several variables subject to constraints in the form of linear equations and linear inequalities.

EXAMPLE 1 Consider a university endowment that needs to invest $100 million. This sum has to be split between three types of investments: stocks, bonds, and cash. The endowment managers expect an annual return of 10%, 7%, and 3% for their stock, bond, and cash investments, respectively. Since stocks are more risky than bonds, the endowment rules require the amount invested in stocks to be no more than one-third of the moneys invested in bonds. In addition, at least 25% of the total amount invested in stocks and bonds must be invested in cash. How should the managers invest the money to maximize the return?

Let us create a mathematical model of this problem. Let x, y, and z be the amounts (in millions of dollars) invested in stocks, bonds, and cash, respectively. By using these variables, we can pose the following optimization problem:

$$\begin{aligned}
\text{maximize} \quad & 0.10x + 0.07y + 0.03z \\
\text{subject to} \quad & x + y + z = 100 \\
& x \le \tfrac{1}{3}y \\
& z \ge 0.25(x + y) \\
& x \ge 0, \quad y \ge 0, \quad z \ge 0.
\end{aligned}$$

■

Although this example is both small and simple, it does show how a problem of optimal decision making can be reduced to an instance of the general linear programming problem

$$\begin{aligned} \text{maximize (or minimize)} \quad & c_1x_1 + \cdots + c_nx_n \\ \text{subject to} \quad & a_{i1}x_1 + \cdots + a_{in}x_n \le \text{(or} \ge \text{or} =\text{)} \; b_i \quad \text{for } i = 1, \ldots, m \\ & x_1 \ge 0, \ldots, x_n \ge 0. \end{aligned}$$

(The last group of constraints—called the nonnegativity constraints—are, strictly speaking, unnecessary because they are special cases of more general constraints $a_{i1}x_1 + \cdots + a_{in}x_n \ge b_i$, but it is convenient to treat them separately.)

Linear programming has proved to be flexible enough to model a wide variety of important applications, such as airline crew scheduling, transportation and communication network planning, oil exploration and refining, and industrial production optimization. In fact, linear programming is considered by many as one of the most important achievements in the history of applied mathematics.

The classic algorithm for this problem is called the ***simplex method*** (Section 10.1). It was discovered by the U.S. mathematician George Dantzig in the 1940s [Dan63]. Although the worst-case efficiency of this algorithm is known to be exponential, it performs very well on typical inputs. Moreover, a more recent algorithm by Narendra Karmarkar [Kar84] not only has a proven polynomial worst-case efficiency but has also performed competitively with the simplex method in empirical tests.

It is important to stress, however, that the simplex method and Karmarkar's algorithm can successfully handle only linear programming problems that do not limit its variables to integer values. When variables of a linear programming problem are required to be integers, the linear programming problem is said to be an ***integer linear programming*** problem. Except for some special cases (e.g., the assignment problem and the problems discussed in Sections 10.2–10.4), integer linear programming problems are much more difficult. There is no known polynomial-time algorithm for solving an arbitrary instance of the general integer linear programming problem and, as we see in Chapter 11, such an algorithm quite possibly does not exist. Other approaches such as the branch-and-bound technique discussed in Section 12.2 are typically used for solving integer linear programming problems.

EXAMPLE 2 Let us see how the knapsack problem can be reduced to a linear programming problem. Recall from Section 3.4 that the knapsack problem can be posed as follows. Given a knapsack of capacity W and n items of weights $w_1, \ldots, w_n$ and values $v_1, \ldots, v_n$, find the most valuable subset of the items that fits into the knapsack. We consider first the ***continuous*** (or ***fractional***) version of the problem, in which any fraction of any item given can be taken into the knapsack. Let x_j, $j = 1, \ldots, n$, be a variable representing a fraction of item j taken into the knapsack. Obviously, x_j must satisfy the inequality $0 \le x_j \le 1$. Then the total weight of the selected items can be expressed by the sum $\sum_{j=1}^{n} w_jx_j$, and their total value by the sum $\sum_{j=1}^{n} v_jx_j$. Thus, the continuous version of the knapsack problem can be posed as the following linear programming problem:

$$\text{maximize} \quad \sum_{j=1}^{n} v_j x_j$$

$$\text{subject to} \quad \sum_{j=1}^{n} w_j x_j \le W$$

$$0 \le x_j \le 1 \quad \text{for } j = 1, \ldots, n.$$

There is no need to apply a general method for solving linear programming problems here: this particular problem can be solved by a simple special algorithm that is introduced in Section 12.3. (But why wait? Try to discover it on your own now.) This reduction of the knapsack problem to an instance of the linear programming problem is still useful, though, to prove the correctness of the algorithm in question.

In the ***discrete*** (or **0-1**) version of the knapsack problem, we are only allowed either to take a whole item or not to take it at all. Hence, we have the following integer linear programming problem for this version:

$$\text{maximize} \quad \sum_{j=1}^{n} v_j x_j$$

$$\text{subject to} \quad \sum_{j=1}^{n} w_j x_j \le W$$

$$x_j \in \{0, 1\} \quad \text{for } j = 1, \ldots, n.$$

This seemingly minor modification makes a drastic difference for the complexity of this and similar problems constrained to take only discrete values in their potential ranges. Despite the fact that the 0-1 version might seem to be easier because it can ignore any subset of the continuous version that has a fractional value of an item, the 0-1 version is, in fact, much more complicated than its continuous counterpart. The reader interested in specific algorithms for solving this problem will find a wealth of literature on the subject, including the monographs [Mar90] and [Kel04]. ■

Reduction to Graph Problems

As we pointed out in Section 1.3, many problems can be solved by a reduction to one of the standard graph problems. This is true, in particular, for a variety of puzzles and games. In these applications, vertices of a graph typically represent possible states of the problem in question, and edges indicate permitted transitions among such states. One of the graph's vertices represents an initial state and another represents a goal state of the problem. (There might be several vertices of the latter kind.) Such a graph is called a ***state-space graph***. Thus, the transformation just described reduces the problem to the question about a path from the initial-state vertex to a goal-state vertex.

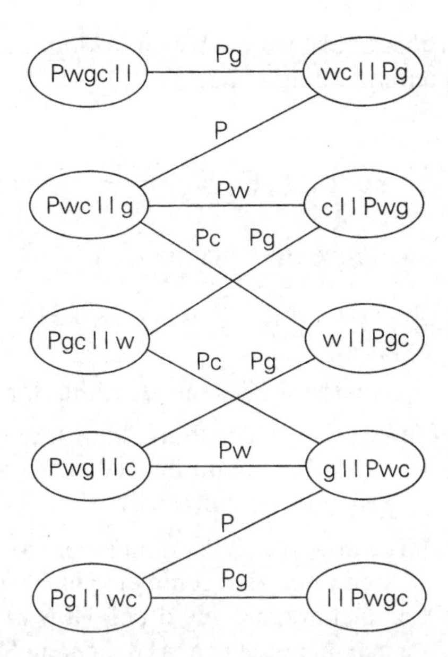

FIGURE 6.18 State-space graph for the peasant, wolf, goat, and cabbage puzzle.

EXAMPLE Let us revisit the classic river-crossing puzzle that was included in the exercises for Section 1.2. A peasant finds himself on a river bank with a wolf, a goat, and a head of cabbage. He needs to transport all three to the other side of the river in his boat. However, the boat has room only for the peasant himself and one other item (either the wolf, the goat, or the cabbage). In his absence, the wolf would eat the goat, and the goat would eat the cabbage. Find a way for the peasant to solve his problem or prove that it has no solution.

The state-space graph for this problem is given in Figure 6.18. Its vertices are labeled to indicate the states they represent: P, w, g, c stand for the peasant, the wolf, the goat, and the cabbage, respectively; the two bars | | denote the river; for convenience, we also label the edges by indicating the boat's occupants for each crossing. In terms of this graph, we are interested in finding a path from the initial-state vertex labeled Pwgc| | to the final-state vertex labeled | |Pwgc.

It is easy to see that there exist two distinct simple paths from the initial-state vertex to the final state vertex (what are they?). If we find them by applying breadth-first search, we get a formal proof that these paths have the smallest number of edges possible. Hence, this puzzle has two solutions requiring seven river crossings, which is the minimum number of crossings needed. ■

Our success in solving this simple puzzle should not lead you to believe that generating and investigating state-space graphs is always a straightforward task. To get a better appreciation of them, consult books on artificial intelligence (AI), the branch of computer science in which state-space graphs are a principal subject.

In this book, we deal with an important special case of state-space graphs in Sections 12.1 and 12.2.

Exercises 6.6

1. **a.** Prove the equality

$$\text{lcm}(m, n) = \frac{m \cdot n}{\gcd(m, n)}$$

that underlies the algorithm for computing $\text{lcm}(m, n)$.

b. Euclid's algorithm is known to be in $O(\log n)$. If it is the algorithm that is used for computing $\gcd(m, n)$, what is the efficiency of the algorithm for computing $\text{lcm}(m, n)$?

2. You are given a list of numbers for which you need to construct a min-heap. (A min-heap is a complete binary tree in which every key is less than or equal to the keys in its children.) How would you use an algorithm for constructing a max-heap (a heap as defined in Section 6.4) to construct a min-heap?

3. Prove that the number of different paths of length $k > 0$ from the ith vertex to the jth vertex in a graph (undirected or directed) equals the (i, j)th element of A^k where A is the adjacency matrix of the graph.

4. **a.** Design an algorithm with a time efficiency better than cubic for checking whether a graph with n vertices contains a cycle of length 3 [Man89].

b. Consider the following algorithm for the same problem. Starting at an arbitrary vertex, traverse the graph by depth-first search and check whether its depth-first search forest has a vertex with a back edge leading to its grandparent. If it does, the graph contains a triangle; if it does not, the graph does not contain a triangle as its subgraph. Is this algorithm correct?

5. Given $n > 3$ points $P_1 = (x_1, y_1), \ldots, P_n = (x_n, y_n)$ in the coordinate plane, design an algorithm to check whether all the points lie within a triangle with its vertices at three of the points given. (You can either design an algorithm from scratch or reduce the problem to another one with a known algorithm.)

6. Consider the problem of finding, for a given positive integer n, the pair of integers whose sum is n and whose product is as large as possible. Design an efficient algorithm for this problem and indicate its efficiency class.

7. The assignment problem introduced in Section 3.4 can be stated as follows: There are n people who need to be assigned to execute n jobs, one person per job. (That is, each person is assigned to exactly one job and each job is assigned to exactly one person.) The cost that would accrue if the ith person is assigned to the jth job is a known quantity $C[i, j]$ for each pair $i, j = 1, \ldots, n$. The problem is to assign the people to the jobs to minimize the total cost of

the assignment. Express the assignment problem as a 0-1 linear programming problem.

8. Solve the instance of the linear programming problem given in Section 6.6:

$$\begin{aligned} \text{maximize} \quad & 0.10x + 0.07y + 0.03z \\ \text{subject to} \quad & x + y + z = 100 \\ & x \le \tfrac{1}{3}y \\ & z \ge 0.25(x + y) \\ & x \ge 0, \quad y \ge 0, \quad z \ge 0. \end{aligned}$$

9. The graph-coloring problem is usually stated as the vertex-coloring problem: Assign the smallest number of colors to vertices of a given graph so that no two adjacent vertices are the same color. Consider the ***edge-coloring*** problem: Assign the smallest number of colors possible to edges of a given graph so that no two edges with the same endpoint are the same color. Explain how the edge-coloring problem can be reduced to a vertex-coloring problem.

10. Consider the two-dimensional ***post office location problem***: given n points $(x_1, y_1), \ldots, (x_n, y_n)$ in the Cartesian plane, find a location (x, y) for a post office that minimizes $\frac{1}{n}\sum_{i=1}^{n}(|x_i - x| + |y_i - y|)$, the average Manhattan distance from the post office to these points. Explain how this problem can be efficiently solved by the problem reduction technique, provided the post office does not have to be located at one of the input points.

11. *Jealous husbands* There are $n \ge 2$ married couples who need to cross a river. They have a boat that can hold no more than two people at a time. To complicate matters, all the husbands are jealous and will not agree on any crossing procedure that would put a wife on the same bank of the river with another woman's husband without the wife's husband being there too, even if there are other people on the same bank. Can they cross the river under such constraints?

a. Solve the problem for $n = 2$.

b. Solve the problem for $n = 3$, which is the classical version of this problem.

c. Does the problem have a solution for $n \ge 4$? If it does, indicate how many river crossings it will take; if it does not, explain why.

12. *Double-n dominoes* Dominoes are small rectangular tiles with dots called spots or pips embossed at both halves of the tiles. A standard "double-six" domino set has 28 tiles: one for each unordered pair of integers from (0, 0) to (6, 6). In general, a "double-n" domino set would consist of domino tiles for each unordered pair of integers from (0, 0) to (n, n). Determine all values of n for which one constructs a ring made up of all the tiles in a double-n domino set.

SUMMARY

- *Transform-and-conquer* is the fourth general algorithm design (and problem-solving) strategy discussed in the book. It is, in fact, a group of techniques based on the idea of transformation to a problem that is easier to solve.
- There are three principal varieties of the transform-and-conquer strategy: *instance simplification*, *representation change*, and *problem reduction*.
- *Instance simplification* is transforming an instance of a problem to an instance of the same problem with some special property that makes the problem easier to solve. List presorting, Gaussian elimination, and rotations in AVL trees are good examples of this strategy.
- *Representation change* implies changing one representation of a problem's instance to another representation of the same instance. Examples discussed in this chapter include representation of a set by a 2-3 tree, heaps and heapsort, Horner's rule for polynomial evaluation, and two binary exponentiation algorithms.
- *Problem reduction* calls for transforming a given problem to another problem that can be solved by a known algorithm. Among examples of applying this idea to algorithmic problem solving (see Section 6.6), reductions to linear programming and reductions to graph problems are especially important.
- Some examples used to illustrate transform-and-conquer happen to be very important data structures and algorithms. They are: heaps and heapsort, AVL and 2-3 trees, Gaussian elimination, and Horner's rule.
- A *heap* is an essentially complete binary tree with keys (one per node) satisfying the parental dominance requirement. Though defined as binary trees, heaps are normally implemented as arrays. Heaps are most important for the efficient implementation of priority queues; they also underlie heapsort.
- *Heapsort* is a theoretically important sorting algorithm based on arranging elements of an array in a heap and then successively removing the largest element from a remaining heap. The algorithm's running time is in $\Theta(n \log n)$ both in the worst case and in the average case; in addition, it is in-place.
- *AVL trees* are binary search trees that are always balanced to the extent possible for a binary tree. The balance is maintained by transformations of four types called *rotations*. All basic operations on AVL trees are in $O(\log n)$; it eliminates the bad worst-case efficiency of classic binary search trees.
- *2-3 trees* achieve a perfect balance in a search tree by allowing a node to contain up to two ordered keys and have up to three children. This idea can be generalized to yield very important *B*-trees, discussed later in the book.

- *Gaussian elimination*—an algorithm for solving systems of linear equations—is a principal algorithm in linear algebra. It solves a system by transforming it to an equivalent system with an upper-triangular coefficient matrix, which is easy to solve by back substitutions. Gaussian elimination requires about $\frac{1}{3}n^3$ multiplications.

- *Horner's rule* is an optimal algorithm for polynomial evaluation without coefficient preprocessing. It requires only n multiplications and n additions to evaluate an n-degree polynomial at a given point. Horner's rule also has a few useful byproducts, such as the synthetic division algorithm.

- Two *binary exponentiation* algorithms for computing a^n are introduced in Section 6.5. Both of them exploit the binary representation of the exponent n, but they process it in the opposite directions: left to right and right to left.

- *Linear programming* concerns optimizing a linear function of several variables subject to constraints in the form of linear equations and linear inequalities. There are efficient algorithms capable of solving very large instances of this problem with many thousands of variables and constraints, provided the variables are not required to be integers. The latter, called *integer linear programming*, constitute a much more difficult class of problems.

7

Space and Time Trade-Offs

Things which matter most must never be at the mercy of things which matter less.

—Johann Wolfgang von Göethe (1749–1832)

Space and time trade-offs in algorithm design are a well-known issue for both theoreticians and practitioners of computing. Consider, as an example, the problem of computing values of a function at many points in its domain. If it is time that is at a premium, we can precompute the function's values and store them in a table. This is exactly what human computers had to do before the advent of electronic computers, in the process burdening libraries with thick volumes of mathematical tables. Though such tables have lost much of their appeal with the widespread use of electronic computers, the underlying idea has proven to be quite useful in the development of several important algorithms for other problems. In somewhat more general terms, the idea is to preprocess the problem's input, in whole or in part, and store the additional information obtained to accelerate solving the problem afterward. We call this approach ***input enhancement***[1] and discuss the following algorithms based on it:

- counting methods for sorting (Section 7.1)
- Boyer-Moore algorithm for string matching and its simplified version suggested by Horspool (Section 7.2)

The other type of technique that exploits space-for-time trade-offs simply uses extra space to facilitate faster and/or more flexible access to the data. We call this approach ***prestructuring***. This name highlights two facets of this variation of the space-for-time trade-off: some processing is done before a problem in question

---

1. The standard terms used synonymously for this technique are ***preprocessing*** and ***preconditioning***. Confusingly, these terms can also be applied to methods that use the idea of preprocessing but do not use extra space (see Chapter 6). Thus, in order to avoid confusion, we use "input enhancement" as a special name for the space-for-time trade-off technique being discussed here.

is actually solved but, unlike the input-enhancement variety, it deals with access structuring. We illustrate this approach by:

- hashing (Section 7.3)
- indexing with B-trees (Section 7.4)

There is one more algorithm design technique related to the space-for-time trade-off idea: ***dynamic programming***. This strategy is based on recording solutions to overlapping subproblems of a given problem in a table from which a solution to the problem in question is then obtained. We discuss this well-developed technique separately, in the next chapter of the book.

Two final comments about the interplay between time and space in algorithm design need to be made. First, the two resources—time and space—do not have to compete with each other in all design situations. In fact, they can align to bring an algorithmic solution that minimizes both the running time and the space consumed. Such a situation arises, in particular, when an algorithm uses a space-efficient data structure to represent a problem's input, which leads, in turn, to a faster algorithm. Consider, as an example, the problem of traversing graphs. Recall that the time efficiency of the two principal traversal algorithms—depth-first search and breadth-first search—depends on the data structure used for representing graphs: it is $\Theta(n^2)$ for the adjacency matrix representation and $\Theta(n+m)$ for the adjacency list representation, where n and m are the numbers of vertices and edges, respectively. If input graphs are sparse, i.e., have few edges relative to the number of vertices (say, $m \in O(n)$), the adjacency list representation may well be more efficient from both the space and the running-time points of view. The same situation arises in the manipulation of sparse matrices and sparse polynomials: if the percentage of zeros in such objects is sufficiently high, we can save both space and time by ignoring zeros in the objects' representation and processing.

Second, one cannot discuss space-time trade-offs without mentioning the hugely important area of data compression. Note, however, that in data compression, size reduction is the goal rather than a technique for solving another problem. We discuss just one data compression algorithm, in the next chapter. The reader interested in this topic will find a wealth of algorithms in such books as [Say05].

7.1 Sorting by Counting

As a first example of applying the input-enhancement technique, we discuss its application to the sorting problem. One rather obvious idea is to count, for each element of a list to be sorted, the total number of elements smaller than this element and record the results in a table. These numbers will indicate the positions of the elements in the sorted list: e.g., if the count is 10 for some element, it should be in the 11th position (with index 10, if we start counting with 0) in the sorted array. Thus, we will be able to sort the list by simply copying its elements to their appropriate positions in a new, sorted list. This algorithm is called ***comparison-counting sort*** (Figure 7.1).

Array $A[0..5]$		62	31	84	96	19	47
Initially	*Count* []	0	0	0	0	0	0
After pass $i = 0$	*Count* []	3	0	1	1	0	0
After pass $i = 1$	*Count* []		1	2	2	0	1
After pass $i = 2$	*Count* []			4	3	0	1
After pass $i = 3$	*Count* []				5	0	1
After pass $i = 4$	*Count* []					0	2
Final state	*Count* []	3	1	4	5	0	2
Array $S[0..5]$		19	31	47	62	84	96

FIGURE 7.1 Example of sorting by comparison counting.

ALGORITHM *ComparisonCountingSort*($A[0..n-1]$)
    //Sorts an array by comparison counting
    //Input: An array $A[0..n-1]$ of orderable elements
    //Output: Array $S[0..n-1]$ of A's elements sorted in nondecreasing order
    for $i \leftarrow 0$ **to** $n-1$ **do** $Count[i] \leftarrow 0$
    for $i \leftarrow 0$ **to** $n-2$ **do**
        for $j \leftarrow i+1$ **to** $n-1$ **do**
            if $A[i] < A[j]$
                $Count[j] \leftarrow Count[j]+1$
            else $Count[i] \leftarrow Count[i]+1$
    for $i \leftarrow 0$ **to** $n-1$ **do** $S[Count[i]] \leftarrow A[i]$
    return S

What is the time efficiency of this algorithm? It should be quadratic because the algorithm considers all the different pairs of an n-element array. More formally, the number of times its basic operation, the comparison $A[i] < A[j]$, is executed is equal to the sum we have encountered several times already:

$$C(n) = \sum_{i=0}^{n-2} \sum_{j=i+1}^{n-1} 1 = \sum_{i=0}^{n-2} [(n-1) - (i+1) + 1] = \sum_{i=0}^{n-2} (n-1-i) = \frac{n(n-1)}{2}.$$

Thus, the algorithm makes the same number of key comparisons as selection sort and in addition uses a linear amount of extra space. On the positive side, the algorithm makes the minimum number of key moves possible, placing each of them directly in their final position in a sorted array.

The counting idea does work productively in a situation in which elements to be sorted belong to a known small set of values. Assume, for example, that we have to sort a list whose values can be either 1 or 2. Rather than applying a general sorting algorithm, we should be able to take advantage of this additional information about values to be sorted. Indeed, we can scan the list to compute the number of 1's and the number of 2's in it and then, on the second pass, simply make the appropriate number of the first elements equal to 1 and the remaining elements equal to 2. More generally, if element values are integers between some lower bound l and upper bound u, we can compute the frequency of each of those values and store them in array $F[0..u-l]$. Then the first $F[0]$ positions in the sorted list must be filled with l, the next $F[1]$ positions with $l+1$, and so on. All this can be done, of course, only if we can overwrite the given elements.

Let us consider a more realistic situation of sorting a list of items with some other information associated with their keys so that we cannot overwrite the list's elements. Then we can copy elements into a new array $S[0..n-1]$ to hold the sorted list as follows. The elements of A whose values are equal to the lowest possible value l are copied into the first $F[0]$ elements of S, i.e., positions 0 through $F[0]-1$; the elements of value $l+1$ are copied to positions from $F[0]$ to $(F[0]+F[1])-1$; and so on. Since such accumulated sums of frequencies are called a distribution in statistics, the method itself is known as ***distribution counting***.

EXAMPLE Consider sorting the array

13	11	12	13	12	12

whose values are known to come from the set {11, 12, 13} and should not be overwritten in the process of sorting. The frequency and distribution arrays are as follows:

Array values	11	12	13
Frequencies	1	3	2
Distribution values	1	4	6

Note that the distribution values indicate the proper positions for the last occurrences of their elements in the final sorted array. If we index array positions from 0 to $n-1$, the distribution values must be reduced by 1 to get corresponding element positions.

It is more convenient to process the input array right to left. For the example, the last element is 12, and, since its distribution value is 4, we place this 12 in position $4-1=3$ of the array S that will hold the sorted list. Then we decrease the 12's distribution value by 1 and proceed to the next (from the right) element in the given array. The entire processing of this example is depicted in Figure 7.2.

	$D[0..2]$			$S[0..5]$					
$A[5] = 12$	1	**4**	6				12		
$A[4] = 12$	1	**3**	6			12			
$A[3] = 13$	1	2	**6**						13
$A[2] = 12$	1	**2**	5		12				
$A[1] = 11$	**1**	1	5	11					
$A[0] = 13$	0	1	**5**					13	

FIGURE 7.2 Example of sorting by distribution counting. The distribution values being decremented are shown in bold. ■

Here is pseudocode of this algorithm.

ALGORITHM *DistributionCountingSort*($A[0..n-1], l, u$)

```
//Sorts an array of integers from a limited range by distribution counting
//Input: An array A[0..n − 1] of integers between l and u (l ≤ u)
//Output: Array S[0..n − 1] of A's elements sorted in nondecreasing order
for j ← 0 to u − l do D[j] ← 0                        //initialize frequencies
for i ← 0 to n − 1 do D[A[i] − l] ← D[A[i] − l] + 1 //compute frequencies
for j ← 1 to u − l do D[j] ← D[j − 1] + D[j]         //reuse for distribution
for i ← n − 1 downto 0 do
    j ← A[i] − l
    S[D[j] − 1] ← A[i]
    D[j] ← D[j] − 1
return S
```

Assuming that the range of array values is fixed, this is obviously a linear algorithm because it makes just two consecutive passes through its input array A. This is a better time-efficiency class than that of the most efficient sorting algorithms—mergesort, quicksort, and heapsort—we have encountered. It is important to remember, however, that this efficiency is obtained by exploiting the specific nature of inputs for which sorting by distribution counting works, in addition to trading space for time.

Exercises 7.1

1. Is it possible to exchange numeric values of two variables, say, u and v, without using any extra storage?
2. Will the comparison-counting algorithm work correctly for arrays with equal values?
3. Assuming that the set of possible list values is $\{a, b, c, d\}$, sort the following list in alphabetical order by the distribution-counting algorithm:

$$b,\ c,\ d,\ c,\ b,\ a,\ a,\ b.$$

4. Is the distribution-counting algorithm stable?
5. Design a one-line algorithm for sorting any array of size n whose values are n distinct integers from 1 to n.
6. The ***ancestry problem*** asks to determine whether a vertex u is an ancestor of vertex v in a given binary (or, more generally, rooted ordered) tree of n vertices. Design a $O(n)$ input-enhancement algorithm that provides sufficient information to solve this problem for any pair of the tree's vertices in constant time.
7. The following technique, known as ***virtual initialization***, provides a time-efficient way to initialize just some elements of a given array $A[0..n-1]$ so that for each of its elements, we can say in constant time whether it has been initialized and, if it has been, with which value. This is done by utilizing a variable *counter* for the number of initialized elements in A and two auxiliary arrays of the same size, say $B[0..n-1]$ and $C[0..n-1]$, defined as follows. $B[0], \ldots, B[counter-1]$ contain the indices of the elements of A that were initialized: $B[0]$ contains the index of the element initialized first, $B[1]$ contains the index of the element initialized second, etc. Furthermore, if $A[i]$ was the kth element $(0 \le k \le counter-1)$ to be initialized, $C[i]$ contains k.
   a. Sketch the state of arrays $A[0..7]$, $B[0..7]$, and $C[0..7]$ after the three assignments

   $$A[3] \leftarrow x; \quad A[7] \leftarrow z; \quad A[1] \leftarrow y.$$

   b. In general, how can we check with this scheme whether $A[i]$ has been initialized and, if it has been, with which value?

8. *Least distance sorting* There are 10 Egyptian stone statues standing in a row in an art gallery hall. A new curator wants to move them so that the statues are ordered by their height. How should this be done to minimize the total distance that the statues are moved? You may assume for simplicity that all the statues have different heights. [Azi10]
9. **a.** Write a program for multiplying two sparse matrices, a $p \times q$ matrix A and a $q \times r$ matrix B.
   b. Write a program for multiplying two sparse polynomials $p(x)$ and $q(x)$ of degrees m and n, respectively.
10. Is it a good idea to write a program that plays the classic game of tic-tac-toe with the human user by storing all possible positions on the game's 3×3 board along with the best move for each of them?

7.2 Input Enhancement in String Matching

In this section, we see how the technique of input enhancement can be applied to the problem of string matching. Recall that the problem of string matching requires finding an occurrence of a given string of m characters called the ***pattern*** in a longer string of n characters called the ***text***. We discussed the brute-force algorithm for this problem in Section 3.2: it simply matches corresponding pairs of characters in the pattern and the text left to right and, if a mismatch occurs, shifts the pattern one position to the right for the next trial. Since the maximum number of such trials is $n - m + 1$ and, in the worst case, m comparisons need to be made on each of them, the worst-case efficiency of the brute-force algorithm is in the $O(nm)$ class. On average, however, we should expect just a few comparisons before a pattern's shift, and for random natural-language texts, the average-case efficiency indeed turns out to be in $O(n + m)$.

Several faster algorithms have been discovered. Most of them exploit the input-enhancement idea: preprocess the pattern to get some information about it, store this information in a table, and then use this information during an actual search for the pattern in a given text. This is exactly the idea behind the two best-known algorithms of this type: the Knuth-Morris-Pratt algorithm [Knu77] and the Boyer-Moore algorithm [Boy77].

The principal difference between these two algorithms lies in the way they compare characters of a pattern with their counterparts in a text: the Knuth-Morris-Pratt algorithm does it left to right, whereas the Boyer-Moore algorithm does it right to left. Since the latter idea leads to simpler algorithms, it is the only one that we will pursue here. (Note that the Boyer-Moore algorithm starts by aligning the pattern against the beginning characters of the text; if the first trial fails, it shifts the pattern to the right. It is comparisons within a trial that the algorithm does right to left, starting with the last character in the pattern.)

Although the underlying idea of the Boyer-Moore algorithm is simple, its actual implementation in a working method is less so. Therefore, we start our discussion with a simplified version of the Boyer-Moore algorithm suggested by R. Horspool [Hor80]. In addition to being simpler, Horspool's algorithm is not necessarily less efficient than the Boyer-Moore algorithm on random strings.

Horspool's Algorithm

Consider, as an example, searching for the pattern BARBER in some text:

$$s_0 \quad \dots \qquad\qquad c \quad \dots \quad s_{n-1}$$
$$\texttt{B A R B E R}$$

Starting with the last R of the pattern and moving right to left, we compare the corresponding pairs of characters in the pattern and the text. If all the pattern's characters match successfully, a matching substring is found. Then the search can be either stopped altogether or continued if another occurrence of the same pattern is desired.

If a mismatch occurs, we need to shift the pattern to the right. Clearly, we would like to make as large a shift as possible without risking the possibility of missing a matching substring in the text. Horspool's algorithm determines the size of such a shift by looking at the character c of the text that is aligned against the last character of the pattern. This is the case even if character c itself matches its counterpart in the pattern.

In general, the following four possibilities can occur.

Case 1 If there are no c's in the pattern—e.g., c is letter S in our example—we can safely shift the pattern by its entire length (if we shift less, some character of the pattern would be aligned against the text's character c that is known not to be in the pattern):

```
s0  ...          S                ...  sn-1
                 ≠
       B A R B E R
                    B A R B E R
```

Case 2 If there are occurrences of character c in the pattern but it is not the last one there—e.g., c is letter B in our example—the shift should align the rightmost occurrence of c in the pattern with the c in the text:

```
s0  ...          B            ...  sn-1
                 ≠
       B A R B E R
             B A R B E R
```

Case 3 If c happens to be the last character in the pattern but there are no c's among its other $m - 1$ characters—e.g., c is letter R in our example—the situation is similar to that of Case 1 and the pattern should be shifted by the entire pattern's length m:

```
s0  ...        M E R              ...  sn-1
               ≠ = =
       L E A D E R
                    L E A D E R
```

Case 4 Finally, if c happens to be the last character in the pattern and there are other c's among its first $m - 1$ characters—e.g., c is letter R in our example—the situation is similar to that of Case 2 and the rightmost occurrence of c among the first $m - 1$ characters in the pattern should be aligned with the text's c:

```
s0  ...          A R          ...  sn-1
                 ≠ =
       R E O R D E R
             R E O R D E R
```

These examples clearly demonstrate that right-to-left character comparisons can lead to farther shifts of the pattern than the shifts by only one position always made by the brute-force algorithm. However, if such an algorithm had to check all the characters of the pattern on every trial, it would lose much of this superiority. Fortunately, the idea of input enhancement makes repetitive comparisons unnecessary. We can precompute shift sizes and store them in a table. The table will be indexed by all possible characters that can be encountered in a text, including, for natural language texts, the space, punctuation symbols, and other special characters. (Note that no other information about the text in which eventual searching will be done is required.) The table's entries will indicate the shift sizes computed by the formula

$$t(c) = \begin{cases} \text{the pattern's length } m, \\ \text{if } c \text{ is not among the first } m-1 \text{ characters of the pattern;} \\ \\ \text{the distance from the rightmost } c \text{ among the first } m-1 \text{ characters} \\ \text{of the pattern to its last character, otherwise.} \end{cases} \quad \textbf{(7.1)}$$

For example, for the pattern BARBER, all the table's entries will be equal to 6, except for the entries for E, B, R, and A, which will be 1, 2, 3, and 4, respectively.

Here is a simple algorithm for computing the shift table entries. Initialize all the entries to the pattern's length m and scan the pattern left to right repeating the following step $m-1$ times: for the jth character of the pattern ($0 \le j \le m-2$), overwrite its entry in the table with $m-1-j$, which is the character's distance to the last character of the pattern. Note that since the algorithm scans the pattern from left to right, the last overwrite will happen for the character's rightmost occurrence—exactly as we would like it to be.

ALGORITHM *ShiftTable*($P[0..m-1]$)
//Fills the shift table used by Horspool's and Boyer-Moore algorithms
//Input: Pattern $P[0..m-1]$ and an alphabet of possible characters
//Output: *Table*$[0..size-1]$ indexed by the alphabet's characters and
//   filled with shift sizes computed by formula (7.1)
for $i \leftarrow 0$ **to** $size-1$ **do** $Table[i] \leftarrow m$
for $j \leftarrow 0$ **to** $m-2$ **do** $Table[P[j]] \leftarrow m-1-j$
return *Table*

Now, we can summarize the algorithm as follows:

Horspool's algorithm

Step 1 For a given pattern of length m and the alphabet used in both the pattern and text, construct the shift table as described above.

Step 2 Align the pattern against the beginning of the text.

Step 3 Repeat the following until either a matching substring is found or the pattern reaches beyond the last character of the text. Starting with the last character in the pattern, compare the corresponding characters in the pattern and text until either all m characters are matched (then stop) or a mismatching pair is encountered. In the latter case, retrieve the entry $t(c)$ from the c's column of the shift table where c is the text's character currently aligned against the last character of the pattern, and shift the pattern by $t(c)$ characters to the right along the text.

Here is pseudocode of Horspool's algorithm.

ALGORITHM *HorspoolMatching*($P[0..m-1]$, $T[0..n-1]$)

```
//Implements Horspool's algorithm for string matching
//Input: Pattern P[0..m − 1] and text T[0..n − 1]
//Output: The index of the left end of the first matching substring
//        or −1 if there are no matches
ShiftTable(P[0..m − 1])        //generate Table of shifts
i ← m − 1                      //position of the pattern's right end
while i ≤ n − 1 do
    k ← 0                      //number of matched characters
    while k ≤ m − 1 and P[m − 1 − k] = T[i − k] do
        k ← k + 1
    if k = m
        return i − m + 1
    else i ← i + Table[T[i]]
return −1
```

EXAMPLE As an example of a complete application of Horspool's algorithm, consider searching for the pattern BARBER in a text that comprises English letters and spaces (denoted by underscores). The shift table, as we mentioned, is filled as follows:

character c	A	B	C	D	E	F	...	R	...	Z	_
shift $t(c)$	4	2	6	6	1	6	6	3	6	6	6

The actual search in a particular text proceeds as follows:

```
J I M _ S A W _ M E _ I N _ A _ B A R B E R S H O P
B A R B E R                 B A R B E R
        B A R B E R             B A R B E R
          B A R B E R                 B A R B E R
```
■

A simple example can demonstrate that the worst-case efficiency of Horspool's algorithm is in $O(nm)$ (Problem 4 in this section's exercises). But for random texts, it is in $\Theta(n)$, and, although in the same efficiency class, Horspool's algorithm is obviously faster on average than the brute-force algorithm. In fact, as mentioned, it is often at least as efficient as its more sophisticated predecessor discovered by R. Boyer and J. Moore.

Boyer-Moore Algorithm

Now we outline the Boyer-Moore algorithm itself. If the first comparison of the rightmost character in the pattern with the corresponding character c in the text fails, the algorithm does exactly the same thing as Horspool's algorithm. Namely, it shifts the pattern to the right by the number of characters retrieved from the table precomputed as explained earlier.

The two algorithms act differently, however, after some positive number k $(0 < k < m)$ of the pattern's characters are matched successfully before a mismatch is encountered:

$$\begin{array}{ccccccccccl} s_0 & \dots & & c & s_{i-k+1} & \dots & s_i & \dots & s_{n-1} & & \text{text} \\ & & & \not\| & \| & & \| & & & & \\ & p_0 & \dots & p_{m-k-1} & p_{m-k} & \dots & p_{m-1} & & & & \text{pattern} \end{array}$$

In this situation, the Boyer-Moore algorithm determines the shift size by considering two quantities. The first one is guided by the text's character c that caused a mismatch with its counterpart in the pattern. Accordingly, it is called the ***bad-symbol shift***. The reasoning behind this shift is the reasoning we used in Horspool's algorithm. If c is not in the pattern, we shift the pattern to just pass this c in the text. Conveniently, the size of this shift can be computed by the formula $t_1(c) - k$ where $t_1(c)$ is the entry in the precomputed table used by Horspool's algorithm (see above) and k is the number of matched characters:

$$\begin{array}{ccccccccccl} s_0 & \dots & & c & s_{i-k+1} & \dots & s_i & & \dots & s_{n-1} & \text{text} \\ & & & \not\| & \| & & \| & & & & \\ & p_0 & \dots & p_{m-k-1} & p_{m-k} & \dots & p_{m-1} & & & & \text{pattern} \\ & & & & p_0 & \dots & & p_{m-1} & & & \end{array}$$

For example, if we search for the pattern BARBER in some text and match the last two characters before failing on letter S in the text, we can shift the pattern by $t_1(\text{S}) - 2 = 6 - 2 = 4$ positions:

$$\begin{array}{ccccccccccccc} s_0 & \dots & & & & \text{S} & \text{E} & \text{R} & & & \dots & s_{n-1} \\ & & & & & \not\| & \| & \| & & & & \\ & & \text{B} & \text{A} & \text{R} & \text{B} & \text{E} & \text{R} & & & & \\ & & & & & & \text{B} & \text{A} & \text{R} & \text{B} & \text{E} & \text{R} \end{array}$$

The same formula can also be used when the mismatching character c of the text occurs in the pattern, provided $t_1(c) - k > 0$. For example, if we search for the pattern BARBER in some text and match the last two characters before failing on letter A, we can shift the pattern by $t_1(\text{A}) - 2 = 4 - 2 = 2$ positions:

$$\begin{array}{ccccccccccc} s_0 & \dots & & & & \text{A} & \text{E} & \text{R} & & \dots & s_{n-1} \\ & & & & & \not\| & \| & \| & & & \\ & & \text{B} & \text{A} & \text{R} & \text{B} & \text{E} & \text{R} & & & \\ & & & & \text{B} & \text{A} & \text{R} & \text{B} & \text{E} & \text{R} & \end{array}$$

If $t_1(c) - k \le 0$, we obviously do not want to shift the pattern by 0 or a negative number of positions. Rather, we can fall back on the brute-force thinking and simply shift the pattern by one position to the right.

To summarize, the bad-symbol shift d_1 is computed by the Boyer-Moore algorithm either as $t_1(c) - k$ if this quantity is positive and as 1 if it is negative or zero. This can be expressed by the following compact formula:

$$d_1 = \max\{t_1(c) - k, 1\}. \tag{7.2}$$

The second type of shift is guided by a successful match of the last $k > 0$ characters of the pattern. We refer to the ending portion of the pattern as its suffix of size k and denote it *suff* (k). Accordingly, we call this type of shift the ***good-suffix shift***. We now apply the reasoning that guided us in filling the bad-symbol shift table, which was based on a single alphabet character c, to the pattern's suffixes of sizes $1, \ldots, m-1$ to fill in the good-suffix shift table.

Let us first consider the case when there is another occurrence of *suff* (k) in the pattern or, to be more accurate, there is another occurrence of *suff* (k) not preceded by the same character as in its rightmost occurrence. (It would be useless to shift the pattern to match another occurrence of *suff* (k) preceded by the same character because this would simply repeat a failed trial.) In this case, we can shift the pattern by the distance d_2 between such a second rightmost occurrence (not preceded by the same character as in the rightmost occurrence) of *suff* (k) and its rightmost occurrence. For example, for the pattern ABCBAB, these distances for $k = 1$ and 2 will be 2 and 4, respectively:

k	**pattern**	d_2
1	$\text{ABC}\overline{\text{B}}\text{A}\underline{\text{B}}$	2
2	$\overline{\text{AB}}\text{CB}\underline{\text{AB}}$	4

What is to be done if there is no other occurrence of *suff* (k) not preceded by the same character as in its rightmost occurrence? In most cases, we can shift the pattern by its entire length m. For example, for the pattern DBCBAB and $k = 3$, we can shift the pattern by its entire length of 6 characters:

```
s0  ...      c B A B            ...  s_{n-1}
             ∦ ‖ ‖ ‖
         D B C B A B
                     D B C B A B
```

Unfortunately, shifting the pattern by its entire length when there is no other occurrence of *suff* (k) not preceded by the same character as in its rightmost occurrence is not always correct. For example, for the pattern ABCBAB and $k = 3$, shifting by 6 could miss a matching substring that starts with the text's AB aligned with the last two characters of the pattern:

```
s0   ...      c  B  A  B  C  B  A  B       ...   sn-1
              ≠  ||  ||  ||
        A  B  C  B  A  B
                           A  B  C  B  A  B
```

Note that the shift by 6 is correct for the pattern DBCBAB but not for ABCBAB, because the latter pattern has the same substring AB as its prefix (beginning part of the pattern) and as its suffix (ending part of the pattern). To avoid such an erroneous shift based on a suffix of size k, for which there is no other occurrence in the pattern not preceded by the same character as in its rightmost occurrence, we need to find the longest prefix of size $l < k$ that matches the suffix of the same size l. If such a prefix exists, the shift size d_2 is computed as the distance between this prefix and the corresponding suffix; otherwise, d_2 is set to the pattern's length m. As an example, here is the complete list of the d_2 values—the good-suffix table of the Boyer-Moore algorithm—for the pattern ABCBAB:

k	**pattern**	d_2
1	ABCBAB	2
2	ABCBAB	4
3	ABCBAB	4
4	ABCBAB	4
5	ABCBAB	4

Now we are prepared to summarize the Boyer-Moore algorithm in its entirety.

The Boyer-Moore algorithm

Step 1 For a given pattern and the alphabet used in both the pattern and the text, construct the bad-symbol shift table as described earlier.

Step 2 Using the pattern, construct the good-suffix shift table as described earlier.

Step 3 Align the pattern against the beginning of the text.

Step 4 Repeat the following step until either a matching substring is found or the pattern reaches beyond the last character of the text. Starting with the last character in the pattern, compare the corresponding characters in the pattern and the text until either all m character pairs are matched (then stop) or a mismatching pair is encountered after $k \geq 0$ character pairs are matched successfully. In the latter case, retrieve the entry $t_1(c)$ from the c's column of the bad-symbol table where c is the text's mismatched character. If $k > 0$, also retrieve the corresponding d_2 entry from the good-suffix table. Shift the pattern to the right by the number of positions computed by the formula

$$d = \begin{cases} d_1 & \text{if } k = 0, \\ \max\{d_1, d_2\} & \text{if } k > 0, \end{cases} \qquad \textbf{(7.3)}$$

where $d_1 = \max\{t_1(c) - k, 1\}$.

Shifting by the maximum of the two available shifts when $k > 0$ is quite logical. The two shifts are based on the observations—the first one about a text's mismatched character, and the second one about a matched group of the pattern's rightmost characters—that imply that shifting by less than d_1 and d_2 characters, respectively, cannot lead to aligning the pattern with a matching substring in the text. Since we are interested in shifting the pattern as far as possible without missing a possible matching substring, we take the maximum of these two numbers.

EXAMPLE As a complete example, let us consider searching for the pattern BAOBAB in a text made of English letters and spaces. The bad-symbol table looks as follows:

c	A	B	C	D	...	O	...	Z	_
$t_1(c)$	1	2	6	6	6	3	6	6	6

The good-suffix table is filled as follows:

k	**pattern**	d_2
1	BAOBAB	2
2	BAOBAB	5
3	BAOBAB	5
4	BAOBAB	5
5	BAOBAB	5

The actual search for this pattern in the text given in Figure 7.3 proceeds as follows. After the last B of the pattern fails to match its counterpart K in the text, the algorithm retrieves $t_1(\text{K}) = 6$ from the bad-symbol table and shifts the pattern by $d_1 = \max\{t_1(\text{K}) - 0, 1\} = 6$ positions to the right. The new try successfully matches two pairs of characters. After the failure of the third comparison on the space character in the text, the algorithm retrieves $t_1(\_) = 6$ from the bad-symbol table and $d_2 = 5$ from the good-suffix table to shift the pattern by $\max\{d_1, d_2\} = \max\{6 - 2, 5\} = 5$. Note that on this iteration it is the good-suffix rule that leads to a farther shift of the pattern.

The next try successfully matches just one pair of B's. After the failure of the next comparison on the space character in the text, the algorithm retrieves $t_1(\_) = 6$ from the bad-symbol table and $d_2 = 2$ from the good-suffix table to shift

```
B  E  S  S  _  K  N  E  W  _  A  B  O  U  T  _  B  A  O  B  A  B  S
B  A  O  B  A  B
d1 = t1(K) − 0 = 6    B  A  O  B  A  B
               d1 = t1(_) − 2 = 4     B  A  O  B  A  B
               d2 = 5                 d1 = t1(_) − 1 = 5
               d = max{4, 5} = 5      d2 = 2
                                      d = max{5, 2} = 5
                                                        B  A  O  B  A  B
```

FIGURE 7.3 Example of string matching with the Boyer-Moore algorithm.

the pattern by $\max\{d_1, d_2\} = \max\{6 - 1, 2\} = 5$. Note that on this iteration it is the bad-symbol rule that leads to a farther shift of the pattern. The next try finds a matching substring in the text after successfully matching all six characters of the pattern with their counterparts in the text. ■

When searching for the first occurrence of the pattern, the worst-case efficiency of the Boyer-Moore algorithm is known to be linear. Though this algorithm runs very fast, especially on large alphabets (relative to the length of the pattern), many people prefer its simplified versions, such as Horspool's algorithm, when dealing with natural-language–like strings.

Exercises 7.2

1. Apply Horspool's algorithm to search for the pattern BAOBAB in the text

   BESS_KNEW_ABOUT_BAOBABS

2. Consider the problem of searching for genes in DNA sequences using Horspool's algorithm. A DNA sequence is represented by a text on the alphabet {A, C, G, T}, and the gene or gene segment is the pattern.

   a. Construct the shift table for the following gene segment of your chromosome 10:

   TCCTATTCTT

   b. Apply Horspool's algorithm to locate the above pattern in the following DNA sequence:

   TTATAGATCTCGTATTCTTTTATAGATCTCCTATTCTT

3. How many character comparisons will be made by Horspool's algorithm in searching for each of the following patterns in the binary text of 1000 zeros?
   a. 00001 **b.** 10000 **c.** 01010

4. For searching in a text of length n for a pattern of length m $(n \geq m)$ with Horspool's algorithm, give an example of
   a. worst-case input. **b.** best-case input.

5. Is it possible for Horspool's algorithm to make more character comparisons than the brute-force algorithm would make in searching for the same pattern in the same text?

6. If Horspool's algorithm discovers a matching substring, how large a shift should it make to search for a next possible match?

7. How many character comparisons will the Boyer-Moore algorithm make in searching for each of the following patterns in the binary text of 1000 zeros?
   a. 00001 **b.** 10000 **c.** 01010

8. **a.** Would the Boyer-Moore algorithm work correctly with just the bad-symbol table to guide pattern shifts?

   b. Would the Boyer-Moore algorithm work correctly with just the good-suffix table to guide pattern shifts?

9. **a.** If the last characters of a pattern and its counterpart in the text do match, does Horspool's algorithm have to check other characters right to left, or can it check them left to right too?

   b. Answer the same question for the Boyer-Moore algorithm.

10. Implement Horspool's algorithm, the Boyer-Moore algorithm, and the brute-force algorithm of Section 3.2 in the language of your choice and run an experiment to compare their efficiencies for matching

    a. random binary patterns in random binary texts.

    b. random natural-language patterns in natural-language texts.

11. You are given two strings S and T, each n characters long. You have to establish whether one of them is a right cyclic shift of the other. For example, PLEA is a right cyclic shift of LEAP, and vice versa. (Formally, T is a right cyclic shift of S if T can be obtained by concatenating the $(n-i)$-character suffix of S and the i-character prefix of S for some $1 \leq i \leq n$.)

    a. Design a space-efficient algorithm for the task. Indicate the space and time efficiencies of your algorithm.

    b. Design a time-efficient algorithm for the task. Indicate the time and space efficiencies of your algorithm.

7.3 Hashing

In this section, we consider a very efficient way to implement dictionaries. Recall that a dictionary is an abstract data type, namely, a set with the operations of searching (lookup), insertion, and deletion defined on its elements. The elements of this set can be of an arbitrary nature: numbers, characters of some alphabet, character strings, and so on. In practice, the most important case is that of records (student records in a school, citizen records in a governmental office, book records in a library).

Typically, records comprise several fields, each responsible for keeping a particular type of information about an entity the record represents. For example, a student record may contain fields for the student's ID, name, date of birth, sex, home address, major, and so on. Among record fields there is usually at least one called a ***key*** that is used for identifying entities represented by the records (e.g., the student's ID). In the discussion below, we assume that we have to implement a dictionary of n records with keys $K_1, K_2, \ldots, K_n$.

Hashing is based on the idea of distributing keys among a one-dimensional array $H[0..m-1]$ called a ***hash table***. The distribution is done by computing, for each of the keys, the value of some predefined function h called the ***hash function***. This function assigns an integer between 0 and $m-1$, called the ***hash address***, to a key.

For example, if keys are nonnegative integers, a hash function can be of the form $h(K) = K \bmod m$; obviously, the remainder of division by m is always between 0 and $m-1$. If keys are letters of some alphabet, we can first assign a letter its position in the alphabet, denoted here $ord(K)$, and then apply the same kind of a function used for integers. Finally, if K is a character string $c_0 c_1 \ldots c_{s-1}$, we can use, as a very unsophisticated option, $(\sum_{i=0}^{s-1} ord(c_i)) \bmod m$. A better option is to compute $h(K)$ as follows:[2]

$$h \leftarrow 0; \quad \textbf{for } i \leftarrow 0 \textbf{ to } s-1 \textbf{ do } h \leftarrow (h * C + ord(c_i)) \bmod m,$$

where C is a constant larger than every $ord(c_i)$.

In general, a hash function needs to satisfy somewhat conflicting requirements:

- A hash table's size should not be excessively large compared to the number of keys, but it should be sufficient to not jeopardize the implementation's time efficiency (see below).
- A hash function needs to distribute keys among the cells of the hash table as evenly as possible. (This requirement makes it desirable, for most applications, to have a hash function dependent on all bits of a key, not just some of them.)
- A hash function has to be easy to compute.

2. This can be obtained by treating $ord(c_i)$ as digits of a number in the C-based system, computing its decimal value by Horner's rule, and finding the remainder of the number after dividing it by m.

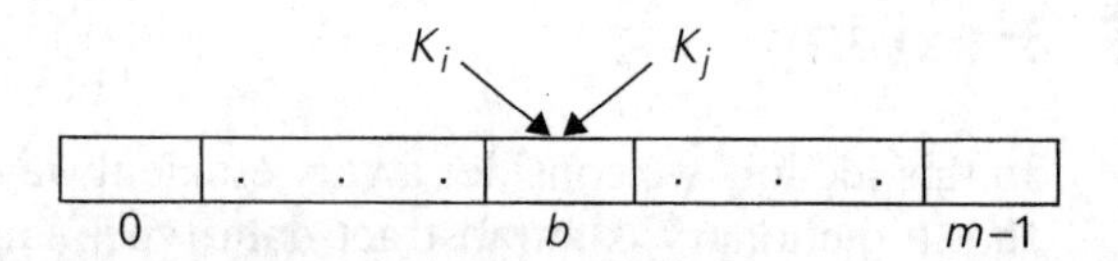

FIGURE 7.4 Collision of two keys in hashing: $h(K_i) = h(K_j)$.

Obviously, if we choose a hash table's size m to be smaller than the number of keys n, we will get ***collisions***—a phenomenon of two (or more) keys being hashed into the same cell of the hash table (Figure 7.4). But collisions should be expected even if m is considerably larger than n (see Problem 5 in this section's exercises). In fact, in the worst case, all the keys could be hashed to the same cell of the hash table. Fortunately, with an appropriately chosen hash table size and a good hash function, this situation happens very rarely. Still, every hashing scheme must have a collision resolution mechanism. This mechanism is different in the two principal versions of hashing: ***open hashing*** (also called ***separate chaining***) and ***closed hashing*** (also called ***open addressing***).

Open Hashing (Separate Chaining)

In open hashing, keys are stored in linked lists attached to cells of a hash table. Each list contains all the keys hashed to its cell. Consider, as an example, the following list of words:

A, FOOL, AND, HIS, MONEY, ARE, SOON, PARTED.

As a hash function, we will use the simple function for strings mentioned above, i.e., we will add the positions of a word's letters in the alphabet and compute the sum's remainder after division by 13.

We start with the empty table. The first key is the word A; its hash value is $h(\text{A}) = 1 \bmod 13 = 1$. The second key—the word FOOL—is installed in the ninth cell since $(6 + 15 + 15 + 12) \bmod 13 = 9$, and so on. The final result of this process is given in Figure 7.5; note a collision of the keys ARE and SOON because $h(\text{ARE}) = (1 + 18 + 5) \bmod 13 = 11$ and $h(\text{SOON}) = (19 + 15 + 15 + 14) \bmod 13 = 11$.

How do we search in a dictionary implemented as such a table of linked lists? We do this by simply applying to a search key the same procedure that was used for creating the table. To illustrate, if we want to search for the key KID in the hash table of Figure 7.5, we first compute the value of the same hash function for the key: $h(\text{KID}) = 11$. Since the list attached to cell 11 is not empty, its linked list may contain the search key. But because of possible collisions, we cannot tell whether this is the case until we traverse this linked list. After comparing the string KID first with the string ARE and then with the string SOON, we end up with an unsuccessful search.

In general, the efficiency of searching depends on the lengths of the linked lists, which, in turn, depend on the dictionary and table sizes, as well as the quality

keys	A	FOOL	AND	HIS	MONEY	ARE	SOON	PARTED
hash addresses	1	9	6	10	7	11	11	12

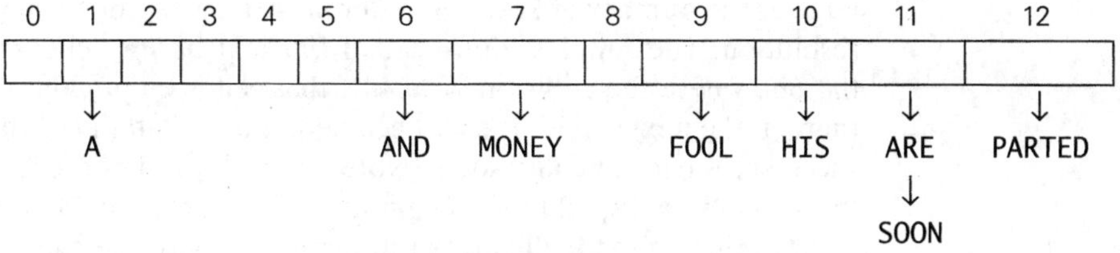

0	1	2	3	4	5	6	7	8	9	10	11	12
	↓					↓	↓		↓	↓	↓	↓
	A					AND	MONEY		FOOL	HIS	ARE	PARTED
											↓	
											SOON	

FIGURE 7.5 Example of a hash table construction with separate chaining.

of the hash function. If the hash function distributes n keys among m cells of the hash table about evenly, each list will be about n/m keys long. The ratio $\alpha = n/m$, called the ***load factor*** of the hash table, plays a crucial role in the efficiency of hashing. In particular, the average number of pointers (chain links) inspected in successful searches, S, and unsuccessful searches, U, turns out to be

$$S \approx 1 + \frac{\alpha}{2} \quad \text{and} \quad U = \alpha, \tag{7.4}$$

respectively, under the standard assumptions of searching for a randomly selected element and a hash function distributing keys uniformly among the table's cells. These results are quite natural. Indeed, they are almost identical to searching sequentially in a linked list; what we have gained by hashing is a reduction in average list size by a factor of m, the size of the hash table.

Normally, we want the load factor to be not far from 1. Having it too small would imply a lot of empty lists and hence inefficient use of space; having it too large would mean longer linked lists and hence longer search times. But if we do have the load factor around 1, we have an amazingly efficient scheme that makes it possible to search for a given key for, on average, the price of one or two comparisons! True, in addition to comparisons, we need to spend time on computing the value of the hash function for a search key, but it is a constant-time operation, independent from n and m. Note that we are getting this remarkable efficiency not only as a result of the method's ingenuity but also at the expense of extra space.

The two other dictionary operations—insertion and deletion—are almost identical to searching. Insertions are normally done at the end of a list (but see Problem 6 in this section's exercises for a possible modification of this rule). Deletion is performed by searching for a key to be deleted and then removing it from its list. Hence, the efficiency of these operations is identical to that of searching, and they are all $\Theta(1)$ in the average case if the number of keys n is about equal to the hash table's size m.

Closed Hashing (Open Addressing)

In closed hashing, all keys are stored in the hash table itself without the use of linked lists. (Of course, this implies that the table size m must be at least as large as the number of keys n.) Different strategies can be employed for collision resolution. The simplest one—called ***linear probing***—checks the cell following the one where the collision occurs. If that cell is empty, the new key is installed there; if the next cell is already occupied, the availability of that cell's immediate successor is checked, and so on. Note that if the end of the hash table is reached, the search is wrapped to the beginning of the table; i.e., it is treated as a circular array. This method is illustrated in Figure 7.6 with the same word list and hash function used above to illustrate separate chaining.

To search for a given key K, we start by computing $h(K)$ where h is the hash function used in the table construction. If the cell $h(K)$ is empty, the search is unsuccessful. If the cell is not empty, we must compare K with the cell's occupant: if they are equal, we have found a matching key; if they are not, we compare K with a key in the next cell and continue in this manner until we encounter either a matching key (a successful search) or an empty cell (unsuccessful search). For example, if we search for the word LIT in the table of Figure 7.6, we will get $h(\text{LIT}) = (12 + 9 + 20) \bmod 13 = 2$ and, since cell 2 is empty, we can stop immediately. However, if we search for KID with $h(\text{KID}) = (11 + 9 + 4) \bmod 13 = 11$, we will have to compare KID with ARE, SOON, PARTED, and A before we can declare the search unsuccessful.

Although the search and insertion operations are straightforward for this version of hashing, deletion is not. For example, if we simply delete the key ARE from the last state of the hash table in Figure 7.6, we will be unable to find the key SOON afterward. Indeed, after computing $h(\text{SOON}) = 11$, the algorithm would find this location empty and report the unsuccessful search result. A simple solution

keys	A	FOOL	AND	HIS	MONEY	ARE	SOON	PARTED
hash addresses	1	9	6	10	7	11	11	12

0	1	2	3	4	5	6	7	8	9	10	11	12
	A											
	A								FOOL			
	A					AND			FOOL			
	A					AND			FOOL	HIS		
	A					AND	MONEY		FOOL	HIS		
	A					AND	MONEY		FOOL	HIS	ARE	
	A					AND	MONEY		FOOL	HIS	ARE	SOON
PARTED	A					AND	MONEY		FOOL	HIS	ARE	SOON

FIGURE 7.6 Example of a hash table construction with linear probing.

is to use "lazy deletion," i.e., to mark previously occupied locations by a special symbol to distinguish them from locations that have not been occupied.

The mathematical analysis of linear probing is a much more difficult problem than that of separate chaining.[3] The simplified versions of these results state that the average number of times the algorithm must access the hash table with the load factor α in successful and unsuccessful searches is, respectively,

$$S \approx \frac{1}{2}(1+\frac{1}{1-\alpha}) \quad \text{and} \quad U \approx \frac{1}{2}(1+\frac{1}{(1-\alpha)^2}) \qquad \textbf{(7.5)}$$

(and the accuracy of these approximations increases with larger sizes of the hash table). These numbers are surprisingly small even for densely populated tables, i.e., for large percentage values of α:

α	$\frac{1}{2}(1+\frac{1}{1-\alpha})$	$\frac{1}{2}(1+\frac{1}{(1-\alpha)^2})$
50%	1.5	2.5
75%	2.5	8.5
90%	5.5	50.5

Still, as the hash table gets closer to being full, the performance of linear probing deteriorates because of a phenomenon called clustering. A ***cluster*** in linear probing is a sequence of contiguously occupied cells (with a possible wrapping). For example, the final state of the hash table of Figure 7.6 has two clusters. Clusters are bad news in hashing because they make the dictionary operations less efficient. As clusters become larger, the probability that a new element will be attached to a cluster increases; in addition, large clusters increase the probability that two clusters will coalesce after a new key's insertion, causing even more clustering.

Several other collision resolution strategies have been suggested to alleviate this problem. One of the most important is ***double hashing***. Under this scheme, we use another hash function, $s(K)$, to determine a fixed increment for the probing sequence to be used after a collision at location $l = h(K)$:

$$(l + s(K)) \bmod m, \qquad (l + 2s(K)) \bmod m, \qquad \ldots . \qquad \textbf{(7.6)}$$

To guarantee that every location in the table is probed by sequence (7.6), the increment $s(k)$ and the table size m must be relatively prime, i.e., their only common divisor must be 1. (This condition is satisfied automatically if m itself is prime.) Some functions recommended in the literature are $s(k) = m - 2 - k \bmod (m-2)$ and $s(k) = 8 - (k \bmod 8)$ for small tables and $s(k) = k \bmod 97 + 1$ for larger ones.

---

3. This problem was solved in 1962 by a young graduate student in mathematics named Donald E. Knuth. Knuth went on to become one of the most important computer scientists of our time. His multivolume treatise *The Art of Computer Programming* [KnuI, KnuII, KnuIII, KnuIV] remains the most comprehensive and influential book on algorithmics ever published.

Mathematical analysis of double hashing has proved to be quite difficult. Some partial results and considerable practical experience with the method suggest that with good hashing functions—both primary and secondary—double hashing is superior to linear probing. But its performance also deteriorates when the table gets close to being full. A natural solution in such a situation is ***rehashing***: the current table is scanned, and all its keys are relocated into a larger table.

It is worthwhile to compare the main properties of hashing with balanced search trees—its principal competitor for implementing dictionaries.

- *Asymptotic time efficiency* With hashing, searching, insertion, and deletion can be implemented to take $\Theta(1)$ time on the average but $\Theta(n)$ time in the very unlikely worst case. For balanced search trees, the average time efficiencies are $\Theta(\log n)$ for both the average and worst cases.
- *Ordering preservation* Unlike balanced search trees, hashing does not assume existence of key ordering and usually does not preserve it. This makes hashing less suitable for applications that need to iterate over the keys in order or require range queries such as counting the number of keys between some lower and upper bounds.

Since its discovery in the 1950s by IBM researchers, hashing has found many important applications. In particular, it has become a standard technique for storing a symbol table—a table of a computer program's symbols generated during compilation. Hashing is quite handy for such AI applications as checking whether positions generated by a chess-playing computer program have already been considered. With some modifications, it has also proved to be useful for storing very large dictionaries on disks; this variation of hashing is called ***extendible hashing***. Since disk access is expensive compared with probes performed in the main memory, it is preferable to make many more probes than disk accesses. Accordingly, a location computed by a hash function in extendible hashing indicates a disk address of a ***bucket*** that can hold up to b keys. When a key's bucket is identified, all its keys are read into main memory and then searched for the key in question. In the next section, we discuss B-trees, a principal alternative for storing large dictionaries.

Exercises 7.3

1. For the input 30, 20, 56, 75, 31, 19 and hash function $h(K) = K \bmod 11$
   a. construct the open hash table.
   b. find the largest number of key comparisons in a successful search in this table.
   c. find the average number of key comparisons in a successful search in this table.
2. For the input 30, 20, 56, 75, 31, 19 and hash function $h(K) = K \bmod 11$
   a. construct the closed hash table.

b. find the largest number of key comparisons in a successful search in this table.

c. find the average number of key comparisons in a successful search in this table.

3. Why is it not a good idea for a hash function to depend on just one letter (say, the first one) of a natural-language word?

4. Find the probability of all n keys being hashed to the same cell of a hash table of size m if the hash function distributes keys evenly among all the cells of the table.

5. *Birthday paradox* The birthday paradox asks how many people should be in a room so that the chances are better than even that two of them will have the same birthday (month and day). Find the quite unexpected answer to this problem. What implication for hashing does this result have?

6. Answer the following questions for the separate-chaining version of hashing.

a. Where would you insert keys if you knew that all the keys in the dictionary are distinct? Which dictionary operations, if any, would benefit from this modification?

b. We could keep keys of the same linked list sorted. Which of the dictionary operations would benefit from this modification? How could we take advantage of this if all the keys stored in the entire table need to be sorted?

7. Explain how to use hashing to check whether all elements of a list are distinct. What is the time efficiency of this application? Compare its efficiency with that of the brute-force algorithm (Section 2.3) and of the presorting-based algorithm (Section 6.1).

8. Fill in the following table with the average-case (as the first entry) and worst-case (as the second entry) efficiency classes for the five implementations of the ADT dictionary:

	unordered array	ordered array	binary search tree	balanced search tree	hashing
search					
insertion					
deletion					

9. We have discussed hashing in the context of techniques based on space–time trade-offs. But it also takes advantage of another general strategy. Which one?

10. Write a computer program that uses hashing for the following problem. Given a natural-language text, generate a list of distinct words with the number of occurrences of each word in the text. Insert appropriate counters in the program to compare the empirical efficiency of hashing with the corresponding theoretical results.

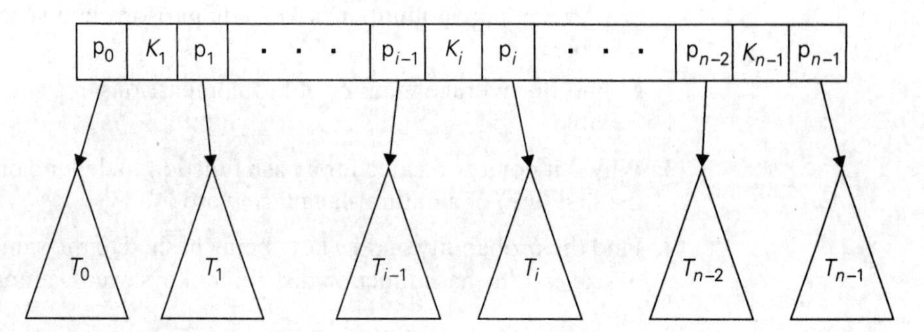

FIGURE 7.7 Parental node of a B-tree.

7.4 B-Trees

The idea of using extra space to facilitate faster access to a given data set is particularly important if the data set in question contains a very large number of records that need to be stored on a disk. A principal device in organizing such data sets is an ***index***, which provides some information about the location of records with indicated key values. For data sets of structured records (as opposed to "unstructured" data such as text, images, sound, and video), the most important index organization is the ***B-tree***, introduced by R. Bayer and E. McGreight [Bay72]. It extends the idea of the 2-3 tree (see Section 6.3) by permitting more than a single key in the same node of a search tree.

In the B-tree version we consider here, all data records (or record keys) are stored at the leaves, in increasing order of the keys. The parental nodes are used for indexing. Specifically, each parental node contains $n-1$ ordered keys $K_1 < \cdots < K_{n-1}$ assumed, for the sake of simplicity, to be distinct. The keys are interposed with n pointers to the node's children so that all the keys in subtree T_0 are smaller than K_1, all the keys in subtree T_1 are greater than or equal to K_1 and smaller than K_2 with K_1 being equal to the smallest key in T_1, and so on, through the last subtree T_{n-1} whose keys are greater than or equal to K_{n-1} with K_{n-1} being equal to the smallest key in T_{n-1} (see Figure 7.7).[4]

In addition, a B-tree of order $m \geq 2$ must satisfy the following structural properties:

- The root is either a leaf or has between 2 and m children.
- Each node, except for the root and the leaves, has between $\lceil m/2 \rceil$ and m children (and hence between $\lceil m/2 \rceil - 1$ and $m-1$ keys).
- The tree is (perfectly) balanced, i.e., all its leaves are at the same level.

4. The node depicted in Figure 7.7 is called the ***n-node***. Thus, all the nodes in a classic binary search tree are 2-nodes; a 2-3 tree introduced in Section 6.3 comprises 2-nodes and 3-nodes.

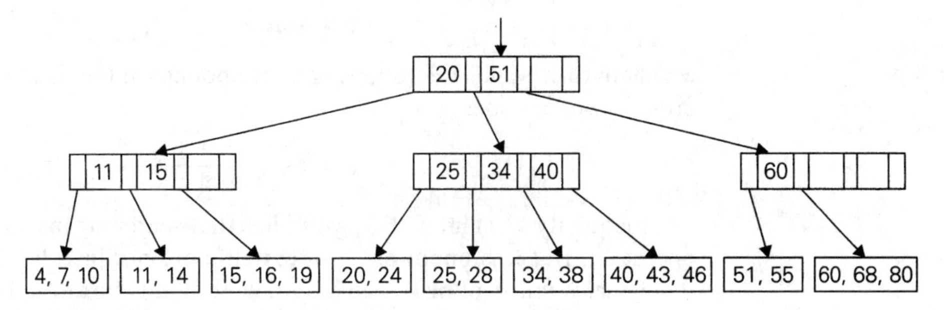

FIGURE 7.8 Example of a B-tree of order 4.

An example of a B-tree of order 4 is given in Figure 7.8.

Searching in a B-tree is very similar to searching in the binary search tree, and even more so in the 2-3 tree. Starting with the root, we follow a chain of pointers to the leaf that may contain the search key. Then we search for the search key among the keys of that leaf. Note that since keys are stored in sorted order, at both parental nodes and leaves, we can use binary search if the number of keys at a node is large enough to make it worthwhile.

It is not the number of key comparisons, however, that we should be concerned about in a typical application of this data structure. When used for storing a large data file on a disk, the nodes of a B-tree normally correspond to the disk pages. Since the time needed to access a disk page is typically several orders of magnitude larger than the time needed to compare keys in the fast computer memory, it is the number of disk accesses that becomes the principal indicator of the efficiency of this and similar data structures.

How many nodes of a B-tree do we need to access during a search for a record with a given key value? This number is, obviously, equal to the height of the tree plus 1. To estimate the height, let us find the smallest number of keys a B-tree of order m and positive height h can have. The root of the tree will contain at least one key. Level 1 will have at least two nodes with at least $\lceil m/2 \rceil - 1$ keys in each of them, for the total minimum number of keys $2(\lceil m/2 \rceil - 1)$. Level 2 will have at least $2\lceil m/2 \rceil$ nodes (the children of the nodes on level 1) with at least $\lceil m/2 \rceil - 1$ in each of them, for the total minimum number of keys $2\lceil m/2 \rceil(\lceil m/2 \rceil - 1)$. In general, the nodes of level i, $1 \le i \le h - 1$, will contain at least $2\lceil m/2 \rceil^{i-1}(\lceil m/2 \rceil - 1)$ keys. Finally, level h, the leaf level, will have at least $2\lceil m/2 \rceil^{h-1}$ nodes with at least one key in each. Thus, for any B-tree of order m with n nodes and height $h > 0$, we have the following inequality:

$$n \ge 1 + \sum_{i=1}^{h-1} 2\lceil m/2 \rceil^{i-1}(\lceil m/2 \rceil - 1) + 2\lceil m/2 \rceil^{h-1}.$$

After a series of standard simplifications (see Problem 2 in this section's exercises), this inequality reduces to

$$n \geq 4\lceil m/2 \rceil^{h-1} - 1,$$

which, in turn, yields the following upper bound on the height h of the B-tree of order m with n nodes:

$$h \leq \lfloor \log_{\lceil m/2 \rceil} \frac{n+1}{4} \rfloor + 1. \quad \textbf{(7.7)}$$

Inequality (7.7) immediately implies that searching in a B-tree is a $O(\log n)$ operation. But it is important to ascertain here not just the efficiency class but the actual number of disk accesses implied by this formula. The following table contains the values of the right-hand-side estimates for a file of 100 million records and a few typical values of the tree's order m:

order m	50	100	250
h's upper bound	6	5	4

Keep in mind that the table's entries are upper estimates for the number of disk accesses. In actual applications, this number rarely exceeds 3, with the B-tree's root and sometimes first-level nodes stored in the fast memory to minimize the number of disk accesses.

The operations of insertion and deletion are less straightforward than searching, but both can also be done in $O(\log n)$ time. Here we outline an insertion algorithm only; a deletion algorithm can be found in the references (e.g., [Aho83], [Cor09]).

The most straightforward algorithm for inserting a new record into a B-tree is quite similar to the algorithm for insertion into a 2-3 tree outlined in Section 6.3. First, we apply the search procedure to the new record's key K to find the appropriate leaf for the new record. If there is room for the record in that leaf, we place it there (in an appropriate position so that the keys remain sorted) and we are done. If there is no room for the record, the leaf is split in half by sending the second half of the records to a new node. After that, the smallest key K' in the new node and the pointer to it are inserted into the old leaf's parent (immediately after the key and pointer to the old leaf). This recursive procedure may percolate up to the tree's root. If the root is already full too, a new root is created with the two halves of the old root's keys split between two children of the new root. As an example, Figure 7.9 shows the result of inserting 65 into the B-tree in Figure 7.8 under the restriction that the leaves cannot contain more than three items.

You should be aware that there are other algorithms for implementing insertions into a B-tree. For example, to avoid the possibility of recursive node splits, we can split full nodes encountered in searching for an appropriate leaf for the new record. Another possibility is to avoid some node splits by moving a key to the node's sibling. For example, inserting 65 into the B-tree in Figure 7.8 can be done by moving 60, the smallest key of the full leaf, to its sibling with keys 51 and 55, and replacing the key value of their parent by 65, the new smallest value in

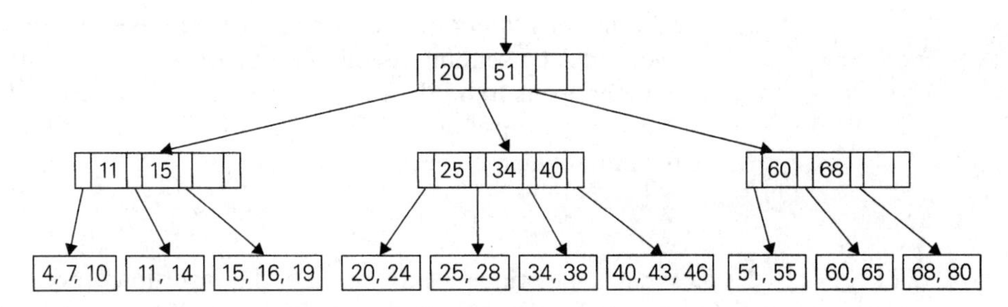

FIGURE 7.9 B-tree obtained after inserting 65 into the B-tree in Figure 7.8.

the second child. This modification tends to save some space at the expense of a slightly more complicated algorithm.

A B-tree does not have to be always associated with the indexing of a large file, and it can be considered as one of several search tree varieties. As with other types of search trees—such as binary search trees, AVL trees, and 2-3 trees—a B-tree can be constructed by successive insertions of data records into the initially empty tree. (The empty tree is considered to be a B-tree, too.) When all keys reside in the leaves and the upper levels are organized as a B-tree comprising an index, the entire structure is usually called, in fact, a ***B^+-tree***.

Exercises 7.4

1. Give examples of using an index in real-life applications that do not involve computers.
2. **a.** Prove the equality

$$1 + \sum_{i=1}^{h-1} 2\lceil m/2\rceil^{i-1}(\lceil m/2\rceil - 1) + 2\lceil m/2\rceil^{h-1} = 4\lceil m/2\rceil^{h-1} - 1,$$

   which was used in the derivation of upper bound (7.7) for the height of a B-tree.

   b. Complete the derivation of inequality (7.7).
3. Find the minimum order of the B-tree that guarantees that the number of disk accesses in searching in a file of 100 million records does not exceed 3. Assume that the root's page is stored in main memory.
4. Draw the B-tree obtained after inserting 30 and then 31 in the B-tree in Figure 7.8. Assume that a leaf cannot contain more than three items.
5. Outline an algorithm for finding the largest key in a B-tree.
6. **a.** A ***top-down 2-3-4 tree*** is a B-tree of order 4 with the following modification of the *insert* operation: Whenever a search for a leaf for a new key

encounters a full node (i.e., a node with three keys), the node is split into two nodes by sending its middle key to the node's parent, or, if the full node happens to be the root, the new root for the middle key is created. Construct a top-down 2-3-4 tree by inserting the following list of keys in the initially empty tree:

10, 6, 15, 31, 20, 27, 50, 44, 18.

b. What is the principal advantage of this insertion procedure compared with the one used for 2-3 trees in Section 6.3? What is its disadvantage?

7. a. Write a program implementing a key insertion algorithm in a B-tree.

b. Write a program for visualization of a key insertion algorithm in a B-tree.

SUMMARY

- Space and time trade-offs in algorithm design are a well-known issue for both theoreticians and practitioners of computing. As an algorithm design technique, trading space for time is much more prevalent than trading time for space.

- *Input enhancement* is one of the two principal varieties of trading space for time in algorithm design. Its idea is to preprocess the problem's input, in whole or in part, and store the additional information obtained in order to accelerate solving the problem afterward. Sorting by distribution counting and several important algorithms for string matching are examples of algorithms based on this technique.

- *Distribution counting* is a special method for sorting lists of elements from a small set of possible values.

- *Horspool's algorithm* for string matching can be considered a simplified version of the *Boyer-Moore algorithm.* Both algorithms are based on the ideas of input enhancement and right-to-left comparisons of a pattern's characters. Both algorithms use the same *bad-symbol shift table*; the Boyer-Moore also uses a second table, called the *good-suffix shift table.*

- *Prestructuring*—the second type of technique that exploits space-for-time trade-offs—uses extra space to facilitate a faster and/or more flexible access to the data. Hashing and B^+-trees are important examples of prestructuring.

- *Hashing* is a very efficient approach to implementing dictionaries. It is based on the idea of mapping keys into a one-dimensional table. The size limitations of such a table make it necessary to employ a *collision resolution* mechanism. The two principal varieties of hashing are *open hashing* or *separate chaining* (with keys stored in linked lists outside of the hash table) and *closed hashing* or *open addressing* (with keys stored inside the table). Both enable searching, insertion, and deletion in $\Theta(1)$ time, on average.

- The *B-tree* is a balanced search tree that generalizes the idea of the 2-3 tree by allowing multiple keys at the same node. Its principal application, called the B^+-tree, is for keeping index-like information about data stored on a disk. By choosing the order of the tree appropriately, one can implement the operations of searching, insertion, and deletion with just a few disk accesses even for extremely large files.

8

Dynamic Programming

An idea, like a ghost . . . must be spoken to a little before it will explain itself.

—Charles Dickens (1812–1870)

Dynamic programming is an algorithm design technique with a rather interesting history. It was invented by a prominent U.S. mathematician, Richard Bellman, in the 1950s as a general method for optimizing multistage decision processes. Thus, the word “programming” in the name of this technique stands for “planning” and does not refer to computer programming. After proving its worth as an important tool of applied mathematics, dynamic programming has eventually come to be considered, at least in computer science circles, as a general algorithm design technique that does not have to be limited to special types of optimization problems. It is from this point of view that we will consider this technique here.

Dynamic programming is a technique for solving problems with overlapping subproblems. Typically, these subproblems arise from a recurrence relating a given problem’s solution to solutions of its smaller subproblems. Rather than solving overlapping subproblems again and again, dynamic programming suggests solving each of the smaller subproblems only once and recording the results in a table from which a solution to the original problem can then be obtained.

This technique can be illustrated by revisiting the Fibonacci numbers discussed in Section 2.5. (If you have not read that section, you will be able to follow the discussion anyway. But it is a beautiful topic, so if you feel a temptation to read it, do succumb to it.) The Fibonacci numbers are the elements of the sequence

$$0,\ 1,\ 1,\ 2,\ 3,\ 5,\ 8,\ 13,\ 21,\ 34,\ldots,$$

which can be defined by the simple recurrence

$$F(n) = F(n-1) + F(n-2) \quad \text{for } n > 1 \tag{8.1}$$

and two initial conditions

$$F(0) = 0, \qquad F(1) = 1. \tag{8.2}$$

If we try to use recurrence (8.1) directly to compute the nth Fibonacci number $F(n)$, we would have to recompute the same values of this function many times (see Figure 2.6 for an example). Note that the problem of computing $F(n)$ is expressed in terms of its smaller and overlapping subproblems of computing $F(n-1)$ and $F(n-2)$. So we can simply fill elements of a one-dimensional array with the $n+1$ consecutive values of $F(n)$ by starting, in view of initial conditions (8.2), with 0 and 1 and using equation (8.1) as the rule for producing all the other elements. Obviously, the last element of this array will contain $F(n)$. Single-loop pseudocode of this very simple algorithm can be found in Section 2.5.

Note that we can, in fact, avoid using an extra array to accomplish this task by recording the values of just the last two elements of the Fibonacci sequence (Problem 8 in Exercises 2.5). This phenomenon is not unusual, and we shall encounter it in a few more examples in this chapter. Thus, although a straightforward application of dynamic programming can be interpreted as a special variety of space-for-time trade-off, a dynamic programming algorithm can sometimes be refined to avoid using extra space.

Certain algorithms compute the nth Fibonacci number without computing all the preceding elements of this sequence (see Section 2.5). It is typical of an algorithm based on the classic bottom-up dynamic programming approach, however, to solve *all* smaller subproblems of a given problem. One variation of the dynamic programming approach seeks to avoid solving unnecessary subproblems. This technique, illustrated in Section 8.2, exploits so-called memory functions and can be considered a top-down variation of dynamic programming.

Whether one uses the classical bottom-up version of dynamic programming or its top-down variation, the crucial step in designing such an algorithm remains the same: deriving a recurrence relating a solution to the problem to solutions to its smaller subproblems. The immediate availability of equation (8.1) for computing the nth Fibonacci number is one of the few exceptions to this rule.

Since a majority of dynamic programming applications deal with optimization problems, we also need to mention a general principle that underlines such applications. Richard Bellman called it the ***principle of optimality***. In terms somewhat different from its original formulation, it says that an optimal solution to any instance of an optimization problem is composed of optimal solutions to its subinstances. The principle of optimality holds much more often than not. (To give a rather rare example, it fails for finding the longest simple path in a graph.) Although its applicability to a particular problem needs to be checked, of course, such a check is usually not a principal difficulty in developing a dynamic programming algorithm.

In the sections and exercises of this chapter are a few standard examples of dynamic programming algorithms. (The algorithms in Section 8.4 were, in fact, invented independently of the discovery of dynamic programming and only later came to be viewed as examples of this technique's applications.) Numerous other applications range from the optimal way of breaking text into lines (e.g., [Baa00]) to image resizing [Avi07] to a variety of applications to sophisticated engineering problems (e.g., [Ber01]).

8.1 Three Basic Examples

The goal of this section is to introduce dynamic programming via three typical examples.

EXAMPLE 1 *Coin-row problem* There is a row of n coins whose values are some positive integers $c_1, c_2, \ldots, c_n$, not necessarily distinct. The goal is to pick up the maximum amount of money subject to the constraint that no two coins adjacent in the initial row can be picked up.

Let $F(n)$ be the maximum amount that can be picked up from the row of n coins. To derive a recurrence for $F(n)$, we partition all the allowed coin selections into two groups: those that include the last coin and those without it. The largest amount we can get from the first group is equal to $c_n + F(n-2)$—the value of the nth coin plus the maximum amount we can pick up from the first $n-2$ coins. The maximum amount we can get from the second group is equal to $F(n-1)$ by the definition of $F(n)$. Thus, we have the following recurrence subject to the obvious initial conditions:

$$\begin{aligned} &F(n) = \max\{c_n + F(n-2), F(n-1)\} \quad \text{for } n > 1, \\ &F(0) = 0, \qquad F(1) = c_1. \end{aligned} \tag{8.3}$$

We can compute $F(n)$ by filling the one-row table left to right in the manner similar to the way it was done for the nth Fibonacci number by Algorithm *Fib(n)* in Section 2.5.

ALGORITHM *CoinRow*($C[1..n]$)

```
//Applies formula (8.3) bottom up to find the maximum amount of money
//that can be picked up from a coin row without picking two adjacent coins
//Input: Array C[1..n] of positive integers indicating the coin values
//Output: The maximum amount of money that can be picked up
F[0] ← 0;  F[1] ← C[1]
for i ← 2 to n do
    F[i] ← max(C[i] + F[i − 2], F[i − 1])
return F[n]
```

The application of the algorithm to the coin row of denominations 5, 1, 2, 10, 6, 2 is shown in Figure 8.1. It yields the maximum amount of 17. It is worth pointing

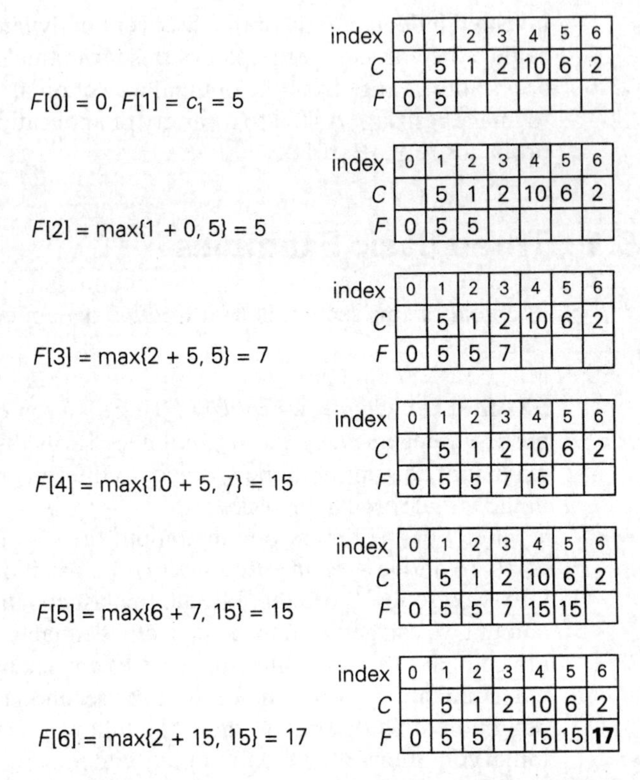

FIGURE 8.1 Solving the coin-row problem by dynamic programming for the coin row 5, 1, 2, 10, 6, 2.

out that, in fact, we also solved the problem for the first i coins in the row given for every $1 \le i \le 6$. For example, for $i = 3$, the maximum amount is $F(3) = 7$.

To find the coins with the maximum total value found, we need to backtrace the computations to see which of the two possibilities—$c_n + F(n-2)$ or $F(n-1)$—produced the maxima in formula (8.3). In the last application of the formula, it was the sum $c_6 + F(4)$, which means that the coin $c_6 = 2$ is a part of an optimal solution. Moving to computing $F(4)$, the maximum was produced by the sum $c_4 + F(2)$, which means that the coin $c_4 = 10$ is a part of an optimal solution as well. Finally, the maximum in computing $F(2)$ was produced by $F(1)$, implying that the coin c_2 is not the part of an optimal solution and the coin $c_1 = 5$ is. Thus, the optimal solution is $\{c_1, c_4, c_6\}$. To avoid repeating the same computations during the backtracing, the information about which of the two terms in (8.3) was larger can be recorded in an extra array when the values of F are computed.

Using the *CoinRow* to find $F(n)$, the largest amount of money that can be picked up, as well as the coins composing an optimal set, clearly takes $\Theta(n)$ time and $\Theta(n)$ space. This is by far superior to the alternatives: the straightforward top-down application of recurrence (8.3) and solving the problem by exhaustive search (Problem 3 in this section's exercises). ■

EXAMPLE 2 *Change-making problem* Consider the general instance of the following well-known problem. Give change for amount n using the minimum number of coins of denominations $d_1 < d_2 < \cdots < d_m$. For the coin denominations used in the United States, as for those used in most if not all other countries, there is a very simple and efficient algorithm discussed in the next chapter. Here, we consider a dynamic programming algorithm for the general case, assuming availability of unlimited quantities of coins for each of the m denominations $d_1 < d_2 < \cdots < d_m$ where $d_1 = 1$.

Let $F(n)$ be the minimum number of coins whose values add up to n; it is convenient to define $F(0) = 0$. The amount n can only be obtained by adding one coin of denomination d_j to the amount $n - d_j$ for $j = 1, 2, \ldots, m$ such that $n \geq d_j$. Therefore, we can consider all such denominations and select the one minimizing $F(n - d_j) + 1$. Since 1 is a constant, we can, of course, find the smallest $F(n - d_j)$ first and then add 1 to it. Hence, we have the following recurrence for $F(n)$:

$$\begin{aligned} F(n) &= \min_{j:n \geq d_j} \{F(n - d_j)\} + 1 \quad \text{for } n > 0, \\ F(0) &= 0. \end{aligned} \tag{8.4}$$

We can compute $F(n)$ by filling a one-row table left to right in the manner similar to the way it was done above for the coin-row problem, but computing a table entry here requires finding the minimum of up to m numbers.

ALGORITHM *ChangeMaking*($D[1..m]$, n)

```
//Applies dynamic programming to find the minimum number of coins
//of denominations d1 < d2 < ... < dm where d1 = 1 that add up to a
//given amount n
//Input: Positive integer n and array D[1..m] of increasing positive
//       integers indicating the coin denominations where D[1] = 1
//Output: The minimum number of coins that add up to n
F[0] ← 0
for i ← 1 to n do
    temp ← ∞; j ← 1
    while j ≤ m and i ≥ D[j] do
        temp ← min(F[i − D[j]], temp)
        j ← j + 1
    F[i] ← temp + 1
return F[n]
```

The application of the algorithm to amount $n = 6$ and denominations 1, 3, 4 is shown in Figure 8.2. The answer it yields is two coins. The time and space efficiencies of the algorithm are obviously $O(nm)$ and $\Theta(n)$, respectively.

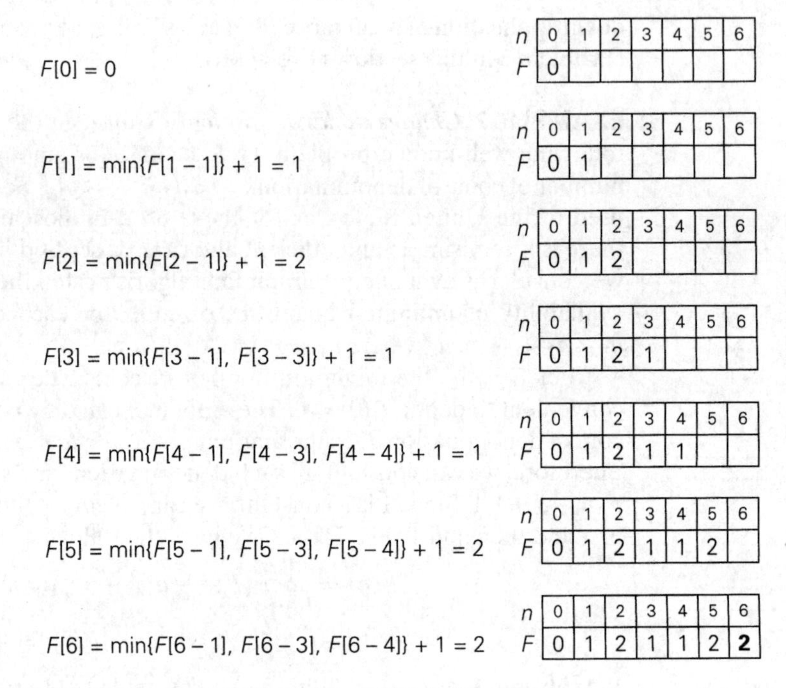

FIGURE 8.2 Application of Algorithm *MinCoinChange* to amount $n = 6$ and coin denominations 1, 3, and 4.

To find the coins of an optimal solution, we need to backtrace the computations to see which of the denominations produced the minima in formula (8.4). For the instance considered, the last application of the formula (for $n = 6$), the minimum was produced by $d_2 = 3$. The second minimum (for $n = 6 - 3$) was also produced for a coin of that denomination. Thus, the minimum-coin set for $n = 6$ is two 3's. ■

EXAMPLE 3 *Coin-collecting problem* Several coins are placed in cells of an $n \times m$ board, no more than one coin per cell. A robot, located in the upper left cell of the board, needs to collect as many of the coins as possible and bring them to the bottom right cell. On each step, the robot can move either one cell to the right or one cell down from its current location. When the robot visits a cell with a coin, it always picks up that coin. Design an algorithm to find the maximum number of coins the robot can collect and a path it needs to follow to do this.

Let $F(i, j)$ be the largest number of coins the robot can collect and bring to the cell (i, j) in the ith row and jth column of the board. It can reach this cell either from the adjacent cell $(i - 1, j)$ above it or from the adjacent cell $(i, j - 1)$ to the left of it. The largest numbers of coins that can be brought to these cells are $F(i - 1, j)$ and $F(i, j - 1)$, respectively. Of course, there are no adjacent cells above the cells in the first row, and there are no adjacent cells to the left of the cells in the first column. For those cells, we assume that $F(i-1, j)$ and $F(i, j-1)$ are equal to 0 for their nonexistent neighbors. Therefore, the largest number of coins the robot can bring to cell (i, j) is the maximum of these two numbers plus one possible coin at cell (i, j) itself. In other words, we have the following formula for $F(i, j)$:

$$\begin{aligned} &F(i, j) = \max\{F(i-1, j), F(i, j-1)\} + c_{ij} \quad \text{for } 1 \le i \le n, \ \ 1 \le j \le m \\ &F(0, j) = 0 \ \text{ for } 1 \le j \le m \quad \text{and} \quad F(i, 0) = 0 \ \text{ for } 1 \le i \le n, \end{aligned} \tag{8.5}$$

where $c_{ij} = 1$ if there is a coin in cell (i, j), and $c_{ij} = 0$ otherwise.

Using these formulas, we can fill in the $n \times m$ table of $F(i, j)$ values either row by row or column by column, as is typical for dynamic programming algorithms involving two-dimensional tables.

```
ALGORITHM RobotCoinCollection(C[1..n, 1..m])
    //Applies dynamic programming to compute the largest number of
    //coins a robot can collect on an n × m board by starting at (1, 1)
    //and moving right and down from upper left to down right corner
    //Input: Matrix C[1..n, 1..m] whose elements are equal to 1 and 0
    //for cells with and without a coin, respectively
    //Output: Largest number of coins the robot can bring to cell (n, m)
    F[1, 1] ← C[1, 1];  for j ← 2 to m do F[1, j] ← F[1, j − 1] + C[1, j]
    for i ← 2 to n do
        F[i, 1] ← F[i − 1, 1] + C[i, 1]
        for j ← 2 to m do
            F[i, j] ← max(F[i − 1, j], F[i, j − 1]) + C[i, j]
    return F[n, m]
```

The algorithm is illustrated in Figure 8.3b for the coin setup in Figure 8.3a. Since computing the value of $F(i, j)$ by formula (8.5) for each cell of the table takes constant time, the time efficiency of the algorithm is $\Theta(nm)$. Its space efficiency is, obviously, also $\Theta(nm)$.

Tracing the computations backward makes it possible to get an optimal path: if $F(i-1, j) > F(i, j-1)$, an optimal path to cell (i, j) must come down from the adjacent cell above it; if $F(i-1, j) < F(i, j-1)$, an optimal path to cell (i, j) must come from the adjacent cell on the left; and if $F(i-1, j) = F(i, j-1)$, it can reach cell (i, j) from either direction. This yields two optimal paths for the instance in Figure 8.3a, which are shown in Figure 8.3c. If ties are ignored, one optimal path can be obtained in $\Theta(n+m)$ time.

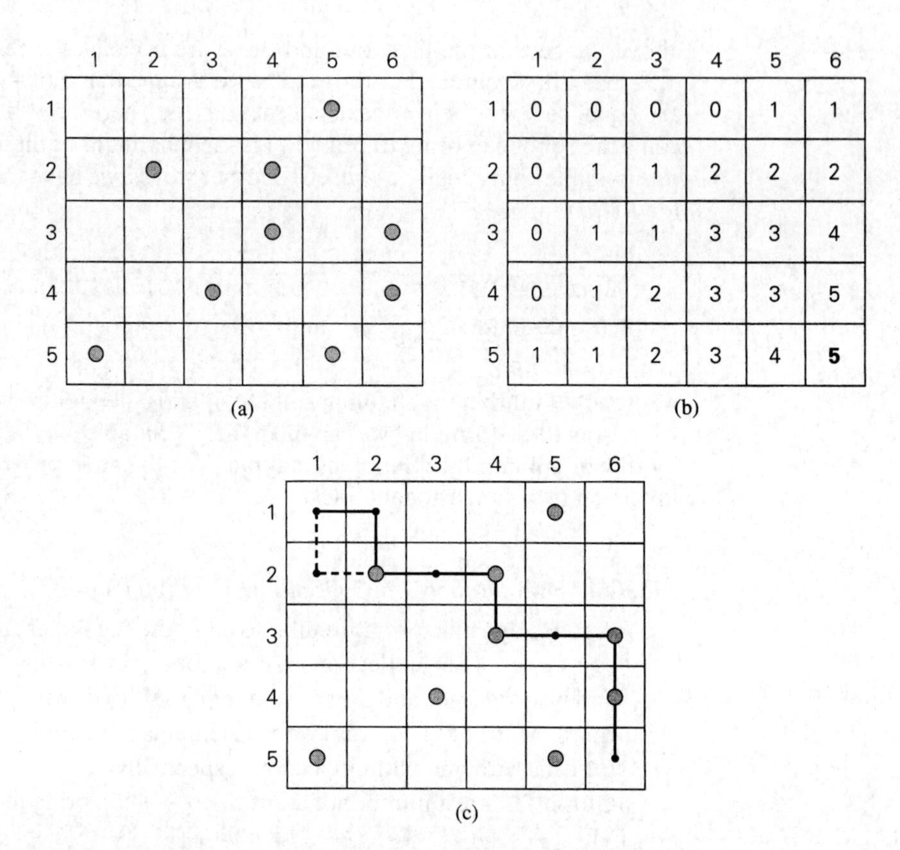

FIGURE 8.3 (a) Coins to collect. (b) Dynamic programming algorithm results. (c) Two paths to collect 5 coins, the maximum number of coins possible. ■

Exercises 8.1

1. What does dynamic programming have in common with divide-and-conquer? What is a principal difference between them?

2. Solve the instance 5, 1, 2, 10, 6 of the coin-row problem.

3. **a.** Show that the time efficiency of solving the coin-row problem by straightforward application of recurrence (8.3) is exponential.

   b. Show that the time efficiency of solving the coin-row problem by exhaustive search is at least exponential.

4. Apply the dynamic programming algorithm to find all the solutions to the change-making problem for the denominations 1, 3, 5 and the amount $n = 9$.

5. How would you modify the dynamic programming algorithm for the coin-collecting problem if some cells on the board are inaccessible for the robot? Apply your algorithm to the board below, where the inaccessible cells are shown by X's. How many optimal paths are there for this board?

	1	2	3	4	5	6
1		X		●		
2	●			X	●	
3		●		X	●	
4				●		●
5	X	X	X		●	

6. *Rod-cutting problem* Design a dynamic programming algorithm for the following problem. Find the maximum total sale price that can be obtained by cutting a rod of n units long into integer-length pieces if the sale price of a piece i units long is p_i for $i = 1, 2, \ldots, n$. What are the time and space efficiencies of your algorithm?

7. *Shortest-path counting* A chess rook can move horizontally or vertically to any square in the same row or in the same column of a chessboard. Find the number of shortest paths by which a rook can move from one corner of a chessboard to the diagonally opposite corner. The length of a path is measured by the number of squares it passes through, including the first and the last squares. Solve the problem

a. by a dynamic programming algorithm.

b. by using elementary combinatorics.

8. *Minimum-sum descent* Some positive integers are arranged in an equilateral triangle with n numbers in its base like the one shown in the figure below for $n = 4$. The problem is to find the smallest sum in a descent from the triangle apex to its base through a sequence of adjacent numbers (shown in the figure by the circles). Design a dynamic programming algorithm for this problem and indicate its time efficiency.

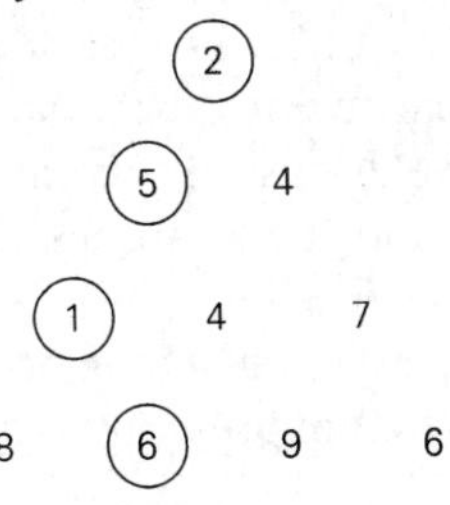

9. *Binomial coefficient* Design an efficient algorithm for computing the binomial coefficient $C(n, k)$ that uses no multiplications. What are the time and space efficiencies of your algorithm?

10. *Longest path in a dag*

    a. Design an efficient algorithm for finding the length of the longest path in a dag. (This problem is important both as a prototype of many other dynamic programming applications and in its own right because it determines the minimal time needed for completing a project comprising precedence-constrained tasks.)

    b. Show how to reduce the coin-row problem discussed in this section to the problem of finding a longest path in a dag.

11. *Maximum square submatrix* Given an $m \times n$ boolean matrix B, find its largest square submatrix whose elements are all zeros. Design a dynamic programming algorithm and indicate its time efficiency. (The algorithm may be useful for, say, finding the largest free square area on a computer screen or for selecting a construction site.)

12. *World Series odds* Consider two teams, A and B, playing a series of games until one of the teams wins n games. Assume that the probability of A winning a game is the same for each game and equal to p, and the probability of A losing a game is $q = 1 - p$. (Hence, there are no ties.) Let $P(i, j)$ be the probability of A winning the series if A needs i more games to win the series and B needs j more games to win the series.

    a. Set up a recurrence relation for $P(i, j)$ that can be used by a dynamic programming algorithm.

    b. Find the probability of team A winning a seven-game series if the probability of it winning a game is 0.4.

    c. Write pseudocode of the dynamic programming algorithm for solving this problem and determine its time and space efficiencies.

8.2 The Knapsack Problem and Memory Functions

We start this section with designing a dynamic programming algorithm for the knapsack problem: given n items of known weights $w_1, \ldots, w_n$ and values $v_1, \ldots, v_n$ and a knapsack of capacity W, find the most valuable subset of the items that fit into the knapsack. (This problem was introduced in Section 3.4, where we discussed solving it by exhaustive search.) We assume here that all the weights and the knapsack capacity are positive integers; the item values do not have to be integers.

To design a dynamic programming algorithm, we need to derive a recurrence relation that expresses a solution to an instance of the knapsack problem in terms of solutions to its smaller subinstances. Let us consider an instance defined by the first i items, $1 \le i \le n$, with weights $w_1, \ldots, w_i$, values $v_1, \ldots, v_i$, and knapsack capacity j, $1 \le j \le W$. Let $F(i, j)$ be the value of an optimal solution to this instance, i.e., the value of the most valuable subset of the first i items that fit into the knapsack of capacity j. We can divide all the subsets of the first i items that fit the knapsack of capacity j into two categories: those that do not include the ith item and those that do. Note the following:

1. Among the subsets that do not include the ith item, the value of an optimal subset is, by definition, $F(i-1, j)$.
2. Among the subsets that do include the ith item (hence, $j - w_i \ge 0$), an optimal subset is made up of this item and an optimal subset of the first $i-1$ items that fits into the knapsack of capacity $j - w_i$. The value of such an optimal subset is $v_i + F(i-1, j-w_i)$.

Thus, the value of an optimal solution among all feasible subsets of the first i items is the maximum of these two values. Of course, if the ith item does not fit into the knapsack, the value of an optimal subset selected from the first i items is the same as the value of an optimal subset selected from the first $i-1$ items. These observations lead to the following recurrence:

$$F(i, j) = \begin{cases} \max\{F(i-1, j),\ v_i + F(i-1, j-w_i)\} & \text{if } j - w_i \ge 0, \\ F(i-1, j) & \text{if } j - w_i < 0. \end{cases} \tag{8.6}$$

It is convenient to define the initial conditions as follows:

$$F(0, j) = 0 \text{ for } j \ge 0 \quad \text{and} \quad F(i, 0) = 0 \text{ for } i \ge 0. \tag{8.7}$$

Our goal is to find $F(n, W)$, the maximal value of a subset of the n given items that fit into the knapsack of capacity W, and an optimal subset itself.

Figure 8.4 illustrates the values involved in equations (8.6) and (8.7). For $i, j > 0$, to compute the entry in the ith row and the jth column, $F(i, j)$, we compute the maximum of the entry in the previous row and the same column and the sum of v_i and the entry in the previous row and w_i columns to the left. The table can be filled either row by row or column by column.

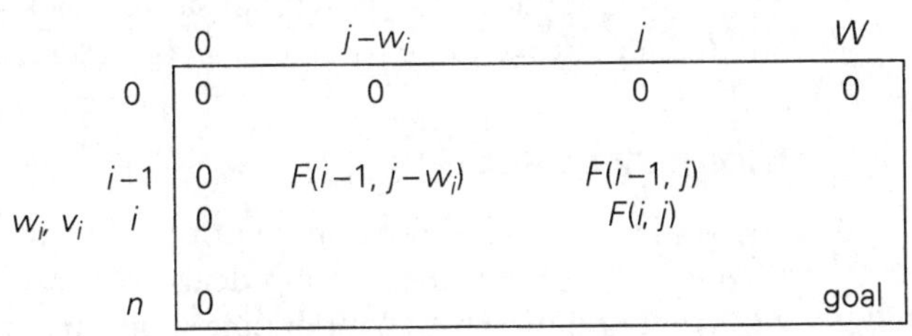

		0	$j-w_i$	j	W
	0	0	0	0	0
	$i-1$	0	$F(i-1, j-w_i)$	$F(i-1, j)$	
w_i, v_i	i	0		$F(i, j)$	
	n	0			goal

FIGURE 8.4 Table for solving the knapsack problem by dynamic programming.

	i	0	1	2	3	4	5
		capacity j					
	0	0	0	0	0	0	0
$w_1=2,\ v_1=12$	1	0	0	12	12	12	12
$w_2=1,\ v_2=10$	2	0	10	12	22	22	22
$w_3=3,\ v_3=20$	3	0	10	12	22	30	32
$w_4=2,\ v_4=15$	4	0	10	15	25	30	**37**

FIGURE 8.5 Example of solving an instance of the knapsack problem by the dynamic programming algorithm.

EXAMPLE 1 Let us consider the instance given by the following data:

item	weight	value
1	2	\$12
2	1	\$10
3	3	\$20
4	2	\$15

capacity $W = 5$.

The dynamic programming table, filled by applying formulas (8.6) and (8.7), is shown in Figure 8.5.

Thus, the maximal value is $F(4, 5) = \$37$. We can find the composition of an optimal subset by backtracing the computations of this entry in the table. Since $F(4, 5) > F(3, 5)$, item 4 has to be included in an optimal solution along with an optimal subset for filling $5 - 2 = 3$ remaining units of the knapsack capacity. The value of the latter is $F(3, 3)$. Since $F(3, 3) = F(2, 3)$, item 3 need not be in an optimal subset. Since $F(2, 3) > F(1, 3)$, item 2 is a part of an optimal selection, which leaves element $F(1, 3 - 1)$ to specify its remaining composition. Similarly, since $F(1, 2) > F(0, 2)$, item 1 is the final part of the optimal solution {item 1, item 2, item 4}. ■

The time efficiency and space efficiency of this algorithm are both in $\Theta(nW)$. The time needed to find the composition of an optimal solution is in $O(n)$. You are asked to prove these assertions in the exercises.

Memory Functions

As we discussed at the beginning of this chapter and illustrated in subsequent sections, dynamic programming deals with problems whose solutions satisfy a recurrence relation with overlapping subproblems. The direct top-down approach to finding a solution to such a recurrence leads to an algorithm that solves common subproblems more than once and hence is very inefficient (typically, exponential or worse). The classic dynamic programming approach, on the other hand, works bottom up: it fills a table with solutions to *all* smaller subproblems, but each of them is solved only once. An unsatisfying aspect of this approach is that solutions to some of these smaller subproblems are often not necessary for getting a solution to the problem given. Since this drawback is not present in the top-down approach, it is natural to try to combine the strengths of the top-down and bottom-up approaches. The goal is to get a method that solves only subproblems that are necessary and does so only once. Such a method exists; it is based on using ***memory functions***.

This method solves a given problem in the top-down manner but, in addition, maintains a table of the kind that would have been used by a bottom-up dynamic programming algorithm. Initially, all the table's entries are initialized with a special "null" symbol to indicate that they have not yet been calculated. Thereafter, whenever a new value needs to be calculated, the method checks the corresponding entry in the table first: if this entry is not "null," it is simply retrieved from the table; otherwise, it is computed by the recursive call whose result is then recorded in the table.

The following algorithm implements this idea for the knapsack problem. After initializing the table, the recursive function needs to be called with $i = n$ (the number of items) and $j = W$ (the knapsack capacity).

ALGORITHM *MFKnapsack*(i, j)

```
//Implements the memory function method for the knapsack problem
//Input: A nonnegative integer i indicating the number of the first
//        items being considered and a nonnegative integer j indicating
//        the knapsack capacity
//Output: The value of an optimal feasible subset of the first i items
//Note: Uses as global variables input arrays Weights[1..n], Values[1..n],
//and table F[0..n, 0..W] whose entries are initialized with −1's except for
//row 0 and column 0 initialized with 0's
if F[i, j] < 0
    if j < Weights[i]
        value ← MFKnapsack(i − 1, j)
    else
        value ← max(MFKnapsack(i − 1, j),
                    Values[i] + MFKnapsack(i − 1, j − Weights[i]))
    F[i, j] ← value
return F[i, j]
```

EXAMPLE 2 Let us apply the memory function method to the instance considered in Example 1. The table in Figure 8.6 gives the results. Only 11 out of 20 nontrivial values (i.e., not those in row 0 or in column 0) have been computed.

	i	0	1	2	3	4	5
				capacity j			
	0	0	0	0	0	0	0
$w_1 = 2, v_1 = 12$	1	0	0	12	12	12	12
$w_2 = 1, v_2 = 10$	2	0	—	12	22	—	22
$w_3 = 3, v_3 = 20$	3	0	—	—	22	—	32
$w_4 = 2, v_4 = 15$	4	0	—	—	—	—	**37**

FIGURE 8.6 Example of solving an instance of the knapsack problem by the memory function algorithm.

Just one nontrivial entry, $V(1, 2)$, is retrieved rather than being recomputed. For larger instances, the proportion of such entries can be significantly larger. ■

In general, we cannot expect more than a constant-factor gain in using the memory function method for the knapsack problem, because its time efficiency class is the same as that of the bottom-up algorithm (why?). A more significant improvement can be expected for dynamic programming algorithms in which a computation of one value takes more than constant time. You should also keep in mind that a memory function algorithm may be less space-efficient than a space-efficient version of a bottom-up algorithm.

Exercises 8.2

1. **a.** Apply the bottom-up dynamic programming algorithm to the following instance of the knapsack problem:

item	weight	value
1	3	\$25
2	2	\$20
3	1	\$15
4	4	\$40
5	5	\$50

capacity $W = 6$.

   b. How many different optimal subsets does the instance of part (a) have?

   c. In general, how can we use the table generated by the dynamic programming algorithm to tell whether there is more than one optimal subset for the knapsack problem's instance?

2. **a.** Write pseudocode of the bottom-up dynamic programming algorithm for the knapsack problem.

   b. Write pseudocode of the algorithm that finds the composition of an optimal subset from the table generated by the bottom-up dynamic programming algorithm for the knapsack problem.

3. For the bottom-up dynamic programming algorithm for the knapsack problem, prove that

   a. its time efficiency is $\Theta(nW)$.

   b. its space efficiency is $\Theta(nW)$.

   c. the time needed to find the composition of an optimal subset from a filled dynamic programming table is $O(n)$.

4. **a.** True or false: A sequence of values in a row of the dynamic programming table for the knapsack problem is always nondecreasing?

   b. True or false: A sequence of values in a column of the dynamic programming table for the knapsack problem is always nondecreasing?

5. Design a dynamic programming algorithm for the version of the knapsack problem in which there are unlimited quantities of copies for each of the n item kinds given. Indicate the time efficiency of the algorithm.

6. Apply the memory function method to the instance of the knapsack problem given in Problem 1. Indicate the entries of the dynamic programming table that are (i) never computed by the memory function method, (ii) retrieved without a recomputation.

7. Prove that the efficiency class of the memory function algorithm for the knapsack problem is the same as that of the bottom-up algorithm (see Problem 3).

8. Explain why the memory function approach is unattractive for the problem of computing a binomial coefficient by the formula $C(n, k) = C(n-1, k-1) + C(n-1, k)$.

9. Write a research report on one of the following well-known applications of dynamic programming:

   a. finding the longest common subsequence in two sequences

   b. optimal string editing

   c. minimal triangulation of a polygon

8.3 Optimal Binary Search Trees

A binary search tree is one of the most important data structures in computer science. One of its principal applications is to implement a dictionary, a set of elements with the operations of searching, insertion, and deletion. If probabilities

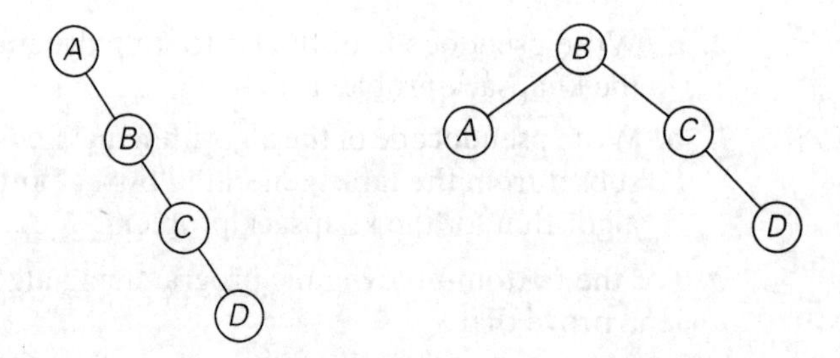

FIGURE 8.7 Two out of 14 possible binary search trees with keys A, B, C, and D.

of searching for elements of a set are known—e.g., from accumulated data about past searches—it is natural to pose a question about an optimal binary search tree for which the average number of comparisons in a search is the smallest possible. For simplicity, we limit our discussion to minimizing the average number of comparisons in a successful search. The method can be extended to include unsuccessful searches as well.

As an example, consider four keys A, B, C, and D to be searched for with probabilities 0.1, 0.2, 0.4, and 0.3, respectively. Figure 8.7 depicts two out of 14 possible binary search trees containing these keys. The average number of comparisons in a successful search in the first of these trees is $0.1 \cdot 1 + 0.2 \cdot 2 + 0.4 \cdot 3 + 0.3 \cdot 4 = 2.9$, and for the second one it is $0.1 \cdot 2 + 0.2 \cdot 1 + 0.4 \cdot 2 + 0.3 \cdot 3 = 2.1$. Neither of these two trees is, in fact, optimal. (Can you tell which binary tree is optimal?)

For our tiny example, we could find the optimal tree by generating all 14 binary search trees with these keys. As a general algorithm, this exhaustive-search approach is unrealistic: the total number of binary search trees with n keys is equal to the nth ***Catalan number***,

$$c(n) = \frac{1}{n+1}\binom{2n}{n} \quad \text{for } n > 0, \quad c(0) = 1,$$

which grows to infinity as fast as $4^n/n^{1.5}$ (see Problem 7 in this section's exercises).

So let $a_1, \ldots, a_n$ be distinct keys ordered from the smallest to the largest and let $p_1, \ldots, p_n$ be the probabilities of searching for them. Let $C(i, j)$ be the smallest average number of comparisons made in a successful search in a binary search tree T_i^j made up of keys $a_i, \ldots, a_j$, where i, j are some integer indices, $1 \le i \le j \le n$. Following the classic dynamic programming approach, we will find values of $C(i, j)$ for all smaller instances of the problem, although we are interested just in $C(1, n)$. To derive a recurrence underlying a dynamic programming algorithm, we will consider all possible ways to choose a root a_k among the keys $a_i, \ldots, a_j$. For such a binary search tree (Figure 8.8), the root contains key a_k, the left subtree T_i^{k-1} contains keys $a_i, \ldots, a_{k-1}$ optimally arranged, and the right subtree T_{k+1}^j

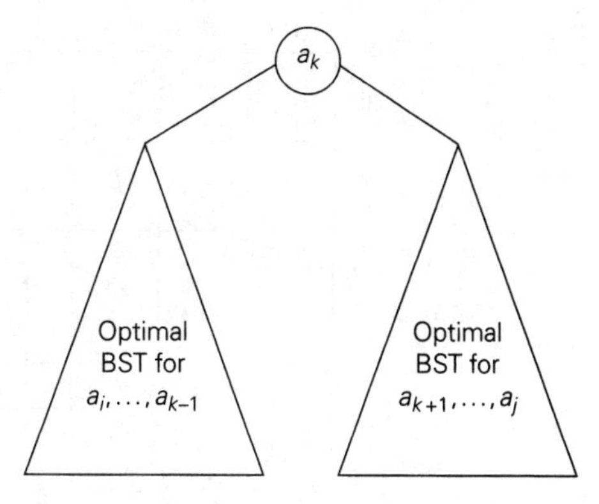

FIGURE 8.8 Binary search tree (BST) with root a_k and two optimal binary search subtrees T_i^{k-1} and T_{k+1}^j.

contains keys $a_{k+1}, \ldots, a_j$ also optimally arranged. (Note how we are taking advantage of the principle of optimality here.)

If we count tree levels starting with 1 to make the comparison numbers equal the keys' levels, the following recurrence relation is obtained:

$$\begin{aligned}
C(i, j) &= \min_{i \le k \le j} \{p_k \cdot 1 + \sum_{s=i}^{k-1} p_s \cdot (\text{level of } a_s \text{ in } T_i^{k-1} + 1) \\
&\qquad + \sum_{s=k+1}^{j} p_s \cdot (\text{level of } a_s \text{ in } T_{k+1}^{j} + 1)\} \\
&= \min_{i \le k \le j} \{\sum_{s=i}^{k-1} p_s \cdot \text{level of } a_s \text{ in } T_i^{k-1} + \sum_{s=k+1}^{j} p_s \cdot \text{level of } a_s \text{ in } T_{k+1}^{j} + \sum_{s=i}^{j} p_s\} \\
&= \min_{i \le k \le j} \{C(i, k-1) + C(k+1, j)\} + \sum_{s=i}^{j} p_s.
\end{aligned}$$

Thus, we have the recurrence

$$C(i, j) = \min_{i \le k \le j} \{C(i, k-1) + C(k+1, j)\} + \sum_{s=i}^{j} p_s \quad \text{for } 1 \le i \le j \le n. \quad \textbf{(8.8)}$$

We assume in formula (8.8) that $C(i, i-1) = 0$ for $1 \le i \le n+1$, which can be interpreted as the number of comparisons in the empty tree. Note that this formula implies that

$$C(i, i) = p_i \quad \text{for } 1 \le i \le n,$$

as it should be for a one-node binary search tree containing a_i.

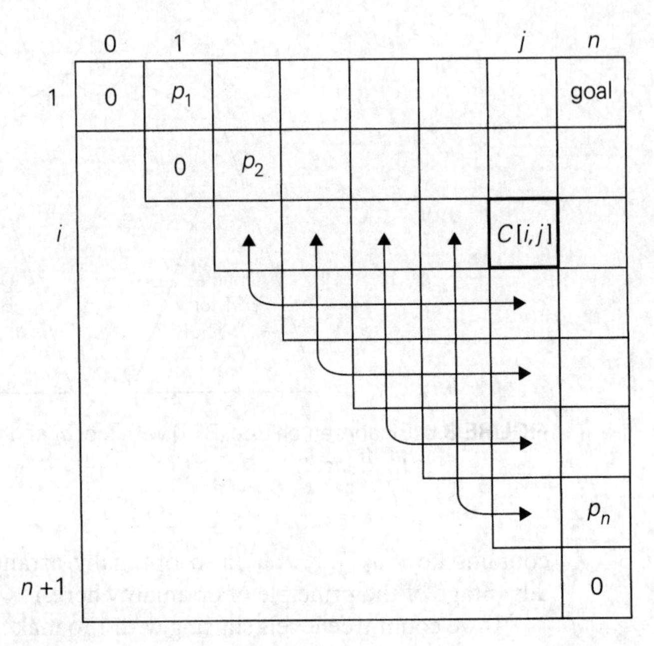

FIGURE 8.9 Table of the dynamic programming algorithm for constructing an optimal binary search tree.

The two-dimensional table in Figure 8.9 shows the values needed for computing $C(i, j)$ by formula (8.8): they are in row i and the columns to the left of column j and in column j and the rows below row i. The arrows point to the pairs of entries whose sums are computed in order to find the smallest one to be recorded as the value of $C(i, j)$. This suggests filling the table along its diagonals, starting with all zeros on the main diagonal and given probabilities p_i, $1 \le i \le n$, right above it and moving toward the upper right corner.

The algorithm we just sketched computes $C(1, n)$—the average number of comparisons for successful searches in the optimal binary tree. If we also want to get the optimal tree itself, we need to maintain another two-dimensional table to record the value of k for which the minimum in (8.8) is achieved. The table has the same shape as the table in Figure 8.9 and is filled in the same manner, starting with entries $R(i, i) = i$ for $1 \le i \le n$. When the table is filled, its entries indicate indices of the roots of the optimal subtrees, which makes it possible to reconstruct an optimal tree for the entire set given.

EXAMPLE Let us illustrate the algorithm by applying it to the four-key set we used at the beginning of this section:

key	A	B	C	D
probability	0.1	0.2	0.4	0.3

The initial tables look like this:

main table

	0	1	2	3	4
1	0	0.1			
2		0	0.2		
3			0	0.4	
4				0	0.3
5					0

root table

	0	1	2	3	4
1		1			
2			2		
3				3	
4					4
5					

Let us compute $C(1, 2)$:

$$C(1, 2) = \min \begin{Bmatrix} k = 1: & C(1, 0) + C(2, 2) + \sum_{s=1}^{2} p_s = 0 + 0.2 + 0.3 = 0.5 \\ k = 2: & C(1, 1) + C(3, 2) + \sum_{s=1}^{2} p_s = 0.1 + 0 + 0.3 = 0.4 \end{Bmatrix}$$
$$= 0.4.$$

Thus, out of two possible binary trees containing the first two keys, A and B, the root of the optimal tree has index 2 (i.e., it contains B), and the average number of comparisons in a successful search in this tree is 0.4.

We will ask you to finish the computations in the exercises. You should arrive at the following final tables:

main table

	0	1	2	3	4
1	0	0.1	0.4	1.1	1.7
2		0	0.2	0.8	1.4
3			0	0.4	1.0
4				0	0.3
5					0

root table

	0	1	2	3	4
1		1	2	3	3
2			2	3	3
3				3	3
4					4
5					

Thus, the average number of key comparisons in the optimal tree is equal to 1.7. Since $R(1, 4) = 3$, the root of the optimal tree contains the third key, i.e., C. Its left subtree is made up of keys A and B, and its right subtree contains just key D (why?). To find the specific structure of these subtrees, we find first their roots by consulting the root table again as follows. Since $R(1, 2) = 2$, the root of the optimal tree containing A and B is B, with A being its left child (and the root of the one-node tree: $R(1, 1) = 1$). Since $R(4, 4) = 4$, the root of this one-node optimal tree is its only key D. Figure 8.10 presents the optimal tree in its entirety.

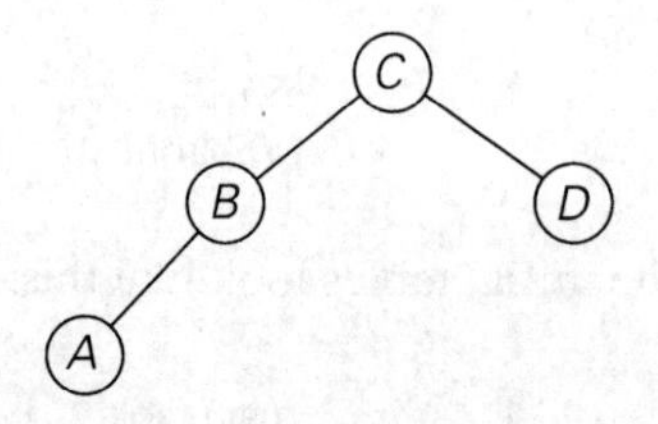

FIGURE 8.10 Optimal binary search tree for the example.

■

Here is pseudocode of the dynamic programming algorithm.

```
ALGORITHM OptimalBST(P[1..n])
    //Finds an optimal binary search tree by dynamic programming
    //Input: An array P[1..n] of search probabilities for a sorted list of n keys
    //Output: Average number of comparisons in successful searches in the
    //        optimal BST and table R of subtrees' roots in the optimal BST
    for i ← 1 to n do
        C[i, i − 1] ← 0
        C[i, i] ← P[i]
        R[i, i] ← i
    C[n + 1, n] ← 0
    for d ← 1 to n − 1 do  //diagonal count
        for i ← 1 to n − d do
            j ← i + d
            minval ← ∞
            for k ← i to j do
                if C[i, k − 1] + C[k + 1, j] < minval
                    minval ← C[i, k − 1] + C[k + 1, j];   kmin ← k
            R[i, j] ← kmin
            sum ← P[i];  for s ← i + 1 to j do sum ← sum + P[s]
            C[i, j] ← minval + sum
    return C[1, n], R
```

The algorithm's space efficiency is clearly quadratic; the time efficiency of this version of the algorithm is cubic (why?). A more careful analysis shows that entries in the root table are always nondecreasing along each row and column. This limits values for $R(i, j)$ to the range $R(i, j-1), \ldots, R(i+1, j)$ and makes it possible to reduce the running time of the algorithm to $\Theta(n^2)$.

Exercises 8.3

1. Finish the computations started in the section's example of constructing an optimal binary search tree.
2. **a.** Why is the time efficiency of algorithm *OptimalBST* cubic?

   b. Why is the space efficiency of algorithm *OptimalBST* quadratic?
3. Write pseudocode for a linear-time algorithm that generates the optimal binary search tree from the root table.
4. Devise a way to compute the sums $\sum_{s=i}^{j} p_s$, which are used in the dynamic programming algorithm for constructing an optimal binary search tree, in constant time (per sum).
5. True or false: The root of an optimal binary search tree always contains the key with the highest search probability?
6. How would you construct an optimal binary search tree for a set of n keys if all the keys are equally likely to be searched for? What will be the average number of comparisons in a successful search in such a tree if $n = 2^k$?
7. **a.** Show that the number of distinct binary search trees $b(n)$ that can be constructed for a set of n orderable keys satisfies the recurrence relation

$$b(n) = \sum_{k=0}^{n-1} b(k)b(n-1-k) \quad \text{for } n > 0, \;\; b(0) = 1.$$

   b. It is known that the solution to this recurrence is given by the Catalan numbers. Verify this assertion for $n = 1, 2, \ldots, 5$.

   c. Find the order of growth of $b(n)$. What implication does the answer to this question have for the exhaustive-search algorithm for constructing an optimal binary search tree?
8. Design a $\Theta(n^2)$ algorithm for finding an optimal binary search tree.
9. Generalize the optimal binary search algorithm by taking into account unsuccessful searches.
10. Write pseudocode of a memory function for the optimal binary search tree problem. You may limit your function to finding the smallest number of key comparisons in a successful search.
11. ***Matrix chain multiplication*** Consider the problem of minimizing the total number of multiplications made in computing the product of n matrices

$$A_1 \cdot A_2 \cdot \ldots \cdot A_n$$

   whose dimensions are $d_0 \times d_1$, $d_1 \times d_2$, $\ldots$, $d_{n-1} \times d_n$, respectively. Assume that all intermediate products of two matrices are computed by the brute-force (definition-based) algorithm.

a. Give an example of three matrices for which the number of multiplications in $(A_1 \cdot A_2) \cdot A_3$ and $A_1 \cdot (A_2 \cdot A_3)$ differ at least by a factor of 1000.

b. How many different ways are there to compute the product of n matrices?

c. Design a dynamic programming algorithm for finding an optimal order of multiplying n matrices.

8.4 Warshall's and Floyd's Algorithms

In this section, we look at two well-known algorithms: Warshall's algorithm for computing the transitive closure of a directed graph and Floyd's algorithm for the all-pairs shortest-paths problem. These algorithms are based on essentially the same idea: exploit a relationship between a problem and its simpler rather than smaller version. Warshall and Floyd published their algorithms without mentioning dynamic programming. Nevertheless, the algorithms certainly have a dynamic programming flavor and have come to be considered applications of this technique.

Warshall's Algorithm

Recall that the adjacency matrix $A = \{a_{ij}\}$ of a directed graph is the boolean matrix that has 1 in its ith row and jth column if and only if there is a directed edge from the ith vertex to the jth vertex. We may also be interested in a matrix containing the information about the existence of directed paths of arbitrary lengths between vertices of a given graph. Such a matrix, called the transitive closure of the digraph, would allow us to determine in constant time whether the jth vertex is reachable from the ith vertex.

Here are a few application examples. When a value in a spreadsheet cell is changed, the spreadsheet software must know all the other cells affected by the change. If the spreadsheet is modeled by a digraph whose vertices represent the spreadsheet cells and edges indicate cell dependencies, the transitive closure will provide such information. In software engineering, transitive closure can be used for investigating data flow and control flow dependencies as well as for inheritance testing of object-oriented software. In electronic engineering, it is used for redundancy identification and test generation for digital circuits.

DEFINITION The ***transitive closure*** of a directed graph with n vertices can be defined as the $n \times n$ boolean matrix $T = \{t_{ij}\}$, in which the element in the ith row and the jth column is 1 if there exists a nontrivial path (i.e., directed path of a positive length) from the ith vertex to the jth vertex; otherwise, t_{ij} is 0.

An example of a digraph, its adjacency matrix, and its transitive closure is given in Figure 8.11.

We can generate the transitive closure of a digraph with the help of depth-first search or breadth-first search. Performing either traversal starting at the ith

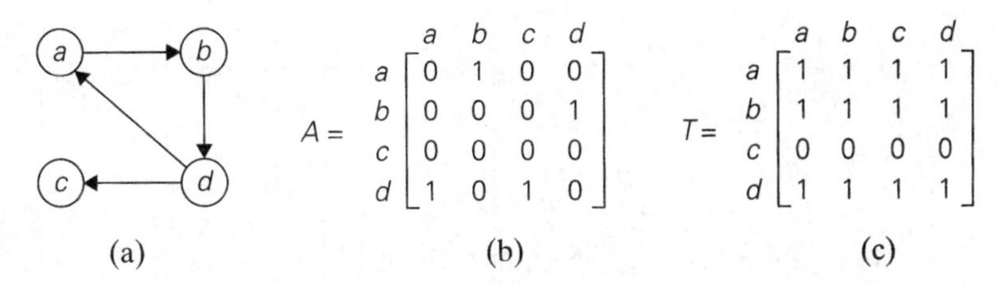

FIGURE 8.11 (a) Digraph. (b) Its adjacency matrix. (c) Its transitive closure.

vertex gives the information about the vertices reachable from it and hence the columns that contain 1's in the ith row of the transitive closure. Thus, doing such a traversal for every vertex as a starting point yields the transitive closure in its entirety.

Since this method traverses the same digraph several times, we should hope that a better algorithm can be found. Indeed, such an algorithm exists. It is called ***Warshall's algorithm*** after Stephen Warshall, who discovered it [War62]. It is convenient to assume that the digraph's vertices and hence the rows and columns of the adjacency matrix are numbered from 1 to n. Warshall's algorithm constructs the transitive closure through a series of $n \times n$ boolean matrices:

$$R^{(0)}, \ldots, R^{(k-1)}, R^{(k)}, \ldots R^{(n)}. \tag{8.9}$$

Each of these matrices provides certain information about directed paths in the digraph. Specifically, the element $r_{ij}^{(k)}$ in the ith row and jth column of matrix $R^{(k)}$ $(i, j = 1, 2, \ldots, n, k = 0, 1, \ldots, n)$ is equal to 1 if and only if there exists a directed path of a positive length from the ith vertex to the jth vertex with each intermediate vertex, if any, numbered not higher than k. Thus, the series starts with $R^{(0)}$, which does not allow any intermediate vertices in its paths; hence, $R^{(0)}$ is nothing other than the adjacency matrix of the digraph. (Recall that the adjacency matrix contains the information about one-edge paths, i.e., paths with no intermediate vertices.) $R^{(1)}$ contains the information about paths that can use the first vertex as intermediate; thus, with more freedom, so to speak, it may contain more 1's than $R^{(0)}$. In general, each subsequent matrix in series (8.9) has one more vertex to use as intermediate for its paths than its predecessor and hence may, but does not have to, contain more 1's. The last matrix in the series, $R^{(n)}$, reflects paths that can use all n vertices of the digraph as intermediate and hence is nothing other than the digraph's transitive closure.

The central point of the algorithm is that we can compute all the elements of each matrix $R^{(k)}$ from its immediate predecessor $R^{(k-1)}$ in series (8.9). Let $r_{ij}^{(k)}$, the element in the ith row and jth column of matrix $R^{(k)}$, be equal to 1. This means that there exists a path from the ith vertex v_i to the jth vertex v_j with each intermediate vertex numbered not higher than k:

$$v_i, \text{ a list of intermediate vertices each numbered not higher than } k, v_j. \tag{8.10}$$

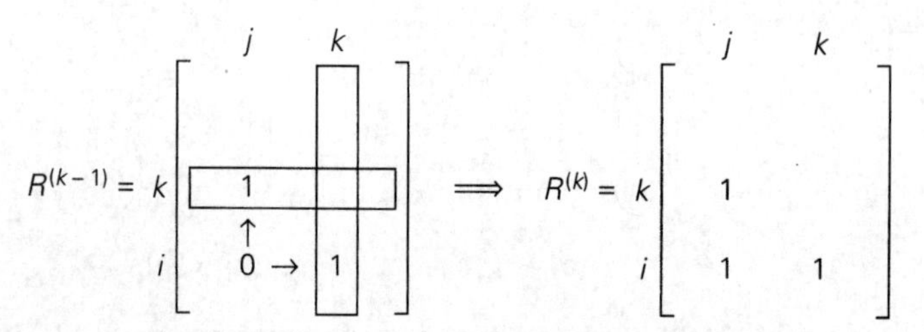

FIGURE 8.12 Rule for changing zeros in Warshall's algorithm.

Two situations regarding this path are possible. In the first, the list of its intermediate vertices does not contain the kth vertex. Then this path from v_i to v_j has intermediate vertices numbered not higher than $k-1$, and therefore $r_{ij}^{(k-1)}$ is equal to 1 as well. The second possibility is that path (8.10) does contain the kth vertex v_k among the intermediate vertices. Without loss of generality, we may assume that v_k occurs only once in that list. (If it is not the case, we can create a new path from v_i to v_j with this property by simply eliminating all the vertices between the first and last occurrences of v_k in it.) With this caveat, path (8.10) can be rewritten as follows:

$$v_i, \text{ vertices numbered} \leq k-1, v_k, \text{ vertices numbered} \leq k-1, v_j.$$

The first part of this representation means that there exists a path from v_i to v_k with each intermediate vertex numbered not higher than $k-1$ (hence, $r_{ik}^{(k-1)}=1$), and the second part means that there exists a path from v_k to v_j with each intermediate vertex numbered not higher than $k-1$ (hence, $r_{kj}^{(k-1)}=1$).

What we have just proved is that if $r_{ij}^{(k)}=1$, then either $r_{ij}^{(k-1)}=1$ or both $r_{ik}^{(k-1)}=1$ and $r_{kj}^{(k-1)}=1$. It is easy to see that the converse of this assertion is also true. Thus, we have the following formula for generating the elements of matrix $R^{(k)}$ from the elements of matrix $R^{(k-1)}$:

$$r_{ij}^{(k)}=r_{ij}^{(k-1)} \quad \text{or} \quad \left(r_{ik}^{(k-1)} \text{ and } r_{kj}^{(k-1)}\right). \tag{8.11}$$

Formula (8.11) is at the heart of Warshall's algorithm. This formula implies the following rule for generating elements of matrix $R^{(k)}$ from elements of matrix $R^{(k-1)}$, which is particularly convenient for applying Warshall's algorithm by hand:

- If an element r_{ij} is 1 in $R^{(k-1)}$, it remains 1 in $R^{(k)}$.
- If an element r_{ij} is 0 in $R^{(k-1)}$, it has to be changed to 1 in $R^{(k)}$ if and only if the element in its row i and column k and the element in its column j and row k are both 1's in $R^{(k-1)}$. This rule is illustrated in Figure 8.12.

As an example, the application of Warshall's algorithm to the digraph in Figure 8.11 is shown in Figure 8.13.

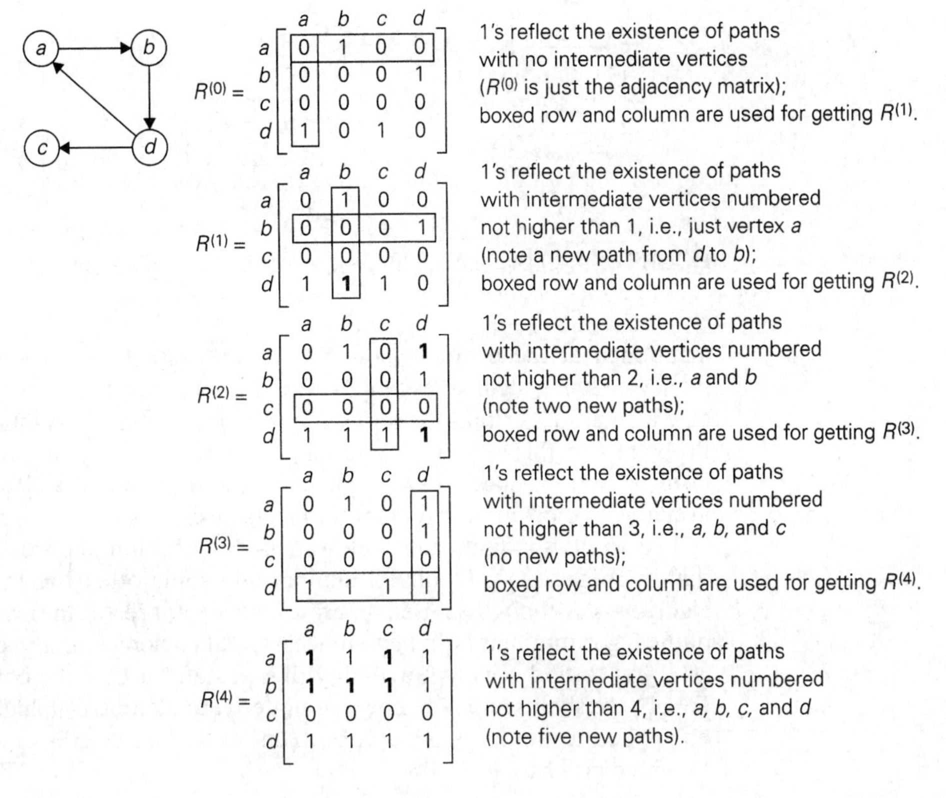

FIGURE 8.13 Application of Warshall's algorithm to the digraph shown. New 1's are in bold.

Here is pseudocode of Warshall's algorithm.

```
ALGORITHM Warshall(A[1..n, 1..n])
    //Implements Warshall's algorithm for computing the transitive closure
    //Input: The adjacency matrix A of a digraph with n vertices
    //Output: The transitive closure of the digraph
    R(0) ← A
    for k ← 1 to n do
        for i ← 1 to n do
            for j ← 1 to n do
                R(k)[i, j] ← R(k−1)[i, j] or (R(k−1)[i, k] and R(k−1)[k, j])
    return R(n)
```

Several observations need to be made about Warshall's algorithm. First, it is remarkably succinct, is it not? Still, its time efficiency is only $\Theta(n^3)$. In fact, for sparse graphs represented by their adjacency lists, the traversal-based algorithm

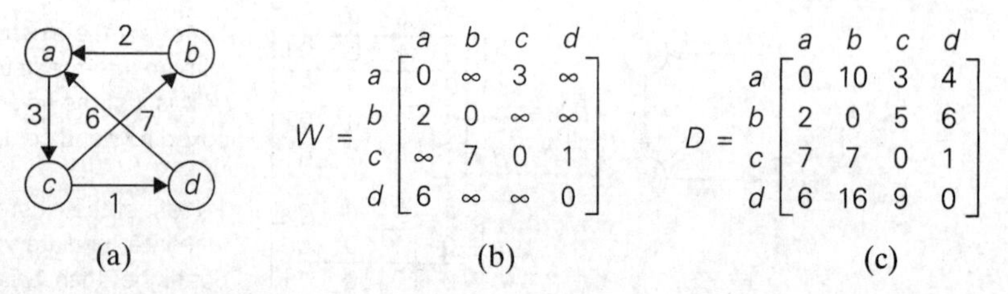

FIGURE 8.14 (a) Digraph. (b) Its weight matrix. (c) Its distance matrix.

mentioned at the beginning of this section has a better asymptotic efficiency than Warshall's algorithm (why?). We can speed up the above implementation of Warshall's algorithm for some inputs by restructuring its innermost loop (see Problem 4 in this section's exercises). Another way to make the algorithm run faster is to treat matrix rows as bit strings and employ the bitwise *or* operation available in most modern computer languages.

As to the space efficiency of Warshall's algorithm, the situation is similar to that of computing a Fibonacci number and some other dynamic programming algorithms. Although we used separate matrices for recording intermediate results of the algorithm, this is, in fact, unnecessary. Problem 3 in this section's exercises asks you to find a way of avoiding this wasteful use of the computer memory. Finally, we shall see below how the underlying idea of Warshall's algorithm can be applied to the more general problem of finding lengths of shortest paths in weighted graphs.

Floyd's Algorithm for the All-Pairs Shortest-Paths Problem

Given a weighted connected graph (undirected or directed), the ***all-pairs shortest-paths problem*** asks to find the distances—i.e., the lengths of the shortest paths—from each vertex to all other vertices. This is one of several variations of the problem involving shortest paths in graphs. Because of its important applications to communications, transportation networks, and operations research, it has been thoroughly studied over the years. Among recent applications of the all-pairs shortest-path problem is precomputing distances for motion planning in computer games.

It is convenient to record the lengths of shortest paths in an $n \times n$ matrix D called the ***distance matrix***: the element d_{ij} in the ith row and the jth column of this matrix indicates the length of the shortest path from the ith vertex to the jth vertex. For an example, see Figure 8.14.

We can generate the distance matrix with an algorithm that is very similar to Warshall's algorithm. It is called ***Floyd's algorithm*** after its co-inventor Robert W. Floyd.[1] It is applicable to both undirected and directed weighted graphs provided

1. Floyd explicitly referenced Warshall's paper in presenting his algorithm [Flo62]. Three years earlier, Bernard Roy published essentially the same algorithm in the proceedings of the French Academy of Sciences [Roy59].

that they do not contain a cycle of a negative length. (The distance between any two vertices in such a cycle can be made arbitrarily small by repeating the cycle enough times.) The algorithm can be enhanced to find not only the lengths of the shortest paths for all vertex pairs but also the shortest paths themselves (Problem 10 in this section's exercises).

Floyd's algorithm computes the distance matrix of a weighted graph with n vertices through a series of $n \times n$ matrices:

$$D^{(0)}, \ldots, D^{(k-1)}, D^{(k)}, \ldots, D^{(n)}. \quad \textbf{(8.12)}$$

Each of these matrices contains the lengths of shortest paths with certain constraints on the paths considered for the matrix in question. Specifically, the element $d_{ij}^{(k)}$ in the ith row and the jth column of matrix $D^{(k)}$ $(i, j = 1, 2, \ldots, n,$ $k = 0, 1, \ldots, n)$ is equal to the length of the shortest path among all paths from the ith vertex to the jth vertex with each intermediate vertex, if any, numbered not higher than k. In particular, the series starts with $D^{(0)}$, which does not allow any intermediate vertices in its paths; hence, $D^{(0)}$ is simply the weight matrix of the graph. The last matrix in the series, $D^{(n)}$, contains the lengths of the shortest paths among all paths that can use all n vertices as intermediate and hence is nothing other than the distance matrix being sought.

As in Warshall's algorithm, we can compute all the elements of each matrix $D^{(k)}$ from its immediate predecessor $D^{(k-1)}$ in series (8.12). Let $d_{ij}^{(k)}$ be the element in the ith row and the jth column of matrix $D^{(k)}$. This means that $d_{ij}^{(k)}$ is equal to the length of the shortest path among all paths from the ith vertex v_i to the jth vertex v_j with their intermediate vertices numbered not higher than k:

$$v_i, \text{ a list of intermediate vertices each numbered not higher than } k, v_j. \quad \textbf{(8.13)}$$

We can partition all such paths into two disjoint subsets: those that do not use the kth vertex v_k as intermediate and those that do. Since the paths of the first subset have their intermediate vertices numbered not higher than $k - 1$, the shortest of them is, by definition of our matrices, of length $d_{ij}^{(k-1)}$.

What is the length of the shortest path in the second subset? If the graph does not contain a cycle of a negative length, we can limit our attention only to the paths in the second subset that use vertex v_k as their intermediate vertex exactly once (because visiting v_k more than once can only increase the path's length). All such paths have the following form:

$$v_i, \text{ vertices numbered} \le k-1, v_k, \text{ vertices numbered} \le k-1, v_j.$$

In other words, each of the paths is made up of a path from v_i to v_k with each intermediate vertex numbered not higher than $k - 1$ and a path from v_k to v_j with each intermediate vertex numbered not higher than $k - 1$. The situation is depicted symbolically in Figure 8.15.

Since the length of the shortest path from v_i to v_k among the paths that use intermediate vertices numbered not higher than $k - 1$ is equal to $d_{ik}^{(k-1)}$ and the length of the shortest path from v_k to v_j among the paths that use intermediate

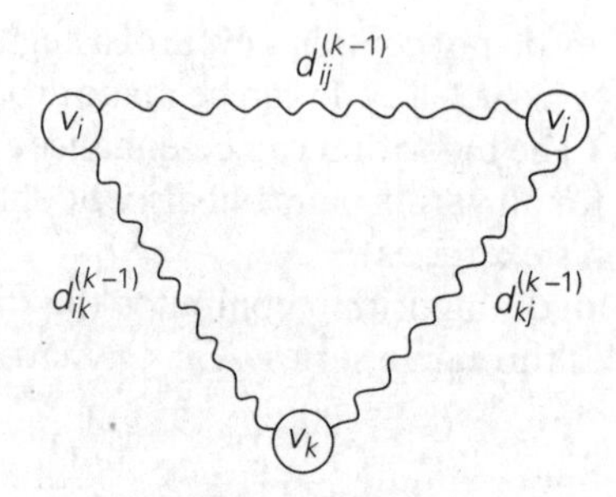

FIGURE 8.15 Underlying idea of Floyd's algorithm.

vertices numbered not higher than $k-1$ is equal to $d_{kj}^{(k-1)}$, the length of the shortest path among the paths that use the kth vertex is equal to $d_{ik}^{(k-1)} + d_{kj}^{(k-1)}$. Taking into account the lengths of the shortest paths in both subsets leads to the following recurrence:

$$d_{ij}^{(k)} = \min\{d_{ij}^{(k-1)},\ d_{ik}^{(k-1)} + d_{kj}^{(k-1)}\} \quad \text{for } k \geq 1,\ d_{ij}^{(0)} = w_{ij}. \qquad \textbf{(8.14)}$$

To put it another way, the element in row i and column j of the current distance matrix $D^{(k-1)}$ is replaced by the sum of the elements in the same row i and the column k and in the same column j and the row k if and only if the latter sum is smaller than its current value.

The application of Floyd's algorithm to the graph in Figure 8.14 is illustrated in Figure 8.16.

Here is pseudocode of Floyd's algorithm. It takes advantage of the fact that the next matrix in sequence (8.12) can be written over its predecessor.

ALGORITHM *Floyd*(W[1..n, 1..n])

```
//Implements Floyd's algorithm for the all-pairs shortest-paths problem
//Input: The weight matrix W of a graph with no negative-length cycle
//Output: The distance matrix of the shortest paths' lengths
D ← W //is not necessary if W can be overwritten
for k ← 1 to n do
    for i ← 1 to n do
        for j ← 1 to n do
            D[i, j] ← min{D[i, j], D[i, k] + D[k, j]}
return D
```

Obviously, the time efficiency of Floyd's algorithm is cubic—as is the time efficiency of Warshall's algorithm. In the next chapter, we examine Dijkstra's algorithm—another method for finding shortest paths.

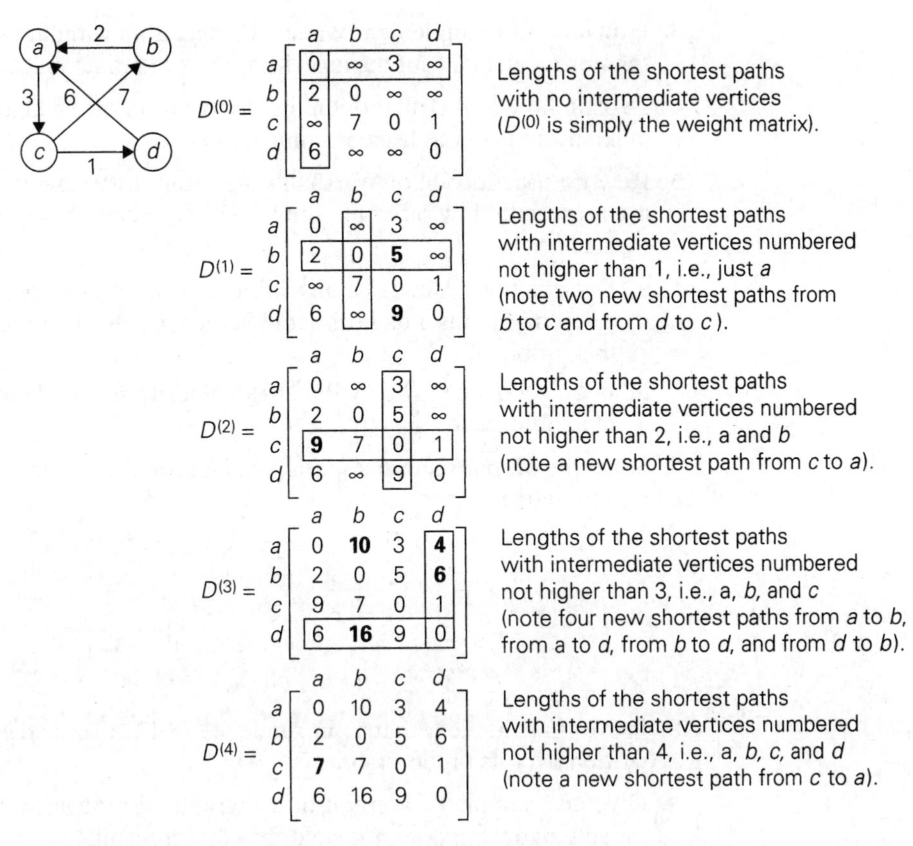

FIGURE 8.16 Application of Floyd's algorithm to the digraph shown. Updated elements are shown in bold.

Exercises 8.4

1. Apply Warshall's algorithm to find the transitive closure of the digraph defined by the following adjacency matrix:

$$\begin{bmatrix} 0 & 1 & 0 & 0 \\ 0 & 0 & 1 & 0 \\ 0 & 0 & 0 & 1 \\ 0 & 0 & 0 & 0 \end{bmatrix}$$

2. **a.** Prove that the time efficiency of Warshall's algorithm is cubic.

   b. Explain why the time efficiency class of Warshall's algorithm is inferior to that of the traversal-based algorithm for sparse graphs represented by their adjacency lists.

3. Explain how to implement Warshall's algorithm without using extra memory for storing elements of the algorithm's intermediate matrices.

4. Explain how to restructure the innermost loop of the algorithm *Warshall* to make it run faster at least on some inputs.

5. Rewrite pseudocode of Warshall's algorithm assuming that the matrix rows are represented by bit strings on which the bitwise *or* operation can be performed.

6. **a.** Explain how Warshall's algorithm can be used to determine whether a given digraph is a dag (directed acyclic graph). Is it a good algorithm for this problem?

   b. Is it a good idea to apply Warshall's algorithm to find the transitive closure of an undirected graph?

7. Solve the all-pairs shortest-path problem for the digraph with the following weight matrix:

$$\begin{bmatrix} 0 & 2 & \infty & 1 & 8 \\ 6 & 0 & 3 & 2 & \infty \\ \infty & \infty & 0 & 4 & \infty \\ \infty & \infty & 2 & 0 & 3 \\ 3 & \infty & \infty & \infty & 0 \end{bmatrix}$$

8. Prove that the next matrix in sequence (8.12) of Floyd's algorithm can be written over its predecessor.

9. Give an example of a graph or a digraph with negative weights for which Floyd's algorithm does not yield the correct result.

10. Enhance Floyd's algorithm so that shortest paths themselves, not just their lengths, can be found.

11. *Jack Straws* In the game of Jack Straws, a number of plastic or wooden "straws" are dumped on the table and players try to remove them one by one without disturbing the other straws. Here, we are only concerned with whether various pairs of straws are connected by a path of touching straws. Given a list of the endpoints for $n > 1$ straws (as if they were dumped on a large piece of graph paper), determine all the pairs of straws that are connected. Note that touching is connecting, but also that two straws can be connected indirectly via other connected straws. [1994 East-Central Regionals of the ACM International Collegiate Programming Contest]

SUMMARY

- *Dynamic programming* is a technique for solving problems with overlapping subproblems. Typically, these subproblems arise from a recurrence relating a solution to a given problem with solutions to its smaller subproblems of the same type. Dynamic programming suggests solving each smaller subproblem once and recording the results in a table from which a solution to the original problem can be then obtained.

- Applicability of dynamic programming to an optimization problem requires the problem to satisfy the *principle of optimality*: an optimal solution to any of its instances must be made up of optimal solutions to its subinstances.
- Among many other problems, the *change-making problem* with arbitrary coin denominations can be solved by dynamic programming.
- Solving a knapsack problem by a dynamic programming algorithm exemplifies an application of this technique to difficult problems of combinatorial optimization.
- The *memory function* technique seeks to combine the strengths of the top-down and bottom-up approaches to solving problems with overlapping subproblems. It does this by solving, in the top-down fashion but only once, just the necessary subproblems of a given problem and recording their solutions in a table.
- Dynamic programming can be used for constructing an *optimal binary search tree* for a given set of keys and known probabilities of searching for them.
- *Warshall's algorithm* for finding the *transitive closure* and *Floyd's algorithm* for the *all-pairs shortest-paths problem* are based on the idea that can be interpreted as an application of the dynamic programming technique.

9

Greedy Technique

Greed, for lack of a better word, is good! Greed is right! Greed works!

—Michael Douglas, US actor in the role of Gordon Gecko,
in the film *Wall Street*, 1987

Let us revisit the ***change-making problem*** faced, at least subconsciously, by millions of cashiers all over the world: give change for a specific amount n with the least number of coins of the denominations $d_1 > d_2 > \cdots > d_m$ used in that locale. (Here, unlike Section 8.1, we assume that the denominations are ordered in decreasing order.) For example, the widely used coin denominations in the United States are $d_1 = 25$ (quarter), $d_2 = 10$ (dime), $d_3 = 5$ (nickel), and $d_4 = 1$ (penny). How would you give change with coins of these denominations of, say, 48 cents? If you came up with the answer 1 quarter, 2 dimes, and 3 pennies, you followed—consciously or not—a logical strategy of making a sequence of best choices among the currently available alternatives. Indeed, in the first step, you could have given one coin of any of the four denominations. "Greedy" thinking leads to giving one quarter because it reduces the remaining amount the most, namely, to 23 cents. In the second step, you had the same coins at your disposal, but you could not give a quarter, because it would have violated the problem's constraints. So your best selection in this step was one dime, reducing the remaining amount to 13 cents. Giving one more dime left you with 3 cents to be given with three pennies.

Is this solution to the instance of the change-making problem optimal? Yes, it is. In fact, one can prove that the greedy algorithm yields an optimal solution for every positive integer amount with these coin denominations. At the same time, it is easy to give an example of coin denominations that do not yield an optimal solution for some amounts—e.g., $d_1 = 25$, $d_2 = 10$, $d_3 = 1$ and $n = 30$.

The approach applied in the opening paragraph to the change-making problem is called ***greedy***. Computer scientists consider it a general design technique despite the fact that it is applicable to optimization problems only. The greedy approach suggests constructing a solution through a sequence of steps, each expanding a partially constructed solution obtained so far, until a complete solution to the problem is reached. On each step—and this is the central point of this technique—the choice made must be:

- *feasible*, i.e., it has to satisfy the problem's constraints
- *locally optimal*, i.e., it has to be the best local choice among all feasible choices available on that step
- *irrevocable*, i.e., once made, it cannot be changed on subsequent steps of the algorithm

These requirements explain the technique's name: on each step, it suggests a "greedy" grab of the best alternative available in the hope that a sequence of locally optimal choices will yield a (globally) optimal solution to the entire problem. We refrain from a philosophical discussion of whether greed is good or bad. (If you have not seen the movie from which the chapter's epigraph is taken, its hero did not end up well.) From our algorithmic perspective, the question is whether such a greedy strategy works or not. As we shall see, there are problems for which a sequence of locally optimal choices does yield an optimal solution for every instance of the problem in question. However, there are others for which this is not the case; for such problems, a greedy algorithm can still be of value if we are interested in or have to be satisfied with an approximate solution.

In the first two sections of the chapter, we discuss two classic algorithms for the minimum spanning tree problem: Prim's algorithm and Kruskal's algorithm. What is remarkable about these algorithms is the fact that they solve the same problem by applying the greedy approach in two different ways, and both of them always yield an optimal solution. In Section 9.3, we introduce another classic algorithm—Dijkstra's algorithm for the shortest-path problem in a weighted graph. Section 9.4 is devoted to Huffman trees and their principal application, Huffman codes—an important data compression method that can be interpreted as an application of the greedy technique. Finally, a few examples of approximation algorithms based on the greedy approach are discussed in Section 12.3.

As a rule, greedy algorithms are both intuitively appealing and simple. Given an optimization problem, it is usually easy to figure out how to proceed in a greedy manner, possibly after considering a few small instances of the problem. What is usually more difficult is to prove that a greedy algorithm yields an optimal solution (when it does). One of the common ways to do this is illustrated by the proof given in Section 9.1: using mathematical induction, we show that a partially constructed solution obtained by the greedy algorithm on each iteration can be extended to an optimal solution to the problem.

The second way to prove optimality of a greedy algorithm is to show that on each step it does at least as well as any other algorithm could in advancing toward the problem's goal. Consider, as an example, the following problem: find the minimum number of moves needed for a chess knight to go from one corner of a 100×100 board to the diagonally opposite corner. (The knight's moves are L-shaped jumps: two squares horizontally or vertically followed by one square in the perpendicular direction.) A greedy solution is clear here: jump as close to the goal as possible on each move. Thus, if its start and finish squares are (1,1) and (100, 100), respectively, a sequence of 66 moves such as

$$(1, 1) - (3, 2) - (4, 4) - \cdots - (97, 97) - (99, 98) - (100, 100)$$

solves the problem. (The number k of two-move advances can be obtained from the equation $1 + 3k = 100$.) Why is this a minimum-move solution? Because if we measure the distance to the goal by the Manhattan distance, which is the sum of the difference between the row numbers and the difference between the column numbers of two squares in question, the greedy algorithm decreases it by 3 on each move—the best the knight can do.

The third way is simply to show that the final result obtained by a greedy algorithm is optimal based on the algorithm's output rather than the way it operates. As an example, consider the problem of placing the maximum number of chips on an 8×8 board so that no two chips are placed on the same or adjacent—vertically, horizontally, or diagonally—squares. To follow the prescription of the greedy strategy, we should place each new chip so as to leave as many available squares as possible for next chips. For example, starting with the upper left corner of the board, we will be able to place 16 chips as shown in Figure 9.1a. Why is this solution optimal? To see why, partition the board into sixteen 4×4 squares as shown in Figure 9.1b. Obviously, it is impossible to place more than one chip in each of these squares, which implies that the total number of nonadjacent chips on the board cannot exceed 16.

As a final comment, we should mention that a rather sophisticated theory has been developed behind the greedy technique, which is based on the abstract combinatorial structure called "matroid." An interested reader can check such books as [Cor09] as well as a variety of Internet resources on the subject.

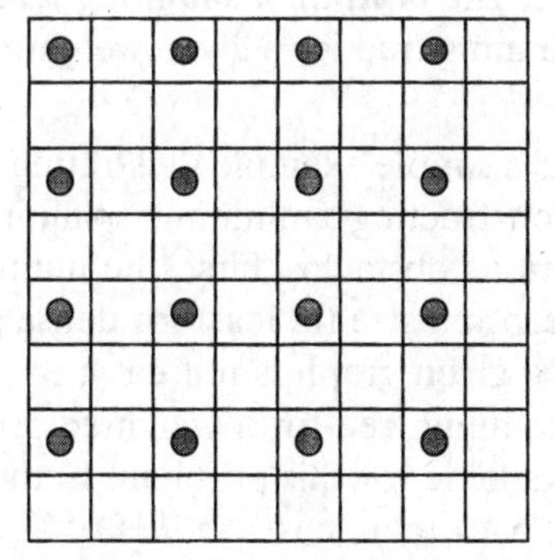

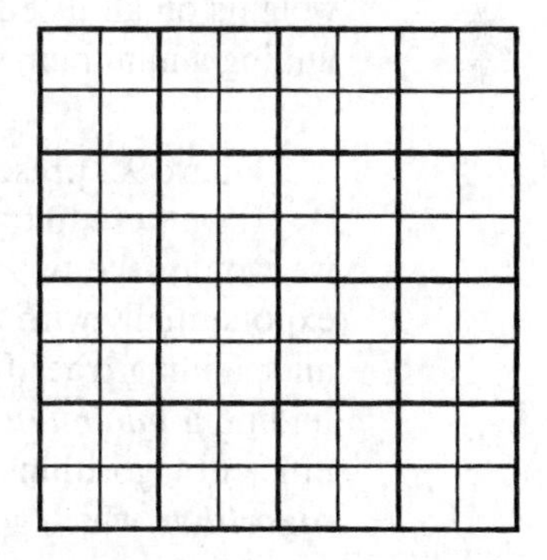

FIGURE 9.1 (a) Placement of 16 chips on non-adjacent squares. (b) Partition of the board proving impossibility of placing more than 16 chips.

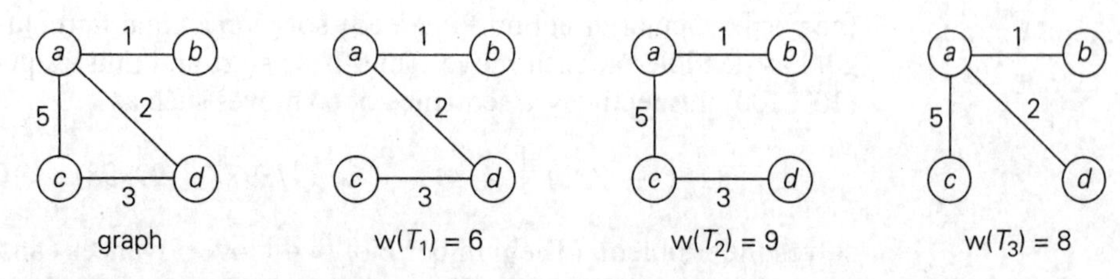

FIGURE 9.2 Graph and its spanning trees, with T_1 being the minimum spanning tree.

9.1 Prim's Algorithm

The following problem arises naturally in many practical situations: given n points, connect them in the cheapest possible way so that there will be a path between every pair of points. It has direct applications to the design of all kinds of networks—including communication, computer, transportation, and electrical—by providing the cheapest way to achieve connectivity. It identifies clusters of points in data sets. It has been used for classification purposes in archeology, biology, sociology, and other sciences. It is also helpful for constructing approximate solutions to more difficult problems such the traveling salesman problem (see Section 12.3).

We can represent the points given by vertices of a graph, possible connections by the graph's edges, and the connection costs by the edge weights. Then the question can be posed as the minimum spanning tree problem, defined formally as follows.

DEFINITION A ***spanning tree*** of an undirected connected graph is its connected acyclic subgraph (i.e., a tree) that contains all the vertices of the graph. If such a graph has weights assigned to its edges, a ***minimum spanning tree*** is its spanning tree of the smallest weight, where the ***weight*** of a tree is defined as the sum of the weights on all its edges. The ***minimum spanning tree problem*** is the problem of finding a minimum spanning tree for a given weighted connected graph.

Figure 9.2 presents a simple example illustrating these notions.

If we were to try constructing a minimum spanning tree by exhaustive search, we would face two serious obstacles. First, the number of spanning trees grows exponentially with the graph size (at least for dense graphs). Second, generating all spanning trees for a given graph is not easy; in fact, it is more difficult than finding a *minimum* spanning tree for a weighted graph by using one of several efficient algorithms available for this problem. In this section, we outline ***Prim's algorithm***, which goes back to at least 1957[1] [Pri57].

1. Robert Prim rediscovered the algorithm published 27 years earlier by the Czech mathematician Vojtěch Jarník in a Czech journal.

Prim's algorithm constructs a minimum spanning tree through a sequence of expanding subtrees. The initial subtree in such a sequence consists of a single vertex selected arbitrarily from the set V of the graph's vertices. On each iteration, the algorithm expands the current tree in the greedy manner by simply attaching to it the nearest vertex not in that tree. (By the nearest vertex, we mean a vertex not in the tree connected to a vertex in the tree by an edge of the smallest weight. Ties can be broken arbitrarily.) The algorithm stops after all the graph's vertices have been included in the tree being constructed. Since the algorithm expands a tree by exactly one vertex on each of its iterations, the total number of such iterations is $n - 1$, where n is the number of vertices in the graph. The tree generated by the algorithm is obtained as the set of edges used for the tree expansions.

Here is pseudocode of this algorithm.

ALGORITHM *Prim(G)*

//Prim's algorithm for constructing a minimum spanning tree
//Input: A weighted connected graph $G = \langle V, E \rangle$
//Output: E_T, the set of edges composing a minimum spanning tree of G
$V_T \leftarrow \{v_0\}$ //the set of tree vertices can be initialized with any vertex
$E_T \leftarrow \varnothing$
for $i \leftarrow 1$ **to** $|V| - 1$ **do**
  find a minimum-weight edge $e^* = (v^*, u^*)$ among all the edges (v, u)
  such that v is in V_T and u is in $V - V_T$
  $V_T \leftarrow V_T \cup \{u^*\}$
  $E_T \leftarrow E_T \cup \{e^*\}$
return E_T

The nature of Prim's algorithm makes it necessary to provide each vertex not in the current tree with the information about the shortest edge connecting the vertex to a tree vertex. We can provide such information by attaching two labels to a vertex: the name of the nearest tree vertex and the length (the weight) of the corresponding edge. Vertices that are not adjacent to any of the tree vertices can be given the ∞ label indicating their "infinite" distance to the tree vertices and a null label for the name of the nearest tree vertex. (Alternatively, we can split the vertices that are not in the tree into two sets, the "fringe" and the "unseen." The fringe contains only the vertices that are not in the tree but are adjacent to at least one tree vertex. These are the candidates from which the next tree vertex is selected. The unseen vertices are all the other vertices of the graph, called "unseen" because they are yet to be affected by the algorithm.) With such labels, finding the next vertex to be added to the current tree $T = \langle V_T, E_T \rangle$ becomes a simple task of finding a vertex with the smallest distance label in the set $V - V_T$. Ties can be broken arbitrarily.

After we have identified a vertex u^* to be added to the tree, we need to perform two operations:

- Move u^* from the set $V - V_T$ to the set of tree vertices V_T.
- For each remaining vertex u in $V - V_T$ that is connected to u^* by a shorter edge than the u's current distance label, update its labels by u^* and the weight of the edge between u^* and u, respectively.[2]

Figure 9.3 demonstrates the application of Prim's algorithm to a specific graph.

Does Prim's algorithm always yield a minimum spanning tree? The answer to this question is yes. Let us prove by induction that each of the subtrees T_i, $i = 0, \ldots, n-1$, generated by Prim's algorithm is a part (i.e., a subgraph) of some minimum spanning tree. (This immediately implies, of course, that the last tree in the sequence, T_{n-1}, is a minimum spanning tree itself because it contains all n vertices of the graph.) The basis of the induction is trivial, since T_0 consists of a single vertex and hence must be a part of any minimum spanning tree. For the inductive step, let us assume that T_{i-1} is part of some minimum spanning tree T. We need to prove that T_i, generated from T_{i-1} by Prim's algorithm, is also a part of a minimum spanning tree. We prove this by contradiction by assuming that no minimum spanning tree of the graph can contain T_i. Let $e_i = (v, u)$ be the minimum weight edge from a vertex in T_{i-1} to a vertex not in T_{i-1} used by Prim's algorithm to expand T_{i-1} to T_i. By our assumption, e_i cannot belong to any minimum spanning tree, including T. Therefore, if we add e_i to T, a cycle must be formed (Figure 9.4).

In addition to edge $e_i = (v, u)$, this cycle must contain another edge (v', u') connecting a vertex $v' \in T_{i-1}$ to a vertex u' that is not in T_{i-1}. (It is possible that v' coincides with v or u' coincides with u but not both.) If we now delete the edge (v', u') from this cycle, we will obtain another spanning tree of the entire graph whose weight is less than or equal to the weight of T since the weight of e_i is less than or equal to the weight of (v', u'). Hence, this spanning tree is a minimum spanning tree, which contradicts the assumption that no minimum spanning tree contains T_i. This completes the correctness proof of Prim's algorithm.

How efficient is Prim's algorithm? The answer depends on the data structures chosen for the graph itself and for the priority queue of the set $V - V_T$ whose vertex priorities are the distances to the nearest tree vertices. (You may want to take another look at the example in Figure 9.3 to see that the set $V - V_T$ indeed operates as a priority queue.) In particular, if a graph is represented by its weight matrix and the priority queue is implemented as an unordered array, the algorithm's running time will be in $\Theta(|V|^2)$. Indeed, on each of the $|V| - 1$ iterations, the array implementing the priority queue is traversed to find and delete the minimum and then to update, if necessary, the priorities of the remaining vertices.

We can also implement the priority queue as a ***min-heap***. A min-heap is a mirror image of the heap structure discussed in Section 6.4. (In fact, it can be implemented by constructing a heap after negating all the key values given.) Namely, a min-heap is a complete binary tree in which every element is less than or equal

2. If the implementation with the fringe/unseen split is pursued, all the unseen vertices adjacent to u^* must also be moved to the fringe.

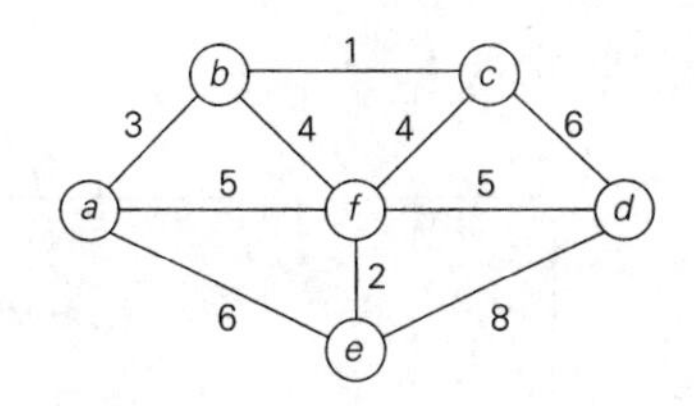

Tree vertices	Remaining vertices	Illustration
a(−, −)	**b**(**a**, **3**) c(−, ∞) d(−, ∞) e(a, 6) f(a, 5)	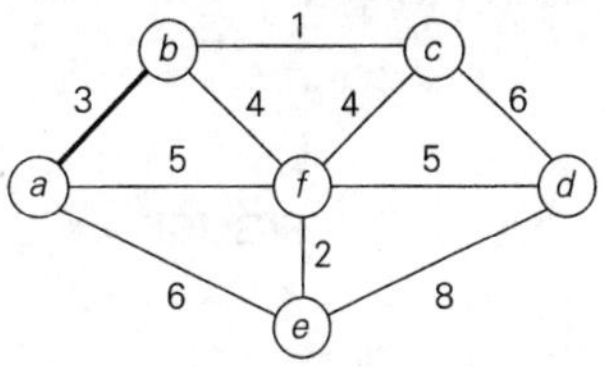
b(a, 3)	**c**(**b**, **1**) d(−, ∞) e(a, 6) f(b, 4)	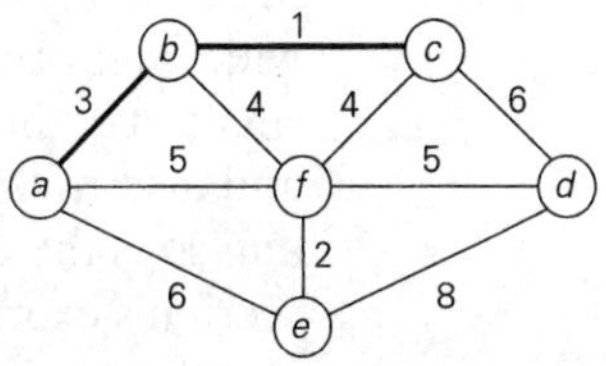
c(b, 1)	d(c, 6) e(a, 6) **f**(**b**, **4**)	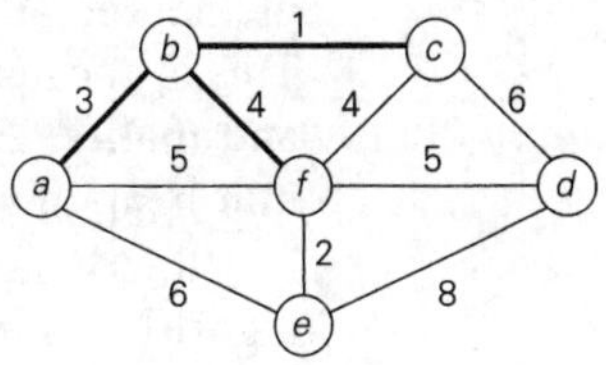
f(b, 4)	d(f, 5) **e**(**f**, **2**)	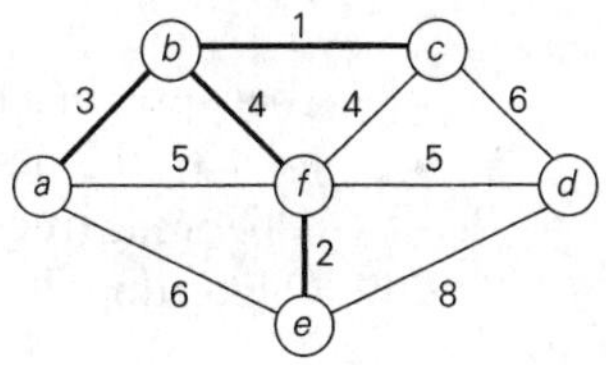
e(f, 2)	**d**(**f**, **5**)	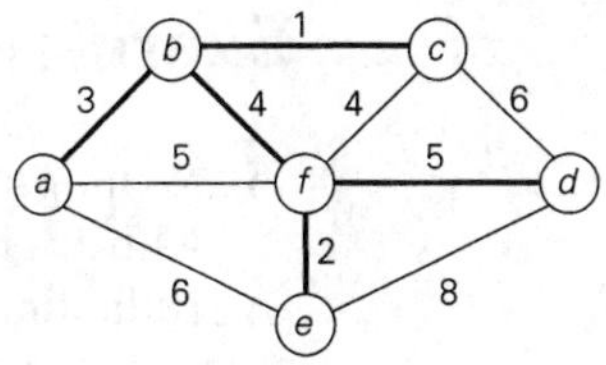
d(f, 5)		

FIGURE 9.3 Application of Prim's algorithm. The parenthesized labels of a vertex in the middle column indicate the nearest tree vertex and edge weight; selected vertices and edges are shown in bold.

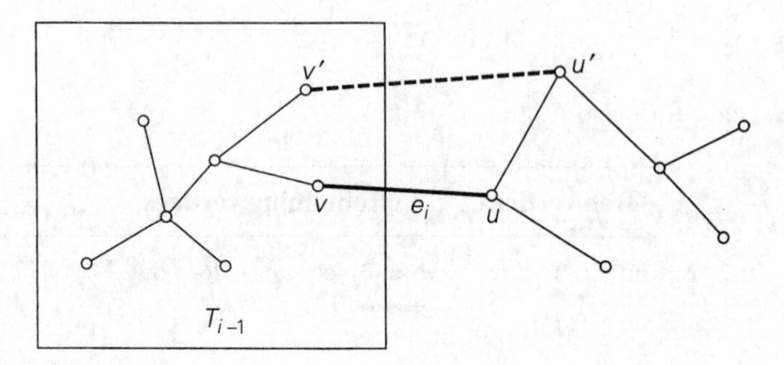

FIGURE 9.4 Correctness proof of Prim's algorithm.

to its children. All the principal properties of heaps remain valid for min-heaps, with some obvious modifications. For example, the root of a min-heap contains the smallest rather than the largest element. Deletion of the smallest element from and insertion of a new element into a min-heap of size n are $O(\log n)$ operations, and so is the operation of changing an element's priority (see Problem 15 in this section's exercises).

If a graph is represented by its adjacency lists and the priority queue is implemented as a min-heap, the running time of the algorithm is in $O(|E| \log |V|)$. This is because the algorithm performs $|V| - 1$ deletions of the smallest element and makes $|E|$ verifications and, possibly, changes of an element's priority in a min-heap of size not exceeding $|V|$. Each of these operations, as noted earlier, is a $O(\log |V|)$ operation. Hence, the running time of this implementation of Prim's algorithm is in

$$(|V| - 1 + |E|)O(\log |V|) = O(|E| \log |V|)$$

because, in a connected graph, $|V| - 1 \leq |E|$.

In the next section, you will find another greedy algorithm for the minimum spanning tree problem, which is "greedy" in a manner different from that of Prim's algorithm.

Exercises 9.1

1. Write pseudocode of the greedy algorithm for the change-making problem, with an amount n and coin denominations $d_1 > d_2 > \cdots > d_m$ as its input. What is the time efficiency class of your algorithm?
2. Design a greedy algorithm for the assignment problem (see Section 3.4). Does your greedy algorithm always yield an optimal solution?
3. *Job scheduling* Consider the problem of scheduling n jobs of known durations $t_1, t_2, \ldots, t_n$ for execution by a single processor. The jobs can be executed in any order, one job at a time. You want to find a schedule that minimizes

the total time spent by all the jobs in the system. (The time spent by one job in the system is the sum of the time spent by this job in waiting plus the time spent on its execution.)

Design a greedy algorithm for this problem. Does the greedy algorithm always yield an optimal solution?

4. *Compatible intervals* Given n open intervals $(a_1, b_1), (a_2, b_2), \ldots, (a_n, b_n)$ on the real line, each representing start and end times of some activity requiring the same resource, the task is to find the largest number of these intervals so that no two of them overlap. Investigate the three greedy algorithms based on

   a. earliest start first.

   b. shortest duration first.

   c. earliest finish first.

   For each of the three algorithms, either prove that the algorithm always yields an optimal solution or give a counterexample showing this not to be the case.

5. *Bridge crossing revisited* Consider the generalization of the bridge crossing puzzle (Problem 2 in Exercises 1.2) in which we have $n > 1$ people whose bridge crossing times are $t_1, t_2, \ldots, t_n$. All the other conditions of the problem remain the same: at most two people at a time can cross the bridge (and they move with the speed of the slower of the two) and they must carry with them the only flashlight the group has.

   Design a greedy algorithm for this problem and find how long it will take to cross the bridge by using this algorithm. Does your algorithm yield a minimum crossing time for every instance of the problem? If it does—prove it; if it does not—find an instance with the smallest number of people for which this happens.

6. *Averaging down* There are $n > 1$ identical vessels, one of them with W pints of water and the others empty. You are allowed to perform the following operation: take two of the vessels and split the total amount of water in them equally between them. The object is to achieve a minimum amount of water in the vessel containing all the water in the initial set up by a sequence of such operations. What is the best way to do this?

7. *Rumor spreading* There are n people, each in possession of a different rumor. They want to share all the rumors with each other by sending electronic messages. Assume that a sender includes all the rumors he or she knows at the time the message is sent and that a message may only have one addressee.

   Design a greedy algorithm that always yields the minimum number of messages they need to send to guarantee that every one of them gets all the rumors.

8. *Bachet's problem of weights* Find an optimal set of n weights $\{w_1, w_2, \ldots, w_n\}$ so that it would be possible to weigh on a balance scale any integer load in the largest possible range from 1 to W, provided

   a. weights can be put only on the free cup of the scale.

b. weights can be put on both cups of the scale.

9. a. Apply Prim's algorithm to the following graph. Include in the priority queue all the vertices not already in the tree.

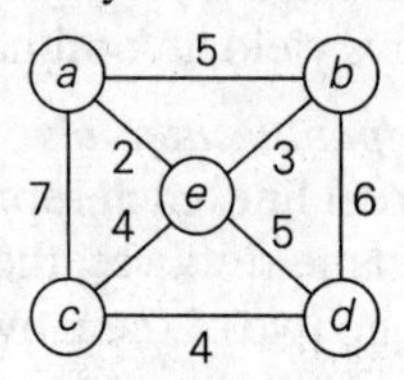

b. Apply Prim's algorithm to the following graph. Include in the priority queue only the fringe vertices (the vertices not in the current tree which are adjacent to at least one tree vertex).

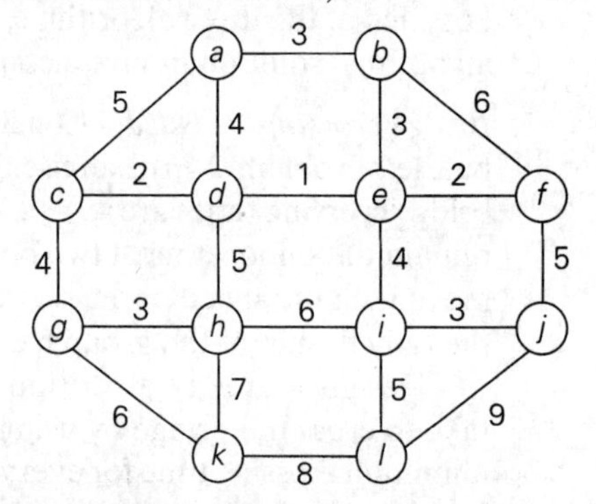

10. The notion of a minimum spanning tree is applicable to a connected weighted graph. Do we have to check a graph's connectivity before applying Prim's algorithm, or can the algorithm do it by itself?

11. Does Prim's algorithm always work correctly on graphs with negative edge weights?

12. Let T be a minimum spanning tree of graph G obtained by Prim's algorithm. Let G_{new} be a graph obtained by adding to G a new vertex and some edges, with weights, connecting the new vertex to some vertices in G. Can we construct a minimum spanning tree of G_{new} by adding one of the new edges to T? If you answer yes, explain how; if you answer no, explain why not.

13. How can one use Prim's algorithm to find a spanning tree of a connected graph with no weights on its edges? Is it a good algorithm for this problem?

14. Prove that any weighted connected graph with distinct weights has exactly one minimum spanning tree.

15. Outline an efficient algorithm for changing an element's value in a min-heap. What is the time efficiency of your algorithm?

9.2 Kruskal's Algorithm

In the previous section, we considered the greedy algorithm that "grows" a minimum spanning tree through a greedy inclusion of the nearest vertex to the vertices already in the tree. Remarkably, there is another greedy algorithm for the minimum spanning tree problem that also always yields an optimal solution. It is named ***Kruskal's algorithm*** after Joseph Kruskal, who discovered this algorithm when he was a second-year graduate student [Kru56]. Kruskal's algorithm looks at a minimum spanning tree of a weighted connected graph $G = \langle V, E\rangle$ as an acyclic subgraph with $|V| - 1$ edges for which the sum of the edge weights is the smallest. (It is not difficult to prove that such a subgraph must be a tree.) Consequently, the algorithm constructs a minimum spanning tree as an expanding sequence of subgraphs that are always acyclic but are not necessarily connected on the intermediate stages of the algorithm.

The algorithm begins by sorting the graph's edges in nondecreasing order of their weights. Then, starting with the empty subgraph, it scans this sorted list, adding the next edge on the list to the current subgraph if such an inclusion does not create a cycle and simply skipping the edge otherwise.

ALGORITHM *Kruskal(G)*

//Kruskal's algorithm for constructing a minimum spanning tree
//Input: A weighted connected graph $G = \langle V, E\rangle$
//Output: E_T, the set of edges composing a minimum spanning tree of G
sort E in nondecreasing order of the edge weights $w(e_{i_1}) \le \cdots \le w(e_{i_{|E|}})$
$E_T \leftarrow \varnothing$; $ecounter \leftarrow 0$ //initialize the set of tree edges and its size
$k \leftarrow 0$ //initialize the number of processed edges
while $ecounter < |V| - 1$ **do**
    $k \leftarrow k + 1$
    if $E_T \cup \{e_{i_k}\}$ is acyclic
        $E_T \leftarrow E_T \cup \{e_{i_k}\}$; $ecounter \leftarrow ecounter + 1$
return E_T

The correctness of Kruskal's algorithm can be proved by repeating the essential steps of the proof of Prim's algorithm given in the previous section. The fact that E_T is actually a tree in Prim's algorithm but generally just an acyclic subgraph in Kruskal's algorithm turns out to be an obstacle that can be overcome.

Figure 9.5 demonstrates the application of Kruskal's algorithm to the same graph we used for illustrating Prim's algorithm in Section 9.1. As you trace the algorithm's operations, note the disconnectedness of some of the intermediate subgraphs.

Applying Prim's and Kruskal's algorithms to the same small graph by hand may create the impression that the latter is simpler than the former. This impression is wrong because, on each of its iterations, Kruskal's algorithm has to check whether the addition of the next edge to the edges already selected would create a

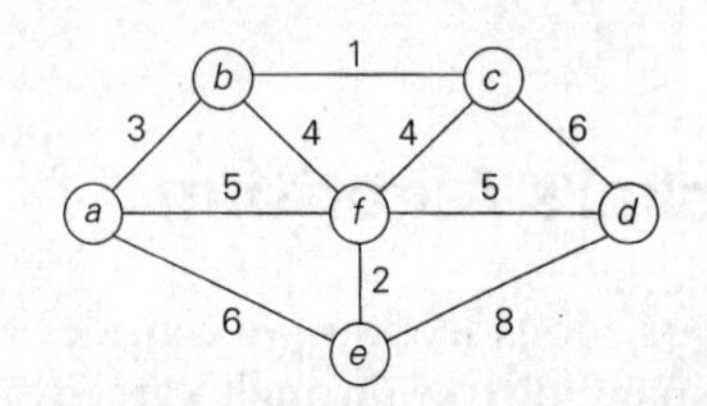

Tree edges	Sorted list of edges	Illustration
	bc ef ab bf cf af df ae cd de 1 2 3 4 4 5 5 6 6 8	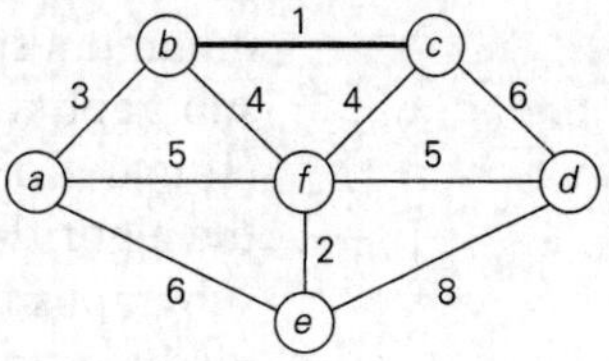
bc 1	bc **ef** ab bf cf af df ae cd de 1 2 3 4 4 5 5 6 6 8	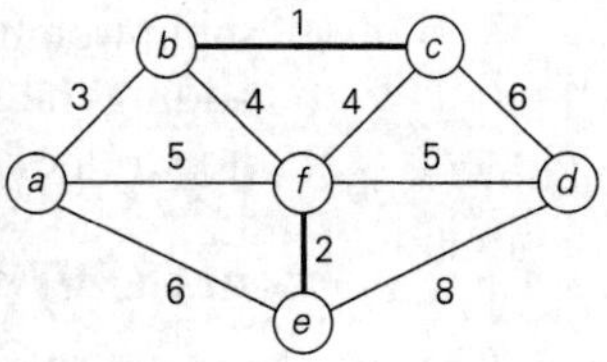
ef 2	bc ef **ab** bf cf af df ae cd de 1 2 3 4 4 5 5 6 6 8	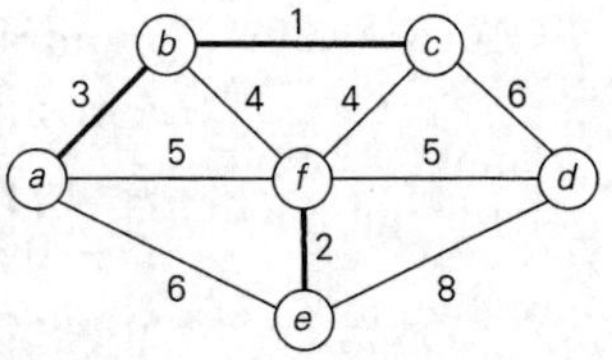
ab 3	bc ef ab **bf** cf af df ae cd de 1 2 3 4 4 5 5 6 6 8	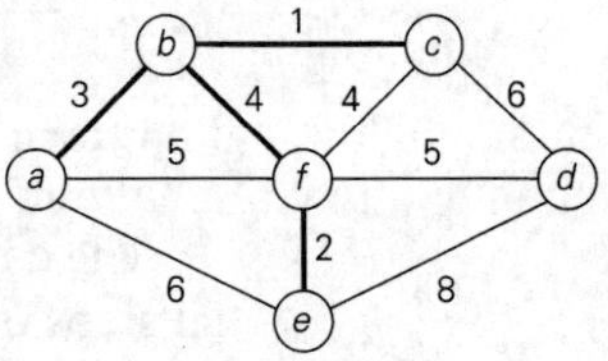
bf 4	bc ef ab bf cf af **df** ae cd de 1 2 3 4 4 5 5 6 6 8	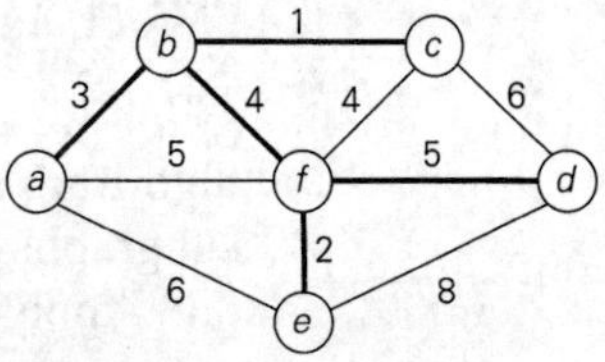
df 5		

FIGURE 9.5 Application of Kruskal's algorithm. Selected edges are shown in bold.

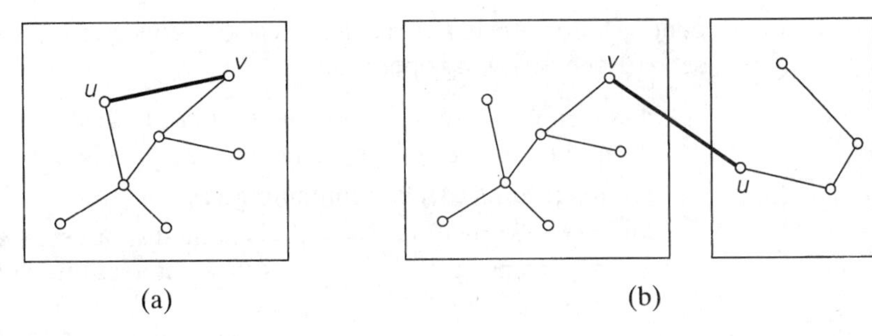

FIGURE 9.6 New edge connecting two vertices may (a) or may not (b) create a cycle.

cycle. It is not difficult to see that a new cycle is created if and only if the new edge connects two vertices already connected by a path, i.e., if and only if the two vertices belong to the same connected component (Figure 9.6). Note also that each connected component of a subgraph generated by Kruskal's algorithm is a tree because it has no cycles.

In view of these observations, it is convenient to use a slightly different interpretation of Kruskal's algorithm. We can consider the algorithm's operations as a progression through a series of forests containing *all* the vertices of a given graph and *some* of its edges. The initial forest consists of $|V|$ trivial trees, each comprising a single vertex of the graph. The final forest consists of a single tree, which is a minimum spanning tree of the graph. On each iteration, the algorithm takes the next edge (u, v) from the sorted list of the graph's edges, finds the trees containing the vertices u and v, and, if these trees are not the same, unites them in a larger tree by adding the edge (u, v).

Fortunately, there are efficient algorithms for doing so, including the crucial check for whether two vertices belong to the same tree. They are called ***union-find*** algorithms. We discuss them in the following subsection. With an efficient union-find algorithm, the running time of Kruskal's algorithm will be dominated by the time needed for sorting the edge weights of a given graph. Hence, with an efficient sorting algorithm, the time efficiency of Kruskal's algorithm will be in $O(|E| \log |E|)$.

Disjoint Subsets and Union-Find Algorithms

Kruskal's algorithm is one of a number of applications that require a dynamic partition of some n element set S into a collection of disjoint subsets $S_1, S_2, \ldots, S_k$. After being initialized as a collection of n one-element subsets, each containing a different element of S, the collection is subjected to a sequence of intermixed union and find operations. (Note that the number of union operations in any such sequence must be bounded above by $n - 1$ because each union increases a subset's size at least by 1 and there are only n elements in the entire set S.) Thus, we are dealing here with an abstract data type of a collection of disjoint subsets of a finite set with the following operations:

makeset(x) creates a one-element set $\{x\}$. It is assumed that this operation can be applied to each of the elements of set S only once.
find(x) returns a subset containing x.
union(x, y) constructs the union of the disjoint subsets S_x and S_y containing x and y, respectively, and adds it to the collection to replace S_x and S_y, which are deleted from it.

For example, let $S = \{1, 2, 3, 4, 5, 6\}$. Then *makeset*(i) creates the set $\{i\}$ and applying this operation six times initializes the structure to the collection of six singleton sets:

$$\{1\},\ \{2\},\ \{3\},\ \{4\},\ \{5\},\ \{6\}.$$

Performing *union*(1, 4) and *union*(5, 2) yields

$$\{1, 4\},\ \{5, 2\},\ \{3\},\ \{6\},$$

and, if followed by *union*(4, 5) and then by *union*(3, 6), we end up with the disjoint subsets

$$\{1, 4, 5, 2\},\ \{3, 6\}.$$

Most implementations of this abstract data type use one element from each of the disjoint subsets in a collection as that subset's ***representative***. Some implementations do not impose any specific constraints on such a representative; others do so by requiring, say, the smallest element of each subset to be used as the subset's representative. Also, it is usually assumed that set elements are (or can be mapped into) integers.

There are two principal alternatives for implementing this data structure. The first one, called the ***quick find***, optimizes the time efficiency of the find operation; the second one, called the ***quick union***, optimizes the union operation.

The quick find uses an array indexed by the elements of the underlying set S; the array's values indicate the representatives of the subsets containing those elements. Each subset is implemented as a linked list whose header contains the pointers to the first and last elements of the list along with the number of elements in the list (see Figure 9.7 for an example).

Under this scheme, the implementation of *makeset*(x) requires assigning the corresponding element in the representative array to x and initializing the corresponding linked list to a single node with the x value. The time efficiency of this operation is obviously in $\Theta(1)$, and hence the initialization of n singleton subsets is in $\Theta(n)$. The efficiency of *find*(x) is also in $\Theta(1)$: all we need to do is to retrieve the x's representative in the representative array. Executing *union*(x, y) takes longer. A straightforward solution would simply append the y's list to the end of the x's list, update the information about their representative for all the elements in the

subset representatives

element index	representative
1	1
2	1
3	3
4	1
5	1
6	3

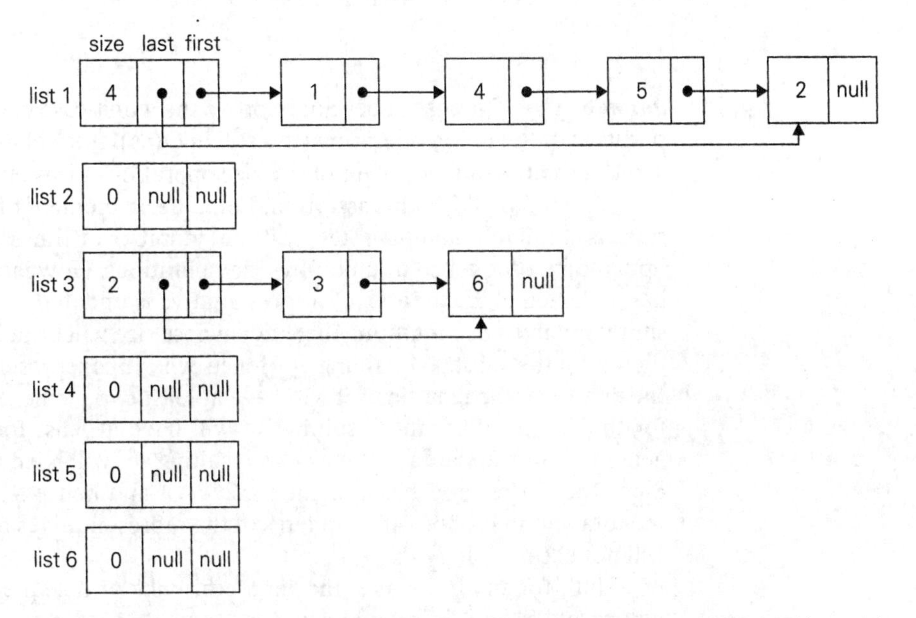

FIGURE 9.7 Linked-list representation of subsets {1, 4, 5, 2} and {3, 6} obtained by quick find after performing *union*(1, 4), *union*(5, 2), *union*(4, 5), and *union*(3, 6). The lists of size 0 are considered deleted from the collection.

y list, and then delete the y's list from the collection. It is easy to verify, however, that with this algorithm the sequence of union operations

$$union(2, 1),\ union(3, 2), \ldots,\ union(i + 1, i), \ldots,\ union(n, n - 1)$$

runs in $\Theta(n^2)$ time, which is slow compared with several known alternatives.

A simple way to improve the overall efficiency of a sequence of union operations is to always append the shorter of the two lists to the longer one, with ties broken arbitrarily. Of course, the size of each list is assumed to be available by, say, storing the number of elements in the list's header. This modification is called the

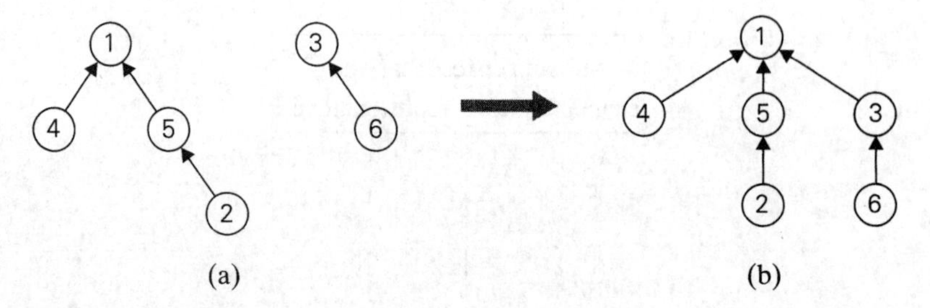

FIGURE 9.8 (a) Forest representation of subsets {1, 4, 5, 2} and {3, 6} used by quick union. (b) Result of *union*(5, 6).

union by size. Though it does not improve the worst-case efficiency of a single application of the union operation (it is still in $\Theta(n)$), the worst-case running time of any legitimate sequence of union-by-size operations turns out to be in $O(n \log n)$.[3]

Here is a proof of this assertion. Let a_i be an element of set S whose disjoint subsets we manipulate, and let A_i be the number of times a_i's representative is updated in a sequence of union-by-size operations. How large can A_i get if set S has n elements? Each time a_i's representative is updated, a_i must be in a smaller subset involved in computing the union whose size will be at least twice as large as the size of the subset containing a_i. Hence, when a_i's representative is updated for the first time, the resulting set will have at least two elements; when it is updated for the second time, the resulting set will have at least four elements; and, in general, if it is updated A_i times, the resulting set will have at least 2^{A_i} elements. Since the entire set S has n elements, $2^{A_i} \le n$ and hence $A_i \le \log_2 n$. Therefore, the total number of possible updates of the representatives for all n elements in S will not exceed $n \log_2 n$.

Thus, for union by size, the time efficiency of a sequence of at most $n - 1$ unions and m finds is in $O(n \log n + m)$.

The ***quick union***—the second principal alternative for implementing disjoint subsets—represents each subset by a rooted tree. The nodes of the tree contain the subset's elements (one per node), with the root's element considered the subset's representative; the tree's edges are directed from children to their parents (Figure 9.8). In addition, a mapping of the set elements to their tree nodes—implemented, say, as an array of pointers—is maintained. This mapping is not shown in Figure 9.8 for the sake of simplicity.

For this implementation, *makeset*(x) requires the creation of a single-node tree, which is a $\Theta(1)$ operation; hence, the initialization of n singleton subsets is in $\Theta(n)$. A *union*(x, y) is implemented by attaching the root of the y's tree to the root of the x's tree (and deleting the y's tree from the collection by making the pointer to its root null). The time efficiency of this operation is clearly $\Theta(1)$. A *find*(x) is

3. This is a specific example of the usefulness of the *amortized efficiency* we mentioned back in Chapter 2.

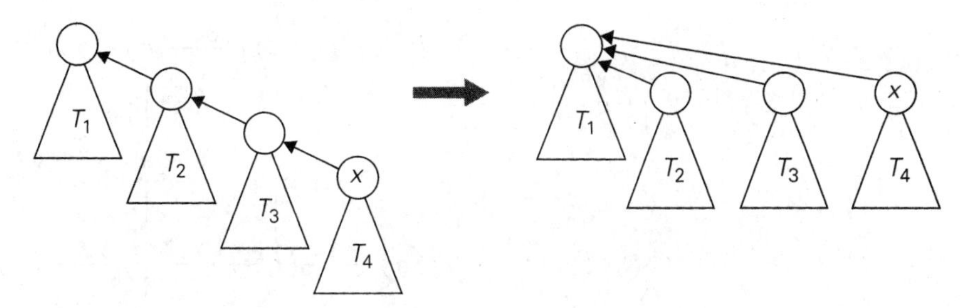

FIGURE 9.9 Path compression.

performed by following the pointer chain from the node containing x to the tree's root whose element is returned as the subset's representative. Accordingly, the time efficiency of a single find operation is in $O(n)$ because a tree representing a subset can degenerate into a linked list with n nodes.

This time bound can be improved. The straightforward way for doing so is to always perform a union operation by attaching a smaller tree to the root of a larger one, with ties broken arbitrarily. The size of a tree can be measured either by the number of nodes (this version is called ***union by size***) or by its height (this version is called ***union by rank***). Of course, these options require storing, for each node of the tree, either the number of node descendants or the height of the subtree rooted at that node, respectively. One can easily prove that in either case the height of the tree will be logarithmic, making it possible to execute each find in $O(\log n)$ time. Thus, for quick union, the time efficiency of a sequence of at most $n - 1$ unions and m finds is in $O(n + m \log n)$.

In fact, an even better efficiency can be obtained by combining either variety of quick union with ***path compression***. This modification makes every node encountered during the execution of a find operation point to the tree's root (Figure 9.9). According to a quite sophisticated analysis that goes beyond the level of this book (see [Tar84]), this and similar techniques improve the efficiency of a sequence of at most $n - 1$ unions and m finds to only slightly worse than linear.

Exercises 9.2

1. Apply Kruskal's algorithm to find a minimum spanning tree of the following graphs.

a.

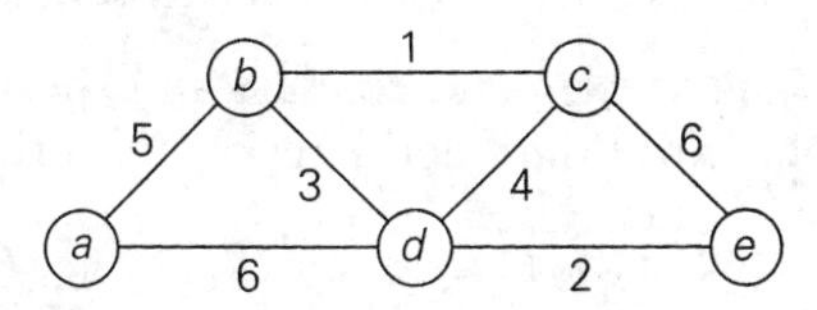

b.

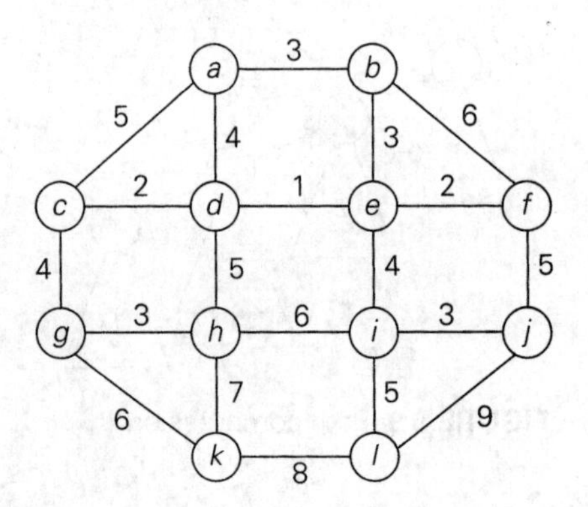

2. Indicate whether the following statements are true or false:

a. If e is a minimum-weight edge in a connected weighted graph, it must be among edges of at least one minimum spanning tree of the graph.

b. If e is a minimum-weight edge in a connected weighted graph, it must be among edges of each minimum spanning tree of the graph.

c. If edge weights of a connected weighted graph are all distinct, the graph must have exactly one minimum spanning tree.

d. If edge weights of a connected weighted graph are not all distinct, the graph must have more than one minimum spanning tree.

3. What changes, if any, need to be made in algorithm *Kruskal* to make it find a ***minimum spanning forest*** for an arbitrary graph? (A minimum spanning forest is a forest whose trees are minimum spanning trees of the graph's connected components.)

4. Does Kruskal's algorithm work correctly on graphs that have negative edge weights?

5. Design an algorithm for finding a ***maximum spanning tree***—a spanning tree with the largest possible edge weight—of a weighted connected graph.

6. Rewrite pseudocode of Kruskal's algorithm in terms of the operations of the disjoint subsets' ADT.

7. Prove the correctness of Kruskal's algorithm.

8. Prove that the time efficiency of $find(x)$ is in $O(\log n)$ for the union-by-size version of quick union.

9. Find at least two Web sites with animations of Kruskal's and Prim's algorithms. Discuss their merits and demerits.

10. Design and conduct an experiment to empirically compare the efficiencies of Prim's and Kruskal's algorithms on random graphs of different sizes and densities.

11. *Steiner tree* Four villages are located at the vertices of a unit square in the Euclidean plane. You are asked to connect them by the shortest network of roads so that there is a path between every pair of the villages along those roads. Find such a network.

12. Write a program generating a random maze based on
    a. Prim's algorithm.
    b. Kruskal's algorithm.

9.3 Dijkstra's Algorithm

In this section, we consider the ***single-source shortest-paths problem***: for a given vertex called the ***source*** in a weighted connected graph, find shortest paths to all its other vertices. It is important to stress that we are not interested here in a single shortest path that starts at the source and visits all the other vertices. This would have been a much more difficult problem (actually, a version of the traveling salesman problem introduced in Section 3.4 and discussed again later in the book). The single-source shortest-paths problem asks for a family of paths, each leading from the source to a different vertex in the graph, though some paths may, of course, have edges in common.

A variety of practical applications of the shortest-paths problem have made the problem a very popular object of study. The obvious but probably most widely used applications are transportation planning and packet routing in communication networks, including the Internet. Multitudes of less obvious applications include finding shortest paths in social networks, speech recognition, document formatting, robotics, compilers, and airline crew scheduling. In the world of entertainment, one can mention pathfinding in video games and finding best solutions to puzzles using their state-space graphs (see Section 6.6 for a very simple example of the latter).

There are several well-known algorithms for finding shortest paths, including Floyd's algorithm for the more general all-pairs shortest-paths problem discussed in Chapter 8. Here, we consider the best-known algorithm for the single-source shortest-paths problem, called ***Dijkstra's algorithm***.[4] This algorithm is applicable to undirected and directed graphs with nonnegative weights only. Since in most applications this condition is satisfied, the limitation has not impaired the popularity of Dijkstra's algorithm.

Dijkstra's algorithm finds the shortest paths to a graph's vertices in order of their distance from a given source. First, it finds the shortest path from the source

4. Edsger W. Dijkstra (1930–2002), a noted Dutch pioneer of the science and industry of computing, discovered this algorithm in the mid-1950s. Dijkstra said about his algorithm: "This was the first graph problem I ever posed myself and solved. The amazing thing was that I didn't publish it. It was not amazing at the time. At the time, algorithms were hardly considered a scientific topic."

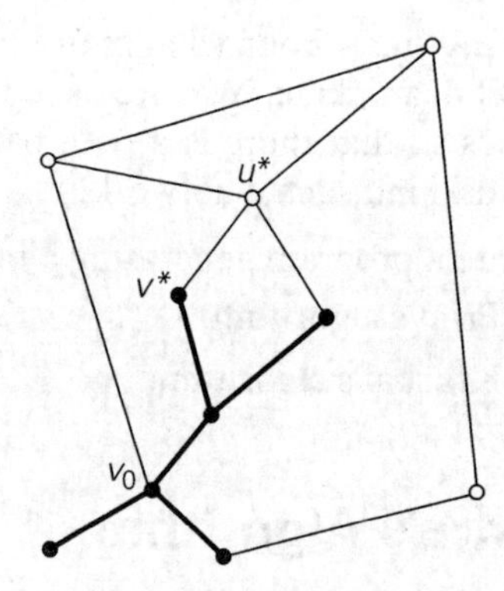

FIGURE 9.10 Idea of Dijkstra's algorithm. The subtree of the shortest paths already found is shown in bold. The next nearest to the source v_0 vertex, u^*, is selected by comparing the lengths of the subtree's paths increased by the distances to vertices adjacent to the subtree's vertices.

to a vertex nearest to it, then to a second nearest, and so on. In general, before its ith iteration commences, the algorithm has already identified the shortest paths to $i - 1$ other vertices nearest to the source. These vertices, the source, and the edges of the shortest paths leading to them from the source form a subtree T_i of the given graph (Figure 9.10). Since all the edge weights are nonnegative, the next vertex nearest to the source can be found among the vertices adjacent to the vertices of T_i. The set of vertices adjacent to the vertices in T_i can be referred to as "fringe vertices"; they are the candidates from which Dijkstra's algorithm selects the next vertex nearest to the source. (Actually, all the other vertices can be treated as fringe vertices connected to tree vertices by edges of infinitely large weights.) To identify the ith nearest vertex, the algorithm computes, for every fringe vertex u, the sum of the distance to the nearest tree vertex v (given by the weight of the edge (v, u)) and the length d_v of the shortest path from the source to v (previously determined by the algorithm) and then selects the vertex with the smallest such sum. The fact that it suffices to compare the lengths of such special paths is the central insight of Dijkstra's algorithm.

To facilitate the algorithm's operations, we label each vertex with two labels. The numeric label d indicates the length of the shortest path from the source to this vertex found by the algorithm so far; when a vertex is added to the tree, d indicates the length of the shortest path from the source to that vertex. The other label indicates the name of the next-to-last vertex on such a path, i.e., the parent of the vertex in the tree being constructed. (It can be left unspecified for the source s and vertices that are adjacent to none of the current tree vertices.) With such labeling, finding the next nearest vertex u^* becomes a simple task of finding a fringe vertex with the smallest d value. Ties can be broken arbitrarily.

After we have identified a vertex u^* to be added to the tree, we need to perform two operations:

- Move u^* from the fringe to the set of tree vertices.
- For each remaining fringe vertex u that is connected to u^* by an edge of weight $w(u^*, u)$ such that $d_{u^*} + w(u^*, u) < d_u$, update the labels of u by u^* and $d_{u^*} + w(u^*, u)$, respectively.

Figure 9.11 demonstrates the application of Dijkstra's algorithm to a specific graph.

The labeling and mechanics of Dijkstra's algorithm are quite similar to those used by Prim's algorithm (see Section 9.1). Both of them construct an expanding subtree of vertices by selecting the next vertex from the priority queue of the remaining vertices. It is important not to mix them up, however. They solve different problems and therefore operate with priorities computed in a different manner: Dijkstra's algorithm compares path lengths and therefore must add edge weights, while Prim's algorithm compares the edge weights as given.

Now we can give pseudocode of Dijkstra's algorithm. It is spelled out—in more detail than Prim's algorithm was in Section 9.1—in terms of explicit operations on two sets of labeled vertices: the set V_T of vertices for which a shortest path has already been found and the priority queue Q of the fringe vertices. (Note that in the following pseudocode, V_T contains a given source vertex and the fringe contains the vertices adjacent to it *after* iteration 0 is completed.)

ALGORITHM *Dijkstra*(G, s)

```
//Dijkstra's algorithm for single-source shortest paths
//Input: A weighted connected graph G = ⟨V, E⟩ with nonnegative weights
//       and its vertex s
//Output: The length d_v of a shortest path from s to v
//        and its penultimate vertex p_v for every vertex v in V
Initialize(Q)  //initialize priority queue to empty
for every vertex v in V
    d_v ← ∞;  p_v ← null
    Insert(Q, v, d_v)  //initialize vertex priority in the priority queue
d_s ← 0;  Decrease(Q, s, d_s)  //update priority of s with d_s
V_T ← ∅
for i ← 0 to |V| − 1 do
    u* ← DeleteMin(Q)  //delete the minimum priority element
    V_T ← V_T ∪ {u*}
    for every vertex u in V − V_T that is adjacent to u* do
        if d_u* + w(u*, u) < d_u
            d_u ← d_u* + w(u*, u);  p_u ← u*
            Decrease(Q, u, d_u)
```

The time efficiency of Dijkstra's algorithm depends on the data structures used for implementing the priority queue and for representing an input graph itself. For the reasons explained in the analysis of Prim's algorithm in Section 9.1, it is

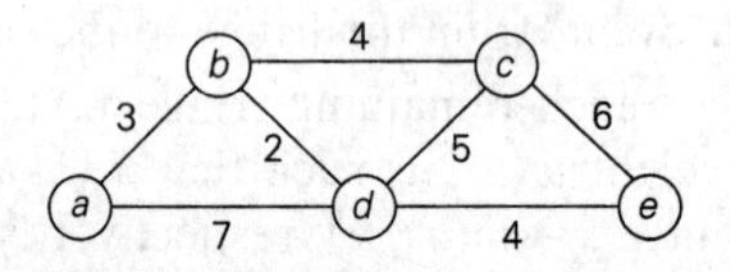

Tree vertices	Remaining vertices	Illustration
a(−, 0)	**b(a, 3)** c(−, ∞) d(a, 7) e(−, ∞)	
b(a, 3)	c(b, 3 + 4) **d(b, 3 + 2)** e(−, ∞)	
d(b, 5)	**c(b, 7)** e(d, 5 + 4)	
c(b, 7)	**e(d, 9)**	
e(d, 9)		

The shortest paths (identified by following nonnumeric labels backward from a destination vertex in the left column to the source) and their lengths (given by numeric labels of the tree vertices) are as follows:

from a to b: $a - b$ of length 3
from a to d: $a - b - d$ of length 5
from a to c: $a - b - c$ of length 7
from a to e: $a - b - d - e$ of length 9

FIGURE 9.11 Application of Dijkstra's algorithm. The next closest vertex is shown in bold.

in $\Theta(|V|^2)$ for graphs represented by their weight matrix and the priority queue implemented as an unordered array. For graphs represented by their adjacency lists and the priority queue implemented as a min-heap, it is in $O(|E| \log |V|)$. A still better upper bound can be achieved for both Prim's and Dijkstra's algorithms if the priority queue is implemented using a sophisticated data structure called the ***Fibonacci heap*** (e.g., [Cor09]). However, its complexity and a considerable overhead make such an improvement primarily of theoretical value.

Exercises 9.3

1. Explain what adjustments if any need to be made in Dijkstra's algorithm and/or in an underlying graph to solve the following problems.
   a. Solve the single-source shortest-paths problem for directed weighted graphs.
   b. Find a shortest path between two given vertices of a weighted graph or digraph. (This variation is called the ***single-pair shortest-path problem.***)
   c. Find the shortest paths to a given vertex from each other vertex of a weighted graph or digraph. (This variation is called the ***single-destination shortest-paths problem.***)
   d. Solve the single-source shortest-paths problem in a graph with nonnegative numbers assigned to its vertices (and the length of a path defined as the sum of the vertex numbers on the path).

2. Solve the following instances of the single-source shortest-paths problem with vertex a as the source:
   a.

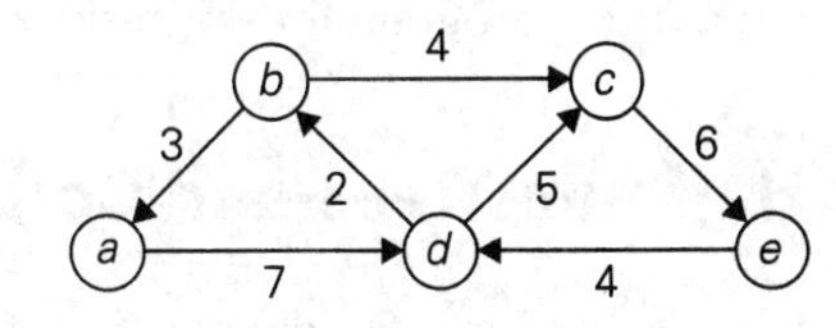

   b.

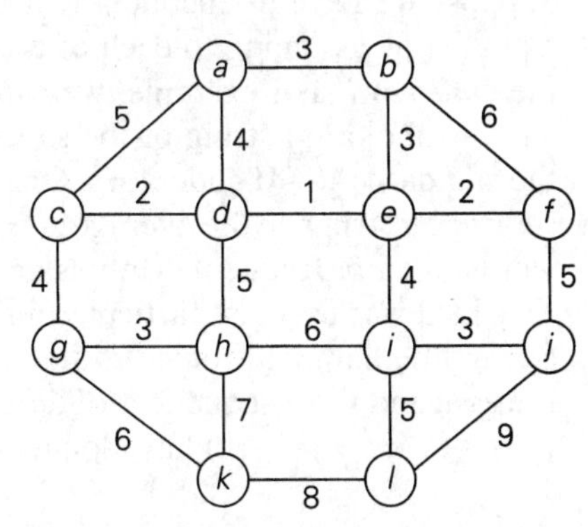

3. Give a counterexample that shows that Dijkstra's algorithm may not work for a weighted connected graph with negative weights.

4. Let T be a tree constructed by Dijkstra's algorithm in the process of solving the single-source shortest-paths problem for a weighted connected graph G.

a. True or false: T is a spanning tree of G?

b. True or false: T is a minimum spanning tree of G?

5. Write pseudocode for a simpler version of Dijkstra's algorithm that finds only the distances (i.e., the lengths of shortest paths but not shortest paths themselves) from a given vertex to all other vertices of a graph represented by its weight matrix.

6. Prove the correctness of Dijkstra's algorithm for graphs with positive weights.

7. Design a linear-time algorithm for solving the single-source shortest-paths problem for dags (directed acyclic graphs) represented by their adjacency lists.

8. Explain how the minimum-sum descent problem (Problem 8 in Exercises 8.1) can be solved by Dijkstra's algorithm.

9. *Shortest-path modeling* Assume you have a model of a weighted connected graph made of balls (representing the vertices) connected by strings of appropriate lengths (representing the edges).

a. Describe how you can solve the single-pair shortest-path problem with this model.

b. Describe how you can solve the single-source shortest-paths problem with this model.

10. Revisit the exercise from Section 1.3 about determining the best route for a subway passenger to take from one designated station to another in a well-developed subway system like those in Washington, DC, or London, UK. Write a program for this task.

9.4 Huffman Trees and Codes

Suppose we have to encode a text that comprises symbols from some n-symbol alphabet by assigning to each of the text's symbols some sequence of bits called the ***codeword***. For example, we can use a ***fixed-length encoding*** that assigns to each symbol a bit string of the same length m ($m \geq \log_2 n$). This is exactly what the standard ASCII code does. One way of getting a coding scheme that yields a shorter bit string on the average is based on the old idea of assigning shorter codewords to more frequent symbols and longer codewords to less frequent symbols. This idea was used, in particular, in the telegraph code invented in the mid-19th century by Samuel Morse. In that code, frequent letters such as e (·) and a (·—) are assigned short sequences of dots and dashes while infrequent letters such as q (— — ·—) and z (— — ··) have longer ones.

Variable-length encoding, which assigns codewords of different lengths to different symbols, introduces a problem that fixed-length encoding does not have. Namely, how can we tell how many bits of an encoded text represent the first (or, more generally, the ith) symbol? To avoid this complication, we can limit ourselves to the so-called ***prefix-free*** (or simply ***prefix***) ***codes***. In a prefix code, no codeword is a prefix of a codeword of another symbol. Hence, with such an encoding, we can simply scan a bit string until we get the first group of bits that is a codeword for some symbol, replace these bits by this symbol, and repeat this operation until the bit string's end is reached.

If we want to create a binary prefix code for some alphabet, it is natural to associate the alphabet's symbols with leaves of a binary tree in which all the left edges are labeled by 0 and all the right edges are labeled by 1. The codeword of a symbol can then be obtained by recording the labels on the simple path from the root to the symbol's leaf. Since there is no simple path to a leaf that continues to another leaf, no codeword can be a prefix of another codeword; hence, any such tree yields a prefix code.

Among the many trees that can be constructed in this manner for a given alphabet with known frequencies of the symbol occurrences, how can we construct a tree that would assign shorter bit strings to high-frequency symbols and longer ones to low-frequency symbols? It can be done by the following greedy algorithm, invented by David Huffman while he was a graduate student at MIT [Huf52].

Huffman's algorithm

Step 1 Initialize n one-node trees and label them with the symbols of the alphabet given. Record the frequency of each symbol in its tree's root to indicate the tree's ***weight***. (More generally, the weight of a tree will be equal to the sum of the frequencies in the tree's leaves.)

Step 2 Repeat the following operation until a single tree is obtained. Find two trees with the smallest weight (ties can be broken arbitrarily, but see Problem 2 in this section's exercises). Make them the left and right subtree of a new tree and record the sum of their weights in the root of the new tree as its weight.

A tree constructed by the above algorithm is called a ***Huffman tree***. It defines—in the manner described above—a ***Huffman code***.

EXAMPLE Consider the five-symbol alphabet {A, B, C, D, _} with the following occurrence frequencies in a text made up of these symbols:

symbol	A	B	C	D	_
frequency	0.35	0.1	0.2	0.2	0.15

The Huffman tree construction for this input is shown in Figure 9.12.

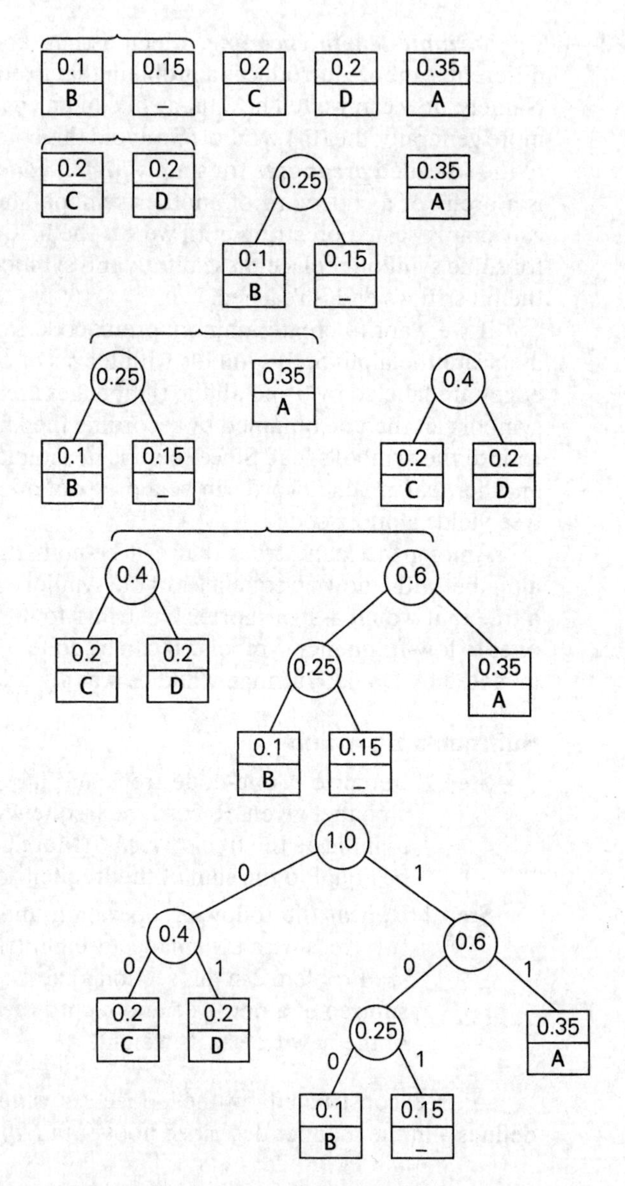

FIGURE 9.12 Example of constructing a Huffman coding tree.

The resulting codewords are as follows:

symbol	A	B	C	D	_
frequency	0.35	0.1	0.2	0.2	0.15
codeword	11	100	00	01	101

Hence, DAD is encoded as 011101, and 10011011011101 is decoded as BAD_AD.

With the occurrence frequencies given and the codeword lengths obtained, the average number of bits per symbol in this code is

$$2 \cdot 0.35 + 3 \cdot 0.1 + 2 \cdot 0.2 + 2 \cdot 0.2 + 3 \cdot 0.15 = 2.25.$$

Had we used a fixed-length encoding for the same alphabet, we would have to use at least 3 bits per each symbol. Thus, for this toy example, Huffman's code achieves the ***compression ratio***—a standard measure of a compression algorithm's effectiveness—of $(3 - 2.25)/3 \cdot 100\% = 25\%$. In other words, Huffman's encoding of the text will use 25% less memory than its fixed-length encoding. (Extensive experiments with Huffman codes have shown that the compression ratio for this scheme typically falls between 20% and 80%, depending on the characteristics of the text being compressed.) ■

Huffman's encoding is one of the most important file-compression methods. In addition to its simplicity and versatility, it yields an optimal, i.e., minimal-length, encoding (provided the frequencies of symbol occurrences are independent and known in advance). The simplest version of Huffman compression calls, in fact, for a preliminary scanning of a given text to count the frequencies of symbol occurrences in it. Then these frequencies are used to construct a Huffman coding tree and encode the text as described above. This scheme makes it necessary, however, to include the coding table into the encoded text to make its decoding possible. This drawback can be overcome by using ***dynamic Huffman encoding***, in which the coding tree is updated each time a new symbol is read from the source text. Further, modern alternatives such as ***Lempel-Ziv*** algorithms (e.g., [Say05]) assign codewords not to individual symbols but to strings of symbols, allowing them to achieve better and more robust compressions in many applications.

It is important to note that applications of Huffman's algorithm are not limited to data compression. Suppose we have n positive numbers $w_1, w_2, \ldots, w_n$ that have to be assigned to n leaves of a binary tree, one per node. If we define the ***weighted path length*** as the sum $\sum_{i=1}^{n} l_i w_i$, where l_i is the length of the simple path from the root to the ith leaf, how can we construct a binary tree with minimum weighted path length? It is this more general problem that Huffman's algorithm actually solves. (For the coding application, l_i and w_i are the length of the codeword and the frequency of the ith symbol, respectively.)

This problem arises in many situations involving decision making. Consider, for example, the game of guessing a chosen object from n possibilities (say, an integer between 1 and n) by asking questions answerable by yes or no. Different strategies for playing this game can be modeled by ***decision trees***[5] such as those depicted in Figure 9.13 for $n = 4$. The length of the simple path from the root to a leaf in such a tree is equal to the number of questions needed to get to the chosen number represented by the leaf. If number i is chosen with probability p_i, the sum

5. Decision trees are discussed in more detail in Section 11.2.

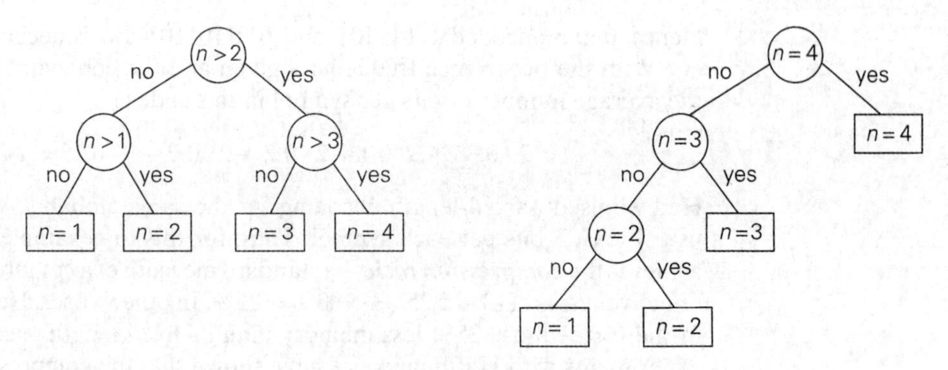

FIGURE 9.13 Two decision trees for guessing an integer between 1 and 4.

$\sum_{i=1}^{n} l_i p_i$, where l_i is the length of the path from the root to the ith leaf, indicates the average number of questions needed to "guess" the chosen number with a game strategy represented by its decision tree. If each of the numbers is chosen with the same probability of $1/n$, the best strategy is to successively eliminate half (or almost half) the candidates as binary search does. This may not be the case for arbitrary p_i's, however. For example, if $n = 4$ and $p_1 = 0.1$, $p_2 = 0.2$, $p_3 = 0.3$, and $p_4 = 0.4$, the minimum weighted path tree is the rightmost one in Figure 9.13. Thus, we need Huffman's algorithm to solve this problem in its general case.

Note that this is the second time we are encountering the problem of constructing an optimal binary tree. In Section 8.3, we discussed the problem of constructing an optimal binary search tree with positive numbers (the search probabilities) assigned to every node of the tree. In this section, given numbers are assigned just to leaves. The latter problem turns out to be easier: it can be solved by the greedy algorithm, whereas the former is solved by the more complicated dynamic programming algorithm.

Exercises 9.4

1. **a.** Construct a Huffman code for the following data:

symbol	A	B	C	D	_
frequency	0.4	0.1	0.2	0.15	0.15

   b. Encode ABACABAD using the code of question (a).

   c. Decode 100010111001010 using the code of question (a).

2. For data transmission purposes, it is often desirable to have a code with a minimum variance of the codeword lengths (among codes of the same average length). Compute the average and variance of the codeword length in two

Huffman codes that result from a different tie breaking during a Huffman code construction for the following data:

symbol	A	B	C	D	E
probability	0.1	0.1	0.2	0.2	0.4

3. Indicate whether each of the following properties is true for every Huffman code.
   a. The codewords of the two least frequent symbols have the same length.
   b. The codeword's length of a more frequent symbol is always smaller than or equal to the codeword's length of a less frequent one.

4. What is the maximal length of a codeword possible in a Huffman encoding of an alphabet of n symbols?

5. **a.** Write pseudocode of the Huffman-tree construction algorithm.
   b. What is the time efficiency class of the algorithm for constructing a Huffman tree as a function of the alphabet size?

6. Show that a Huffman tree can be constructed in linear time if the alphabet symbols are given in a sorted order of their frequencies.

7. Given a Huffman coding tree, which algorithm would you use to get the codewords for all the symbols? What is its time-efficiency class as a function of the alphabet size?

8. Explain how one can generate a Huffman code without an explicit generation of a Huffman coding tree.

9. **a.** Write a program that constructs a Huffman code for a given English text and encode it.
   b. Write a program for decoding of an English text which has been encoded with a Huffman code.
   c. Experiment with your encoding program to find a range of typical compression ratios for Huffman's encoding of English texts of, say, 1000 words.
   d. Experiment with your encoding program to find out how sensitive the compression ratios are to using standard estimates of frequencies instead of actual frequencies of symbol occurrences in English texts.

10. *Card guessing* Design a strategy that minimizes the expected number of questions asked in the following game [Gar94]. You have a deck of cards that consists of one ace of spades, two deuces of spades, three threes, and on up to nine nines, making 45 cards in all. Someone draws a card from the shuffled deck, which you have to identify by asking questions answerable with yes or no.

SUMMARY

- The *greedy technique* suggests constructing a solution to an optimization problem through a sequence of steps, each expanding a partially constructed solution obtained so far, until a complete solution to the problem is reached. On each step, the choice made must be *feasible*, *locally optimal*, and *irrevocable.*

- *Prim's algorithm* is a greedy algorithm for constructing a *minimum spanning tree* of a weighted connected graph. It works by attaching to a previously constructed subtree a vertex closest to the vertices already in the tree.

- *Kruskal's algorithm* is another greedy algorithm for the minimum spanning tree problem. It constructs a minimum spanning tree by selecting edges in nondecreasing order of their weights provided that the inclusion does not create a cycle. Checking the latter condition efficiently requires an application of one of the so-called *union-find algorithms.*

- *Dijkstra's algorithm* solves the *single-source shortest-path problem* of finding shortest paths from a given vertex (the source) to all the other vertices of a weighted graph or digraph. It works as Prim's algorithm but compares path lengths rather than edge lengths. Dijkstra's algorithm always yields a correct solution for a graph with nonnegative weights.

- A *Huffman tree* is a binary tree that minimizes the weighted path length from the root to the leaves of predefined weights. The most important application of Huffman trees is Huffman codes.

- A *Huffman code* is an optimal prefix-free variable-length encoding scheme that assigns bit strings to symbols based on their frequencies in a given text. This is accomplished by a greedy construction of a binary tree whose leaves represent the alphabet symbols and whose edges are labeled with 0's and 1's.

10

Iterative Improvement

The most successful men in the end are those whose success is the result of steady accretion.

—Alexander Graham Bell (1835–1910)

The greedy strategy, considered in the preceding chapter, constructs a solution to an optimization problem piece by piece, always adding a locally optimal piece to a partially constructed solution. In this chapter, we discuss a different approach to designing algorithms for optimization problems. It starts with some feasible solution (a solution that satisfies all the constraints of the problem) and proceeds to improve it by repeated applications of some simple step. This step typically involves a small, localized change yielding a feasible solution with an improved value of the objective function. When no such change improves the value of the objective function, the algorithm returns the last feasible solution as optimal and stops.

There can be several obstacles to the successful implementation of this idea. First, we need an initial feasible solution. For some problems, we can always start with a trivial solution or use an approximate solution obtained by some other (e.g., greedy) algorithm. But for others, finding an initial solution may require as much effort as solving the problem after a feasible solution has been identified. Second, it is not always clear what changes should be allowed in a feasible solution so that we can check efficiently whether the current solution is locally optimal and, if not, replace it with a better one. Third—and this is the most fundamental difficulty—is an issue of local versus global extremum (maximum or minimum). Think about the problem of finding the highest point in a hilly area with no map on a foggy day. A logical thing to do would be to start walking "up the hill" from the point you are at until it becomes impossible to do so because no direction would lead up. You will have reached a local highest point, but because of a limited feasibility, there will be no simple way to tell whether the point is the highest (global maximum you are after) in the entire area.

Fortunately, there are important problems that can be solved by iterative-improvement algorithms. The most important of them is linear programming.

We have already encountered this topic in Section 6.6. Here, in Section 10.1, we introduce the simplex method, the classic algorithm for linear programming. Discovered by the U.S. mathematician George B. Dantzig in 1947, this algorithm has proved to be one of the most consequential achievements in the history of algorithmics.

In Section 10.2, we consider the important problem of maximizing the amount of flow that can be sent through a network with links of limited capacities. This problem is a special case of linear programming. However, its special structure makes it possible to solve the problem by algorithms that are more efficient than the simplex method. We outline the classic iterative-improvement algorithm for this problem, discovered by the American mathematicians L. R. Ford, Jr., and D. R. Fulkerson in the 1950s.

The last two sections of the chapter deal with bipartite matching. This is the problem of finding an optimal pairing of elements taken from two disjoint sets. Examples include matching workers and jobs, high school graduates and colleges, and men and women for marriage. Section 10.3 deals with the problem of maximizing the number of matched pairs; Section 10.4 is concerned with the matching stability.

We also discuss several iterative-improvement algorithms in Section 12.3, where we consider approximation algorithms for the traveling salesman and knapsack problems. Other examples of iterative-improvement algorithms can be found in the algorithms textbook by Moret and Shapiro [Mor91], books on continuous and discrete optimization (e.g., [Nem89]), and the literature on heuristic search (e.g., [Mic10]).

10.1 The Simplex Method

We have already encountered linear programming (see Section 6.6)—the general problem of optimizing a linear function of several variables subject to a set of linear constraints:

$$\begin{aligned} \text{maximize (or minimize)} \quad & c_1x_1 + \cdots + c_nx_n \\ \text{subject to} \quad & a_{i1}x_1 + \cdots + a_{in}x_n \le (\text{or} \ge \text{or} =)\ b_i \quad \text{for } i = 1, \ldots, m \\ & x_1 \ge 0, \ldots, x_n \ge 0. \end{aligned} \qquad \textbf{(10.1)}$$

We mentioned there that many important practical problems can be modeled as instances of linear programming. Two researchers, L. V. Kantorovich of the former Soviet Union and the Dutch-American T. C. Koopmans, were even awarded the Nobel Prize in 1975 for their contributions to linear programming theory and its applications to economics. Apparently because there is no Nobel Prize in mathematics, the Royal Swedish Academy of Sciences failed to honor the U.S. mathematician G. B. Dantzig, who is universally recognized as the father of linear programming in its modern form and the inventor of the simplex method, the classic algorithm for solving such problems.[1]

Geometric Interpretation of Linear Programming

Before we introduce a general method for solving linear programming problems, let us consider a small example, which will help us to see the fundamental properties of such problems.

EXAMPLE 1 Consider the following linear programming problem in two variables:

$$\begin{aligned} \text{maximize} \quad & 3x + 5y \\ \text{subject to} \quad & x + \ \ y \le 4 \\ & x + 3y \le 6 \\ & x \ge 0, \quad y \ge 0. \end{aligned} \qquad \textbf{(10.2)}$$

By definition, a ***feasible solution*** to this problem is any point (x, y) that satisfies all the constraints of the problem; the problem's ***feasible region*** is the set of all its feasible points. It is instructive to sketch the feasible region in the Cartesian plane. Recall that any equation $ax + by = c$, where coefficients a and b are not both equal to zero, defines a straight line. Such a line divides the plane into two half-planes: for all the points in one of them, $ax + by < c$, while for all the points in the other, $ax + by > c$. (It is easy to determine which of the two half-planes is which: take any point (x_0, y_0) not on the line $ax + by = c$ and check which of the two inequalities hold, $ax_0 + by_0 > c$ or $ax_0 + by_0 < c$.) In particular, the set of points defined by inequality $x + y \le 4$ comprises the points on and below the line $x + y = 4$, and the set of points defined by inequality $x + 3y \le 6$ comprises the points on and below the line $x + 3y = 6$. Since the points of the feasible region must satisfy all the constraints of the problem, the feasible region is obtained by the intersection of these two half-planes and the first quadrant of the Cartesian plane defined by the nonnegativity constraints $x \ge 0$, $y \ge 0$ (see Figure 10.1). Thus, the feasible region for problem (10.2) is the convex polygon with the vertices (0, 0), (4, 0), (0, 2), and (3, 1). (The last point, which is the point of intersection of the lines $x + y = 4$ and $x + 3y = 6$, is obtained by solving the system of these two linear equations.) Our task is to find an ***optimal solution***, a point in the feasible region with the largest value of the ***objective function*** $z = 3x + 5y$.

Are there feasible solutions for which the value of the objective function equals, say, 20? The points (x, y) for which the objective function $z = 3x + 5y$ is equal to 20 form the line $3x + 5y = 20$. Since this line does not have common points

1. George B. Dantzig (1914–2005) has received many honors, including the National Medal of Science presented by the president of the United States in 1976. The citation states that the National Medal was awarded "for inventing linear programming and discovering methods that led to wide-scale scientific and technical applications to important problems in logistics, scheduling, and network optimization, and to the use of computers in making efficient use of the mathematical theory."

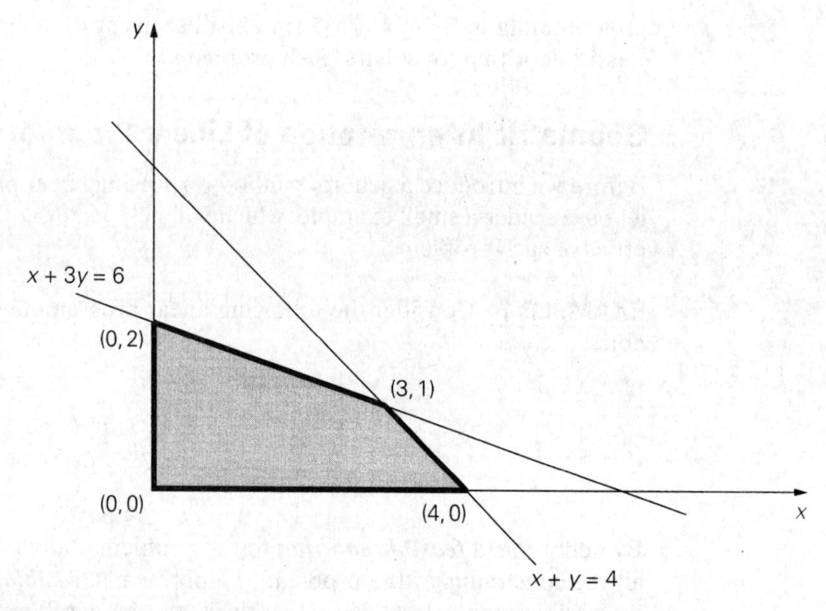

FIGURE 10.1 Feasible region of problem (10.2).

with the feasible region—see Figure 10.2—the answer to the posed question is no. On the other hand, there are infinitely many feasible points for which the objective function is equal to, say, 10: they are the intersection points of the line $3x + 5y = 10$ with the feasible region. Note that the lines $3x + 5y = 20$ and $3x + 5y = 10$ have the same slope, as would any line defined by equation $3x + 5y = z$ where z is some constant. Such lines are called ***level lines*** of the objective function. Thus, our problem can be restated as finding the largest value of the parameter z for which the level line $3x + 5y = z$ has a common point with the feasible region.

We can find this line either by shifting, say, the line $3x + 5y = 20$ south-west (without changing its slope!) toward the feasible region until it hits the region for the first time or by shifting, say, the line $3x + 5y = 10$ north-east until it hits the feasible region for the last time. Either way, it will happen at the point (3, 1) with the corresponding z value $3 \cdot 3 + 5 \cdot 1 = 14$. This means that the optimal solution to the linear programming problem in question is $x = 3$, $y = 1$, with the maximal value of the objective function equal to 14.

Note that if we had to maximize $z = 3x + 3y$ as the objective function in problem (10.2), the level line $3x + 3y = z$ for the largest value of z would coincide with the boundary line segment that has the same slope as the level lines (draw this line in Figure 10.2). Consequently, all the points of the line segment between vertices (3, 1) and (4, 0), including the vertices themselves, would be optimal solutions, yielding, of course, the same maximal value of the objective function. ■

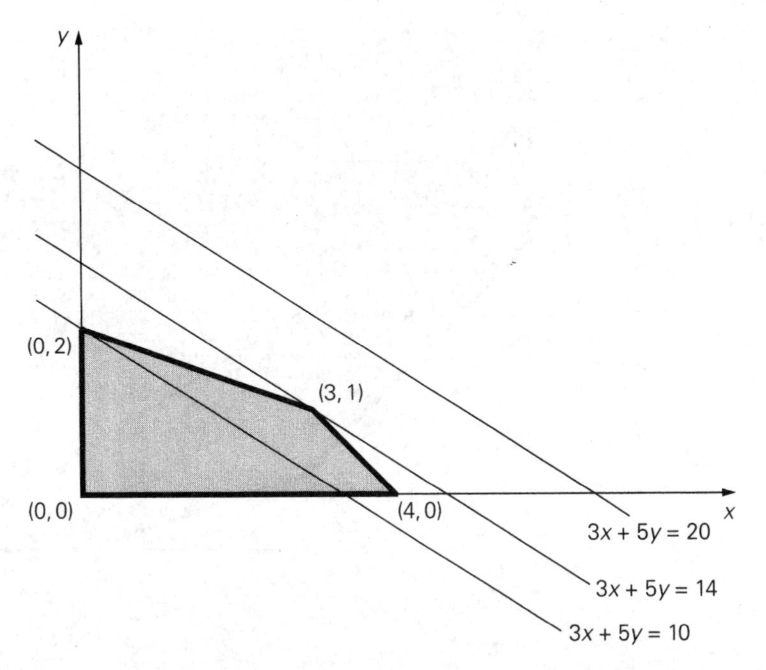

FIGURE 10.2 Solving a two-dimensional linear programming problem geometrically.

Does every linear programming problem have an optimal solution that can be found at a vertex of its feasible region? Without appropriate qualifications, the answer to this question is no. To begin with, the feasible region of a linear programming problem can be empty. For example, if the constraints include two contradictory requirements, such as $x + y \leq 1$ and $x + y \geq 2$, there can be no points in the problem's feasible region. Linear programming problems with the empty feasible region are called ***infeasible***. Obviously, infeasible problems do not have optimal solutions.

Another complication may arise if the problem's feasible region is unbounded, as the following example demonstrates.

EXAMPLE 2 If we reverse the inequalities in problem (10.2) to $x + y \geq 4$ and $x + 3y \geq 6$, the feasible region of the new problem will become unbounded (see Figure 10.3). If the feasible region of a linear programming problem is unbounded, its objective function may or may not attain a finite optimal value on it. For example, the problem of maximizing $z = 3x + 5y$ subject to the constraints $x + y \geq 4$, $x + 3y \geq 6$, $x \geq 0$, $y \geq 0$ has no optimal solution, because there are points in the feasible region making $3x + 5y$ as large as we wish. Such problems are called ***unbounded***. On the other hand, the problem of minimizing $z = 3x + 5y$ subject to the same constraints has an optimal solution (which?). ■

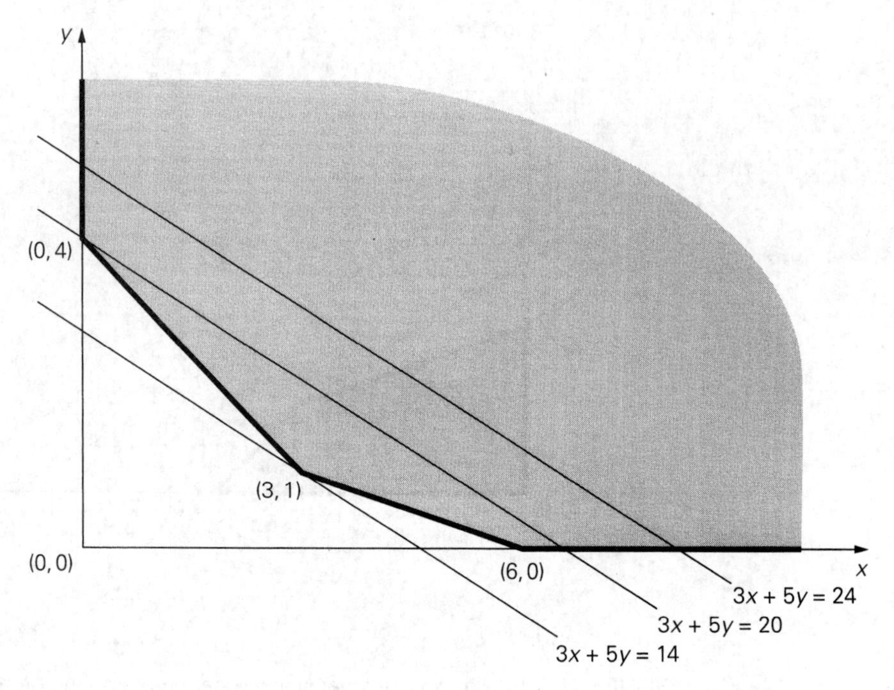

FIGURE 10.3 Unbounded feasible region of a linear programming problem with constraints $x + y \geq 4$, $x + 3y \geq 6$, $x \geq 0$, $y \geq 0$, and three level lines of the function $3x + 5y$.

Fortunately, the most important features of the examples we considered above hold for problems with more than two variables. In particular, a feasible region of a typical linear programming problem is in many ways similar to convex polygons in the two-dimensional Cartesian plane. Specifically, it always has a finite number of vertices, which mathematicians prefer to call ***extreme points*** (see Section 3.3). Furthermore, an optimal solution to a linear programming problem can be found at one of the extreme points of its feasible region. We reiterate these properties in the following theorem.

THEOREM (*Extreme Point Theorem*) Any linear programming problem with a nonempty bounded feasible region has an optimal solution; moreover, an optimal solution can always be found at an extreme point of the problem's feasible region.[2]

This theorem implies that to solve a linear programming problem, at least in the case of a bounded feasible region, we can ignore all but a finite number of

2. Except for some degenerate instances (such as maximizing $z = x + y$ subject to $x + y = 1$), if a linear programming problem with an unbounded feasible region has an optimal solution, it can also be found at an extreme point of the feasible region.

points in its feasible region. In principle, we can solve such a problem by computing the value of the objective function at each extreme point and selecting the one with the best value. There are two major obstacles to implementing this plan, however. The first lies in the need for a mechanism for generating the extreme points of the feasible region. As we are going to see below, a rather straightforward algebraic procedure for this task has been discovered. The second obstacle lies in the number of extreme points a typical feasible region has. Here, the news is bad: the number of extreme points is known to grow exponentially with the size of the problem. This makes the exhaustive inspection of extreme points unrealistic for most linear programming problems of nontrivial sizes.

Fortunately, it turns out that there exists an algorithm that typically inspects only a small fraction of the extreme points of the feasible region before reaching an optimal one. This famous algorithm is called the ***simplex method***. The idea of this algorithm can be described in geometric terms as follows. Start by identifying an extreme point of the feasible region. Then check whether one can get an improved value of the objective function by going to an adjacent extreme point. If it is not the case, the current point is optimal—stop; if it is the case, proceed to an adjacent extreme point with an improved value of the objective function. After a finite number of steps, the algorithm will either reach an extreme point where an optimal solution occurs or determine that no optimal solution exists.

An Outline of the Simplex Method

Our task now is to "translate" the geometric description of the simplex method into the more algorithmically precise language of algebra. To begin with, before we can apply the simplex method to a linear programming problem, it has to be represented in a special form called the ***standard form***. The standard form has the following requirements:

- It must be a maximization problem.
- All the constraints (except the nonnegativity constraints) must be in the form of linear equations with nonnegative right-hand sides.
- All the variables must be required to be nonnegative.

Thus, the general linear programming problem in standard form with m constraints and n unknowns ($n \geq m$) is

$$\begin{aligned} \text{maximize} \quad & c_1x_1 + \cdots + c_nx_n \\ \text{subject to} \quad & a_{i1}x_1 + \cdots + a_{in}x_n = b_i, \quad \text{where } b_i \geq 0 \text{ for } i = 1, 2, \ldots, m \\ & x_1 \geq 0, \ldots, x_n \geq 0. \end{aligned} \tag{10.3}$$

It can also be written in compact matrix notations:

$$\begin{aligned} \text{maximize} \quad & cx \\ \text{subject to} \quad & Ax = b \\ & x \geq 0, \end{aligned}$$

where

$$c = [c_1\, c_2\, \dots\, c_n],\quad x = \begin{bmatrix} x_1 \\ x_2 \\ \vdots \\ x_n \end{bmatrix},\quad A = \begin{bmatrix} a_{11} & a_{12} & \dots & a_{1n} \\ \vdots & \vdots & & \vdots \\ a_{m1} & a_{m2} & \dots & a_{mn} \end{bmatrix},\quad b = \begin{bmatrix} b_1 \\ b_2 \\ \vdots \\ b_m \end{bmatrix}.$$

Any linear programming problem can be transformed into an equivalent problem in standard form. If an objective function needs to be minimized, it can be replaced by the equivalent problem of maximizing the same objective function with all its coefficients c_j replaced by $-c_j$, $j = 1, 2, \dots, n$ (see Section 6.6 for a more general discussion of such transformations). If a constraint is given as an inequality, it can be replaced by an equivalent equation by adding a ***slack variable*** representing the difference between the two sides of the original inequality. For example, the two inequalities of problem (10.2) can be transformed, respectively, into the following equations:

$$x + y + u = 4 \text{ where } u \geq 0 \quad \text{and} \quad x + 3y + v = 6 \text{ where } v \geq 0.$$

Finally, in most linear programming problems, the variables are required to be nonnegative to begin with because they represent some physical quantities. If this is not the case in an initial statement of a problem, an unconstrained variable x_j can be replaced by the difference between two new nonnegative variables: $x_j = x_j' - x_j''$, $x_j' \geq 0$, $x_j'' \geq 0$.

Thus, problem (10.2) in standard form is the following linear programming problem in four variables:

$$\begin{aligned} \text{maximize} \quad & 3x + 5y + 0u + 0v \\ \text{subject to} \quad & x + y + u = 4 \\ & x + 3y + + v = 6 \\ & x, y, u, v \geq 0. \end{aligned} \qquad \textbf{(10.4)}$$

It is easy to see that if we find an optimal solution (x^*, y^*, u^*, v^*) to problem (10.4), we can obtain an optimal solution to problem (10.2) by simply ignoring its last two coordinates.

The principal advantage of the standard form lies in the simple mechanism it provides for identifying extreme points of the feasible region. To do this for problem (10.4), for example, we need to set two of the four variables in the constraint equations to zero to get a system of two linear equations in two unknowns and solve this system. For the general case of a problem with m equations in n unknowns ($n \geq m$), $n - m$ variables need to be set to zero to get a system of m equations in m unknowns. If the system obtained has a unique solution—as any nondegenerate system of linear equations with the number of equations equal to the number of unknowns does—we have a ***basic solution***; its coordinates set to zero before solving the system are called ***nonbasic***, and its coordinates obtained by solving the system are called ***basic***. (This terminology comes from linear algebra.

Specifically, we can rewrite the system of constraint equations of (10.4) as

$$x\begin{bmatrix}1\\1\end{bmatrix}+y\begin{bmatrix}1\\3\end{bmatrix}+u\begin{bmatrix}1\\0\end{bmatrix}+v\begin{bmatrix}0\\1\end{bmatrix}=\begin{bmatrix}4\\6\end{bmatrix}.$$

A basis in the two-dimensional vector space is composed of any two vectors that are not proportional to each other; once a basis is chosen, any vector can be uniquely expressed as a sum of multiples of the basis vectors. Basic and nonbasic variables indicate which of the given vectors are, respectively, included and excluded in a particular basis choice.)

If all the coordinates of a basic solution are nonnegative, the basic solution is called a ***basic feasible solution***. For example, if we set to zero variables x and y and solve the resulting system for u and v, we obtain the basic feasible solution (0, 0, 4, 6); if we set to zero variables x and u and solve the resulting system for y and v, we obtain the basic solution (0, 4, 0, −6), which is not feasible. The importance of basic feasible solutions lies in the one-to-one correspondence between them and the extreme points of the feasible region. For example, (0, 0, 4, 6) is an extreme point of the feasible region of problem (10.4) (with the point (0, 0) in Figure 10.1 being its projection on the x, y plane). Incidentally, (0, 0, 4, 6) is a natural starting point for the simplex method's application to this problem.

As mentioned above, the simplex method progresses through a series of adjacent extreme points (basic feasible solutions) with increasing values of the objective function. Each such point can be represented by a ***simplex tableau***, a table storing the information about the basic feasible solution corresponding to the extreme point. For example, the simplex tableau for (0, 0, 4, 6) of problem (10.4) is presented below:

	x	y	u	v	
u	1	1	1	0	4
← v	1	3	0	1	6
	−3	−5	0	0	0
		↑			

(10.5)

In general, a simplex tableau for a linear programming problem in standard form with n unknowns and m linear equality constraints ($n \geq m$) has $m+1$ rows and $n+1$ columns. Each of the first m rows of the table contains the coefficients of a corresponding constraint equation, with the last column's entry containing the equation's right-hand side. The columns, except the last one, are labeled by the names of the variables. The rows are labeled by the basic variables of the basic feasible solution the tableau represents; the values of the basic variables of this solution are in the last column. Also note that the columns labeled by the basic variables form the $m \times m$ identity matrix.

The last row of a simplex tableau is called the ***objective row***. It is initialized by the coefficients of the objective function with their signs reversed (in the first n columns) and the value of the objective function at the initial point (in the last column). On subsequent iterations, the objective row is transformed the same way as all the other rows. The objective row is used by the simplex method to check whether the current tableau represents an optimal solution: it does if all the entries in the objective row—except, possibly, the one in the last column—are nonnegative. If this is not the case, any of the negative entries indicates a nonbasic variable that can become basic in the next tableau.

For example, according to this criterion, the basic feasible solution (0, 0, 4, 6) represented by tableau (10.5) is not optimal. The negative value in the x-column signals the fact that we can increase the value of the objective function $z = 3x + 5y + 0u + 0v$ by increasing the value of the x-coordinate in the current basic feasible solution (0, 0, 4, 6). Indeed, since the coefficient for x in the objective function is positive, the larger the x value, the larger the value of this function. Of course, we will need to "compensate" an increase in x by adjusting the values of the basic variables u and v so that the new point is still feasible. For this to be the case, both conditions

$$x + u = 4 \quad \text{where } u \geq 0$$
$$x + v = 6 \quad \text{where } v \geq 0$$

must be satisfied, which means that

$$x \leq \min\{4, 6\} = 4.$$

Note that if we increase the value of x from 0 to 4, the largest amount possible, we will find ourselves at the point (4, 0, 0, 2), an adjacent to (0, 0, 4, 6) extreme point of the feasible region, with $z = 12$.

Similarly, the negative value in the y-column of the objective row signals the fact that we can also increase the value of the objective function by increasing the value of the y-coordinate in the initial basic feasible solution (0, 0, 4, 6). This requires

$$y + u = 4 \quad \text{where } u \geq 0$$
$$3y + v = 6 \quad \text{where } v \geq 0,$$

which means that

$$y \leq \min\{\frac{4}{1}, \frac{6}{3}\} = 2.$$

If we increase the value of y from 0 to 2, the largest amount possible, we will find ourselves at the point (0, 2, 2, 0), another adjacent to (0, 0, 4, 6) extreme point, with $z = 10$.

If there are several negative entries in the objective row, a commonly used rule is to select the most negative one, i.e., the negative number with the largest absolute value. This rule is motivated by the observation that such a choice yields the largest increase in the objective function's value per unit of change in a variable's value. (In our example, an increase in the x-value from 0 to 1 at (0, 0, 4, 6) changes the value of $z = 3x + 5y + 0u + 0v$ from 0 to 3, while an increase in the y-value from 0 to 1 at (0, 0, 4, 6) changes z from 0 to 5.) Note, however, that the feasibility constraints impose different limits on how much each of the variables may increase. In our example, in particular, the choice of the y-variable over the x-variable leads to a smaller increase in the value of the objective function. Still, we will employ this commonly used rule and select variable y as we continue with our example. A new basic variable is called the ***entering variable***, while its column is referred to as the ***pivot column***; we mark the pivot column by $\uparrow$.

Now we will explain how to choose a ***departing variable***, i.e., a basic variable to become nonbasic in the next tableau. (The total number of basic variables in any basic solution must be equal to m, the number of the equality constraints.) As we saw above, to get to an adjacent extreme point with a larger value of the objective function, we need to increase the entering variable by the largest amount possible to make one of the old basic variables zero while preserving the nonnegativity of all the others. We can translate this observation into the following rule for choosing a departing variable in a simplex tableau: for each *positive* entry in the pivot column, compute the θ-***ratio*** by dividing the row's last entry by the entry in the pivot column. For the example of tableau (10.5), these θ-ratios are

$$\theta_u = \frac{4}{1} = 4, \qquad \theta_v = \frac{6}{3} = 2.$$

The row with the smallest θ-ratio determines the departing variable, i.e., the variable to become nonbasic. Ties may be broken arbitrarily. For our example, it is variable v. We mark the row of the departing variable, called the ***pivot row***, by $\longleftarrow$ and denote it $\overleftarrow{\text{row}}$. Note that if there are no positive entries in the pivot column, no θ-ratio can be computed, which indicates that the problem is unbounded and the algorithm stops.

Finally, the following steps need to be taken to transform a current tableau into the next one. (This transformation, called ***pivoting***, is similar to the principal step of the Gauss-Jordan elimination algorithm for solving systems of linear equations—see Problem 8 in Exercises 6.2.) First, divide all the entries of the pivot row by the ***pivot***, its entry in the pivot column, to obtain $\overleftarrow{\text{row}}_{new}$. For tableau (10.5), we obtain

$$\overleftarrow{\text{row}}_{\text{new}}: \quad \frac{1}{3} \quad 1 \quad 0 \quad \frac{1}{3} \quad 2.$$

Then, replace each of the other rows, including the objective row, by the difference

$$\text{row} - c \cdot \overleftarrow{\text{row}}_{\text{new}},$$

where c is the row's entry in the pivot column. For tableau (10.5), this yields

$$\text{row 1} - 1 \cdot \overleftarrow{\text{row}}_{\text{new}}: \quad \frac{2}{3} \quad 0 \quad 1 \quad -\frac{1}{3} \quad 2,$$

$$\text{row 3} - (-5) \cdot \overleftarrow{\text{row}}_{\text{new}}: \quad -\frac{4}{3} \quad 0 \quad 0 \quad \frac{5}{3} \quad 10.$$

Thus, the simplex method transforms tableau (10.5) into the following tableau:

	x	y	u	v	
← u	$\frac{2}{3}$	0	1	$-\frac{1}{3}$	2
y	$\frac{1}{3}$	1	0	$\frac{1}{3}$	2
	$-\frac{4}{3}$ ↑	0	0	$\frac{5}{3}$	10

(10.6)

Tableau (10.6) represents the basic feasible solution (0, 2, 2, 0) with an increased value of the objective function, which is equal to 10. It is not optimal, however (why?).

The next iteration—do it yourself as a good exercise!—yields tableau (10.7):

	x	y	u	v	
x	1	0	$\frac{3}{2}$	$-\frac{1}{2}$	3
y	0	1	$-\frac{1}{2}$	$\frac{1}{2}$	1
	0	0	2	1	14

(10.7)

This tableau represents the basic feasible solution (3, 1, 0, 0). It is optimal because all the entries in the objective row of tableau (10.7) are nonnegative. The maximal value of the objective function is equal to 14, the last entry in the objective row.

Let us summarize the steps of the simplex method.

Summary of the simplex method

Step 0 *Initialization* Present a given linear programming problem in standard form and set up an initial tableau with nonnegative entries in the rightmost column and m other columns composing the $m \times m$ identity matrix. (Entries in the objective row are to be disregarded in verifying these requirements.) These m columns define the basic variables of the initial basic feasible solution, used as the labels of the tableau's rows.

Step 1 *Optimality test* If all the entries in the objective row (except, possibly, the one in the rightmost column, which represents the value of the objective function) are nonnegative—stop: the tableau represents an optimal solution whose basic variables' values are in the rightmost column and the remaining, nonbasic variables' values are zeros.

Step 2 *Finding the entering variable* Select a negative entry from among the first n elements of the objective row. (A commonly used rule is to select the negative entry with the largest absolute value, with ties broken arbitrarily.) Mark its column to indicate the entering variable and the pivot column.

Step 3 *Finding the departing variable* For each positive entry in the pivot column, calculate the θ-ratio by dividing that row's entry in the rightmost column by its entry in the pivot column. (If all the entries in the pivot column are negative or zero, the problem is unbounded—stop.) Find the row with the smallest θ-ratio (ties may be broken arbitrarily), and mark this row to indicate the departing variable and the pivot row.

Step 4 *Forming the next tableau* Divide all the entries in the pivot row by its entry in the pivot column. Subtract from each of the other rows, including the objective row, the new pivot row multiplied by the entry in the pivot column of the row in question. (This will make all the entries in the pivot column 0's except for 1 in the pivot row.) Replace the label of the pivot row by the variable's name of the pivot column and go back to Step 1.

Further Notes on the Simplex Method

Formal proofs of validity of the simplex method steps can be found in books devoted to a detailed discussion of linear programming (e.g., [Dan63]). A few important remarks about the method still need to be made, however. Generally speaking, an iteration of the simplex method leads to an extreme point of the problem's feasible region with a greater value of the objective function. In degenerate cases, which arise when one or more basic variables are equal to zero, the simplex method can only guarantee that the value of the objective function at the new extreme point is greater than or equal to its value at the previous point. In turn, this opens the door to the possibility not only that the objective function's values "stall" for several iterations in a row but that the algorithm might cycle back to a previously considered point and hence never terminate. The latter phenomenon is called ***cycling***. Although it rarely if ever happens in practice, specific examples of problems where cycling does occur have been constructed. A simple modification of Steps 2 and 3 of the simplex method, called ***Bland's rule***, eliminates even the theoretical possibility of cycling. Assuming that the variables are denoted by a subscripted letter (e.g., $x_1, x_2, \ldots, x_n$), this rule can be stated as follows:

Step 2 modified Among the columns with a negative entry in the objective row, select the column with the smallest subscript.

Step 3 modified Resolve a tie among the smallest θ-ratios by selecting the row labeled by the basic variable with the smallest subscript.

Another caveat deals with the assumptions made in Step 0. They are automatically satisfied if a problem is given in the form where all the constraints imposed on nonnegative variables are inequalities $a_{i1}x_1 + \cdots + a_{in}x_n \le b_i$ with $b_i \ge 0$ for $i = 1, 2, \ldots, m$. Indeed, by adding a nonnegative slack variable x_{n+i} into the ith constraint, we obtain the equality $a_{i1}x_1 + \cdots + a_{in}x_n + x_{n+i} = b_i$, and all the requirements imposed on an initial tableau of the simplex method are satisfied for the obvious basic feasible solution $x_1 = \cdots = x_n = 0$, $x_{n+1} = \cdots = x_{n+m} = 1$. But if a problem is not given in such a form, finding an initial basic feasible solution may present a nontrivial obstacle. Moreover, for problems with an empty feasible region, no initial basic feasible solution exists, and we need an algorithmic way to identify such problems. One of the ways to address these issues is to use an extension to the classic simplex method called the ***two-phase simplex method*** (see, e.g., [Kol95]). In a nutshell, this method adds a set of artificial variables to the equality constraints of a given problem so that the new problem has an obvious basic feasible solution. It then solves the linear programming problem of minimizing the sum of the artificial variables by the simplex method. The optimal solution to this problem either yields an initial tableau for the original problem or indicates that the feasible region of the original problem is empty.

How efficient is the simplex method? Since the algorithm progresses through a sequence of adjacent points of a feasible region, one should probably expect bad news because the number of extreme points is known to grow exponentially with the problem size. Indeed, the worst-case efficiency of the simplex method has been shown to be exponential as well. Fortunately, more than half a century of practical experience with the algorithm has shown that the number of iterations in a typical application ranges between m and $3m$, with the number of operations per iteration proportional to mn, where m and n are the numbers of equality constraints and variables, respectively.

Since its discovery in 1947, the simplex method has been a subject of intensive study by many researchers. Some of them have worked on improvements to the original algorithm and details of its efficient implementation. As a result of these efforts, programs implementing the simplex method have been polished to the point that very large problems with hundreds of thousands of constraints and variables can be solved in a routine manner. In fact, such programs have evolved into sophisticated software packages. These packages enable the user to enter a problem's constraints and obtain a solution in a user-friendly form. They also provide tools for investigating important properties of the solution, such as its sensitivity to changes in the input data. Such investigations are very important for many applications, including those in economics. At the other end of the spectrum, linear programming problems of a moderate size can nowadays be solved on a desktop using a standard spreadsheet facility or by taking advantage of specialized software available on the Internet.

Researchers have also tried to find algorithms for solving linear programming problems with polynomial-time efficiency in the worst case. An important milestone in the history of such algorithms was the proof by L. G. Khachian [Kha79] showing that the ***ellipsoid method*** can solve any linear programming problem in polynomial time. Although the ellipsoid method was much slower than the simplex method in practice, its better worst-case efficiency encouraged a search for alternatives to the simplex method. In 1984, Narendra Karmarkar published an algorithm that not only had a polynomial worst-case efficiency but also was competitive with the simplex method in empirical tests as well. Although we are not going to discuss ***Karmarkar's algorithm*** [Kar84] here, it is worth pointing out that it is also based on the iterative-improvement idea. However, Karmarkar's algorithm generates a sequence of feasible solutions that lie within the feasible region rather than going through a sequence of adjacent extreme points as the simplex method does. Such algorithms are called ***interior-point methods*** (see, e.g., [Arb93]).

Exercises 10.1

1. Consider the following version of the post office location problem (Problem 3 in Exercises 3.3): Given n integers $x_1, x_2, \ldots, x_n$ representing coordinates of n villages located along a straight road, find a location for a post office that minimizes the average distance between the villages. The post office may be, but is not required to be, located at one of the villages. Devise an iterative-improvement algorithm for this problem. Is this an efficient way to solve this problem?

2. Solve the following linear programming problems geometrically.

a.

$$\begin{aligned} \text{maximize} \quad & 3x + y \\ \text{subject to} \quad & -x + y \le 1 \\ & 2x + y \le 4 \\ & x \ge 0,\ y \ge 0 \end{aligned}$$

b.

$$\begin{aligned} \text{maximize} \quad & x + 2y \\ \text{subject to} \quad & 4x \ge y \\ & y \le 3 + x \\ & x \ge 0,\ y \ge 0 \end{aligned}$$

3. Consider the linear programming problem

$$\begin{aligned} \text{minimize} \quad & c_1x + c_2y \\ \text{subject to} \quad & x + y \ge 4 \\ & x + 3y \ge 6 \\ & x \ge 0,\ y \ge 0 \end{aligned}$$

where c_1 and c_2 are some real numbers not both equal to zero.

a. Give an example of the coefficient values c_1 and c_2 for which the problem has a unique optimal solution.

b. Give an example of the coefficient values c_1 and c_2 for which the problem has infinitely many optimal solutions.

c. Give an example of the coefficient values c_1 and c_2 for which the problem does not have an optimal solution.

4. Would the solution to problem (10.2) be different if its inequality constraints were strict, i.e., $x + y < 4$ and $x + 3y < 6$, respectively?

5. Trace the simplex method on

a. the problem of Exercise 2a.

b. the problem of Exercise 2b.

6. Trace the simplex method on the problem of Example 1 in Section 6.6

a. by hand.

b. by using one of the implementations available on the Internet.

7. Determine how many iterations the simplex method needs to solve the problem

$$\text{maximize} \quad \sum_{j=1}^{n} x_j$$
$$\text{subject to} \quad 0 \le x_j \le b_j, \quad \text{where } b_j > 0 \text{ for } j = 1, 2, \ldots, n.$$

8. Can we apply the simplex method to solve the knapsack problem (see Example 2 in Section 6.6)? If you answer yes, indicate whether it is a good algorithm for the problem in question; if you answer no, explain why not.

9. Prove that no linear programming problem can have exactly $k \ge 1$ optimal solutions unless $k = 1$.

10. If a linear programming problem

$$\text{maximize} \quad \sum_{j=1}^{n} c_j x_j$$
$$\text{subject to} \quad \sum_{j=1}^{n} a_{ij} x_j \le b_i \quad \text{for } i = 1, 2, \ldots, m$$
$$x_1, x_2, \ldots, x_n \ge 0$$

is considered as ***primal***, then its ***dual*** is defined as the linear programming problem

$$\text{minimize} \quad \sum_{i=1}^{m} b_i y_i$$
$$\text{subject to} \quad \sum_{i=1}^{m} a_{ij} y_i \ge c_j \quad \text{for } j = 1, 2, \ldots, n$$
$$y_1, y_2, \ldots, y_m \ge 0.$$

a. Express the primal and dual problems in matrix notations.

b. Find the dual of the linear programming problem

$$\begin{aligned} \text{maximize } & x_1 + 4x_2 - x_3 \\ \text{subject to } & x_1 + x_2 + x_3 \le 6 \\ & x_1 - x_2 - 2x_3 \le 2 \\ & x_1, x_2, x_3 \ge 0. \end{aligned}$$

c. Solve the primal and dual problems and compare the optimal values of their objective functions.

10.2 The Maximum-Flow Problem

In this section, we consider the important problem of maximizing the flow of a material through a transportation network (pipeline system, communication system, electrical distribution system, and so on). We will assume that the transportation network in question can be represented by a connected weighted digraph with n vertices numbered from 1 to n and a set of edges E, with the following properties:

- It contains exactly one vertex with no entering edges; this vertex is called the ***source*** and assumed to be numbered 1.
- It contains exactly one vertex with no leaving edges; this vertex is called the ***sink*** and assumed to be numbered n.
- The weight u_{ij} of each directed edge (i, j) is a positive integer, called the edge ***capacity***. (This number represents the upper bound on the amount of the material that can be sent from i to j through a link represented by this edge.)

A digraph satisfying these properties is called a ***flow network*** or simply a ***network***.[3] A small instance of a network is given in Figure 10.4.

It is assumed that the source and the sink are the only source and destination of the material, respectively; all the other vertices can serve only as points where a flow can be redirected without consuming or adding any amount of the material. In other words, the total amount of the material entering an intermediate vertex must be equal to the total amount of the material leaving the vertex. This condition is called the ***flow-conservation requirement***. If we denote the amount sent through edge (i, j) by x_{ij}, then for any intermediate vertex i, the flow-conservation requirement can be expressed by the following equality constraint:

$$\sum_{j:(j,i)\in E} x_{ji} = \sum_{j:(i,j)\in E} x_{ij} \quad \text{for } i = 2, 3, \ldots, n-1, \tag{10.8}$$

3. In a slightly more general model, one can consider a network with several sources and sinks and allow capacities u_{ij} to be infinitely large.

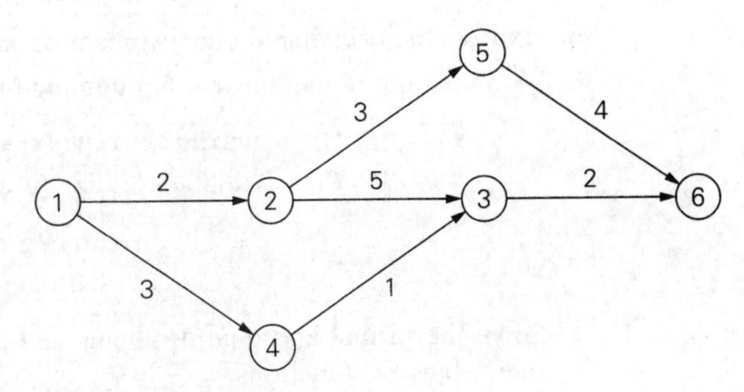

FIGURE 10.4 Example of a network graph. The vertex numbers are vertex "names"; the edge numbers are edge capacities.

where the sums in the left- and right-hand sides express the total inflow and outflow entering and leaving vertex i, respectively.

Since no amount of the material can change by going through intermediate vertices of the network, it stands to reason that the total amount of the material leaving the source must end up at the sink. (This observation can also be derived formally from equalities (10.8), a task you will be asked to do in the exercises.) Thus, we have the following equality:

$$\sum_{j:\,(1,j)\in E} x_{1j} = \sum_{j:\,(j,n)\in E} x_{jn}. \tag{10.9}$$

This quantity, the total outflow from the source—or, equivalently, the total inflow into the sink—is called the ***value*** of the flow. We denote it by v. It is this quantity that we will want to maximize over all possible flows in a network.

Thus, a (feasible) ***flow*** is an assignment of real numbers x_{ij} to edges (i, j) of a given network that satisfy flow-conservation constraints (10.8) and the ***capacity constraints***

$$0 \le x_{ij} \le u_{ij} \quad \text{for every edge } (i, j) \in E. \tag{10.10}$$

The ***maximum-flow problem*** can be stated formally as the following optimization problem:

$$\begin{aligned} &\text{maximize} \quad v = \sum_{j:\,(1,j)\in E} x_{1j} \\ &\text{subject to} \quad \sum_{j:\,(j,i)\in E} x_{ji} - \sum_{j:\,(i,j)\in E} x_{ij} = 0 \quad \text{for } i = 2, 3, \ldots, n-1 \\ &\qquad\qquad 0 \le x_{ij} \le u_{ij} \quad \text{for every edge } (i, j) \in E. \end{aligned} \tag{10.11}$$

We can solve linear programming problem (10.11) by the simplex method or by another algorithm for general linear programming problems (see Section 10.1). However, the special structure of problem (10.11) can be exploited to design faster algorithms. In particular, it is quite natural to employ the iterative-improvement idea as follows. We can always start with the zero flow (i.e., set $x_{ij} = 0$ for every edge (i, j) in the network). Then, on each iteration, we can try to find a path from source to sink along which some additional flow can be sent. Such a path is called ***flow augmenting***. If a flow-augmenting path is found, we adjust the flow along the edges of this path to get a flow of an increased value and try to find an augmenting path for the new flow. If no flow-augmenting path can be found, we conclude that the current flow is optimal. This general template for solving the maximum-flow problem is called the ***augmenting-path method***, also known as the ***Ford-Fulkerson method*** after L. R. Ford, Jr., and D. R. Fulkerson, who discovered it (see [For57]).

An actual implementation of the augmenting path idea is, however, not quite straightforward. To see this, let us consider the network in Figure 10.4. We start with the zero flow shown in Figure 10.5a. (In that figure, the zero amounts sent through each edge are separated from the edge capacities by the slashes; we will use this notation in the other examples as well.) It is natural to search for a flow-augmenting path from source to sink by following directed edges (i, j) for which the current flow x_{ij} is less than the edge capacity u_{ij}. Among several possibilities, let us assume that we identify the augmenting path 1→2→3→6 first. We can increase the flow along this path by a maximum of 2 units, which is the smallest unused capacity of its edges. The new flow is shown in Figure 10.5b. This is as far as our simpleminded idea about flow-augmenting paths will be able to take us. Unfortunately, the flow shown in Figure 10.5b is not optimal: its value can still be increased along the path 1→4→3←2→5→6 by increasing the flow by 1 on edges (1, 4), (4, 3), (2, 5), and (5, 6) and *decreasing* it by 1 on edge (2, 3). The flow obtained as the result of this augmentation is shown in Figure 10.5c. It is indeed maximal. (Can you tell why?)

Thus, to find a flow-augmenting path for a flow x, we need to consider paths from source to sink in the underlying *undirected* graph in which any two consecutive vertices i, j are either

i. connected by a directed edge from i to j with some positive unused capacity $r_{ij} = u_{ij} - x_{ij}$ (so that we can increase the flow through that edge by up to r_{ij} units), or

ii. connected by a directed edge from j to i with some positive flow x_{ji} (so that we can decrease the flow through that edge by up to x_{ji} units).

Edges of the first kind are called ***forward edges*** because their tail is listed before their head in the vertex list $1 \rightarrow \cdots i \rightarrow j \cdots \rightarrow n$ defining the path; edges of the second kind are called ***backward edges*** because their tail is listed after their head in the path list $1 \rightarrow \cdots i \leftarrow j \cdots \rightarrow n$. To illustrate, for the path 1→4→3←2→5→6 of the last example, (1, 4), (4, 3), (2, 5), and (5, 6) are the forward edges, and (3, 2) is the backward edge.

For a given flow-augmenting path, let r be the minimum of all the unused capacities r_{ij} of its forward edges and all the flows x_{ji} of its backward edges. It is easy to see that if we increase the current flow by r on each forward edge and decrease it by this amount on each backward edge, we will obtain a feasible

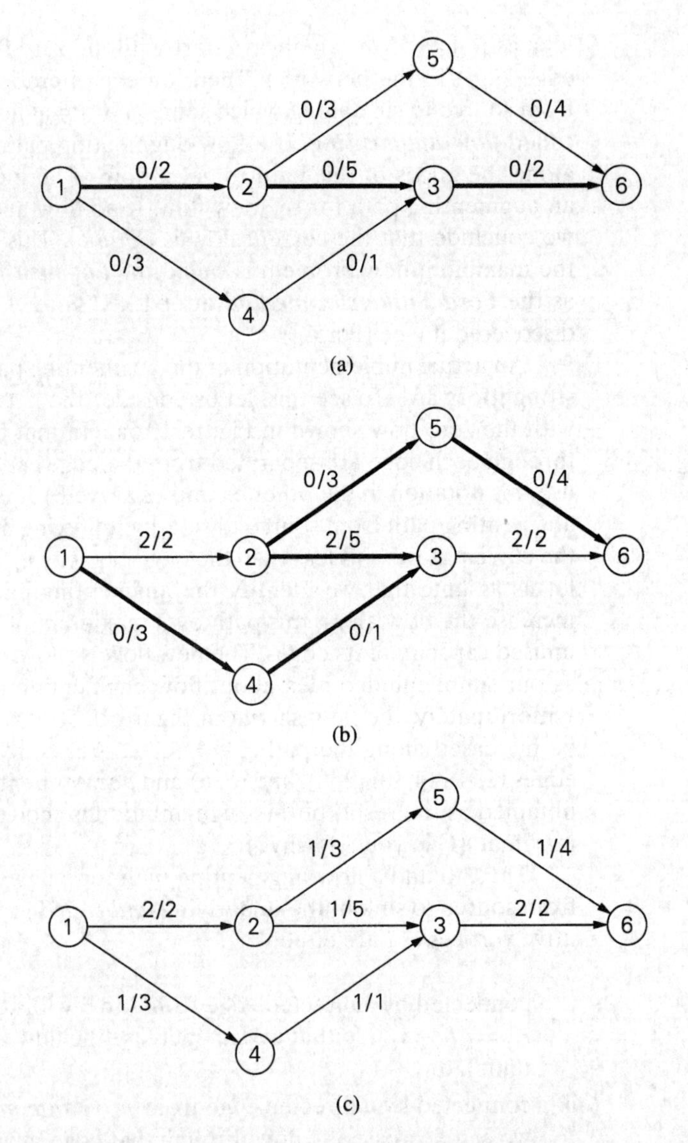

FIGURE 10.5 Illustration of the augmenting-path method. Flow-augmenting paths are shown in bold. The flow amounts and edge capacities are indicated by the numbers before and after the slash, respectively.

flow whose value is r units greater than the value of its predecessor. Indeed, let i be an intermediate vertex on a flow-augmenting path. There are four possible combinations of forward and backward edges incident to vertex i:

$$\xrightarrow{+r} i \xrightarrow{+r}, \qquad \xrightarrow{+r} i \xleftarrow{-r}, \qquad \xleftarrow{-r} i \xrightarrow{+r}, \qquad \xleftarrow{-r} i \xleftarrow{-r}.$$

For each of them, the flow-conservation requirement for vertex i will still hold after the flow adjustments indicated above the edge arrows. Further, since r is the minimum among all the positive unused capacities on the forward edges and all the positive flows on the backward edges of the flow-augmenting path, the new flow will satisfy the capacity constraints as well. Finally, adding r to the flow on the first edge of the augmenting path will increase the value of the flow by r.

Under the assumption that all the edge capacities are integers, r will be a positive integer too. Hence, the flow value increases at least by 1 on each iteration of the augmenting-path method. Since the value of a maximum flow is bounded above (e.g., by the sum of the capacities of the source edges), the augmenting-path method has to stop after a finite number of iterations.[4] Surprisingly, the final flow always turns out to be maximal, irrespective of a sequence of augmenting paths. This remarkable result stems from the proof of the Max-Flow Min-Cut Theorem (see, e.g., [For62]), which we replicate later in this section.

The augmenting-path method—as described above in its general form—does not indicate a specific way for generating flow-augmenting paths. A bad sequence of such paths may, however, have a dramatic impact on the method's efficiency. Consider, for example, the network in Figure 10.6a, in which U stands for some large positive integer. If we augment the zero flow along the path $1\to2\to3\to4$, we shall obtain the flow of value 1 shown in Figure 10.6b. Augmenting that flow along the path $1\to3\leftarrow2\to4$ will increase the flow value to 2 (Figure 10.6c). If we continue selecting this pair of flow-augmenting paths, we will need a total of $2U$ iterations to reach the maximum flow of value $2U$ (Figure 10.6d). Of course, we can obtain the maximum flow in just two iterations by augmenting the initial zero flow along the path $1\to2\to4$ followed by augmenting the new flow along the path $1\to3\to4$. The dramatic difference between $2U$ and 2 iterations makes the point.

Fortunately, there are several ways to generate flow-augmenting paths efficiently and avoid the degradation in performance illustrated by the previous example. The simplest of them uses breadth-first search to generate augmenting paths with the least number of edges (see Section 3.5). This version of the augmenting-path method, called ***shortest-augmenting-path*** or ***first-labeled-first-scanned algorithm***, was suggested by J. Edmonds and R. M. Karp [Edm72]. The labeling refers to marking a new (unlabeled) vertex with two labels. The first label indicates the amount of additional flow that can be brought from the source to the vertex being labeled. The second label is the name of the vertex from which the vertex being labeled was reached. (It can be left undefined for the source.) It is also convenient to add the + or – sign to the second label to indicate whether the vertex was reached via a forward or backward edge, respectively. The source can be always labeled with $\infty, -$. For the other vertices, the labels are computed as follows.

4. If capacity upper bounds are irrational numbers, the augmenting-path method may not terminate (see, e.g., [Chv83, pp. 387–388], for a cleverly devised example demonstrating such a situation). This limitation is only of theoretical interest because we cannot store irrational numbers in a computer, and rational numbers can be transformed into integers by changing the capacity measurement unit.

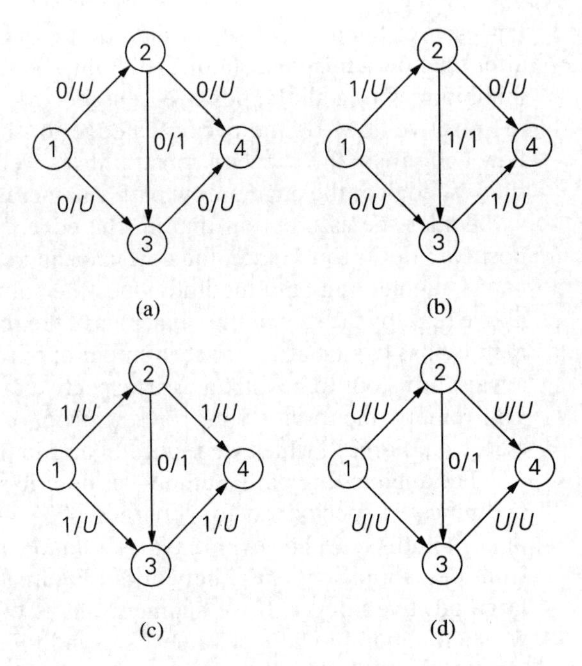

FIGURE 10.6 Efficiency degradation of the augmenting-path method.

If unlabeled vertex j is connected to the front vertex i of the traversal queue by a directed edge from i to j with positive unused capacity $r_{ij} = u_{ij} - x_{ij}$, then vertex j is labeled with l_j, i^+, where $l_j = \min\{l_i, r_{ij}\}$.

If unlabeled vertex j is connected to the front vertex i of the traversal queue by a directed edge from j to i with positive flow x_{ji}, then vertex j is labeled with l_j, i^-, where $l_j = \min\{l_i, x_{ji}\}$.

If this labeling-enhanced traversal ends up labeling the sink, the current flow can be augmented by the amount indicated by the sink's first label. The augmentation is performed along the augmenting path traced by following the vertex second labels from sink to source: the current flow quantities are increased on the forward edges and decreased on the backward edges of this path. If, on the other hand, the sink remains unlabeled after the traversal queue becomes empty, the algorithm returns the current flow as maximum and stops.

ALGORITHM *ShortestAugmentingPath(G)*

```
//Implements the shortest-augmenting-path algorithm
//Input: A network with single source 1, single sink n, and
//        positive integer capacities u_ij on its edges (i, j)
//Output: A maximum flow x
assign x_ij = 0 to every edge (i, j) in the network
label the source with ∞, − and add the source to the empty queue Q
```

```
while not Empty(Q) do
    i ← Front(Q);  Dequeue(Q)
    for every edge from i to j do   //forward edges
        if j is unlabeled
            r_ij ← u_ij − x_ij
            if r_ij > 0
                l_j ← min{l_i, r_ij};  label j with l_j, i+
                Enqueue(Q, j)
    for every edge from j to i do   //backward edges
        if j is unlabeled
            if x_ji > 0
                l_j ← min{l_i, x_ji};  label j with l_j, i−
                Enqueue(Q, j)
    if the sink has been labeled
        //augment along the augmenting path found
        j ← n   //start at the sink and move backwards using second labels
        while j ≠ 1  //the source hasn't been reached
            if the second label of vertex j is i+
                x_ij ← x_ij + l_n
            else  //the second label of vertex j is i−
                x_ji ← x_ji − l_n
            j ← i;  i ← the vertex indicated by i's second label
        erase all vertex labels except the ones of the source
        reinitialize Q with the source
return x //the current flow is maximum
```

An application of this algorithm to the network in Figure 10.4 is illustrated in Figure 10.7.

The optimality of a final flow obtained by the augmenting-path method stems from a theorem that relates network flows to network cuts. A ***cut*** induced by partitioning vertices of a network into some subset X containing the source and $\bar{X}$, the complement of X, containing the sink is the set of all the edges with a tail in X and a head in $\bar{X}$. We denote a cut $C(X, \bar{X})$ or simply C. For example, for the network in Figure 10.4:

if $X = \{1\}$ and hence $\bar{X} = \{2, 3, 4, 5, 6\}$, $C(X, \bar{X}) = \{(1, 2), (1, 4)\}$;

if $X = \{1, 2, 3, 4, 5\}$ and hence $\bar{X} = \{6\}$, $C(X, \bar{X}) = \{(3, 6), (5, 6)\}$;

if $X = \{1, 2, 4\}$ and hence $\bar{X} = \{3, 5, 6\}$, $C(X, \bar{X}) = \{(2, 3), (2, 5), (4, 3)\}$.

The name "cut" stems from the following property: if all the edges of a cut were deleted from the network, there would be no directed path from source to sink. Indeed, let $C(X, \bar{X})$ be a cut. Consider a directed path from source to sink. If v_i is the first vertex of that path which belongs to $\bar{X}$ (the set of such vertices is not

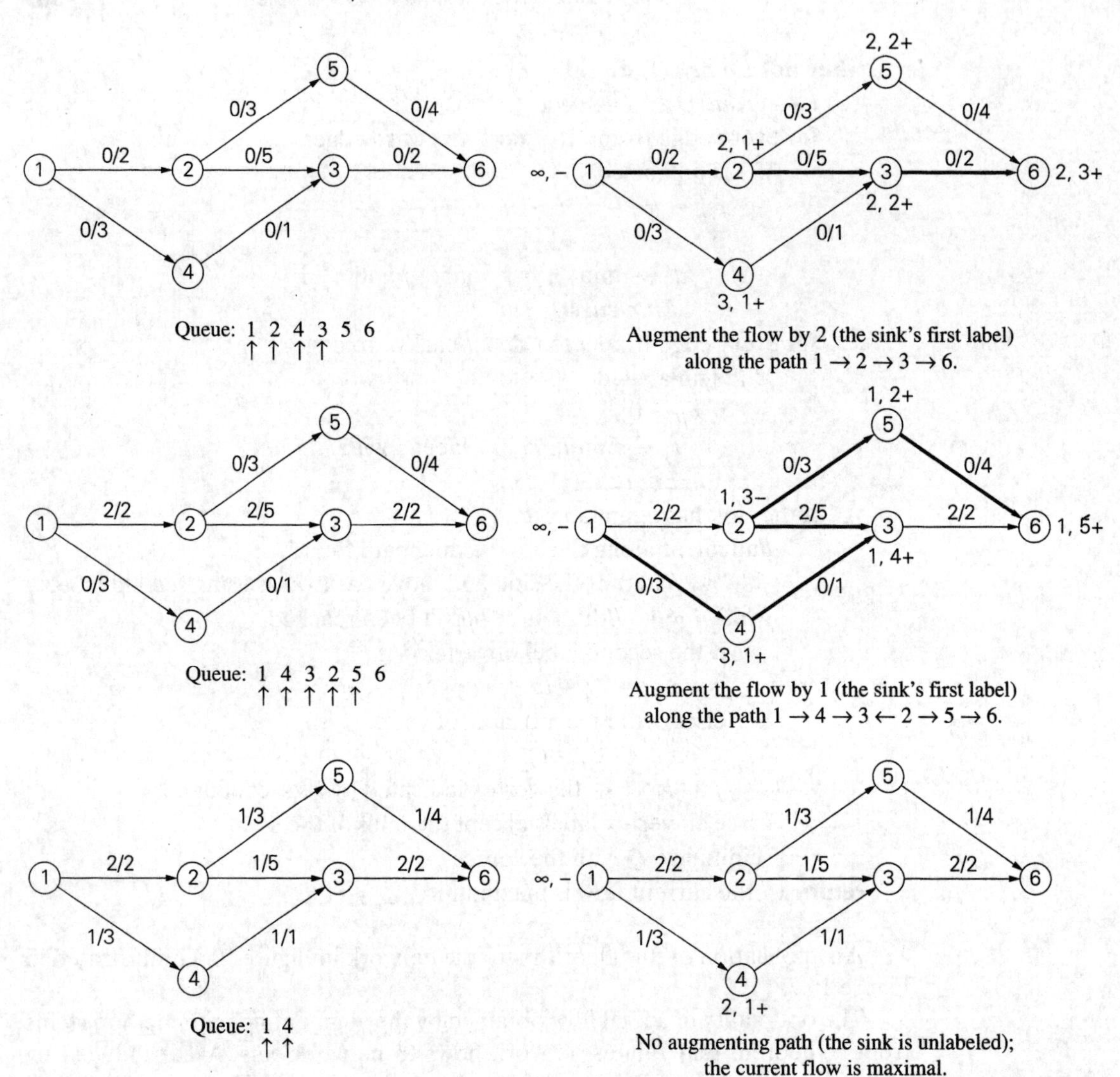

FIGURE 10.7 Illustration of the shortest-augmenting-path algorithm. The diagrams on the left show the current flow before the next iteration begins; the diagrams on the right show the results of the vertex labeling on that iteration, the augmenting path found (in bold), and the flow before its augmentation. Vertices deleted from the queue are indicated by the ↑ symbol.

empty, because it contains the sink), then v_i is not the source and its immediate predecessor v_{i-1} on that path belongs to X. Hence, the edge from v_{i-1} to v_i must be an element of the cut $C(X, \bar{X})$. This proves the property in question.

The ***capacity*** of a cut $C(X, \bar{X})$, denoted $c(X, \bar{X})$, is defined as the sum of capacities of the edges that compose the cut. For the three examples of cuts given above, the capacities are equal to 5, 6, and 9, respectively. Since the number of different cuts in a network is nonempty and finite (why?), there always exists a ***minimum cut***, i.e., a cut with the smallest capacity. (What is a minimum cut in the network of Figure 10.4?) The following theorem establishes an important relationship between the notions of maximum flow and minimum cut.

THEOREM (*Max-Flow Min-Cut Theorem*) The value of a maximum flow in a network is equal to the capacity of its minimum cut.

PROOF First, let x be a feasible flow of value v and let $C(X, \bar{X})$ be a cut of capacity c in the same network. Consider the flow across this cut defined as the difference between the sum of the flows on the edges from X to $\bar{X}$ and the sum of the flows on the edges from $\bar{X}$ to X. It is intuitively clear and can be formally derived from the equations expressing the flow-conservation requirement and the definition of the flow value (Problem 6b in this section's exercises) that the flow across the cut $C(X, \bar{X})$ is equal to v, the value of the flow:

$$v = \sum_{i \in X,\, j \in \bar{X}} x_{ij} - \sum_{j \in \bar{X},\, i \in X} x_{ji}. \quad \textbf{(10.12)}$$

Since the second sum is nonnegative and the flow x_{ij} on any edge (i, j) cannot exceed the edge capacity u_{ij}, equality (10.12) implies that

$$v \le \sum_{i \in X,\, j \in \bar{X}} x_{ij} \le \sum_{i \in X,\, j \in \bar{X}} u_{ij},$$

i.e.,

$$v \le c. \quad \textbf{(10.13)}$$

Thus, the value of any feasible flow in a network cannot exceed the capacity of any cut in that network.

Let v^* be the value of a final flow x^* obtained by the augmenting-path method. If we now find a cut whose capacity is equal to v^*, we will have to conclude, in view of inequality (10.13), that (i) the value v^* of the final flow is maximal among all feasible flows, (ii) the cut's capacity is minimal among all cuts in the network, and (iii) the maximum-flow value is equal to the minimum-cut capacity.

To find such a cut, consider the set of vertices X^* that can be reached from the source by following an undirected path composed of forward edges with positive unused capacities (with respect to the final flow x^*) and backward edges with positive flows on them. This set contains the source but does not contain the sink: if it did, we would have an augmenting path for the flow x^*, which would contradict the assumption that the flow x^* is final. Consider the cut $C(X^*, \overline{X^*})$. By the definition of set X^*, each edge (i, j) from X^* to $\overline{X^*}$ has zero unused capacity, i.e., $x^*_{ij} = u_{ij}$, and each edge (j, i) from $\overline{X^*}$ to X^* has the zero flow on it (otherwise, j would be in X^*). Applying equality (10.12) to the final flow x^* and the set X^* defined above, we obtain

$$v^* = \sum_{i \in X^*, j \in \overline{X^*}} x^*_{ij} - \sum_{j \in \overline{X^*}, i \in X^*} x^*_{ji} = \sum_{i \in X^*, j \in \overline{X^*}} u_{ij} - 0 = c(X^*, \overline{X^*}),$$

which proves the theorem. ■

The proof outlined above accomplishes more than proving the equality of the maximum-flow value and the minimum-cut capacity. It also implies that when the augmenting-path method terminates, it yields both a maximum flow and a minimum cut. If labeling of the kind utilized in the shortest-augmenting-path algorithm is used, a minimum cut is formed by the edges from the labeled to unlabeled vertices on the last iteration of the method. Finally, the proof implies that all such edges must be full (i.e., the flows must be equal to the edge capacities), and all the edges from unlabeled vertices to labeled, if any, must be empty (i.e., have zero flows on them). In particular, for the network in Figure 10.7, the algorithm finds the cut $\{(1, 2), (4, 3)\}$ of minimum capacity 3, both edges of which are full as required.

Edmonds and Karp proved in their paper [Edm72] that the number of augmenting paths needed by the shortest-augmenting-path algorithm never exceeds $nm/2$, where n and m are the number of vertices and edges, respectively. Since the time required to find a shortest augmenting path by breadth-first search is in $O(n + m) = O(m)$ for networks represented by their adjacency lists, the time efficiency of the shortest-augmenting-path algorithm is in $O(nm^2)$.

More efficient algorithms for the maximum-flow problem are known (see the monograph [Ahu93], as well as appropriate chapters in such books as [Cor09] and [Kle06]). Some of them implement the augmenting-path idea in a more efficient manner. Others are based on the concept of preflows. A ***preflow*** is a flow that satisfies the capacity constraints but not the flow-conservation requirement. Any vertex is allowed to have more flow entering the vertex than leaving it. A preflow-push algorithm moves the excess flow toward the sink until the flow-conservation requirement is reestablished for all intermediate vertices of the network. Faster algorithms of this kind have worst-case efficiency close to $O(nm)$. Note that preflow-push algorithms fall outside the iterative-improvement paradigm because they do not generate a sequence of improving solutions that satisfy *all* the constraints of the problem.

To conclude this section, it is worth pointing out that although the initial interest in studying network flows was caused by transportation applications, this model has also proved to be useful for many other areas. We discuss one of them in the next section.

Exercises 10.2

1. Since maximum-flow algorithms require processing edges in both directions, it is convenient to modify the adjacency matrix representation of a network as follows. If there is a directed edge from vertex i to vertex j of capacity u_{ij}, then the element in the ith row and the jth column is set to u_{ij}, and the element in the jth row and the ith column is set to $-u_{ij}$; if there is no edge between vertices i and j, both these elements are set to zero. Outline a simple algorithm for identifying a source and a sink in a network presented by such a matrix and indicate its time efficiency.

2. Apply the shortest-augmenting path algorithm to find a maximum flow and a minimum cut in the following networks.

a.

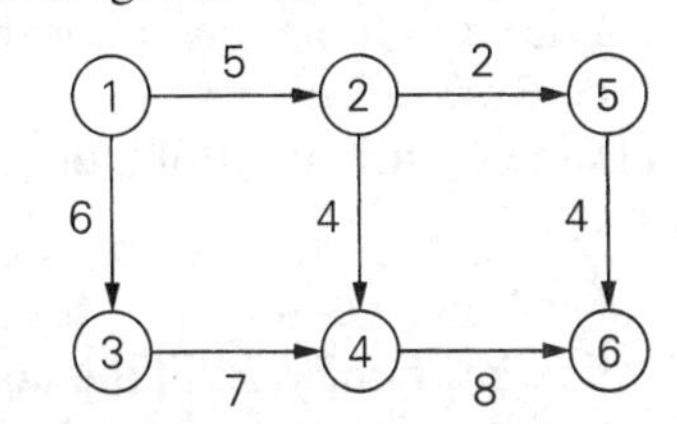

b.

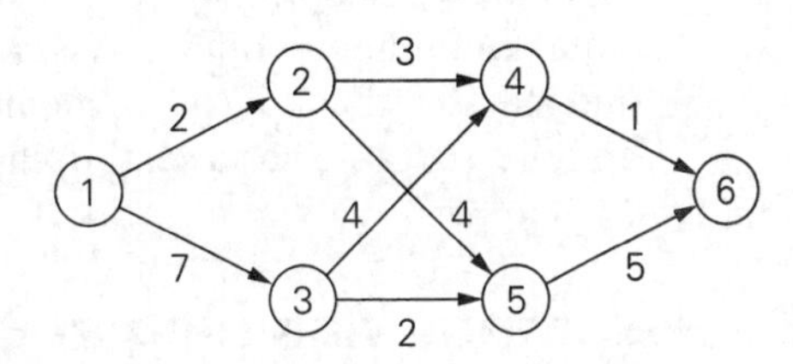

3. **a.** Does the maximum-flow problem always have a unique solution? Would your answer be different for networks with different capacities on all their edges?

   b. Answer the same questions for the minimum-cut problem of finding a cut of the smallest capacity in a given network.

4. **a.** Explain how the maximum-flow problem for a network with several sources and sinks can be transformed into the same problem for a network with a single source and a single sink.

   b. Some networks have capacity constraints on the flow amounts that can flow through their intermediate vertices. Explain how the maximum-flow problem for such a network can be transformed to the maximum-flow problem for a network with edge capacity constraints only.

5. Consider a network that is a rooted tree, with the root as its source, the leaves as its sinks, and all the edges directed along the paths from the root to the leaves. Design an efficient algorithm for finding a maximum flow in such a network. What is the time efficiency of your algorithm?

6. **a.** Prove equality (10.9).

b. Prove that for any flow in a network and any cut in it, the value of the flow is equal to the flow across the cut (see equality (10.12)). Explain the relationship between this property and equality (10.9).

7. a. Express the maximum-flow problem for the network in Figure 10.4 as a linear programming problem.

b. Solve this linear programming problem by the simplex method.

8. As an alternative to the shortest-augmenting-path algorithm, Edmonds and Karp [Edm72] suggested the maximum-capacity-augmenting-path algorithm, in which a flow is augmented along the path that increases the flow by the largest amount. Implement both these algorithms in the language of your choice and perform an empirical investigation of their relative efficiency.

9. Write a report on a more advanced maximum-flow algorithm such as (i) Dinitz's algorithm, (ii) Karzanov's algorithm, (iii) Malhotra-Kamar-Maheshwari algorithm, or (iv) Goldberg-Tarjan algorithm.

10. *Dining problem* Several families go out to dinner together. To increase their social interaction, they would like to sit at tables so that no two members of the same family are at the same table. Show how to find a seating arrangement that meets this objective (or prove that no such arrangement exists) by using a maximum-flow problem. Assume that the dinner contingent has p families and that the ith family has a_i members. Also assume that q tables are available and the jth table has a seating capacity of b_j. [Ahu93]

10.3 Maximum Matching in Bipartite Graphs

In many situations we are faced with a problem of pairing elements of two sets. The traditional example is boys and girls for a dance, but you can easily think of more serious applications. It is convenient to represent elements of two given sets by vertices of a graph, with edges between vertices that can be paired. A ***matching*** in a graph is a subset of its edges with the property that no two edges share a vertex. A ***maximum matching***—more precisely, a ***maximum cardinality matching***—is a matching with the largest number of edges. (What is it for the graph in Figure 10.8? Is it unique?) The maximum-matching problem is the problem of finding a maximum matching in a given graph. For an arbitrary graph, this is a rather difficult problem. It was solved in 1965 by Jack Edmonds [Edm65]. (See [Gal86] for a good survey and more recent references.)

We limit our discussion in this section to the simpler case of bipartite graphs. In a ***bipartite graph***, all the vertices can be partitioned into two disjoint sets V and U, not necessarily of the same size, so that every edge connects a vertex in one of these sets to a vertex in the other set. In other words, a graph is bipartite if its vertices can be colored in two colors so that every edge has its vertices colored in different colors; such graphs are also said to be ***2-colorable***. The graph in Figure 10.8 is bipartite. It is not difficult to prove that a graph is bipartite if and only if it does not have a cycle of an odd length. We will assume for the rest of this section that

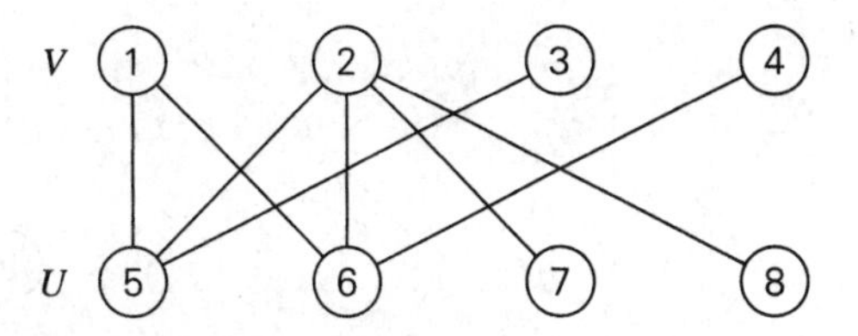

FIGURE 10.8 Example of a bipartite graph.

the vertex set of a given bipartite graph has been already partitioned into sets V and U as required by the definition (see Problem 8 in Exercises 3.5).

Let us apply the iterative-improvement technique to the maximum-cardinality-matching problem. Let M be a matching in a bipartite graph $G = \langle V, U, E \rangle$. How can we improve it, i.e., find a new matching with more edges? Obviously, if every vertex in either V or U is ***matched*** (has a ***mate***), i.e., serves as an endpoint of an edge in M, this cannot be done and M is a maximum matching. Therefore, to have a chance at improving the current matching, both V and U must contain ***unmatched*** (also called ***free***) ***vertices***, i.e., vertices that are not incident to any edge in M. For example, for the matching $M_a = \{(4, 8), (5, 9)\}$ in the graph in Figure 10.9a, vertices 1, 2, 3, 6, 7, and 10 are free, and vertices 4, 5, 8, and 9 are matched.

Another obvious observation is that we can immediately increase a current matching by adding an edge between two free vertices. For example, adding (1, 6) to the matching $M_a = \{(4, 8), (5, 9)\}$ in the graph in Figure 10.9a yields a larger matching $M_b = \{(1, 6), (4, 8), (5, 9)\}$ (Figure 10.9b). Let us now try to find a matching larger than M_b by matching vertex 2. The only way to do this would be to include the edge (2, 6) in a new matching. This inclusion requires removal of (1, 6), which can be compensated by inclusion of (1, 7) in the new matching. This new matching $M_c = \{(1, 7), (2, 6), (4, 8), (5, 9)\}$ is shown in Figure 10.9c.

In general, we increase the size of a current matching M by constructing a simple path from a free vertex in V to a free vertex in U whose edges are alternately in $E - M$ and in M. That is, the first edge of the path does not belong to M, the second one does, and so on, until the last edge that does not belong to M. Such a path is called ***augmenting*** with respect to the matching M. For example, the path 2, 6, 1, 7 is an augmenting path with respect to the matching M_b in Figure 10.9b. Since the length of an augmenting path is always odd, adding to the matching M the path's edges in the odd-numbered positions and deleting from it the path's edges in the even-numbered positions yields a matching with one more edge than in M. Such a matching adjustment is called ***augmentation***. Thus, in Figure 10.9, the matching M_b was obtained by augmentation of the matching M_a along the augmenting path 1, 6, and the matching M_c was obtained by augmentation of the matching M_b along the augmenting path 2, 6, 1, 7. Moving further, 3, 8, 4, 9, 5, 10 is an augmenting path for the matching M_c (Figure 10.9c). After adding to M_c the edges (3, 8), (4, 9), and (5, 10) and deleting (4, 8) and (5, 9), we obtain the matching $M_d = \{(1, 7), (2, 6), (3, 8), (4, 9), (5, 10)\}$ shown in Figure 10.9d. The

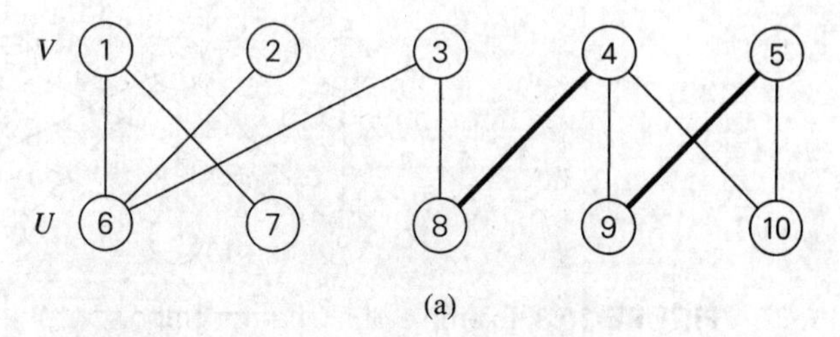

(a)

Augmenting path: 1, 6

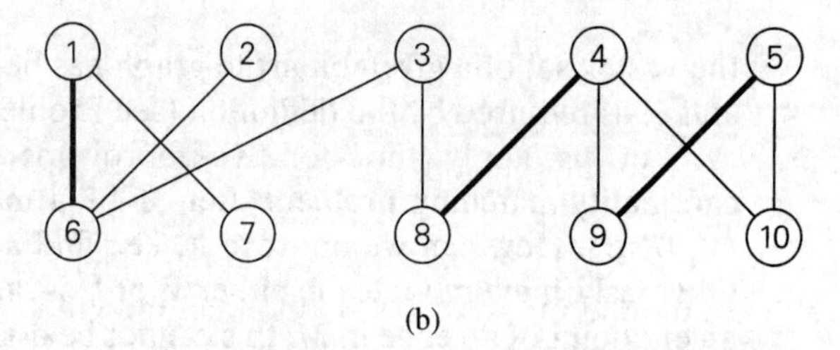

(b)

Augmenting path: 2, 6, 1, 7

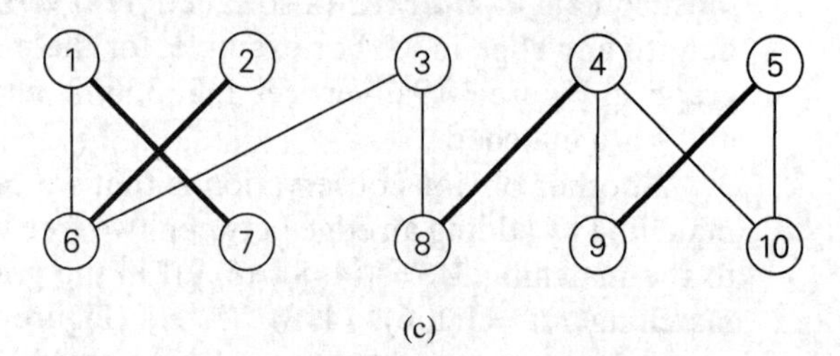

(c)

Augmenting path: 3, 8, 4, 9, 5, 10

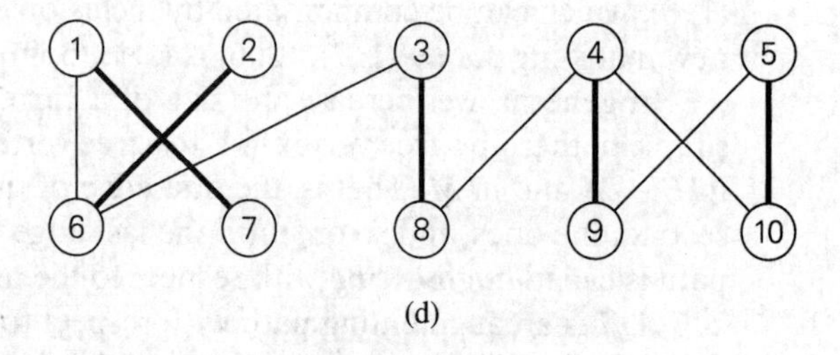

(d)

Maximum matching

FIGURE 10.9 Augmenting paths and matching augmentations.

matching M_d is not only a maximum matching but also ***perfect***, i.e., a matching that matches all the vertices of the graph.

Before we discuss an algorithm for finding an augmenting path, let us settle the issue of what nonexistence of such a path means. According to the theorem discovered by the French mathematician Claude Berge, it means the current matching is maximal.

THEOREM A matching M is a maximum matching if and only if there exists no augmenting path with respect to M.

PROOF If an augmenting path with respect to a matching M exists, then the size of the matching can be increased by augmentation. Let us prove the more difficult part: if no augmenting path with respect to a matching M exists, then the matching is a maximum matching. Assume that, on the contrary, this is not the case for a certain matching M in a graph G. Let M^* be a maximum matching in G; by our assumption, the number of edges in M^* is at least one more than the number of edges in M, i.e., $|M^*| > |M|$. Consider the edges in the symmetric difference $M \oplus M^* = (M - M^*) \cup (M^* - M)$, the set of all the edges that are either in M or in M^* but not in both. Note that $|M^* - M| > |M - M^*|$ because $|M^*| > |M|$ by assumption. Let G' be the subgraph of G made up of all the edges in $M \oplus M^*$ and their endpoints. By definition of a matching, any vertex in $G' \subseteq G$ can be incident to no more than one edge in M and no more than one edge in M^*. Hence, each of the vertices in G' has degree 2 or less, and therefore every connected component of G' is either a path or an even-length cycle of alternating edges from $M - M^*$ and $M^* - M$. Since $|M^* - M| > |M - M^*|$ and the number of edges from $M - M^*$ and $M^* - M$ is the same for any even-length cycle of alternating edges in G', there must exist at least one path of alternating edges that starts and ends with an edge from $M^* - M$. Hence, this is an augmenting path for the matching M, which contradicts the assumption that no such path exists. ■

Our discussion of augmenting paths leads to the following general method for constructing a maximum matching in a bipartite graph. Start with some initial matching (e.g., the empty set). Find an augmenting path and augment the current matching along this path. When no augmenting path can be found, terminate the algorithm and return the last matching, which is maximum.

We now give a specific algorithm implementing this general template. We will search for an augmenting path for a matching M by a BFS-like traversal of the graph that starts simultaneously at all the free vertices in one of the sets V and U, say, V. (It would be logical to select the smaller of the two vertex sets, but we will ignore this observation in the pseudocode below.) Recall that an augmenting path, if it exists, is an odd-length path that connects a free vertex in V with a free vertex in U and which, unless it consists of a single edge, "zigs" from a vertex in V to another vertex' mate in U, then "zags" back to V along the uniquely defined edge from M, and so on until a free vertex in U is reached. (Draw augmenting paths for the matchings in Figure 10.9, for example.) Hence, any candidate to be such a path must have its edges alternate in the pattern just described. This motivates the following rules for labeling vertices during the BFS-like traversal of the graph.

Case 1 (the queue's front vertex w is in V) If u is a free vertex adjacent to w, it is used as the other endpoint of an augmenting path; so the labeling stops and augmentation of the matching commences. The augmenting path in question is obtained by moving backward along the vertex labels (see below) to alternately add and delete its edges to and from the current matching. If u is not free and connected to w by an edge not in M, label u with w unless it has been already labeled.

Case 2 (the front vertex w is in U) In this case, w must be matched and we label its mate in V with w.

Here is pseudocode of the algorithm in its entirety.

```
ALGORITHM  MaximumBipartiteMatching(G)
    //Finds a maximum matching in a bipartite graph by a BFS-like traversal
    //Input: A bipartite graph G = ⟨V, U, E⟩
    //Output: A maximum-cardinality matching M in the input graph
    initialize set M of edges with some valid matching (e.g., the empty set)
    initialize queue Q with all the free vertices in V (in any order)
    while not Empty(Q) do
        w ← Front(Q);  Dequeue(Q)
        if w ∈ V
            for every vertex u adjacent to w do
                if u is free
                    //augment
                    M ← M ∪ (w, u)
                    v ← w
                    while v is labeled do
                        u ← vertex indicated by v's label;  M ← M − (v, u)
                        v ← vertex indicated by u's label;  M ← M ∪ (v, u)
                    remove all vertex labels
                    reinitialize Q with all free vertices in V
                    break  //exit the for loop
                else  //u is matched
                    if (w, u) ∉ M and u is unlabeled
                        label u with w
                        Enqueue(Q, u)
        else  //w ∈ U (and matched)
            label the mate v of w with w
            Enqueue(Q, v)
    return M  //current matching is maximum
```

An application of this algorithm to the matching in Figure 10.9a is shown in Figure 10.10. Note that the algorithm finds a maximum matching that differs from the one in Figure 10.9d.

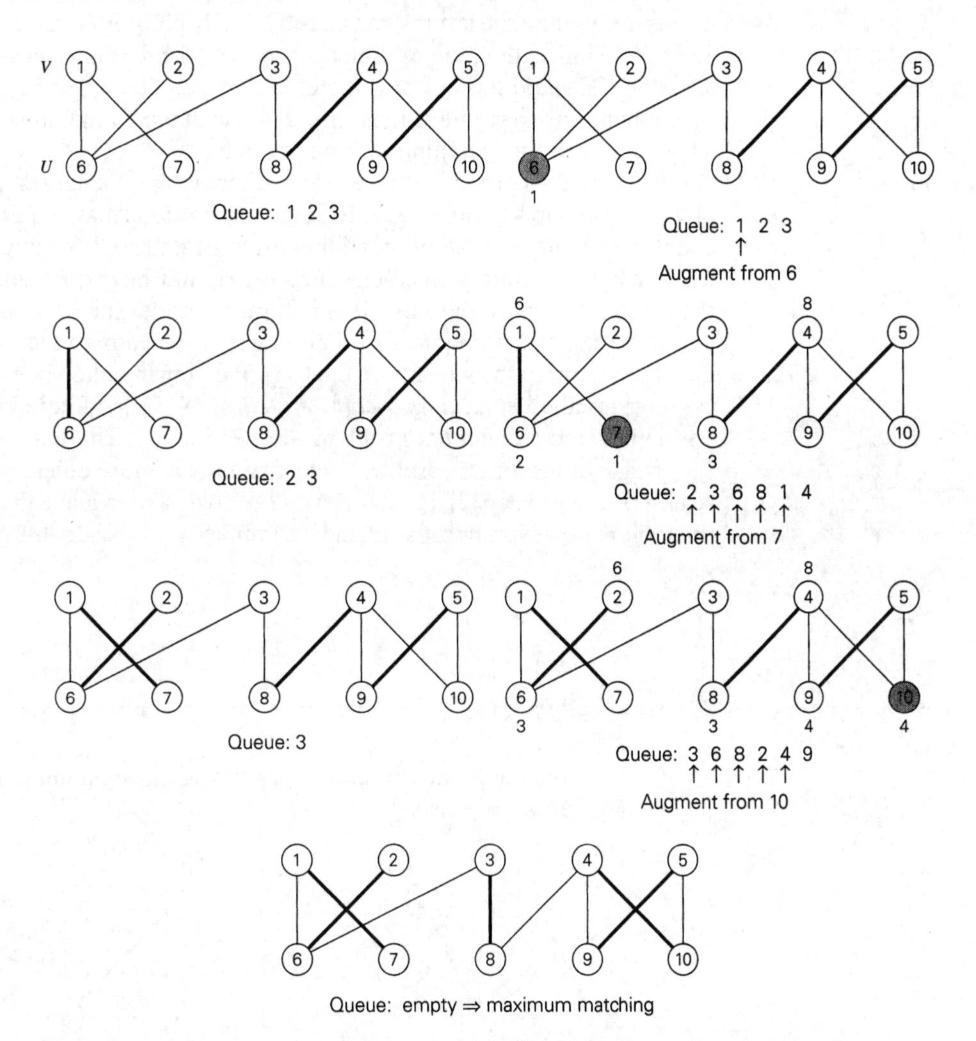

FIGURE 10.10 Application of the maximum-cardinality-matching algorithm. The left column shows a current matching and initialized queue at the next iteration's start; the right column shows the vertex labeling generated by the algorithm before augmentation is performed. Matching edges are shown in bold. Vertex labels indicate the vertices from which the labeling is done. The discovered endpoint of an augmenting path is shaded and labeled for clarity. Vertices deleted from the queue are indicated by ↑.

How efficient is the maximum-matching algorithm? Each iteration except the last one matches two previously free vertices—one from each of the sets V and U. Therefore, the total number of iterations cannot exceed $\lfloor n/2 \rfloor + 1$, where $n = |V| + |U|$ is the number of vertices in the graph. The time spent on each iteration is in $O(n + m)$, where $m = |E|$ is the number of edges in the graph. (This assumes that the information about the status of each vertex—free or matched and the vertex' mate if the latter—can be retrieved in constant time, e.g., by storing it in an array.) Hence, the time efficiency of the algorithm is in $O(n(n + m))$. Hopcroft and Karp [Hop73] showed how the efficiency can be improved to $O(\sqrt{n}(n + m))$ by combining several iterations into a single stage to maximize the number of edges added to the matching with one search.

We were concerned in this section with matching the largest possible number of vertex pairs in a bipartite graph. Some applications may require taking into account the quality or cost of matching different pairs. For example, workers may execute jobs with different efficiencies, or girls may have different preferences for their potential dance partners. It is natural to model such situations by bipartite graphs with weights assigned to their edges. This leads to the problem of maximizing the sum of the weights on edges connecting matched pairs of vertices. This problem is called ***maximum-weight matching***. We encountered it under a different name—the assignment problem—in Section 3.4. There are several sophisticated algorithms for this problem, which are much more efficient than exhaustive search (see, e.g., [Pap82], [Gal86], [Ahu93]). We have to leave them outside of our discussion, however, because of their complexity, especially for general graphs.

Exercises 10.3

1. For each matching shown below in bold, find an augmentation or explain why no augmentation exists.

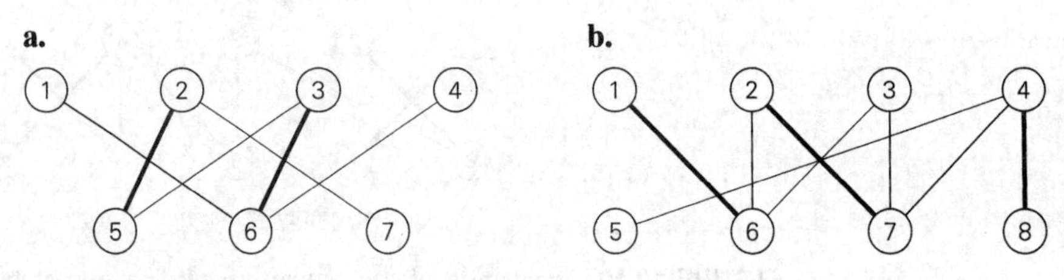

2. Apply the maximum-matching algorithm to the following bipartite graph:

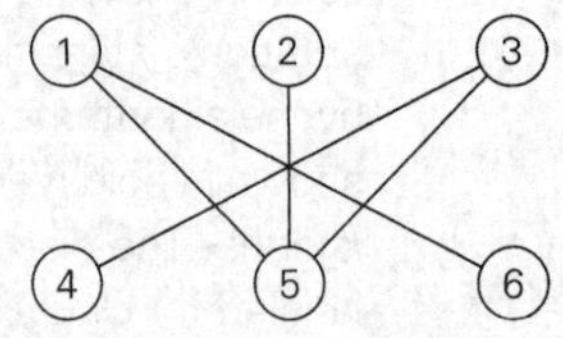

3. a. What is the largest and what is the smallest possible cardinality of a matching in a bipartite graph $G = \langle V, U, E \rangle$ with n vertices in each vertex set V and U and at least n edges?

b. What is the largest and what is the smallest number of distinct solutions the maximum-cardinality-matching problem can have for a bipartite graph $G = \langle V, U, E \rangle$ with n vertices in each vertex set V and U and at least n edges?

4. a. ***Hall's Marriage Theorem*** asserts that a bipartite graph $G = \langle V, U, E \rangle$ has a matching that matches all vertices of the set V if and only if for each subset $S \subseteq V$, $|R(S)| \geq |S|$ where $R(S)$ is the set of all vertices adjacent to a vertex in S. Check this property for the following graph with (i) $V = \{1, 2, 3, 4\}$ and (ii) $V = \{5, 6, 7\}$.

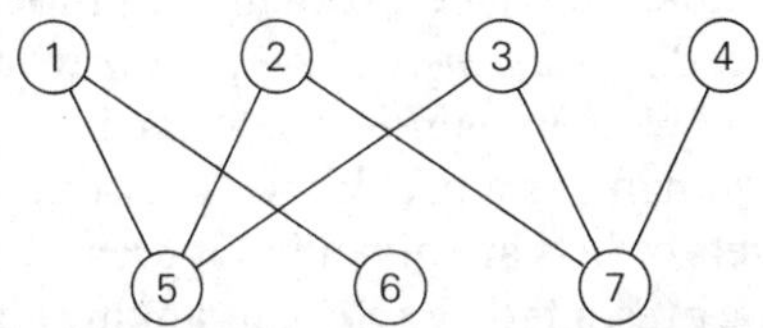

b. You have to devise an algorithm that returns yes if there is a matching in a bipartite graph $G = \langle V, U, E \rangle$ that matches all vertices in V and returns no otherwise. Would you base your algorithm on checking the condition of Hall's Marriage Theorem?

5. Suppose there are five committees A, B, C, D, and E composed of six persons a, b, c, d, e, and f as follows: committee A's members are b and e; committee B's members are b, d, and e; committee C's members are a, c, d, e, and f; committee D's members are b, d, and e; committee E's members are b and e. Is there a ***system of distinct representatives***, i.e., is it possible to select a representative from each committee so that all the selected persons are distinct?

6. Show how the maximum-cardinality-matching problem for a bipartite graph can be reduced to the maximum-flow problem discussed in Section 10.2.

7. Consider the following greedy algorithm for finding a maximum matching in a bipartite graph $G = \langle V, U, E \rangle$. Sort all the vertices in nondecreasing order of their degrees. Scan this sorted list to add to the current matching (initially empty) the edge from the list's free vertex to an adjacent free vertex of the lowest degree. If the list's vertex is matched or if there are no adjacent free vertices for it, the vertex is simply skipped. Does this algorithm always produce a maximum matching in a bipartite graph?

8. Design a linear-time algorithm for finding a maximum matching in a tree.

9. Implement the maximum-matching algorithm of this section in the language of your choice. Experiment with its performance on bipartite graphs with n vertices in each of the vertex sets and randomly generated edges (in both dense and sparse modes) to compare the observed running time with the algorithm's theoretical efficiency.

10. *Domino puzzle* A domino is a 2×1 tile that can be oriented either horizontally or vertically. A tiling of a given board composed of 1×1 squares is covering it with dominoes exactly and without overlap. Is it possible to tile with dominoes an 8×8 board without two unit squares at its diagonally opposite corners?

10.4 The Stable Marriage Problem

In this section, we consider an interesting version of bipartite matching called the stable marriage problem. Consider a set $Y = \{m_1, m_2, \ldots, m_n\}$ of n men and a set $X = \{w_1, w_2, \ldots, w_n\}$ of n women. Each man has a preference list ordering the women as potential marriage partners with no ties allowed. Similarly, each woman has a preference list of the men, also with no ties. Examples of these two sets of lists are given in Figures 10.11a and 10.11b. The same information can also be presented by an $n \times n$ ranking matrix (see Figure 10.11c). The rows and columns of the matrix represent the men and women of the two sets, respectively. A cell in row m and column w contains two rankings: the first is the position (ranking) of w in the m's preference list; the second is the position (ranking) of m in the w's preference list. For example, the pair 3, 1 in Jim's row and Ann's column in the matrix in Figure 10.11c indicates that Ann is Jim's third choice while Jim is Ann's first. Which of these two ways to represent such information is better depends on the task at hand. For example, it is easier to specify a match of the sets' elements by using the ranking matrix, whereas the preference lists might be a more efficient data structure for implementing a matching algorithm.

A ***marriage matching*** M is a set of n (m, w) pairs whose members are selected from disjoint n-element sets Y and X in a one-one fashion, i.e., each man m from Y is paired with exactly one woman w from X and vice versa. (If we represent Y and X as vertices of a complete bipartite graph with edges connecting possible marriage partners, then a marriage matching is a perfect matching in such a graph.)

men's preferences

	1st	2nd	3rd
Bob:	Lea	Ann	Sue
Jim:	Lea	Sue	Ann
Tom:	Sue	Lea	Ann

(a)

women's preferences

	1st	2nd	3rd
Ann:	Jim	Tom	Bob
Lea:	Tom	Bob	Jim
Sue:	Jim	Tom	Bob

(b)

ranking matrix

	Ann	Lea	Sue
Bob	2,3	1,2	3,3
Jim	3,1	1,3	2,1
Tom	3,2	2,1	1,2

(c)

FIGURE 10.11 Data for an instance of the stable marriage problem. (a) Men's preference lists; (b) women's preference lists. (c) Ranking matrix (with the boxed cells composing an unstable matching).

A pair (m, w), where $m \in Y$, $w \in X$, is said to be a ***blocking pair*** for a marriage matching M if man m and woman w are not matched in M but they prefer each other to their mates in M. For example, (Bob, Lea) is a blocking pair for the marriage matching $M = \{$(Bob, Ann), (Jim, Lea), (Tom, Sue)$\}$ (Figure 10.11c) because they are not matched in M while Bob prefers Lea to Ann and Lea prefers Bob to Jim. A marriage matching M is called ***stable*** if there is no blocking pair for it; otherwise, M is called ***unstable***. According to this definition, the marriage matching in Figure 10.11c is unstable because Bob and Lea can drop their designated mates to join in a union they both prefer. The ***stable marriage problem*** is to find a stable marriage matching for men's and women's given preferences.

Surprisingly, this problem always has a solution. (Can you find it for the instance in Figure 10.11?) It can be found by the following algorithm.

Stable marriage algorithm

Input: A set of n men and a set of n women along with rankings of the women by each man and rankings of the men by each woman with no ties allowed in the rankings

Output: A stable marriage matching

Step 0 Start with all the men and women being free.

Step 1 While there are free men, arbitrarily select one of them and do the following:

Proposal The selected free man m proposes to w, the next woman on his preference list (who is the highest-ranked woman who has not rejected him before).

Response If w is free, she accepts the proposal to be matched with m. If she is not free, she compares m with her current mate. If she prefers m to him, she accepts m's proposal, making her former mate free; otherwise, she simply rejects m's proposal, leaving m free.

Step 2 Return the set of n matched pairs.

Before we analyze this algorithm, it is useful to trace it on some input. Such an example is presented in Figure 10.12.

Let us discuss properties of the stable marriage algorithm.

THEOREM The stable marriage algorithm terminates after no more than n^2 iterations with a stable marriage output.

PROOF The algorithm starts with n men having the total of n^2 women on their ranking lists. On each iteration, one man makes a proposal to a woman. This reduces the total number of women to whom the men can still propose in the future because no man proposes to the same woman more than once. Hence, the algorithm must stop after no more than n^2 iterations.

Free men: Bob, Jim, Tom

	Ann	Lea	Sue
Bob	2, 3	1,2	3, 3
Jim	3, 1	1, 3	2, 1
Tom	3, 2	2, 1	1, 2

Bob proposed to Lea
Lea accepted

Free men: Jim, Tom

	Ann	Lea	Sue
Bob	2, 3	1,2	3, 3
Jim	3, 1	1, 3	2, 1
Tom	3, 2	2, 1	1, 2

Jim proposed to Lea
Lea rejected

Free men: Jim, Tom

	Ann	Lea	Sue
Bob	2, 3	1,2	3, 3
Jim	3, 1	1, 3	2,1
Tom	3, 2	2, 1	1, 2

Jim proposed to Sue
Sue accepted

Free men: Tom

	Ann	Lea	Sue
Bob	2, 3	1,2	3, 3
Jim	3, 1	1, 3	2,1
Tom	3, 2	2, 1	1, 2

Tom proposed to Sue
Sue rejected

Free men: Tom

	Ann	Lea	Sue
Bob	2, 3	1, 2	3, 3
Jim	3, 1	1, 3	2,1
Tom	3, 2	2,1	1, 2

Tom proposed to Lea
Lea replaced Bob with Tom

Free men: Bob

	Ann	Lea	Sue
Bob	2,3	1, 2	3, 3
Jim	3, 1	1, 3	2,1
Tom	3, 2	2,1	1, 2

Bob proposed to Ann
Ann accepted

FIGURE 10.12 Application of the stable marriage algorithm. An accepted proposal is indicated by a boxed cell; a rejected proposal is shown by an underlined cell.

Let us now prove that the final matching M is a stable marriage matching. Since the algorithm stops after all the n men are one-one matched to the n women, the only thing that needs to be proved is the stability of M. Suppose, on the contrary, that M is unstable. Then there exists a blocking pair of a man m and a woman w who are unmatched in M and such that both m and w prefer each other to the persons they are matched with in M. Since m proposes to every woman on his ranking list in decreasing order of preference and w precedes m's match in M, m must have proposed to w on some iteration. Whether w refused m's proposal or accepted it but replaced him on a subsequent iteration with a higher-ranked match, w's mate in M must be higher on w's preference list than m because the rankings of the men matched to a given woman may only improve on each iteration of the algorithm. This contradicts the assumption that w prefers m to her final match in M. ■

The stable marriage algorithm has a notable shortcoming. It is not "gender neutral." In the form presented above, it favors men's preferences over women's preferences. We can easily see this by tracing the algorithm on the following instance of the problem:

	woman 1	woman 2
man 1	1, 2	2, 1
man 2	2, 1	1, 2

The algorithm obviously yields the stable matching $M = \{(\text{man } 1, \text{woman } 1), (\text{man } 2, \text{woman } 2)\}$. In this matching, both men are matched to their first choices, which is not the case for the women. One can prove that the algorithm always yields a stable matching that is ***man-optimal***: it assigns to each man the highest-ranked woman possible under any stable marriage. Of course, this gender bias can be reversed, but not eliminated, by reversing the roles played by men and women in the algorithm, i.e., by making women propose and men accept or reject their proposals.

There is another important corollary to the fact that the stable marriage algorithm always yields a gender-optimal stable matching. It is easy to prove that a man (woman)-optimal matching is unique for a given set of participant preferences. Therefore the algorithm's output does not depend on the order in which the free men (women) make their proposals. Consequently, we can use any data structure we might prefer—e.g., a queue or a stack—for representing this set with no impact on the algorithm's outcome.

The notion of the stable matching as well as the algorithm discussed above was introduced by D. Gale and L. S. Shapley in the paper titled "College Admissions and the Stability of Marriage" [Gal62]. I do not know which of the two applications mentioned in the title you would consider more important. The point is that stability is a matching property that can be desirable in a variety of applications. For example, it has been used for many years in the United States for matching medical-school graduates with hospitals for residency training. For a brief history of this application and an in-depth discussion of the stable marriage problem and its extensions, see the monograph by Gusfield and Irwing [Gus89].

Exercises 10.4

1. Consider an instance of the stable marriage problem given by the following ranking matrix:

	A	B	C
α	1, 3	2, 2	3, 1
β	3, 1	1, 3	2, 2
γ	2, 2	3, 1	1, 3

For each of its marriage matchings, indicate whether it is stable or not. For the unstable matchings, specify a blocking pair. For the stable matchings, indicate whether they are man-optimal, woman-optimal, or neither. (Assume that the Greek and Roman letters denote the men and women, respectively.)

2. Design a simple algorithm for checking whether a given marriage matching is stable and determine its time efficiency class.
3. Find a stable marriage matching for the instance given in Problem 1 by applying the stable marriage algorithm
   a. in its men-proposing version.
   b. in its women-proposing version.
4. Find a stable marriage matching for the instance defined by the following ranking matrix:

	A	B	C	D
α	1, 3	2, 3	3, 2	4, 3
β	1, 4	4, 1	3, 4	2, 2
γ	2, 2	1, 4	3, 3	4, 1
δ	4, 1	2, 2	3, 1	1, 4

5. Determine the time-efficiency class of the stable marriage algorithm
   a. in the worst case.
   b. in the best case.
6. Prove that a man-optimal stable marriage set is always unique. Is it also true for a woman-optimal stable marriage matching?
7. Prove that in the man-optimal stable matching, each woman has the worst partner that she can have in any stable marriage matching.
8. Implement the stable-marriage algorithm given in Section 10.4 so that its running time is in $O(n^2)$. Run an experiment to ascertain its average-case efficiency.
9. Write a report on the ***college admission problem*** (residents-hospitals assignment) that generalizes the stable marriage problem in that a college can accept "proposals" from more than one applicant.
10. Consider the ***problem of the roommates***, which is related to but more difficult than the stable marriage problem: "An even number of boys wish to divide up into pairs of roommates. A set of pairings is called stable if under it there are no two boys who are not roommates and who prefer each other to their actual roommates." [Gal62] Give an instance of this problem that does *not* have a stable pairing.

SUMMARY

- The *iterative-improvement technique* involves finding a solution to an optimization problem by generating a sequence of feasible solutions with improving values of the problem's objective function. Each subsequent solution in such a sequence typically involves a small, localized change in the previous feasible solution. When no such change improves the value of the objective function, the algorithm returns the last feasible solution as optimal and stops.

- Important problems that can be solved exactly by iterative-improvement algorithms include linear programming, maximizing the flow in a network, and matching the maximum possible number of vertices in a graph.
- The *simplex method* is the classic method for solving the general linear programming problem. It works by generating a sequence of adjacent extreme points of the problem's feasible region with improving values of the objective function.
- The *maximum-flow problem* asks to find the maximum flow possible in a network, a weighted directed graph with a source and a sink.
- The *Ford-Fulkerson method* is a classic template for solving the maximum-flow problem by the iterative-improvement approach. The *shortest-augmenting-path method* implements this idea by labeling network vertices in the breadth-first search manner.
- The Ford-Fulkerson method also finds a *minimum cut* in a given network.
- A *maximum cardinality matching* is the largest subset of edges in a graph such that no two edges share the same vertex. For a bipartite graph, it can be found by a sequence of augmentations of previously obtained matchings.
- The *stable marriage problem* is to find a *stable matching* for elements of two n-element sets based on given matching preferences. This problem always has a solution that can be found by the *Gale-Shapley algorithm*.

11

Limitations of Algorithm Power

Intellect distinguishes between the possible and the impossible; reason distinguishes between the sensible and the senseless. Even the possible can be senseless.

—Max Born (1882–1970), *My Life and My Views*, 1968

In the preceding chapters of this book, we encountered dozens of algorithms for solving a variety of different problems. A fair assessment of algorithms as problem-solving tools is inescapable: they are very powerful instruments, especially when they are executed by modern computers. But the power of algorithms is not unlimited, and its limits are the subject of this chapter. As we shall see, some problems cannot be solved by any algorithm. Other problems can be solved algorithmically but not in polynomial time. And even when a problem can be solved in polynomial time by some algorithms, there are usually lower bounds on their efficiency.

We start, in Section 11.1, with methods for obtaining lower bounds, which are estimates on a minimum amount of work needed to solve a problem. In general, obtaining a nontrivial lower bound even for a simple-sounding problem is a very difficult task. As opposed to ascertaining the efficiency of a particular algorithm, the task here is to establish a limit on the efficiency of *any* algorithm, known or unknown. This also necessitates a careful description of the operations such algorithms are allowed to perform. If we fail to define carefully the "rules of the game," so to speak, our claims may end up in the large dustbin of impossibility-related statements as, for example, the one made by the celebrated British physicist Lord Kelvin in 1895: "Heavier-than-air flying machines are impossible."

Section 11.2 discusses decision trees. This technique allows us, among other applications, to establish lower bounds on the efficiency of comparison-based algorithms for sorting and for searching in sorted arrays. As a result, we will be able to answer such questions as whether it is possible to invent a faster sorting algorithm than mergesort and whether binary search is the fastest algorithm for searching in a sorted array. (What does your intuition tell you the answers to these questions will turn out to be?) Incidentally, decision trees are also a great vehicle for directing us to a solution of some puzzles, such as the coin-weighing problem discussed in Section 4.4.

Section 11.3 deals with the question of intractability: which problems can and cannot be solved in polynomial time. This well-developed area of theoretical computer science is called computational complexity theory. We present the basic elements of this theory and discuss informally such fundamental notions as P, NP, and NP-complete problems, including the most important unresolved question of theoretical computer science about the relationship between P and NP problems.

The last section of this chapter deals with numerical analysis. This branch of computer science concerns algorithms for solving problems of "continuous" mathematics—solving equations and systems of equations, evaluating such functions as $\sin x$ and $\ln x$, computing integrals, and so on. The nature of such problems imposes two types of limitations. First, most cannot be solved exactly. Second, solving them even approximately requires dealing with numbers that can be represented in a digital computer with only a limited level of precision. Manipulating approximate numbers without proper care can lead to very inaccurate results. We will see that even solving a basic quadratic equation on a computer poses significant difficulties that require a modification of the canonical formula for the equation's roots.

11.1 Lower-Bound Arguments

We can look at the efficiency of an algorithm two ways. We can establish its asymptotic efficiency class (say, for the worst case) and see where this class stands with respect to the hierarchy of efficiency classes outlined in Section 2.2. For example, selection sort, whose efficiency is quadratic, is a reasonably fast algorithm, whereas the algorithm for the Tower of Hanoi problem is very slow because its efficiency is exponential. We can argue, however, that this comparison is akin to the proverbial comparison of apples to oranges because these two algorithms solve different problems. The alternative and possibly "fairer" approach is to ask how efficient a particular algorithm is with respect to other algorithms for the same problem. Seen in this light, selection sort has to be considered slow because there are $O(n \log n)$ sorting algorithms; the Tower of Hanoi algorithm, on the other hand, turns out to be the fastest possible for the problem it solves.

When we want to ascertain the efficiency of an algorithm with respect to other algorithms for the same problem, it is desirable to know the best possible efficiency *any* algorithm solving the problem may have. Knowing such a ***lower bound*** can tell us how much improvement we can hope to achieve in our quest for a better algorithm for the problem in question. If such a bound is ***tight***, i.e., we already know an algorithm in the same efficiency class as the lower bound, we can hope for a constant-factor improvement at best. If there is a gap between the efficiency of the fastest algorithm and the best lower bound known, the door for possible improvement remains open: either a faster algorithm matching the lower bound could exist or a better lower bound could be proved.

In this section, we present several methods for establishing lower bounds and illustrate them with specific examples. As we did in analyzing the efficiency of specific algorithms in the preceding chapters, we should distinguish between a lower-bound class and a minimum number of times a particular operation needs to be executed. As a rule, the second problem is more difficult than the first. For example, we can immediately conclude that any algorithm for finding the median of n numbers must be in $\Omega(n)$ (why?), but it is not simple at all to prove that any comparison-based algorithm for this problem must do at least $3(n-1)/2$ comparisons in the worst case (for odd n).

Trivial Lower Bounds

The simplest method of obtaining a lower-bound class is based on counting the number of items in the problem's input that must be processed and the number of output items that need to be produced. Since any algorithm must at least "read" all the items it needs to process and "write" all its outputs, such a count yields a ***trivial lower bound***. For example, any algorithm for generating all permutations of n distinct items must be in $\Omega(n!)$ because the size of the output is $n!$. And this bound is tight because good algorithms for generating permutations spend a constant time on each of them except the initial one (see Section 4.3).

As another example, consider the problem of evaluating a polynomial of degree n

$$p(x) = a_n x^n + a_{n-1} x^{n-1} + \cdots + a_0$$

at a given point x, given its coefficients $a_n, a_{n-1}, \ldots, a_0$. It is easy to see that all the coefficients have to be processed by any polynomial-evaluation algorithm. Indeed, if it were not the case, we could change the value of an unprocessed coefficient, which would change the value of the polynomial at a nonzero point x. This means that any such algorithm must be in $\Omega(n)$. This lower bound is tight because both the right-to-left evaluation algorithm (Problem 2 in Exercises 6.5) and Horner's rule (Section 6.5) are both linear.

In a similar vein, a trivial lower bound for computing the product of two $n \times n$ matrices is $\Omega(n^2)$ because any such algorithm has to process $2n^2$ elements in the input matrices and generate n^2 elements of the product. It is still unknown, however, whether this bound is tight.

Trivial lower bounds are often too low to be useful. For example, the trivial bound for the traveling salesman problem is $\Omega(n^2)$, because its input is $n(n-1)/2$ intercity distances and its output is a list of $n+1$ cities making up an optimal tour. But this bound is all but useless because there is no known algorithm with the running time being a polynomial function of any degree.

There is another obstacle to deriving a meaningful lower bound by this method. It lies in determining which part of an input must be processed by any algorithm solving the problem in question. For example, searching for an element of a given value in a sorted array does not require processing all its elements (why?). As another example, consider the problem of determining connectivity of an undirected graph defined by its adjacency matrix. It is plausible to expect that any such algorithm would have to check the existence of each of the $n(n-1)/2$ potential edges, but the proof of this fact is not trivial.

Information-Theoretic Arguments

While the approach outlined above takes into account the size of a problem's output, the information-theoretical approach seeks to establish a lower bound based on the amount of information it has to produce. Consider, as an example, the well-known game of deducing a positive integer between 1 and n selected by somebody by asking that person questions with yes/no answers. The amount of uncertainty that any algorithm solving this problem has to resolve can be measured by $\lceil \log_2 n \rceil$, the number of bits needed to specify a particular number among the n possibilities. We can think of each question (or, to be more accurate, an answer to each question) as yielding at most 1 bit of information about the algorithm's output, i.e., the selected number. Consequently, any such algorithm will need at least $\lceil \log_2 n \rceil$ such steps before it can determine its output in the worst case.

The approach we just exploited is called the ***information-theoretic argument*** because of its connection to information theory. It has proved to be quite useful for finding the so-called ***information-theoretic lower bounds*** for many problems involving comparisons, including sorting and searching. Its underlying idea can be realized much more precisely through the mechanism of ***decision trees***. Because of the importance of this technique, we discuss it separately and in more detail in Section 11.2.

Adversary Arguments

Let us revisit the same game of "guessing" a number used to introduce the idea of an information-theoretic argument. We can prove that any algorithm that solves this problem must ask at least $\lceil \log_2 n \rceil$ questions in its worst case by playing the role of a hostile adversary who wants to make an algorithm ask as many questions as possible. The adversary starts by considering each of the numbers between 1 and n as being potentially selected. (This is cheating, of course, as far as the game is concerned, but not as a way to prove our assertion.) After each question, the adversary gives an answer that leaves him with the largest set of numbers consistent with this and all the previously given answers. This strategy leaves him with at least one-half of the numbers he had before his last answer. If an algorithm stops before the size of the set is reduced to 1, the adversary can exhibit a number that could be a legitimate input the algorithm failed to identify. It is a simple technical matter now to show that one needs $\lceil \log_2 n \rceil$ iterations to shrink an n-element set to a one-element set by halving and rounding up the size of the remaining set. Hence, at least $\lceil \log_2 n \rceil$ questions need to be asked by any algorithm in the worst case.

This example illustrates the ***adversary method*** for establishing lower bounds. It is based on following the logic of a malevolent but honest adversary: the malevolence makes him push the algorithm down the most time-consuming path, and his honesty forces him to stay consistent with the choices already made. A lower bound is then obtained by measuring the amount of work needed to shrink a set of potential inputs to a single input along the most time-consuming path.

As another example, consider the problem of merging two sorted lists of size n

$$a_1 < a_2 < \cdots < a_n \quad \text{and} \quad b_1 < b_2 < \cdots < b_n$$

into a single sorted list of size $2n$. For simplicity, we assume that all the a's and b's are distinct, which gives the problem a unique solution. We encountered this problem when discussing mergesort in Section 5.1. Recall that we did merging by repeatedly comparing the first elements in the remaining lists and outputting the smaller among them. The number of key comparisons in the worst case for this algorithm for merging is $2n - 1$.

Is there an algorithm that can do merging faster? The answer turns out to be no. Knuth [KnuIII, p. 198] quotes the following adversary method for proving that $2n - 1$ is a lower bound on the number of key comparisons made by any comparison-based algorithm for this problem. The adversary will employ the following rule: reply true to the comparison $a_i < b_j$ if and only if $i < j$. This will force any correct merging algorithm to produce the only combined list consistent with this rule:

$$b_1 < a_1 < b_2 < a_2 < \cdots < b_n < a_n.$$

To produce this combined list, any correct algorithm will have to explicitly compare $2n - 1$ adjacent pairs of its elements, i.e., b_1 to a_1, a_1 to b_2, and so on. If one of these comparisons has not been made, e.g., a_1 has not been compared to b_2, we can transpose these keys to get

$$b_1 < b_2 < a_1 < a_2 < \cdots < b_n < a_n,$$

which is consistent with all the comparisons made but cannot be distinguished from the correct configuration given above. Hence, $2n - 1$ is, indeed, a lower bound for the number of key comparisons needed for any merging algorithm.

Problem Reduction

We have already encountered the problem-reduction approach in Section 6.6. There, we discussed getting an algorithm for problem P by reducing it to another problem Q solvable with a known algorithm. A similar reduction idea can be used for finding a lower bound. To show that problem P is at least as hard as another problem Q with a known lower bound, we need to reduce Q to P (not P to Q!). In other words, we should show that an arbitrary instance of problem Q can be transformed (in a reasonably efficient fashion) to an instance of problem P, so any algorithm solving P would solve Q as well. Then a lower bound for Q will be a lower bound for P. Table 11.1 lists several important problems that are often used for this purpose.

TABLE 11.1 Problems often used for establishing lower bounds by problem reduction

Problem	Lower bound	Tightness
sorting	$\Omega(n \log n)$	yes
searching in a sorted array	$\Omega(\log n)$	yes
element uniqueness problem	$\Omega(n \log n)$	yes
multiplication of n-digit integers	$\Omega(n)$	unknown
multiplication of $n \times n$ matrices	$\Omega(n^2)$	unknown

We will establish the lower bounds for sorting and searching in the next section. The element uniqueness problem asks whether there are duplicates among n given numbers. (We encountered this problem in Sections 2.3 and 6.1.) The proof of the lower bound for this seemingly simple problem is based on a very sophisticated mathematical analysis that is well beyond the scope of this book (see, e.g., [Pre85] for a rather elementary exposition). As to the last two algebraic problems in Table 11.1, the lower bounds quoted are trivial, but whether they can be improved remains unknown.

As an example of establishing a lower bound by reduction, let us consider the ***Euclidean minimum spanning tree problem***: given n points in the Cartesian plane, construct a tree of minimum total length whose vertices are the given points. As a problem with a known lower bound, we use the element uniqueness problem. We can transform any set $x_1, x_2, \ldots, x_n$ of n real numbers into a set of n points in the Cartesian plane by simply adding 0 as the points' y coordinate: $(x_1, 0), (x_2, 0), \ldots, (x_n, 0)$. Let T be a minimum spanning tree found for this set of points. Since T must contain a shortest edge, checking whether T contains a zero-length edge will answer the question about uniqueness of the given numbers. This reduction implies that $\Omega(n \log n)$ is a lower bound for the Euclidean minimum spanning tree problem, too.

Since the final results about the complexity of many problems are not known, the reduction technique is often used to compare the relative complexity of problems. For example, the formulas

$$x \cdot y = \frac{(x+y)^2 - (x-y)^2}{4} \quad \text{and} \quad x^2 = x \cdot x$$

show that the problems of computing the product of two n-digit integers and squaring an n-digit integer belong to the same complexity class, despite the latter being seemingly simpler than the former.

There are several similar results for matrix operations. For example, multiplying two symmetric matrices turns out to be in the same complexity class as multiplying two arbitrary square matrices. This result is based on the observation that not only is the former problem a special case of the latter one, but also that we can reduce the problem of multiplying two arbitrary square matrices of order n, say, A and B, to the problem of multiplying two symmetric matrices

$$X = \begin{bmatrix} 0 & A \\ A^T & 0 \end{bmatrix} \quad \text{and} \quad Y = \begin{bmatrix} 0 & B^T \\ B & 0 \end{bmatrix},$$

where A^T and B^T are the transpose matrices of A and B (i.e., $A^T[i, j] = A[j, i]$ and $B^T[i, j] = B[j, i]$), respectively, and 0 stands for the $n \times n$ matrix whose elements are all zeros. Indeed,

$$XY = \begin{bmatrix} 0 & A \\ A^T & 0 \end{bmatrix} \begin{bmatrix} 0 & B^T \\ B & 0 \end{bmatrix} = \begin{bmatrix} AB & 0 \\ 0 & A^T B^T \end{bmatrix},$$

from which the needed product AB can be easily extracted. (True, we will have to multiply matrices twice the original size, but this is just a minor technical complication with no impact on the complexity classes.)

Though such results are interesting, we will encounter even more important applications of the reduction approach to comparing problem complexity in Section 11.3.

Exercises 11.1

1. Prove that any algorithm solving the alternating-disk puzzle (Problem 14 in Exercises 3.1) must make at least $n(n+1)/2$ moves to solve it. Is this lower bound tight?

2. Prove that the classic recursive algorithm for the Tower of Hanoi puzzle (Section 2.4) makes the minimum number of disk moves needed to solve the problem.

3. Find a trivial lower-bound class for each of the following problems and indicate, if you can, whether this bound is tight.
   a. finding the largest element in an array
   b. checking completeness of a graph represented by its adjacency matrix
   c. generating all the subsets of an n-element set
   d. determining whether n given real numbers are all distinct

4. Consider the problem of identifying a lighter fake coin among n identical-looking coins with the help of a balance scale. Can we use the same information-theoretic argument as the one in the text for the number of questions in the guessing game to conclude that any algorithm for identifying the fake will need at least $\lceil \log_2 n \rceil$ weighings in the worst case?

5. Prove that any comparison-based algorithm for finding the largest element of an n-element set of real numbers must make $n-1$ comparisons in the worst case.

6. Find a tight lower bound for sorting an array by exchanging its adjacent elements.

7. Give an adversary-argument proof that the time efficiency of any algorithm that checks connectivity of a graph with n vertices is in $\Omega(n^2)$, provided the only operation allowed for an algorithm is to inquire about the presence of an edge between two vertices of the graph. Is this lower bound tight?

8. What is the minimum number of comparisons needed for a comparison-based sorting algorithm to merge any two sorted lists of sizes n and $n+1$ elements, respectively? Prove the validity of your answer.

9. Find the product of matrices A and B through a transformation to a product of two symmetric matrices if

$$A = \begin{bmatrix} 1 & -1 \\ 2 & 3 \end{bmatrix} \quad \text{and} \quad B = \begin{bmatrix} 0 & 1 \\ -1 & 2 \end{bmatrix}.$$

10. **a.** Can one use this section's formulas that indicate the complexity equivalence of multiplication and squaring of integers to show the complexity equivalence of multiplication and squaring of square matrices?

    b. Show that multiplication of two matrices of order n can be reduced to squaring a matrix of order $2n$.

11. Find a tight lower-bound class for the problem of finding two closest numbers among n real numbers $x_1, x_2, \ldots, x_n$.

12. Find a tight lower-bound class for the number placement problem (Problem 9 in Exercises 6.1).

11.2 Decision Trees

Many important algorithms, especially those for sorting and searching, work by comparing items of their inputs. We can study the performance of such algorithms with a device called a ***decision tree***. As an example, Figure 11.1 presents a decision tree of an algorithm for finding a minimum of three numbers. Each internal node of a binary decision tree represents a key comparison indicated in the node, e.g., $k < k'$. The node's left subtree contains the information about subsequent comparisons made if $k < k'$, and its right subtree does the same for the case of $k > k'$. (For the sake of simplicity, we assume throughout this section that all input items are distinct.) Each leaf represents a possible outcome of the algorithm's run on some input of size n. Note that the number of leaves can be greater than the number of outcomes because, for some algorithms, the same outcome can be arrived at through a different chain of comparisons. (This happens to be the case for the decision tree in Figure 11.1.) An important point is that the number of leaves must be at least as large as the number of possible outcomes. The algorithm's work on a particular input of size n can be traced by a path from the root to a leaf in its decision tree, and the number of comparisons made by the algorithm on such

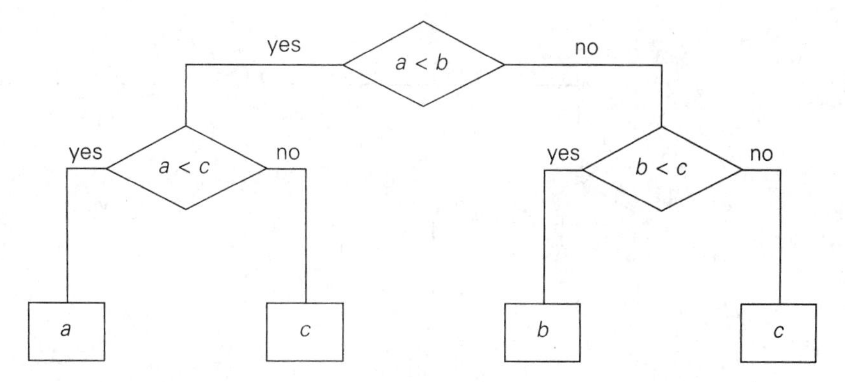

FIGURE 11.1 Decision tree for finding a minimum of three numbers.

a run is equal to the length of this path. Hence, the number of comparisons in the worst case is equal to the height of the algorithm's decision tree.

The central idea behind this model lies in the observation that a tree with a given number of leaves, which is dictated by the number of possible outcomes, has to be tall enough to have that many leaves. Specifically, it is not difficult to prove that for any binary tree with l leaves and height h,

$$h \geq \lceil \log_2 l \rceil. \tag{11.1}$$

Indeed, a binary tree of height h with the largest number of leaves has all its leaves on the last level (why?). Hence, the largest number of leaves in such a tree is 2^h. In other words, $2^h \geq l$, which immediately implies (11.1).

Inequality (11.1) puts a lower bound on the heights of binary decision trees and hence the worst-case number of comparisons made by any comparison-based algorithm for the problem in question. Such a bound is called the ***information-theoretic lower bound*** (see Section 11.1). We illustrate this technique below on two important problems: sorting and searching in a sorted array.

Decision Trees for Sorting

Most sorting algorithms are comparison based, i.e., they work by comparing elements in a list to be sorted. By studying properties of decision trees for such algorithms, we can derive important lower bounds on their time efficiencies.

We can interpret an outcome of a sorting algorithm as finding a permutation of the element indices of an input list that puts the list's elements in ascending order. Consider, as an example, a three-element list a, b, c of orderable items such as real numbers or strings. For the outcome $a < c < b$ obtained by sorting this list (see Figure 11.2), the permutation in question is 1, 3, 2. In general, the number of possible outcomes for sorting an arbitrary n-element list is equal to $n!$.

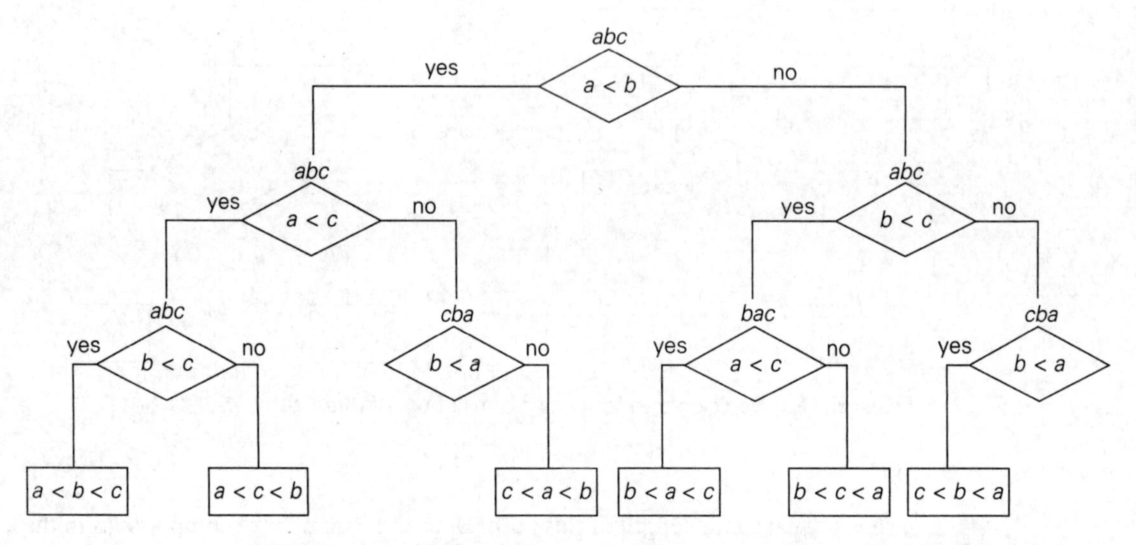

FIGURE 11.2 Decision tree for the tree-element selection sort. A triple above a node indicates the state of the array being sorted. Note two redundant comparisons $b < a$ with a single possible outcome because of the results of some previously made comparisons.

Inequality (11.1) implies that the height of a binary decision tree for any comparison-based sorting algorithm and hence the worst-case number of comparisons made by such an algorithm cannot be less than $\lceil \log_2 n! \rceil$:

$$C_{worst}(n) \geq \lceil \log_2 n! \rceil. \tag{11.2}$$

Using Stirling's formula for $n!$, we get

$$\lceil \log_2 n! \rceil \approx \log_2 \sqrt{2\pi n}(n/e)^n = n \log_2 n - n \log_2 e + \frac{\log_2 n}{2} + \frac{\log_2 2\pi}{2} \approx n \log_2 n.$$

In other words, about $n \log_2 n$ comparisons are necessary in the worst case to sort an arbitrary n-element list by any comparison-based sorting algorithm. Note that mergesort makes about this number of comparisons in its worst case and hence is asymptotically optimal. This also implies that the asymptotic lower bound $n \log_2 n$ is tight and therefore cannot be substantially improved. We should point out, however, that the lower bound of $\lceil \log_2 n! \rceil$ can be improved for some values of n. For example, $\lceil \log_2 12! \rceil = 29$, but it has been proved that 30 comparisons are necessary (and sufficient) to sort an array of 12 elements in the worst case.

We can also use decision trees for analyzing the average-case efficiencies of comparison-based sorting algorithms. We can compute the average number of comparisons for a particular algorithm as the average depth of its decision tree's leaves, i.e., as the average path length from the root to the leaves. For example, for

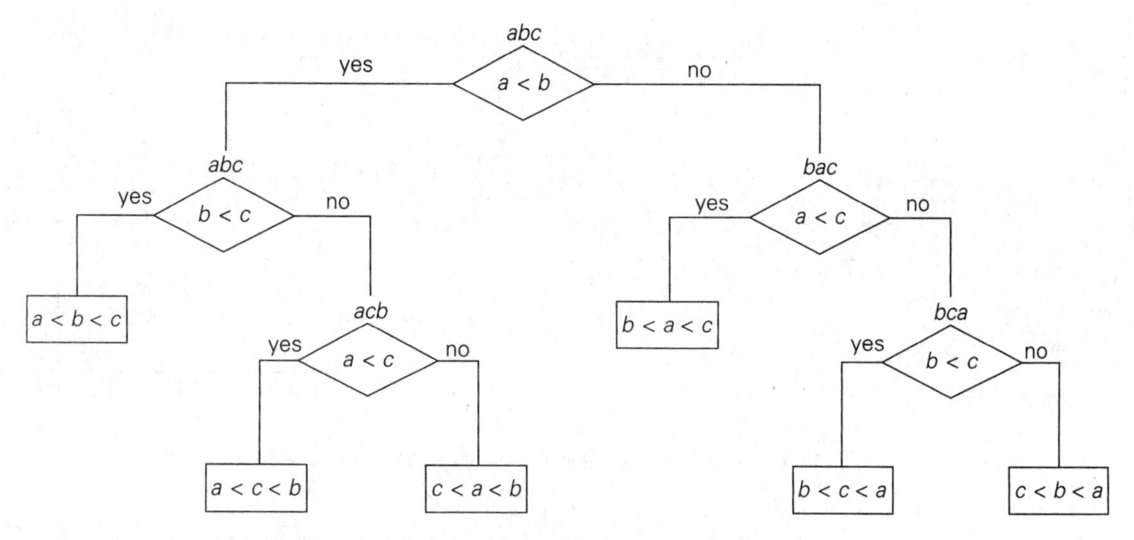

FIGURE 11.3 Decision tree for the three-element insertion sort.

the three-element insertion sort whose decision tree is given in Figure 11.3, this number is $(2+3+3+2+3+3)/6 = 2\frac{2}{3}$.

Under the standard assumption that all $n!$ outcomes of sorting are equally likely, the following lower bound on the average number of comparisons C_{avg} made by any comparison-based algorithm in sorting an n-element list has been proved:

$$C_{avg}(n) \geq \log_2 n!. \qquad \textbf{(11.3)}$$

As we saw earlier, this lower bound is about $n \log_2 n$. You might be surprised that the lower bounds for the average and worst cases are almost identical. Remember, however, that these bounds are obtained by maximizing the number of comparisons made in the average and worst cases, respectively. For a particular sorting algorithm, the average-case efficiency can, of course, be significantly better than their worst-case efficiency.

Decision Trees for Searching a Sorted Array

In this section, we shall see how decision trees can be used for establishing lower bounds on the number of key comparisons in searching a sorted array of n keys: $A[0] < A[1] < \cdots < A[n-1]$. The principal algorithm for this problem is binary search. As we saw in Section 4.4, the number of comparisons made by binary search in the worst case, $C^{bs}_{worst}(n)$, is given by the formula

$$C^{bs}_{worst}(n) = \lfloor \log_2 n \rfloor + 1 = \lceil \log_2(n+1) \rceil. \qquad \textbf{(11.4)}$$

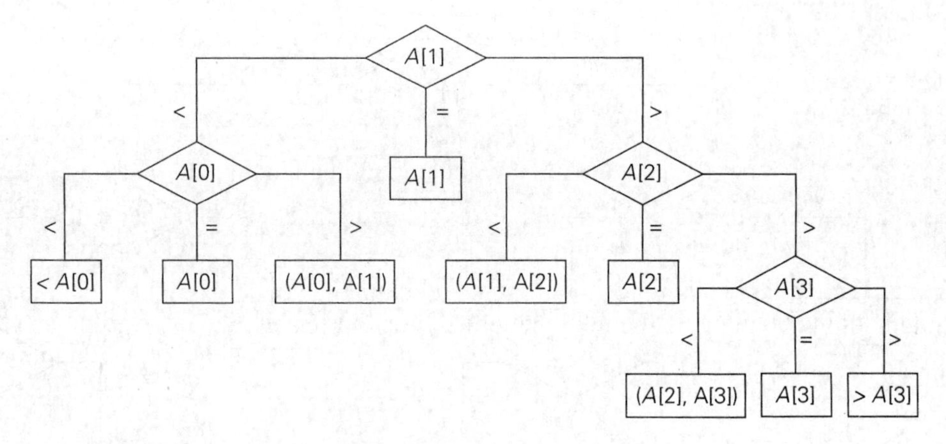

FIGURE 11.4 Ternary decision tree for binary search in a four-element array.

We will use decision trees to determine whether this is the smallest possible number of comparisons.

Since we are dealing here with three-way comparisons in which search key K is compared with some element $A[i]$ to see whether $K < A[i]$, $K = A[i]$, or $K > A[i]$, it is natural to try using ternary decision trees. Figure 11.4 presents such a tree for the case of $n = 4$. The internal nodes of that tree indicate the array's elements being compared with the search key. The leaves indicate either a matching element in the case of a successful search or a found interval that the search key belongs to in the case of an unsuccessful search.

We can represent any algorithm for searching a sorted array by three-way comparisons with a ternary decision tree similar to that in Figure 11.4. For an array of n elements, all such decision trees will have $2n + 1$ leaves (n for successful searches and $n + 1$ for unsuccessful ones). Since the minimum height h of a ternary tree with l leaves is $\lceil \log_3 l \rceil$, we get the following lower bound on the number of worst-case comparisons:

$$C_{worst}(n) \geq \lceil \log_3(2n + 1) \rceil.$$

This lower bound is smaller than $\lceil \log_2(n + 1) \rceil$, the number of worst-case comparisons for binary search, at least for large values of n (and smaller than or equal to $\lceil \log_2(n + 1) \rceil$ for every positive integer n—see Problem 7 in this section's exercises). Can we prove a better lower bound, or is binary search far from being optimal? The answer turns out to be the former. To obtain a better lower bound, we should consider binary rather than ternary decision trees, such as the one in Figure 11.5. Internal nodes in such a tree correspond to the same three-way comparisons as before, but they also serve as terminal nodes for successful searches. Leaves therefore represent only unsuccessful searches, and there are $n + 1$ of them for searching an n-element array.

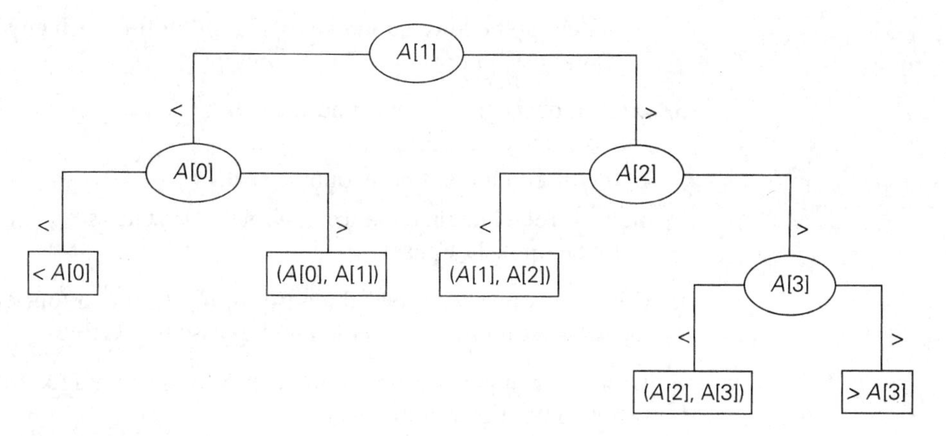

FIGURE 11.5 Binary decision tree for binary search in a four-element array.

As comparison of the decision trees in Figures 11.4 and 11.5 illustrates, the binary decision tree is simply the ternary decision tree with all the middle subtrees eliminated. Applying inequality (11.1) to such binary decision trees immediately yields

$$C_{worst}(n) \geq \lceil \log_2(n+1) \rceil. \qquad \textbf{(11.5)}$$

This inequality closes the gap between the lower bound and the number of worst-case comparisons made by binary search, which is also $\lceil \log_2(n+1) \rceil$. A much more sophisticated analysis (see, e.g., [KnuIII, Section 6.2.1]) shows that under the standard assumptions about searches, binary search makes the smallest number of comparisons on the average, as well. The average number of comparisons made by this algorithm turns out to be about $\log_2 n - 1$ and $\log_2(n+1)$ for successful and unsuccessful searches, respectively.

Exercises 11.2

1. Prove by mathematical induction that

a. $h \geq \lceil \log_2 l \rceil$ for any binary tree with height h and the number of leaves l.

b. $h \geq \lceil \log_3 l \rceil$ for any ternary tree with height h and the number of leaves l.

2. Consider the problem of finding the median of a three-element set $\{a, b, c\}$ of orderable items.

a. What is the information-theoretic lower bound for comparison-based algorithms solving this problem?

b. Draw a decision tree for an algorithm solving this problem.

c. If the worst-case number of comparisons in your algorithm is greater than the information-theoretic lower bound, do you think an algorithm matching the lower bound exists? (Either find such an algorithm or prove its impossibility.)

3. Draw a decision tree and find the number of key comparisons in the worst and average cases for

a. the three-element basic bubble sort.

b. the three-element enhanced bubble sort (which stops if no swaps have been made on its last pass).

4. Design a comparison-based algorithm for sorting a four-element array with the smallest number of element comparisons possible.

5. Design a comparison-based algorithm for sorting a five-element array with seven comparisons in the worst case.

6. Draw a binary decision tree for searching a four-element sorted list by sequential search.

7. Compare the two lower bounds for searching a sorted array—$\lceil \log_3(2n+1) \rceil$ and $\lceil \log_2(n+1) \rceil$—to show that

a. $\lceil \log_3(2n+1) \rceil \le \lceil \log_2(n+1) \rceil$ for every positive integer n.

b. $\lceil \log_3(2n+1) \rceil < \lceil \log_2(n+1) \rceil$ for every positive integer $n \ge n_0$.

8. What is the information-theoretic lower bound for finding the maximum of n numbers by comparison-based algorithms? Is this bound tight?

9. A ***tournament tree*** is a complete binary tree reflecting results of a "knockout tournament": its leaves represent n players entering the tournament, and each internal node represents a winner of a match played by the players represented by the node's children. Hence, the winner of the tournament is represented by the root of the tree.

a. What is the total number of games played in such a tournament?

b. How many rounds are there in such a tournament?

c. Design an efficient algorithm to determine the second-best player using the information produced by the tournament. How many extra games does your algorithm require?

10. *Advanced fake-coin problem* There are $n \ge 3$ coins identical in appearance; either all are genuine or exactly one of them is fake. It is unknown whether the fake coin is lighter or heavier than the genuine one. You have a balance scale with which you can compare any two sets of coins. That is, by tipping to the left, to the right, or staying even, the balance scale will tell whether the sets weigh the same or which of the sets is heavier than the other, but not by how much. The problem is to find whether all the coins are genuine and, if not, to find the fake coin and establish whether it is lighter or heavier than the genuine ones.

a. Prove that any algorithm for this problem must make at least $\lceil \log_3(2n + 1) \rceil$ weighings in the worst case.

b. Draw a decision tree for an algorithm that solves the problem for $n = 3$ coins in two weighings.

c. Prove that there exists no algorithm that solves the problem for $n = 4$ coins in two weighings.

d. Draw a decision tree for an algorithm that solves the problem for $n = 4$ coins in two weighings by using an extra coin known to be genuine.

e. Draw a decision tree for an algorithm that solves the classic version of the problem—that for $n = 12$ coins in three weighings (with no extra coins being used).

11. *Jigsaw puzzle* A jigsaw puzzle contains n pieces. A "section" of the puzzle is a set of one or more pieces that have been connected to each other. A "move" consists of connecting two sections. What algorithm will minimize the number of moves required to complete the puzzle?

11.3 *P*, *NP*, and *NP*-Complete Problems

In the study of the computational complexity of problems, the first concern of both computer scientists and computing professionals is whether a given problem can be solved in polynomial time by some algorithm.

DEFINITION 1 We say that an algorithm solves a problem in polynomial time if its worst-case time efficiency belongs to $O(p(n))$ where $p(n)$ is a polynomial of the problem's input size n. (Note that since we are using big-oh notation here, problems solvable in, say, logarithmic time are solvable in polynomial time as well.) Problems that can be solved in polynomial time are called ***tractable***, and problems that cannot be solved in polynomial time are called ***intractable***.

There are several reasons for drawing the intractability line in this way. First, the entries of Table 2.1 and their discussion in Section 2.1 imply that we cannot solve arbitrary instances of intractable problems in a reasonable amount of time unless such instances are very small. Second, although there might be a huge difference between the running times in $O(p(n))$ for polynomials of drastically different degrees, there are very few useful polynomial-time algorithms with the degree of a polynomial higher than three. In addition, polynomials that bound running times of algorithms do not usually have extremely large coefficients. Third, polynomial functions possess many convenient properties; in particular, both the sum and composition of two polynomials are always polynomials too. Fourth, the choice of this class has led to a development of an extensive theory called ***computational complexity***, which seeks to classify problems according to their inherent difficulty. And according to this theory, a problem's intractability remains the same for all principal models of computations and all reasonable input-encoding schemes for the problem under consideration.

We just touch on some basic notions and ideas of complexity theory in this section. If you are interested in a more formal treatment of this theory, you will have no trouble finding a wealth of textbooks devoted to the subject (e.g., [Sip05], [Aro09]).

P and *NP* Problems

Most problems discussed in this book can be solved in polynomial time by some algorithm. They include computing the product and the greatest common divisor of two integers, sorting a list, searching for a key in a list or for a pattern in a text string, checking connectivity and acyclicity of a graph, and finding a minimum spanning tree and shortest paths in a weighted graph. (You are invited to add more examples to this list.) Informally, we can think about problems that can be solved in polynomial time as the set that computer science theoreticians call P. A more formal definition includes in P only ***decision problems***, which are problems with yes/no answers.

DEFINITION 2 Class P is a class of decision problems that can be solved in polynomial time by (deterministic) algorithms. This class of problems is called ***polynomial***.

The restriction of P to decision problems can be justified by the following reasons. First, it is sensible to exclude problems not solvable in polynomial time because of their exponentially large output. Such problems do arise naturally—e.g., generating subsets of a given set or all the permutations of n distinct items—but it is apparent from the outset that they cannot be solved in polynomial time. Second, many important problems that are not decision problems in their most natural formulation can be reduced to a series of decision problems that are easier to study. For example, instead of asking about the minimum number of colors needed to color the vertices of a graph so that no two adjacent vertices are colored the same color, we can ask whether there exists such a coloring of the graph's vertices with no more than m colors for $m = 1, 2, \ldots$. (The latter is called the ***m*-coloring problem**.) The first value of m in this series for which the decision problem of m-coloring has a solution solves the optimization version of the graph-coloring problem as well.

It is natural to wonder whether *every* decision problem can be solved in polynomial time. The answer to this question turns out to be no. In fact, some decision problems cannot be solved at all by any algorithm. Such problems are called ***undecidable***, as opposed to ***decidable*** problems that can be solved by an algorithm. A famous example of an undecidable problem was given by Alan

Turing in 1936.[1] The problem in question is called the ***halting problem***: given a computer program and an input to it, determine whether the program will halt on that input or continue working indefinitely on it.

Here is a surprisingly short proof of this remarkable fact. By way of contradiction, assume that A is an algorithm that solves the halting problem. That is, for any program P and input I,

$$A(P, I) = \begin{cases} 1, & \text{if program } P \text{ halts on input } I; \\ 0, & \text{if program } P \text{ does not halt on input } I. \end{cases}$$

We can consider program P as an input to itself and use the output of algorithm A for pair (P, P) to construct a program Q as follows:

$$Q(P) = \begin{cases} \text{halts,} & \text{if } A(P, P) = 0, \text{ i.e., if program } P \text{ does not halt on input } P; \\ \text{does not halt,} & \text{if } A(P, P) = 1, \text{ i.e., if program } P \text{ halts on input } P. \end{cases}$$

Then on substituting Q for P, we obtain

$$Q(Q) = \begin{cases} \text{halts,} & \text{if } A(Q, Q) = 0, \text{ i.e., if program } Q \text{ does not halt on input } Q; \\ \text{does not halt,} & \text{if } A(Q, Q) = 1, \text{ i.e., if program } Q \text{ halts on input } Q. \end{cases}$$

This is a contradiction because neither of the two outcomes for program Q is possible, which completes the proof.

Are there decidable but intractable problems? Yes, there are, but the number of *known* examples is surprisingly small, especially of those that arise naturally rather than being constructed for the sake of a theoretical argument.

There are many important problems, however, for which no polynomial-time algorithm has been found, nor has the impossibility of such an algorithm been proved. The classic monograph by M. Garey and D. Johnson [Gar79] contains a list of several hundred such problems from different areas of computer science, mathematics, and operations research. Here is just a small sample of some of the best-known problems that fall into this category:

Hamiltonian circuit problem Determine whether a given graph has a Hamiltonian circuit—a path that starts and ends at the same vertex and passes through all the other vertices exactly once.

Traveling salesman problem Find the shortest tour through n cities with known positive integer distances between them (find the shortest Hamiltonian circuit in a complete graph with positive integer weights).

---

1. This was just one of many breakthrough contributions to theoretical computer science made by the English mathematician and computer science pioneer Alan Turing (1912–1954). In recognition of this, the ACM—the principal society of computing professionals and researchers—has named after him an award given for outstanding contributions to theoretical computer science. A lecture given on such an occasion by Richard Karp [Kar86] provides an interesting historical account of the development of complexity theory.

Knapsack problem Find the most valuable subset of n items of given positive integer weights and values that fit into a knapsack of a given positive integer capacity.

Partition problem Given n positive integers, determine whether it is possible to partition them into two disjoint subsets with the same sum.

Bin-packing problem Given n items whose sizes are positive rational numbers not larger than 1, put them into the smallest number of bins of size 1.

Graph-coloring problem For a given graph, find its chromatic number, which is the smallest number of colors that need to be assigned to the graph's vertices so that no two adjacent vertices are assigned the same color.

Integer linear programming problem Find the maximum (or minimum) value of a linear function of several integer-valued variables subject to a finite set of constraints in the form of linear equalities and inequalities.

Some of these problems are decision problems. Those that are not have decision-version counterparts (e.g., the m-coloring problem for the graph-coloring problem). What all these problems have in common is an exponential (or worse) growth of choices, as a function of input size, from which a solution needs to be found. Note, however, that some problems that also fall under this umbrella can be solved in polynomial time. For example, the Eulerian circuit problem—the problem of the existence of a cycle that traverses all the edges of a given graph exactly once—can be solved in $O(n^2)$ time by checking, in addition to the graph's connectivity, whether all the graph's vertices have even degrees. This example is particularly striking: it is quite counterintuitive to expect that the problem about cycles traversing all the edges exactly once (Eulerian circuits) can be so much easier than the seemingly similar problem about cycles visiting all the vertices exactly once (Hamiltonian circuits).

Another common feature of a vast majority of decision problems is the fact that although solving such problems can be computationally difficult, checking whether a proposed solution actually solves the problem is computationally easy, i.e., it can be done in polynomial time. (We can think of such a proposed solution as being randomly generated by somebody leaving us with the task of verifying its validity.) For example, it is easy to check whether a proposed list of vertices is a Hamiltonian circuit for a given graph with n vertices. All we need to check is that the list contains $n + 1$ vertices of the graph in question, that the first n vertices are distinct whereas the last one is the same as the first, and that every consecutive pair of the list's vertices is connected by an edge. This general observation about decision problems has led computer scientists to the notion of a nondeterministic algorithm.

DEFINITION 3 A ***nondeterministic algorithm*** is a two-stage procedure that takes as its input an instance I of a decision problem and does the following.

Nondeterministic ("guessing") stage: An arbitrary string S is generated that can be thought of as a candidate solution to the given instance I (but may be complete gibberish as well).

Deterministic ("verification") stage: A deterministic algorithm takes both I and S as its input and outputs yes if S represents a solution to instance I. (If S is not a solution to instance I, the algorithm either returns no or is allowed not to halt at all.)

We say that a nondeterministic algorithm solves a decision problem if and only if for every yes instance of the problem it returns yes on some execution. (In other words, we require a nondeterministic algorithm to be capable of "guessing" a solution at least once and to be able to verify its validity. And, of course, we do not want it to ever output a yes answer on an instance for which the answer should be no.) Finally, a nondeterministic algorithm is said to be ***nondeterministic polynomial*** if the time efficiency of its verification stage is polynomial.

Now we can define the class of *NP* problems.

DEFINITION 4 Class *NP* is the class of decision problems that can be solved by nondeterministic polynomial algorithms. This class of problems is called ***nondeterministic polynomial***.

Most decision problems are in *NP*. First of all, this class includes all the problems in *P*:

$$P \subseteq NP.$$

This is true because, if a problem is in *P*, we can use the deterministic polynomial-time algorithm that solves it in the verification-stage of a nondeterministic algorithm that simply ignores string S generated in its nondeterministic ("guessing") stage. But *NP* also contains the Hamiltonian circuit problem, the partition problem, decision versions of the traveling salesman, the knapsack, graph coloring, and many hundreds of other difficult combinatorial optimization problems cataloged in [Gar79]. The halting problem, on the other hand, is among the rare examples of decision problems that are known not to be in *NP*.

This leads to the most important open question of theoretical computer science: Is *P* a proper subset of *NP*, or are these two classes, in fact, the same? We can put this symbolically as

$$P \stackrel{?}{=} NP.$$

Note that $P = NP$ would imply that each of many hundreds of difficult combinatorial decision problems can be solved by a polynomial-time algorithm, although computer scientists have failed to find such algorithms despite their persistent efforts over many years. Moreover, many well-known decision problems are known to be "*NP*-complete" (see below), which seems to cast more doubts on the possibility that $P = NP$.

NP-Complete Problems

Informally, an *NP*-complete problem is a problem in *NP* that is as difficult as any other problem in this class because, by definition, any other problem in *NP* can be reduced to it in polynomial time (shown symbolically in Figure 11.6).

Here are more formal definitions of these concepts.

DEFINITION 5 A decision problem D_1 is said to be ***polynomially reducible*** to a decision problem D_2, if there exists a function t that transforms instances of D_1 to instances of D_2 such that:

1. t maps all yes instances of D_1 to yes instances of D_2 and all no instances of D_1 to no instances of D_2
2. t is computable by a polynomial time algorithm

This definition immediately implies that if a problem D_1 is polynomially reducible to some problem D_2 that can be solved in polynomial time, then problem D_1 can also be solved in polynomial time (why?).

DEFINITION 6 A decision problem D is said to be ***NP-complete*** if:

1. it belongs to class *NP*
2. every problem in *NP* is polynomially reducible to D

The fact that closely related decision problems are polynomially reducible to each other is not very surprising. For example, let us prove that the Hamiltonian circuit problem is polynomially reducible to the decision version of the traveling

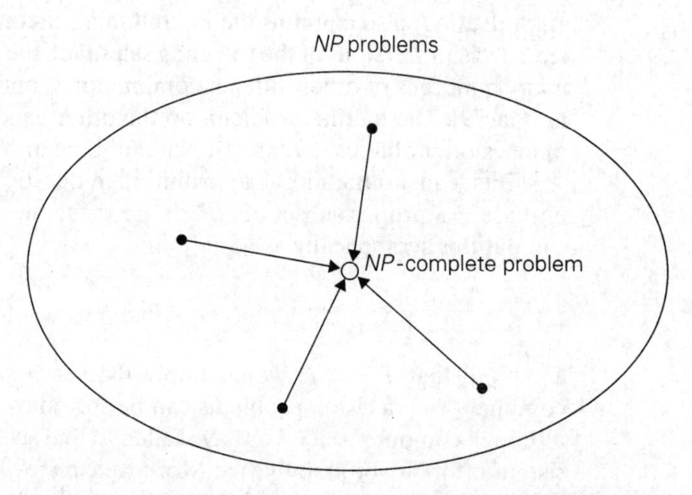

FIGURE 11.6 Notion of an *NP*-complete problem. Polynomial-time reductions of *NP* problems to an *NP*-complete problem are shown by arrows.

salesman problem. The latter can be stated as the existence problem of a Hamiltonian circuit not longer than a given positive integer m in a given complete graph with positive integer weights. We can map a graph G of a given instance of the Hamiltonian circuit problem to a complete weighted graph G' representing an instance of the traveling salesman problem by assigning 1 as the weight to each edge in G and adding an edge of weight 2 between any pair of nonadjacent vertices in G. As the upper bound m on the Hamiltonian circuit length, we take $m = n$, where n is the number of vertices in G (and G'). Obviously, this transformation can be done in polynomial time.

Let G be a yes instance of the Hamiltonian circuit problem. Then G has a Hamiltonian circuit, and its image in G' will have length n, making the image a yes instance of the decision traveling salesman problem. Conversely, if we have a Hamiltonian circuit of the length not larger than n in G', then its length must be exactly n (why?) and hence the circuit must be made up of edges present in G, making the inverse image of the yes instance of the decision traveling salesman problem be a yes instance of the Hamiltonian circuit problem. This completes the proof.

The notion of *NP*-completeness requires, however, polynomial reducibility of *all* problems in *NP*, both known and unknown, to the problem in question. Given the bewildering variety of decision problems, it is nothing short of amazing that specific examples of *NP*-complete problems have been actually found. Nevertheless, this mathematical feat was accomplished independently by Stephen Cook in the United States and Leonid Levin in the former Soviet Union.[2] In his 1971 paper, Cook [Coo71] showed that the so-called ***CNF-satisfiability problem*** is *NP*-complete. The CNF-satisfiability problem deals with boolean expressions. Each boolean expression can be represented in conjunctive normal form, such as the following expression involving three boolean variables x_1, x_2, and x_3 and their negations denoted $\bar{x}_1$, $\bar{x}_2$, and $\bar{x}_3$, respectively:

$$(x_1 \vee \bar{x}_2 \vee \bar{x}_3)\&(\bar{x}_1 \vee x_2)\&(\bar{x}_1 \vee \bar{x}_2 \vee \bar{x}_3).$$

The CNF-satisfiability problem asks whether or not one can assign values *true* and *false* to variables of a given boolean expression in its CNF form to make the entire expression *true*. (It is easy to see that this can be done for the above formula: if $x_1 = true$, $x_2 = true$, and $x_3 = false$, the entire expression is *true*.)

Since the Cook-Levin discovery of the first known *NP*-complete problems, computer scientists have found many hundreds, if not thousands, of other examples. In particular, the well-known problems (or their decision versions) mentioned above—Hamiltonian circuit, traveling salesman, partition, bin packing, and graph coloring—are all *NP*-complete. It is known, however, that if $P \neq NP$ there must exist *NP* problems that neither are in P nor are *NP*-complete.

2. As it often happens in the history of science, breakthrough discoveries are made independently and almost simultaneously by several scientists. In fact, Levin introduced a more general notion than *NP*-completeness, which was not limited to decision problems, but his paper [Lev73] was published two years after Cook's.

For a while, the leading candidate to be such an example was the problem of determining whether a given integer is prime or composite. But in an important theoretical breakthrough, Professor Manindra Agrawal and his students Neeraj Kayal and Nitin Saxena of the Indian Institute of Technology in Kanpur announced in 2002 a discovery of a deterministic polynomial-time algorithm for primality testing [Agr04]. Their algorithm does not solve, however, the related problem of factoring large composite integers, which lies at the heart of the widely used encryption method called the ***RSA algorithm*** [Riv78].

Showing that a decision problem is *NP*-complete can be done in two steps. First, one needs to show that the problem in question is in *NP*; i.e., a randomly generated string can be checked in polynomial time to determine whether or not it represents a solution to the problem. Typically, this step is easy. The second step is to show that every problem in *NP* is reducible to the problem in question in polynomial time. Because of the transitivity of polynomial reduction, this step can be done by showing that a known *NP*-complete problem can be transformed to the problem in question in polynomial time (see Figure 11.7). Although such a transformation may need to be quite ingenious, it is incomparably simpler than proving the existence of a transformation for every problem in *NP*. For example, if we already know that the Hamiltonian circuit problem is *NP*-complete, its polynomial reducibility to the decision traveling salesman problem implies that the latter is also *NP*-complete (after an easy check that the decision traveling salesman problem is in class *NP*).

The definition of *NP*-completeness immediately implies that if there exists a deterministic polynomial-time algorithm for just one *NP*-complete problem, then every problem in *NP* can be solved in polynomial time by a deterministic algorithm, and hence $P = NP$. In other words, finding a polynomial-time algorithm

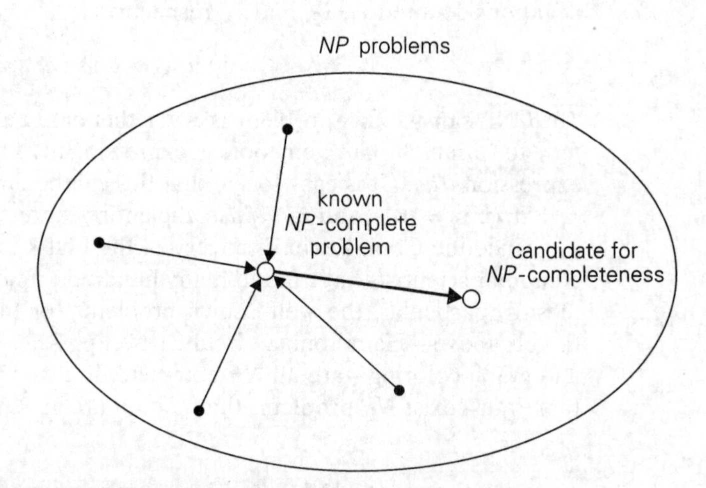

FIGURE 11.7 Proving *NP*-completeness by reduction.

for one *NP*-complete problem would mean that there is no qualitative difference between the complexity of checking a proposed solution and finding it in polynomial time for the vast majority of decision problems of all kinds. Such implications make most computer scientists believe that $P \neq NP$, although nobody has been successful so far in finding a mathematical proof of this intriguing conjecture. Surprisingly, in interviews with the authors of a book about the lives and discoveries of 15 prominent computer scientists [Sha98], Cook seemed to be uncertain about the eventual resolution of this dilemma whereas Levin contended that we should expect the $P = NP$ outcome.

Whatever the eventual answer to the $P \stackrel{?}{=} NP$ question proves to be, knowing that a problem is *NP*-complete has important practical implications for today. It means that faced with a problem known to be *NP*-complete, we should probably not aim at gaining fame and fortune[3] by designing a polynomial-time algorithm for solving all its instances. Rather, we should concentrate on several approaches that seek to alleviate the intractability of such problems. These approaches are outlined in the next chapter of the book.

Exercises 11.3

1. A game of chess can be posed as the following decision problem: given a legal positioning of chess pieces and information about which side is to move, determine whether that side can win. Is this decision problem decidable?

2. A certain problem can be solved by an algorithm whose running time is in $O(n^{\log_2 n})$. Which of the following assertions is true?
   a. The problem is tractable.
   b. The problem is intractable.
   c. Impossible to tell.

3. Give examples of the following graphs or explain why such examples cannot exist.
   a. graph with a Hamiltonian circuit but without an Eulerian circuit
   b. graph with an Eulerian circuit but without a Hamiltonian circuit
   c. graph with both a Hamiltonian circuit and an Eulerian circuit
   d. graph with a cycle that includes all the vertices but with neither a Hamiltonian circuit nor an Eulerian circuit

3. In 2000, The Clay Mathematics Institute (CMI) of Cambridge, Massachusetts, designated a $1 million prize for the solution to this problem.

4. For each of the following graphs, find its chromatic number.

a.

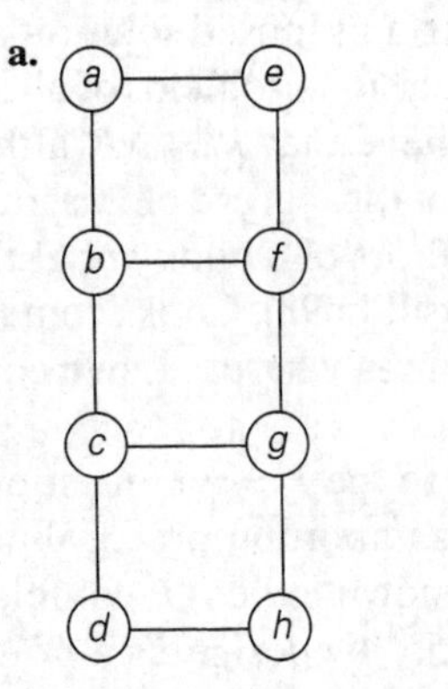

b.

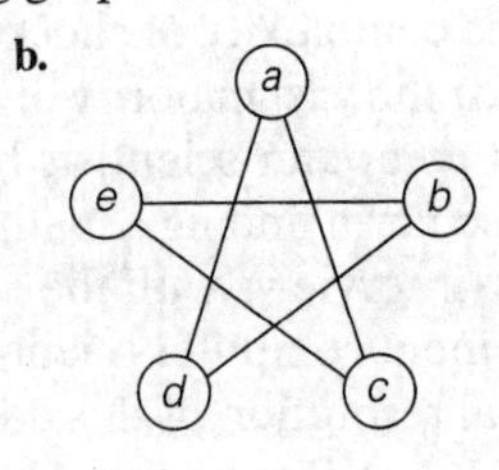

c.

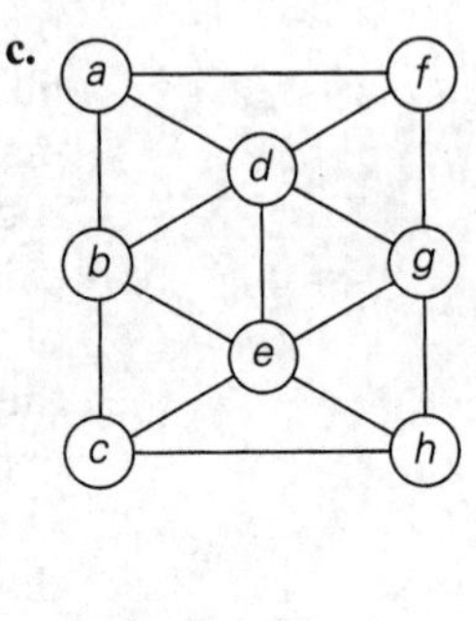

5. Design a polynomial-time algorithm for the graph 2-coloring problem: determine whether vertices of a given graph can be colored in no more than two colors so that no two adjacent vertices are colored the same color.

6. Consider the following brute-force algorithm for solving the composite number problem: Check successive integers from 2 to $\lfloor n/2 \rfloor$ as possible divisors of n. If one of them divides n evenly, return yes (i.e., the number is composite); if none of them does, return no. Why does this algorithm not put the problem in class P?

7. State the decision version for each of the following problems and outline a polynomial-time algorithm that verifies whether or not a proposed solution solves the problem. (You may assume that a proposed solution represents a legitimate input to your verification algorithm.)

a. knapsack problem **b.** bin packing problem

8. Show that the partition problem is polynomially reducible to the decision version of the knapsack problem.

9. Show that the following three problems are polynomially reducible to each other.

(i) Determine, for a given graph $G = \langle V, E \rangle$ and a positive integer $m \leq |V|$, whether G contains a ***clique*** of size m or more. (A clique of size k in a graph is its complete subgraph of k vertices.)

(ii) Determine, for a given graph $G = \langle V, E \rangle$ and a positive integer $m \leq |V|$, whether there is a ***vertex cover*** of size m or less for G. (A vertex cover of size k for a graph $G = \langle V, E \rangle$ is a subset $V' \subseteq V$ such that $|V'| = k$ and, for each edge $(u, v) \in E$, at least one of u and v belongs to V'.)

(iii) Determine, for a given graph $G = \langle V, E \rangle$ and a positive integer $m \leq |V|$, whether G contains an ***independent set*** of size m or more. (An independent set of size k for a graph $G = \langle V, E \rangle$ is a subset $V' \subseteq V$ such that $|V'| = k$ and for all $u, v \in V'$, vertices u and v are *not* adjacent in G.)

10. Determine whether the following problem is *NP*-complete. Given several sequences of uppercase and lowercase letters, is it possible to select a letter from each sequence without selecting both the upper- and lowercase versions of any letter? For example, if the sequences are Abc, BC, aB, and ac, it is possible to choose A from the first sequence, B from the second and third, and c from the fourth. An example where there is no way to make the required selections is given by the four sequences AB, Ab, aB, and ab. [Kar86]

11. Which of the following diagrams do not contradict the current state of our knowledge about the complexity classes *P*, *NP*, and *NPC* (*NP*-complete problems)?

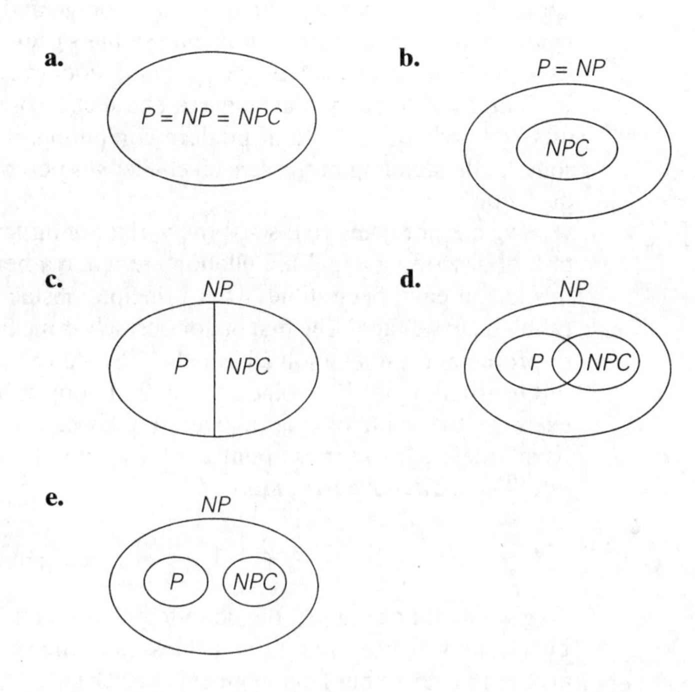

12. King Arthur expects 150 knights for an annual dinner at Camelot. Unfortunately, some of the knights quarrel with each other, and Arthur knows who quarrels with whom. Arthur wants to seat his guests around a table so that no two quarreling knights sit next to each other.

a. Which standard problem can be used to model King Arthur's task?

b. As a research project, find a proof that Arthur's problem has a solution if each knight does not quarrel with at least 75 other knights.

11.4 Challenges of Numerical Algorithms

Numerical analysis is usually described as the branch of computer science concerned with algorithms for solving mathematical problems. This description needs an important clarification: the problems in question are problems of "continuous" mathematics—solving equations and systems of equations, evaluating such functions as $\sin x$ and $\ln x$, computing integrals, and so on—as opposed to problems of discrete mathematics dealing with such structures as graphs, trees, permutations, and combinations. Our interest in efficient algorithms for mathematical problems stems from the fact that these problems arise as models of many real-life phenomena both in the natural world and in the social sciences. In fact, numerical analysis used to be the main area of research, study, and application of computer science. With the rapid proliferation of computers in business and everyday-life applications, which deal primarily with storage and retrieval of information, the *relative* importance of numerical analysis has shrunk in the last 30 years. However, its applications, enhanced by the power of modern computers, continue to expand in all areas of fundamental research and technology. Thus, wherever one's interests lie in the wide world of modern computing, it is important to have at least some understanding of the special challenges posed by continuous mathematical problems.

We are not going to discuss the variety of difficulties posed by modeling, the task of describing a real-life phenomenon in mathematical terms. Assuming that this has already been done, what principal obstacles to solving a mathematical problem do we face? The first major obstacle is the fact that most numerical analysis problems cannot be solved exactly.[4] They have to be solved approximately, and this is usually done by replacing an infinite object by a finite approximation. For example, the value of e^x at a given point x can be computed by approximating its infinite Taylor's series about $x = 0$ by a finite sum of its first terms, called the nth-degree ***Taylor polynomial***:

$$e^x \approx 1 + x + \frac{x^2}{2!} + \cdots + \frac{x^n}{n!}. \qquad \textbf{(11.6)}$$

To give another example, the definite integral of a function can be approximated by a finite weighted sum of its values, as in the ***composite trapezoidal rule*** that you might remember from your calculus class:

$$\int_a^b f(x)dx \approx \frac{h}{2}[f(a) + 2\sum_{i=1}^{n-1} f(x_i) + f(b)], \qquad \textbf{(11.7)}$$

where $h = (b - a)/n$, $x_i = a + ih$ for $i = 0, 1, \ldots, n$ (Figure 11.8).

The errors of such approximations are called ***truncation errors***. One of the major tasks in numerical analysis is to estimate the magnitudes of truncation

4. Solving a system of linear equations and polynomial evaluation, discussed in Sections 6.2 and 6.5, respectively, are rare exceptions to this rule.

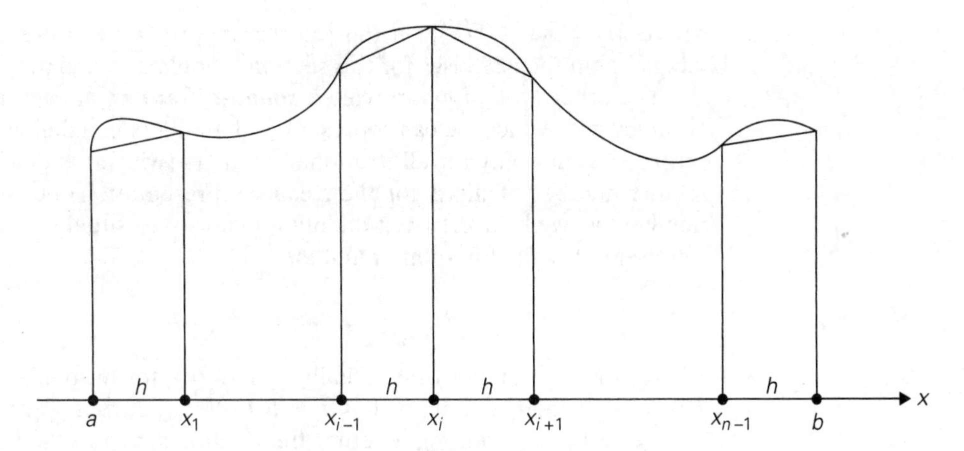

FIGURE 11.8 Composite trapezoidal rule.

errors. This is typically done by using calculus tools, from elementary to quite advanced. For example, for approximation (11.6) we have

$$|e^x - [1 + x + \frac{x^2}{2!} + \cdots + \frac{x^n}{n!}]| \le \frac{M}{(n+1)!}|x|^{n+1}, \tag{11.8}$$

where $M = \max e^{\xi}$ on the segment with the endpoints at 0 and x. This formula makes it possible to determine the degree of Taylor's polynomial needed to guarantee a predefined accuracy level of approximation (11.6).

For example, if we want to compute $e^{0.5}$ by formula (11.6) and guarantee the truncation error to be smaller than 10^{-4}, we can proceed as follows. First, we estimate M of formula (11.8):

$$M = \max_{0 \le \xi \le 0.5} e^{\xi} \le e^{0.5} < 2.$$

Using this bound and the desired accuracy level of 10^{-4}, we obtain from (11.8)

$$\frac{M}{(n+1)!}|0.5|^{n+1} < \frac{2}{(n+1)!}0.5^{n+1} < 10^{-4}.$$

To solve the last inequality, we can compute the first few values of

$$\frac{2}{(n+1)!}0.5^{n+1} = \frac{2^{-n}}{(n+1)!}$$

to see that the smallest value of n for which this inequality holds is 5.

Similarly, for approximation (11.7), the standard bound of the truncation error is given by the inequality

$$|\int_a^b f(x)dx - \frac{h}{2}[f(a) + 2\sum_{i=1}^{n-1} f(x_i) + f(b)]| \le \frac{(b-a)h^2}{12}M_2, \tag{11.9}$$

where $M_2 = \max |f''(x)|$ on the interval $a \le x \le b$. You are asked to use this inequality in the exercises for this section (Problems 5 and 6).

The other type of errors, called ***round-off errors***, are caused by the limited accuracy with which we can represent real numbers in a digital computer. These errors arise not only for all irrational numbers (which, by definition, require an infinite number of digits for their exact representation) but for many rational numbers as well. In the overwhelming majority of situations, real numbers are represented as floating-point numbers,

$$\pm .d_1 d_2 \dots d_p \cdot B^E, \qquad \textbf{(11.10)}$$

where B is the number base, usually 2 or 16 (or, for unsophisticated calculators, 10); $d_1, d_2, \dots, d_p$ are digits ($0 \le d_i < B$ for $i = 1, 2, \dots, p$ and $d_1 > 0$ unless the number is 0) representing together the fractional part of the number and called its ***mantissa***; and E is an integer ***exponent*** with the range of values approximately symmetric about 0.

The accuracy of the floating-point representation depends on the number of ***significant digits*** p in representation (11.10). Most computers permit two or even three levels of precision: ***single precision*** (typically equivalent to between 6 and 7 significant decimal digits), ***double precision*** (13 to 14 significant decimal digits), and ***extended precision*** (19 to 20 significant decimal digits). Using higher-precision arithmetic slows computations but may help to overcome some of the problems caused by round-off errors. Higher precision may need to be used only for a particular step of the algorithm in question.

As with an approximation of any kind, it is important to distinguish between the ***absolute error*** and the ***relative error*** of representing a number α^* by its approximation α:

$$\text{absolute error} = |\alpha - \alpha^*|, \qquad \textbf{(11.11)}$$

$$\text{relative error} = \frac{|\alpha - \alpha^*|}{|\alpha^*|}. \qquad \textbf{(11.12)}$$

(The relative error is undefined if $\alpha^* = 0$.)

Very large and very small numbers cannot be represented in floating-point arithmetic because of the phenomena called ***overflow*** and ***underflow***, respectively. An overflow happens when an arithmetic operation yields a result outside the range of the computer's floating-point numbers. Typical examples of overflow arise from the multiplication of large numbers or division by a very small number. Sometimes we can eliminate this problem by making a simple change in the order in which an expression is evaluated (e.g., $(10^{29} \cdot 11^{30})/12^{30} = 10^{29} \cdot (11/12)^{30}$), by replacing an expression with an equal one (e.g., computing $\binom{100}{2}$ not as $100!/(2!(100-2)!)$ but as $(100 \cdot 99)/2$), or by computing a logarithm of an expression instead of the expression itself.

Underflow occurs when the result of an operation is a nonzero fraction of such a small magnitude that it cannot be represented as a nonzero floating-point number. Usually, underflow numbers are replaced by zero, but a special signal is generated by hardware to indicate such an event has occurred.

It is important to remember that, in addition to inaccurate representation of numbers, the arithmetic operations performed in a computer are not always exact, either. In particular, subtracting two nearly equal floating-point numbers may cause a large increase in relative error. This phenomenon is called ***subtractive cancellation***.

EXAMPLE 1 Consider two irrational numbers

$$\alpha^* = \pi = 3.14159265\ldots \quad \text{and} \quad \beta^* = \pi - 6 \cdot 10^{-7} = 3.14159205\ldots$$

represented by floating-point numbers $\alpha = 0.3141593 \cdot 10^1$ and $\beta = 0.3141592 \cdot 10^1$, respectively. The relative errors of these approximations are small:

$$\frac{|\alpha - \alpha^*|}{\alpha^*} = \frac{0.0000003\ldots}{\pi} < \frac{4}{3}10^{-7}$$

and

$$\frac{|\beta - \beta^*|}{\beta^*} = \frac{0.00000005\ldots}{\pi - 6 \cdot 10^{-7}} < \frac{1}{3}10^{-7},$$

respectively. The relative error of representing the difference $\gamma^* = \alpha^* - \beta^*$ by the difference of the floating-point representations $\gamma = \alpha - \beta$ is

$$\frac{|\gamma - \gamma^*|}{\gamma^*} = \frac{10^{-6} - 6 \cdot 10^{-7}}{6 \cdot 10^{-7}} = \frac{2}{3},$$

which is very large for a relative error despite quite accurate approximations for both α and β. ■

Note that we may get a significant magnification of round-off error if a low-accuracy difference is used as a divisor. (We already encountered this problem in discussing Gaussian elimination in Section 6.2. Our solution there was to use partial pivoting.) Many numerical algorithms involve thousands or even millions of arithmetic operations for typical inputs. For such algorithms, the propagation of round-off errors becomes a major concern from both the practical and theoretical standpoints. For some algorithms, round-off errors can propagate through the algorithm's operations with increasing effect. This highly undesirable property of a numerical algorithm is called ***instability***. Some problems exhibit such a high level of sensitivity to changes in their input that it is all but impossible to design a stable algorithm to solve them. Such problems are called ***ill-conditioned***.

EXAMPLE 2 Consider the following system of two linear equations in two unknowns:

$$1.001x + 0.999y = 2$$
$$0.999x + 1.001y = 2.$$

Its only solution is $x = 1$, $y = 1$. To see how sensitive this system is to small changes to its right-hand side, consider the system with the same coefficient matrix but slightly different right-hand side values:

$$\begin{aligned} 1.001x + 0.999y &= 2.002 \\ 0.999x + 1.001y &= 1.998. \end{aligned}$$

The only solution to this system is $x = 2$, $y = 0$, which is quite far from the solution to the previous system. Note that the coefficient matrix of this system is close to being singular (why?). Hence, a minor change in its coefficients may yield a system with either no solutions or infinitely many solutions, depending on its right-hand-side values. You can find a more formal and detailed discussion of how we can measure the degree of ill-condition of the coefficient matrix in numerical analysis textbooks (e.g., [Ger03]). ■

We conclude with a well-known problem of finding real roots of the quadratic equation

$$ax^2 + bx + c = 0 \tag{11.13}$$

for any real coefficients a, b, and c ($a \neq 0$). According to secondary-school algebra, equation (11.13) has real roots if and only if its discriminant $D = b^2 - 4ac$ is nonnegative, and these roots can be found by the following formula

$$x_{1,2} = \frac{-b \pm \sqrt{b^2 - 4ac}}{2a}. \tag{11.14}$$

Although formula (11.14) provides a complete solution to the posed problem as far as a mathematician is concerned, it is far from being a complete solution for an algorithm designer. The first major obstacle is evaluating the square root. Even for most positive integers D, $\sqrt{D}$ is an irrational number that can be computed only approximately. There is a method of computing square roots that is much better than the one commonly taught in secondary school. (It follows from ***Newton's method***, a very important algorithm for solving equations, which we discuss in Section 12.4.) This method generates the sequence $\{x_n\}$ of approximations to $\sqrt{D}$, where D is a given nonnegative number, according to the formula

$$x_{n+1} = \frac{1}{2}(x_n + \frac{D}{x_n}) \quad \text{for } n = 0, 1, \ldots, \tag{11.15}$$

where the initial approximation x_0 can be chosen, among other possibilities, as $x_0 = (1 + D)/2$. It is not difficult to prove that sequence (11.15) is decreasing (if $D \neq 1$) and always converges to $\sqrt{D}$. We can stop generating its elements either when the difference between its two consecutive elements is less than a predefined error tolerance $\epsilon > 0$

$$x_n - x_{n+1} < \epsilon$$

or when x_{n+1}^2 is sufficiently close to D. Approximation sequence (11.15) converges very fast to $\sqrt{D}$ for most values of D. In particular, one can prove that if $0.25 \le D < 1$, then no more than four iterations are needed to guarantee that

$$|x_n - \sqrt{D}| < 4 \cdot 10^{-15},$$

and we can always scale a given value of d to one in the interval [0.25, 1) by the formula $d = D2^p$, where p is an even integer.

EXAMPLE 3 Let us apply Newton's algorithm to compute $\sqrt{2}$. (For simplicity, we ignore scaling.) We will round off the numbers to six decimal places and use the standard numerical analysis notation $\doteq$ to indicate the round-offs.

$$x_0 = \frac{1}{2}(1+2) = 1.500000,$$

$$x_1 = \frac{1}{2}(x_0 + \frac{2}{x_0}) \doteq 1.416667,$$

$$x_2 = \frac{1}{2}(x_1 + \frac{2}{x_1}) \doteq 1.414216,$$

$$x_3 = \frac{1}{2}(x_2 + \frac{2}{x_2}) \doteq 1.414214,$$

$$x_4 = \frac{1}{2}(x_3 + \frac{2}{x_3}) \doteq 1.414214.$$

At this point we have to stop because $x_4 = x_3 \doteq 1.414214$ and hence all other approximations will be the same. The exact value of $\sqrt{2}$ is 1.41421356 . . . . ■

With the issue of computing square roots squared away (I do not know whether or not the pun was intended), are we home free to write a program based on formula (11.14)? The answer is no because of the possible impact of round-off errors. Among other obstacles, we are faced here with the menace of subtractive cancellation. If b^2 is much larger than $4ac$, $\sqrt{b^2 - 4ac}$ will be very close to $|b|$, and a root computed by formula (11.14) might have a large relative error.

EXAMPLE 4 Let us follow a paper by George Forsythe[5] [For69] and consider the equation

$$x^2 - 10^5 x + 1 = 0.$$

Its true roots to 11 significant digits are

$$x_1^* \doteq 99999.999990$$

5. George E. Forsythe (1917–1972), a noted numerical analyst, played a leading role in establishing computer science as a separate academic discipline in the United States. It is his words that are used as the epigraph to this book's preface.

and

$$x_2^* \doteq 0.000010000000001.$$

If we use formula (11.14) and perform all the computations in decimal floating-point arithmetic with, say, seven significant digits, we obtain

$$\begin{aligned}
(-b)^2 &= 0.1000000 \cdot 10^{11}, \\
4ac &= 0.4000000 \cdot 10^{1}, \\
D &\doteq 0.1000000 \cdot 10^{11}, \\
\sqrt{D} &\doteq 0.1000000 \cdot 10^{6}, \\
x_1 &\doteq \frac{-b + \sqrt{D}}{2a} \doteq 0.1000000 \cdot 10^{6}, \\
x_2 &\doteq \frac{-b - \sqrt{D}}{2a} \doteq 0.
\end{aligned}$$

And although the relative error of approximating x_1^* by x_1 is very small, for the second root it is very large:

$$\frac{|x_2 - x_2^*|}{x_2^*} = 1 \quad \text{(i.e., 100\%)}$$ ■

To avoid the possibility of subtractive cancellation in formula (11.14), we can use instead another formula, obtained as follows:

$$\begin{aligned}
x_1 &= \frac{-b + \sqrt{b^2 - 4ac}}{2a} \\
&= \frac{(-b + \sqrt{b^2 - 4ac})(-b - \sqrt{b^2 - 4ac})}{2a(-b - \sqrt{b^2 - 4ac})} \\
&= \frac{2c}{-b - \sqrt{b^2 - 4ac}},
\end{aligned}$$

with no danger of subtractive cancellation in the denominator if $b > 0$. As to x_2, it can be computed by the standard formula

$$x_2 = \frac{-b - \sqrt{b^2 - 4ac}}{2a},$$

with no danger of cancellation either for a positive value of b.

The case of $b < 0$ is symmetric: we can use the formulas

$$x_1 = \frac{-b + \sqrt{b^2 - 4ac}}{2a}$$

and

$$x_2 = \frac{2c}{-b + \sqrt{b^2 - 4ac}}.$$

(The case of $b = 0$ can be considered with either of the other two cases.)

There are several other obstacles to applying formula (11.14), which are related to limitations of floating-point arithmetic: if a is very small, division by a can cause an overflow; there seems to be no way to fight the danger of subtractive cancellation in computing $b^2 - 4ac$ other than calculating it with double precision; and so on. These problems have been overcome by William Kahan of the University of Toronto (see [For69]), and his algorithm is considered to be a significant achievement in the history of numerical analysis.

Hopefully, this brief overview has piqued your interest enough for you to seek more information in the many books devoted exclusively to numerical algorithms. In this book, we discuss one more topic in the next chapter: three classic methods for solving equations in one unknown.

Exercises 11.4

1. Some textbooks define the number of significant digits in the approximation of number α^* by number α as the largest nonnegative integer k for which

$$\frac{|\alpha - \alpha^*|}{|\alpha^*|} < 5 \cdot 10^{-k}.$$

   According to this definition, how many significant digits are there in the approximation of π by

   a. 3.1415? **b.** 3.1417?

2. If $\alpha = 1.5$ is known to approximate some number α^* with the absolute error not exceeding 10^{-2}, find
   a. the range of possible values of α^*.
   b. the range of the relative errors of these approximations.

3. Find the approximate value of $\sqrt{e} = 1.648721\ldots$ obtained by the fifth-degree Taylor's polynomial about 0 and compute the truncation error of this approximation. Does the result agree with the theoretical prediction made in the section?

4. Derive formula (11.7) of the composite trapezoidal rule.

5. Use the composite trapezoidal rule with $n = 4$ to approximate the following definite integrals. Find the truncation error of each approximation and compare it with the one given by formula (11.9).

   a. $\int_0^1 x^2 dx$ **b.** $\int_1^3 x^{-1} dx$

6. If $\int_0^1 e^{\sin x} dx$ is to be computed by the composite trapezoidal rule, how large should the number of subintervals be to guarantee a truncation error smaller than 10^{-4}? Smaller than 10^{-6}?

7. Solve the two systems of linear equations and indicate whether they are ill-conditioned.

a. $\begin{aligned} 2x + 5y &= 7 \\ 2x + 5.000001y &= 7.000001 \end{aligned}$ **b.** $\begin{aligned} 2x + 5y &= 7 \\ 2x + 4.999999y &= 7.000002 \end{aligned}$

8. Write a computer program to solve the equation $ax^2 + bx + c = 0$.

9. **a.** Prove that for any nonnegative number D, the sequence of Newton's method for computing $\sqrt{D}$ is strictly decreasing and converges to $\sqrt{D}$ for any value of the initial approximation $x_0 > \sqrt{D}$.

   b. Prove that if $0.25 \le D < 1$ and $x_0 = (1 + D)/2$, no more than four iterations of Newton's method are needed to guarantee that

$$|x_n - \sqrt{D}| < 4 \cdot 10^{-15}.$$

10. Apply four iterations of Newton's method to compute $\sqrt{3}$ and estimate the absolute and relative errors of this approximation.

SUMMARY

- Given a class of algorithms for solving a particular problem, a *lower bound* indicates the best possible efficiency *any* algorithm from this class can have.

- A *trivial lower bound* is based on counting the number of items in the problem's input that must be processed and the number of output items that need to be produced.

- An *information-theoretic lower bound* is usually obtained through a mechanism of *decision trees*. This technique is particularly useful for comparison-based algorithms for sorting and searching. Specifically:
  - Any general comparison-based sorting algorithm must perform at least $\lceil \log_2 n! \rceil \approx n \log_2 n$ key comparisons in the worst case.
  - Any general comparison-based algorithm for searching a sorted array must perform at least $\lceil \log_2(n + 1) \rceil$ key comparisons in the worst case.

- The *adversary method* for establishing lower bounds is based on following the logic of a malevolent adversary who forces the algorithm into the most time-consuming path.

- A lower bound can also be established by *reduction*, i.e., by reducing a problem with a known lower bound to the problem in question.

- *Complexity theory* seeks to classify problems according to their computational complexity. The principal split is between *tractable* and *intractable* problems—

problems that can and cannot be solved in polynomial time, respectively. For purely technical reasons, complexity theory concentrates on *decision problems*, which are problems with yes/no answers.

- The *halting problem* is an example of an *undecidable* decision problem; i.e., it cannot be solved by any algorithm.
- P is the class of all decision problems that can be solved in polynomial time. NP is the class of all decision problems whose randomly guessed solutions can be verified in polynomial time.
- Many important problems in NP (such as the Hamiltonian circuit problem) are known to be NP-complete: all other problems in NP are reducible to such a problem in polynomial time. The first proof of a problem's NP-completeness was published by S. Cook for the *CNF-satisfiability problem.*
- It is not known whether $P = NP$ or P is just a proper subset of NP. This question is the most important unresolved issue in theoretical computer science. A discovery of a polynomial-time algorithm for any of the thousands of known NP-complete problems would imply that $P = NP$.
- *Numerical analysis* is a branch of computer science dealing with solving continuous mathematical problems. Two types of errors occur in solving a majority of such problems: truncation error and round-off error. *Truncation errors* stem from replacing infinite objects by their finite approximations. *Round-off errors* are due to inaccuracies of representing numbers in a digital computer.
- *Subtractive cancellation* happens as a result of subtracting two near-equal floating-point numbers. It may lead to a sharp increase in the relative round-off error and therefore should be avoided (by either changing the expression's form or by using a higher precision in computing such a difference).
- Writing a general computer program for solving quadratic equations $ax^2 + bx + c = 0$ is a difficult task. The problem of computing square roots can be solved by utilizing *Newton's method*; the problem of subtractive cancellation can be dealt with by using different formulas depending on whether coefficient b is positive or negative and by computing the discriminant $b^2 - 4ac$ with double precision.

12

Coping with the Limitations of Algorithm Power

Keep on the lookout for novel ideas that others have used successfully. Your idea has to be original only in its adaptation to the problem you're working on.

—Thomas Edison (1847–1931)

As we saw in the previous chapter, there are problems that are difficult to solve algorithmically. At the same time, some of them are so important that we cannot just sigh in resignation and do nothing. This chapter outlines several ways of dealing with such difficult problems.

Sections 12.1 and 12.2 introduce two algorithm design techniques—***backtracking*** and ***branch-and-bound***—that often make it possible to solve at least some large instances of difficult combinatorial problems. Both strategies can be considered an improvement over exhaustive search, discussed in Section 3.4. Unlike exhaustive search, they construct candidate solutions one component at a time and evaluate the partially constructed solutions: if no potential values of the remaining components can lead to a solution, the remaining components are not generated at all. This approach makes it possible to solve some large instances of difficult combinatorial problems, though, in the worst case, we still face the same curse of exponential explosion encountered in exhaustive search.

Both backtracking and branch-and-bound are based on the construction of a ***state-space tree*** whose nodes reflect specific choices made for a solution's components. Both techniques terminate a node as soon as it can be guaranteed that no solution to the problem can be obtained by considering choices that correspond to the node's descendants. The techniques differ in the nature of problems they can be applied to. Branch-and-bound is applicable only to optimization problems because it is based on computing a bound on possible values of the problem's objective function. Backtracking is not constrained by this demand, but more often than not, it applies to nonoptimization problems. The other distinction between backtracking and branch-and-bound lies in the order in which nodes of the state-space tree are generated. For backtracking, this tree is usually developed depth-first (i.e., similar to DFS). Branch-and-bound can generate nodes according to several rules: the most natural one is the so-called best-first rule explained in Section 12.2.

Section 12.3 takes a break from the idea of solving a problem exactly. The algorithms presented there solve problems approximately but fast. Specifically, we consider a few approximation algorithms for the traveling salesman and knapsack problems. For the traveling salesman problem, we discuss basic theoretical results and pertinent empirical data for several well-known approximation algorithms. For the knapsack problem, we first introduce a greedy algorithm and then a parametric family of polynomial-time algorithms that yield arbitrarily good approximations.

Section 12.4 is devoted to algorithms for solving nonlinear equations. After a brief discussion of this very important problem, we examine three classic methods for approximate root finding: the bisection method, the method of false position, and Newton's method.

12.1 Backtracking

Throughout the book (see in particular Sections 3.4 and 11.3), we have encountered problems that require finding an element with a special property in a domain that grows exponentially fast (or faster) with the size of the problem's input: a Hamiltonian circuit among all permutations of a graph's vertices, the most valuable subset of items for an instance of the knapsack problem, and the like. We addressed in Section 11.3 the reasons for believing that many such problems might not be solvable in polynomial time. Also recall that we discussed in Section 3.4 how such problems can be solved, at least in principle, by exhaustive search. The exhaustive-search technique suggests generating all candidate solutions and then identifying the one (or the ones) with a desired property.

Backtracking is a more intelligent variation of this approach. The principal idea is to construct solutions one component at a time and evaluate such partially constructed candidates as follows. If a partially constructed solution can be developed further without violating the problem's constraints, it is done by taking the first remaining legitimate option for the next component. If there is no legitimate option for the next component, no alternatives for *any* remaining component need to be considered. In this case, the algorithm backtracks to replace the last component of the partially constructed solution with its next option.

It is convenient to implement this kind of processing by constructing a tree of choices being made, called the ***state-space tree***. Its root represents an initial state before the search for a solution begins. The nodes of the first level in the tree represent the choices made for the first component of a solution, the nodes of the second level represent the choices for the second component, and so on. A node in a state-space tree is said to be ***promising*** if it corresponds to a partially constructed solution that may still lead to a complete solution; otherwise, it is called ***nonpromising***. Leaves represent either nonpromising dead ends or complete solutions found by the algorithm. In the majority of cases, a state-space tree for a backtracking algorithm is constructed in the manner of depth-first search. If the current node is promising, its child is generated by adding the first remaining legitimate option for the next component of a solution, and the processing moves to this child. If the current node turns out to be nonpromising, the algorithm backtracks to the node's parent to consider the next possible option for its last component; if there is no such option, it backtracks one more level up the tree, and so on. Finally, if the algorithm reaches a complete solution to the problem, it either stops (if just one solution is required) or continues searching for other possible solutions.

n-Queens Problem

As our first example, we use a perennial favorite of textbook writers: the ***n*-queens problem**. The problem is to place n queens on an $n \times n$ chessboard so that no two queens attack each other by being in the same row or in the same column or on the same diagonal. For $n = 1$, the problem has a trivial solution, and it is easy to see that there is no solution for $n = 2$ and $n = 3$. So let us consider the four-queens problem and solve it by the backtracking technique. Since each of the four queens has to be placed in its own row, all we need to do is to assign a column for each queen on the board presented in Figure 12.1.

We start with the empty board and then place queen 1 in the first possible position of its row, which is in column 1 of row 1. Then we place queen 2, after trying unsuccessfully columns 1 and 2, in the first acceptable position for it, which is square (2, 3), the square in row 2 and column 3. This proves to be a dead end because there is no acceptable position for queen 3. So, the algorithm backtracks and puts queen 2 in the next possible position at (2, 4). Then queen 3 is placed at (3, 2), which proves to be another dead end. The algorithm then backtracks all the way to queen 1 and moves it to (1, 2). Queen 2 then goes to (2, 4), queen 3 to (3, 1), and queen 4 to (4, 3), which is a solution to the problem. The state-space tree of this search is shown in Figure 12.2.

If other solutions need to be found (how many of them are there for the four-queens problem?), the algorithm can simply resume its operations at the leaf at which it stopped. Alternatively, we can use the board's symmetry for this purpose.

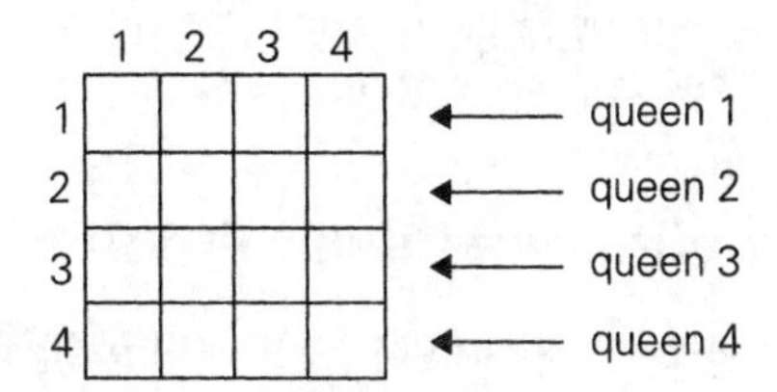

FIGURE 12.1 Board for the four-queens problem.

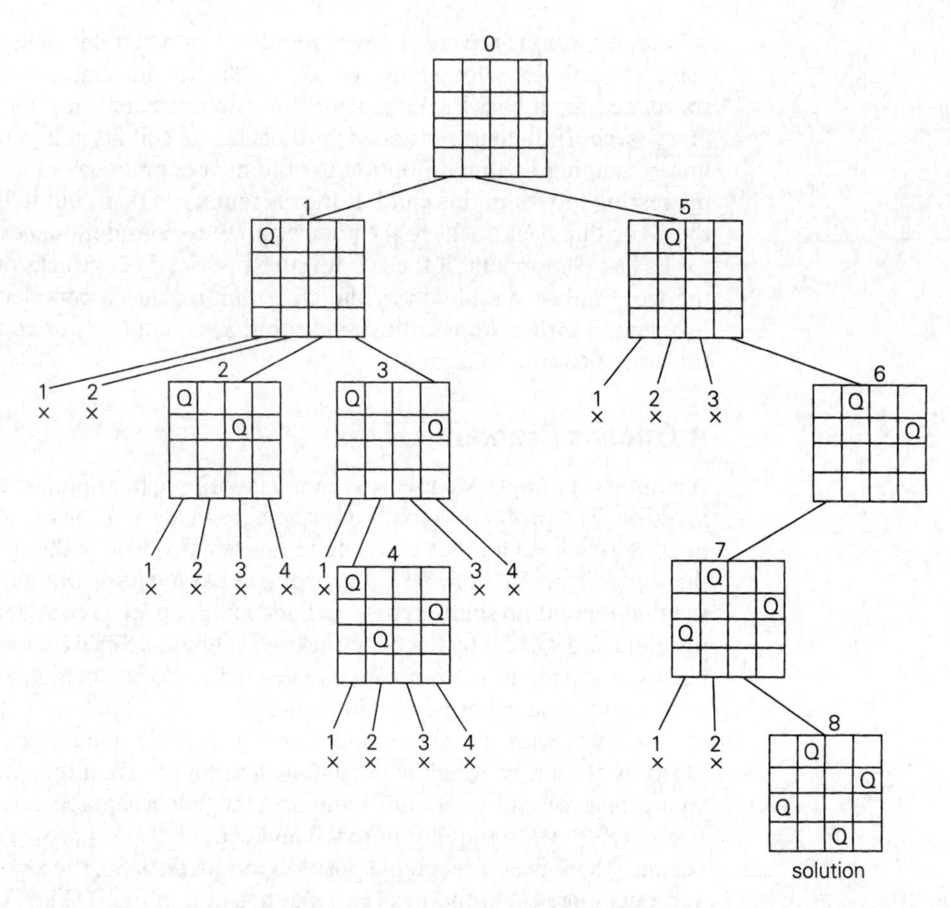

FIGURE 12.2 State-space tree of solving the four-queens problem by backtracking. × denotes an unsuccessful attempt to place a queen in the indicated column. The numbers above the nodes indicate the order in which the nodes are generated.

Finally, it should be pointed out that a single solution to the n-queens problem for any $n \geq 4$ can be found in linear time. In fact, over the last 150 years mathematicians have discovered several alternative formulas for nonattacking positions of n queens [Bel09]. Such positions can also be found by applying some general algorithm design strategies (Problem 4 in this section's exercises).

Hamiltonian Circuit Problem

As our next example, let us consider the problem of finding a Hamiltonian circuit in the graph in Figure 12.3a.

Without loss of generality, we can assume that if a Hamiltonian circuit exists, it starts at vertex a. Accordingly, we make vertex a the root of the state-space

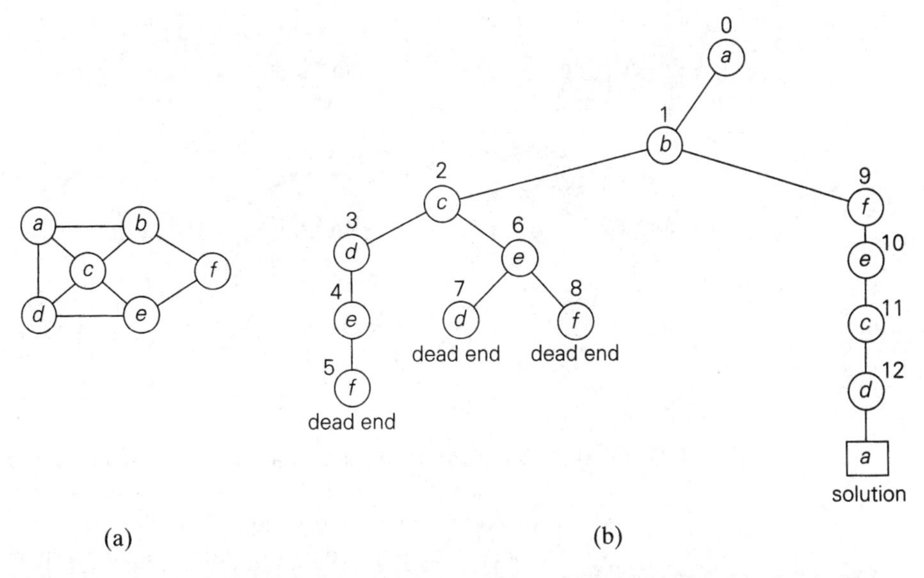

FIGURE 12.3 (a) Graph. (b) State-space tree for finding a Hamiltonian circuit. The numbers above the nodes of the tree indicate the order in which the nodes are generated.

tree (Figure 12.3b). The first component of our future solution, if it exists, is a first intermediate vertex of a Hamiltonian circuit to be constructed. Using the alphabet order to break the three-way tie among the vertices adjacent to a, we select vertex b. From b, the algorithm proceeds to c, then to d, then to e, and finally to f, which proves to be a dead end. So the algorithm backtracks from f to e, then to d, and then to c, which provides the first alternative for the algorithm to pursue. Going from c to e eventually proves useless, and the algorithm has to backtrack from e to c and then to b. From there, it goes to the vertices f, e, c, and d, from which it can legitimately return to a, yielding the Hamiltonian circuit a, b, f, e, c, d, a. If we wanted to find another Hamiltonian circuit, we could continue this process by backtracking from the leaf of the solution found.

Subset-Sum Problem

As our last example, we consider the ***subset-sum problem***: find a subset of a given set $A = \{a_1, \ldots, a_n\}$ of n positive integers whose sum is equal to a given positive integer d. For example, for $A = \{1, 2, 5, 6, 8\}$ and $d = 9$, there are two solutions: $\{1, 2, 6\}$ and $\{1, 8\}$. Of course, some instances of this problem may have no solutions.

It is convenient to sort the set's elements in increasing order. So, we will assume that

$$a_1 < a_2 < \cdots < a_n.$$

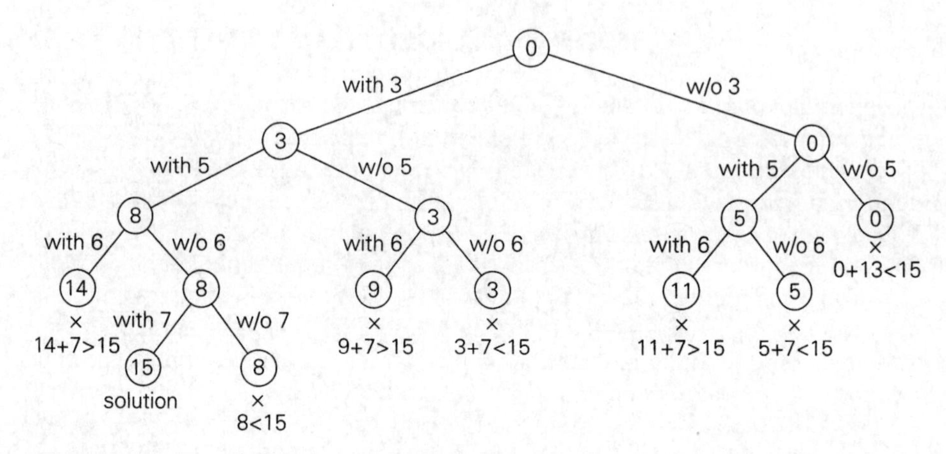

FIGURE 12.4 Complete state-space tree of the backtracking algorithm applied to the instance $A = \{3, 5, 6, 7\}$ and $d = 15$ of the subset-sum problem. The number inside a node is the sum of the elements already included in the subsets represented by the node. The inequality below a leaf indicates the reason for its termination.

The state-space tree can be constructed as a binary tree like that in Figure 12.4 for the instance $A = \{3, 5, 6, 7\}$ and $d = 15$. The root of the tree represents the starting point, with no decisions about the given elements made as yet. Its left and right children represent, respectively, inclusion and exclusion of a_1 in a set being sought. Similarly, going to the left from a node of the first level corresponds to inclusion of a_2 while going to the right corresponds to its exclusion, and so on. Thus, a path from the root to a node on the ith level of the tree indicates which of the first i numbers have been included in the subsets represented by that node.

We record the value of s, the sum of these numbers, in the node. If s is equal to d, we have a solution to the problem. We can either report this result and stop or, if all the solutions need to be found, continue by backtracking to the node's parent. If s is not equal to d, we can terminate the node as nonpromising if either of the following two inequalities holds:

$$s + a_{i+1} > d \quad \text{(the sum } s \text{ is too large)},$$

$$s + \sum_{j=i+1}^{n} a_j < d \quad \text{(the sum } s \text{ is too small)}.$$

General Remarks

From a more general perspective, most backtracking algorithms fit the following description. An output of a backtracking algorithm can be thought of as an n-tuple $(x_1, x_2, \ldots, x_n)$ where each coordinate x_i is an element of some finite linearly ordered set S_i. For example, for the n-queens problem, each S_i is the set of integers (column numbers) 1 through n. The tuple may need to satisfy some additional constraints (e.g., the nonattacking requirements in the n-queens problem). Depending on the problem, all solution tuples can be of the same length (the n-queens and the Hamiltonian circuit problem) and of different lengths (the subset-sum problem). A backtracking algorithm generates, explicitly or implicitly, a state-space tree; its nodes represent partially constructed tuples with the first i coordinates defined by the earlier actions of the algorithm. If such a tuple $(x_1, x_2, \ldots, x_i)$ is not a solution, the algorithm finds the next element in S_{i+1} that is consistent with the values of $(x_1, x_2, \ldots, x_i)$ and the problem's constraints, and adds it to the tuple as its $(i+1)$st coordinate. If such an element does not exist, the algorithm backtracks to consider the next value of x_i, and so on.

To start a backtracking algorithm, the following pseudocode can be called for $i = 0$; $X[1..0]$ represents the empty tuple.

```
ALGORITHM  Backtrack(X[1..i])
    //Gives a template of a generic backtracking algorithm
    //Input: X[1..i] specifies first i promising components of a solution
    //Output: All the tuples representing the problem's solutions
    if X[1..i] is a solution write X[1..i]
    else    //see Problem 9 in this section's exercises
        for each element x ∈ S_{i+1} consistent with X[1..i] and the constraints do
            X[i + 1] ← x
            Backtrack(X[1..i + 1])
```

Our success in solving small instances of three difficult problems earlier in this section should not lead you to the false conclusion that backtracking is a very efficient technique. In the worst case, it may have to generate all possible candidates in an exponentially (or faster) growing state space of the problem at hand. The hope, of course, is that a backtracking algorithm will be able to prune enough branches of its state-space tree before running out of time or memory or both. The success of this strategy is known to vary widely, not only from problem to problem but also from one instance to another of the same problem.

There are several tricks that might help reduce the size of a state-space tree. One is to exploit the symmetry often present in combinatorial problems. For example, the board of the n-queens problem has several symmetries so that some solutions can be obtained from others by reflection or rotation. This implies, in particular, that we need not consider placements of the first queen in the last $\lfloor n/2 \rfloor$ columns, because any solution with the first queen in square $(1, i)$, $\lceil n/2 \rceil \le i \le n$, can be obtained by reflection (which?) from a solution with the first queen in square $(1, n - i + 1)$. This observation cuts the size of the tree by about half. Another trick is to preassign values to one or more components of a solution, as we did in the Hamiltonian circuit example. Data presorting in the subset-sum example demonstrates potential benefits of yet another opportunity: rearrange data of an instance given.

It would be highly desirable to be able to estimate the size of the state-space tree of a backtracking algorithm. As a rule, this is too difficult to do analytically, however. Knuth [Knu75] suggested generating a random path from the root to a leaf and using the information about the number of choices available during the path generation for estimating the size of the tree. Specifically, let c_1 be the number of values of the first component x_1 that are consistent with the problem's constraints. We randomly select one of these values (with equal probability $1/c_1$) to move to one of the root's c_1 children. Repeating this operation for c_2 possible values for x_2 that are consistent with x_1 and the other constraints, we move to one of the c_2 children of that node. We continue this process until a leaf is reached after randomly selecting values for $x_1, x_2, \ldots, x_n$. By assuming that the nodes on level i have c_i children on average, we estimate the number of nodes in the tree as

$$1 + c_1 + c_1c_2 + \cdots + c_1c_2 \cdots c_n.$$

Generating several such estimates and computing their average yields a useful estimation of the actual size of the tree, although the standard deviation of this random variable can be large.

In conclusion, three things on behalf of backtracking need to be said. First, it is typically applied to difficult combinatorial problems for which no efficient algorithms for finding exact solutions possibly exist. Second, unlike the exhaustive-search approach, which is doomed to be extremely slow for all instances of a problem, backtracking at least holds a hope for solving some instances of nontrivial sizes in an acceptable amount of time. This is especially true for optimization problems, for which the idea of backtracking can be further enhanced by evaluating the quality of partially constructed solutions. How this can be done is explained in the next section. Third, even if backtracking does not eliminate any elements of a problem's state space and ends up generating all its elements, it provides a specific technique for doing so, which can be of value in its own right.

Exercises 12.1

1. **a.** Continue the backtracking search for a solution to the four-queens problem, which was started in this section, to find the second solution to the problem.

   b. Explain how the board's symmetry can be used to find the second solution to the four-queens problem.

2. **a.** Which is the *last* solution to the five-queens problem found by the backtracking algorithm?

   b. Use the board's symmetry to find at least four other solutions to the problem.

3. a. Implement the backtracking algorithm for the n-queens problem in the language of your choice. Run your program for a sample of n values to get the numbers of nodes in the algorithm's state-space trees. Compare these numbers with the numbers of candidate solutions generated by the exhaustive-search algorithm for this problem (see Problem 9 in Exercises 3.4).

   b. For each value of n for which you run your program in part (a), estimate the size of the state-space tree by the method described in Section 12.1 and compare the estimate with the actual number of nodes you obtained.

4. Design a linear-time algorithm that finds a solution to the n-queens problem for any $n \geq 4$.

5. Apply backtracking to the problem of finding a Hamiltonian circuit in the following graph.

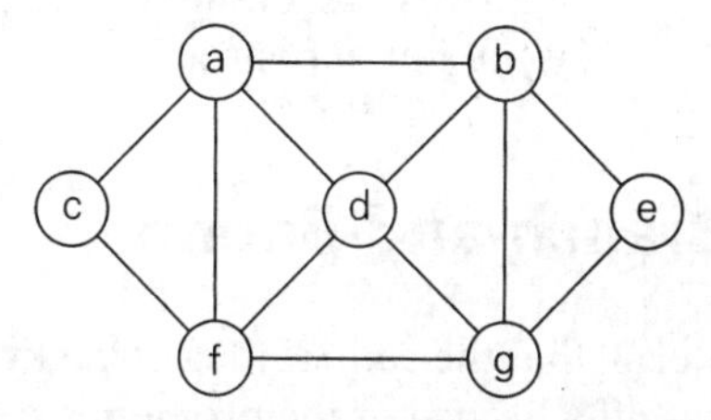

6. Apply backtracking to solve the 3-coloring problem for the graph in Figure 12.3a.

7. Generate all permutations of {1, 2, 3, 4} by backtracking.

8. a. Apply backtracking to solve the following instance of the subset sum problem: $A = \{1, 3, 4, 5\}$ and $d = 11$.

   b. Will the backtracking algorithm work correctly if we use just one of the two inequalities to terminate a node as nonpromising?

9. The general template for backtracking algorithms, which is given in the section, works correctly only if no solution is a prefix to another solution to the problem. Change the template's pseudocode to work correctly without this restriction.

10. Write a program implementing a backtracking algorithm for

    a. the Hamiltonian circuit problem.

    b. the m-coloring problem.

11. *Puzzle pegs* This puzzle-like game is played on a board with 15 small holes arranged in an equilateral triangle. In an initial position, all but one of the holes are occupied by pegs, as in the example shown below. A legal move is a jump of a peg over its immediate neighbor into an empty square opposite; the jump removes the jumped-over neighbor from the board.

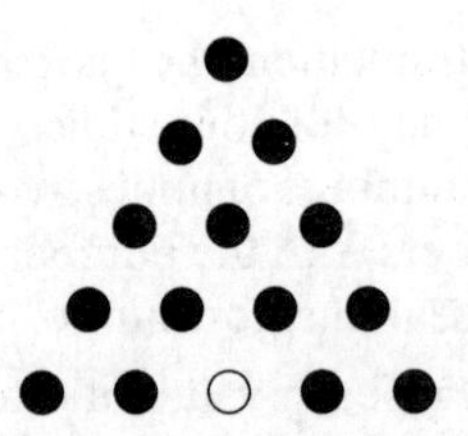

Design and implement a backtracking algorithm for solving the following versions of this puzzle.

a. Starting with a given location of the empty hole, find a shortest sequence of moves that eliminates 14 pegs with no limitations on the final position of the remaining peg.

b. Starting with a given location of the empty hole, find a shortest sequence of moves that eliminates 14 pegs with the remaining peg at the empty hole of the initial board.

12.2 Branch-and-Bound

Recall that the central idea of backtracking, discussed in the previous section, is to cut off a branch of the problem's state-space tree as soon as we can deduce that it cannot lead to a solution. This idea can be strengthened further if we deal with an optimization problem. An optimization problem seeks to minimize or maximize some objective function (a tour length, the value of items selected, the cost of an assignment, and the like), usually subject to some constraints. Note that in the standard terminology of optimization problems, a ***feasible solution*** is a point in the problem's search space that satisfies all the problem's constraints (e.g., a Hamiltonian circuit in the traveling salesman problem or a subset of items whose total weight does not exceed the knapsack's capacity in the knapsack problem), whereas an ***optimal solution*** is a feasible solution with the best value of the objective function (e.g., the shortest Hamiltonian circuit or the most valuable subset of items that fit the knapsack).

Compared to backtracking, branch-and-bound requires two additional items:

- a way to provide, for every node of a state-space tree, a bound on the best value of the objective function[1] on any solution that can be obtained by adding further components to the partially constructed solution represented by the node
- the value of the best solution seen so far

If this information is available, we can compare a node's bound value with the value of the best solution seen so far. If the bound value is not better than the value of the best solution seen so far—i.e., not smaller for a minimization problem

1. This bound should be a lower bound for a minimization problem and an upper bound for a maximization problem.

and not larger for a maximization problem—the node is nonpromising and can be terminated (some people say the branch is "pruned"). Indeed, no solution obtained from it can yield a better solution than the one already available. This is the principal idea of the branch-and-bound technique.

In general, we terminate a search path at the current node in a state-space tree of a branch-and-bound algorithm for any one of the following three reasons:

- The value of the node's bound is not better than the value of the best solution seen so far.
- The node represents no feasible solutions because the constraints of the problem are already violated.
- The subset of feasible solutions represented by the node consists of a single point (and hence no further choices can be made)—in this case, we compare the value of the objective function for this feasible solution with that of the best solution seen so far and update the latter with the former if the new solution is better.

Assignment Problem

Let us illustrate the branch-and-bound approach by applying it to the problem of assigning n people to n jobs so that the total cost of the assignment is as small as possible. We introduced this problem in Section 3.4, where we solved it by exhaustive search. Recall that an instance of the assignment problem is specified by an $n \times n$ cost matrix C so that we can state the problem as follows: select one element in each row of the matrix so that no two selected elements are in the same column and their sum is the smallest possible. We will demonstrate how this problem can be solved using the branch-and-bound technique by considering the same small instance of the problem that we investigated in Section 3.4:

$$
\begin{array}{cc}
 & \begin{array}{cccc} \text{job } 1 & \text{job } 2 & \text{job } 3 & \text{job } 4 \end{array} \\
C = & \begin{bmatrix} 9 & 2 & 7 & 8 \\ 6 & 4 & 3 & 7 \\ 5 & 8 & 1 & 8 \\ 7 & 6 & 9 & 4 \end{bmatrix}
\end{array}
\begin{array}{l} \\ \text{person } a \\ \text{person } b \\ \text{person } c \\ \text{person } d \end{array}
$$

How can we find a lower bound on the cost of an optimal selection without actually solving the problem? We can do this by several methods. For example, it is clear that the cost of any solution, including an optimal one, cannot be smaller than the sum of the smallest elements in each of the matrix's rows. For the instance here, this sum is $2 + 3 + 1 + 4 = 10$. It is important to stress that this is not the cost of any legitimate selection (3 and 1 came from the same column of the matrix); it is just a lower bound on the cost of any legitimate selection. We can and will apply the same thinking to partially constructed solutions. For example, for any legitimate selection that selects 9 from the first row, the lower bound will be $9 + 3 + 1 + 4 = 17$.

One more comment is in order before we embark on constructing the problem's state-space tree. It deals with the order in which the tree nodes will be generated. Rather than generating a single child of the last promising node as we did in backtracking, we will generate all the children of the most promising node among nonterminated leaves in the current tree. (Nonterminated, i.e., still promising, leaves are also called ***live***.) How can we tell which of the nodes is most promising? We can do this by comparing the lower bounds of the live nodes. It is sensible to consider a node with the best bound as most promising, although this does not, of course, preclude the possibility that an optimal solution will ultimately belong to a different branch of the state-space tree. This variation of the strategy is called the ***best-first branch-and-bound***.

So, returning to the instance of the assignment problem given earlier, we start with the root that corresponds to no elements selected from the cost matrix. As we already discussed, the lower-bound value for the root, denoted lb, is 10. The nodes on the first level of the tree correspond to selections of an element in the first row of the matrix, i.e., a job for person a (Figure 12.5).

So we have four live leaves—nodes 1 through 4—that may contain an optimal solution. The most promising of them is node 2 because it has the smallest lower-bound value. Following our best-first search strategy, we branch out from that node first by considering the three different ways of selecting an element from the second row and not in the second column—the three different jobs that can be assigned to person b (Figure 12.6).

Of the six live leaves—nodes 1, 3, 4, 5, 6, and 7—that may contain an optimal solution, we again choose the one with the smallest lower bound, node 5. First, we consider selecting the third column's element from c's row (i.e., assigning person c to job 3); this leaves us with no choice but to select the element from the fourth column of d's row (assigning person d to job 4). This yields leaf 8 (Figure 12.7), which corresponds to the feasible solution $\{a \to 2, b \to 1, c \to 3, d \to 4\}$ with the total cost of 13. Its sibling, node 9, corresponds to the feasible solution $\{a \to 2, b \to 1, c \to 4, d \to 3\}$ with the total cost of 25. Since its cost is larger than the cost of the solution represented by leaf 8, node 9 is simply terminated. (Of course, if

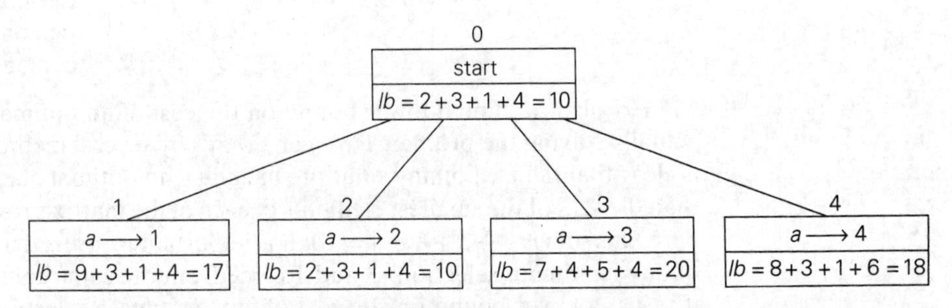

FIGURE 12.5 Levels 0 and 1 of the state-space tree for the instance of the assignment problem being solved with the best-first branch-and-bound algorithm. The number above a node shows the order in which the node was generated. A node's fields indicate the job number assigned to person a and the lower bound value, lb, for this node.

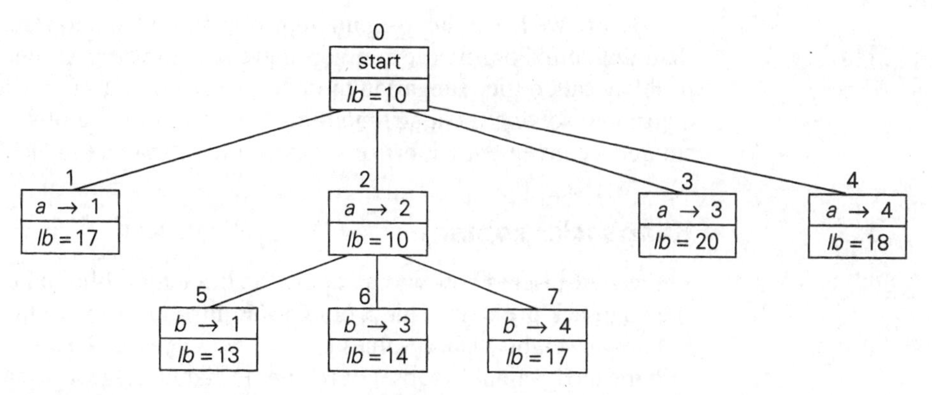

FIGURE 12.6 Levels 0, 1, and 2 of the state-space tree for the instance of the assignment problem being solved with the best-first branch-and-bound algorithm.

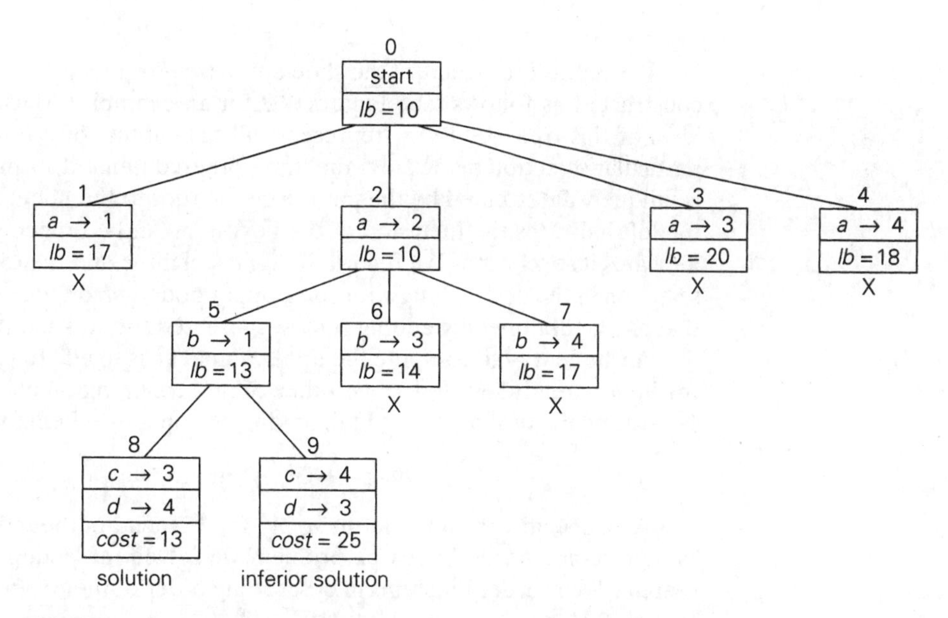

FIGURE 12.7 Complete state-space tree for the instance of the assignment problem solved with the best-first branch-and-bound algorithm.

its cost were smaller than 13, we would have to replace the information about the best solution seen so far with the data provided by this node.)

Now, as we inspect each of the live leaves of the last state-space tree—nodes 1, 3, 4, 6, and 7 in Figure 12.7—we discover that their lower-bound values are not smaller than 13, the value of the best selection seen so far (leaf 8). Hence, we terminate all of them and recognize the solution represented by leaf 8 as the optimal solution to the problem.

Before we leave the assignment problem, we have to remind ourselves again that, unlike for our next examples, there is a polynomial-time algorithm for this problem called the Hungarian method (e.g., [Pap82]). In the light of this efficient algorithm, solving the assignment problem by branch-and-bound should be considered a convenient educational device rather than a practical recommendation.

Knapsack Problem

Let us now discuss how we can apply the branch-and-bound technique to solving the knapsack problem. This problem was introduced in Section 3.4: given n items of known weights w_i and values v_i, $i = 1, 2, \ldots, n$, and a knapsack of capacity W, find the most valuable subset of the items that fit in the knapsack. It is convenient to order the items of a given instance in descending order by their value-to-weight ratios. Then the first item gives the best payoff per weight unit and the last one gives the worst payoff per weight unit, with ties resolved arbitrarily:

$$v_1/w_1 \geq v_2/w_2 \geq \cdots \geq v_n/w_n.$$

It is natural to structure the state-space tree for this problem as a binary tree constructed as follows (see Figure 12.8 for an example). Each node on the ith level of this tree, $0 \leq i \leq n$, represents all the subsets of n items that include a particular selection made from the first i ordered items. This particular selection is uniquely determined by the path from the root to the node: a branch going to the left indicates the inclusion of the next item, and a branch going to the right indicates its exclusion. We record the total weight w and the total value v of this selection in the node, along with some upper bound ub on the value of any subset that can be obtained by adding zero or more items to this selection.

A simple way to compute the upper bound ub is to add to v, the total value of the items already selected, the product of the remaining capacity of the knapsack $W - w$ and the best per unit payoff among the remaining items, which is v_{i+1}/w_{i+1}:

$$ub = v + (W - w)(v_{i+1}/w_{i+1}). \qquad \textbf{(12.1)}$$

As a specific example, let us apply the branch-and-bound algorithm to the same instance of the knapsack problem we solved in Section 3.4 by exhaustive search. (We reorder the items in descending order of their value-to-weight ratios, though.)

item	weight	value	$\frac{\text{value}}{\text{weight}}$
1	4	\$40	10
2	7	\$42	6
3	5	\$25	5
4	3	\$12	4

The knapsack's capacity W is 10.

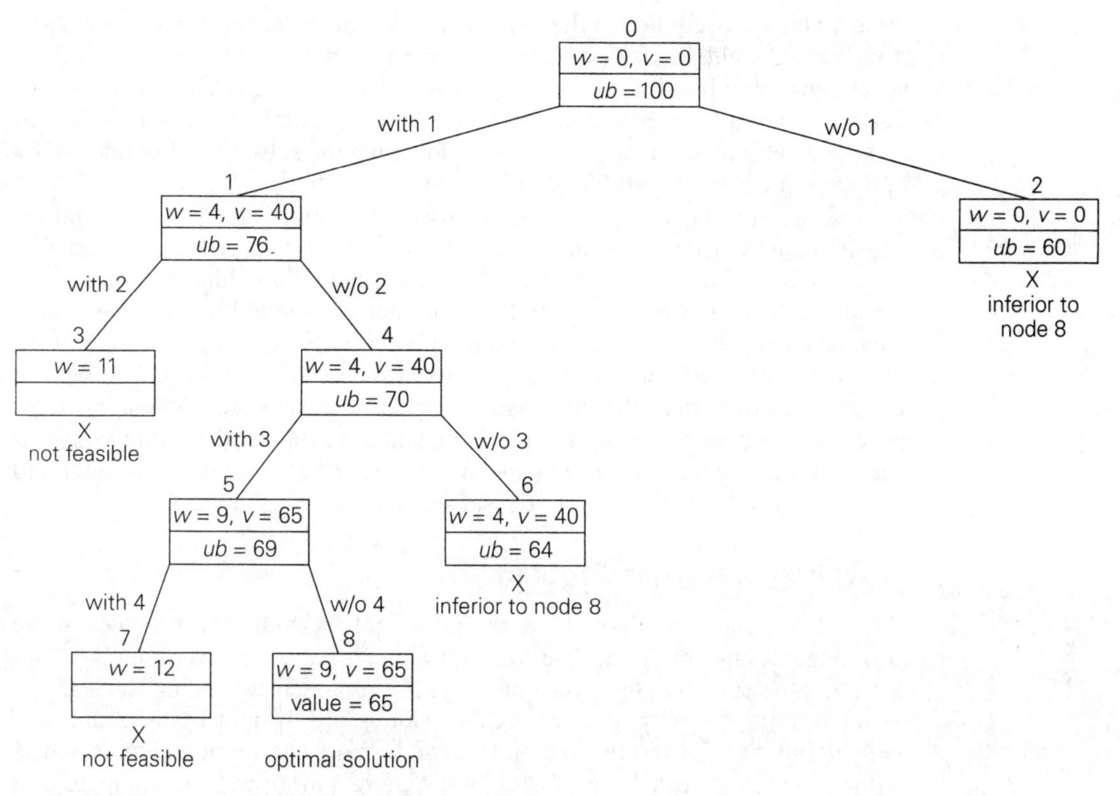

FIGURE 12.8 State-space tree of the best-first branch-and-bound algorithm for the instance of the knapsack problem.

At the root of the state-space tree (see Figure 12.8), no items have been selected as yet. Hence, both the total weight of the items already selected w and their total value v are equal to 0. The value of the upper bound computed by formula (12.1) is \$100. Node 1, the left child of the root, represents the subsets that include item 1. The total weight and value of the items already included are 4 and \$40, respectively; the value of the upper bound is $40 + (10 - 4) * 6 = \$76$. Node 2 represents the subsets that do not include item 1. Accordingly, $w = 0$, $v = \$0$, and $ub = 0 + (10 - 0) * 6 = \$60$. Since node 1 has a larger upper bound than the upper bound of node 2, it is more promising for this maximization problem, and we branch from node 1 first. Its children—nodes 3 and 4—represent subsets with item 1 and with and without item 2, respectively. Since the total weight w of every subset represented by node 3 exceeds the knapsack's capacity, node 3 can be terminated immediately. Node 4 has the same values of w and v as its parent; the upper bound ub is equal to $40 + (10 - 4) * 5 = \$70$. Selecting node 4 over node 2 for the next branching (why?), we get nodes 5 and 6 by respectively including and excluding item 3. The total weights and values as well as the upper bounds for these nodes are computed in the same way as for the preceding nodes. Branching from node 5 yields node 7, which represents no feasible solutions, and node 8, which represents just a single subset {1, 3} of value $65. The remaining live nodes 2 and 6 have smaller upper-bound values than the value of the solution represented by node 8. Hence, both can be terminated making the subset {1, 3} of node 8 the optimal solution to the problem.

Solving the knapsack problem by a branch-and-bound algorithm has a rather unusual characteristic. Typically, internal nodes of a state-space tree do not define a point of the problem's search space, because some of the solution's components remain undefined. (See, for example, the branch-and-bound tree for the assignment problem discussed in the preceding subsection.) For the knapsack problem, however, every node of the tree represents a subset of the items given. We can use this fact to update the information about the best subset seen so far after generating each new node in the tree. If we had done this for the instance investigated above, we could have terminated nodes 2 and 6 before node 8 was generated because they both are inferior to the subset of value $65 of node 5.

Traveling Salesman Problem

We will be able to apply the branch-and-bound technique to instances of the traveling salesman problem if we come up with a reasonable lower bound on tour lengths. One very simple lower bound can be obtained by finding the smallest element in the intercity distance matrix D and multiplying it by the number of cities n. But there is a less obvious and more informative lower bound for instances with symmetric matrix D, which does not require a lot of work to compute. It is not difficult to show (Problem 8 in this section's exercises) that we can compute a lower bound on the length l of any tour as follows. For each city i, $1 \le i \le n$, find the sum s_i of the distances from city i to the two nearest cities; compute the sum s of these n numbers, divide the result by 2, and, if all the distances are integers, round up the result to the nearest integer:

$$lb = \lceil s/2 \rceil. \tag{12.2}$$

For example, for the instance in Figure 12.9a, formula (12.2) yields

$$lb = \lceil [(1+3) + (3+6) + (1+2) + (3+4) + (2+3)]/2 \rceil = 14.$$

Moreover, for any subset of tours that must include particular edges of a given graph, we can modify lower bound (12.2) accordingly. For example, for all the Hamiltonian circuits of the graph in Figure 12.9a that must include edge (a, d), we get the following lower bound by summing up the lengths of the two shortest edges incident with each of the vertices, with the required inclusion of edges (a, d) and (d, a):

$$\lceil [(1+5) + (3+6) + (1+2) + (3+5) + (2+3)]/2 \rceil = 16.$$

We now apply the branch-and-bound algorithm, with the bounding function given by formula (12.2), to find the shortest Hamiltonian circuit for the graph in

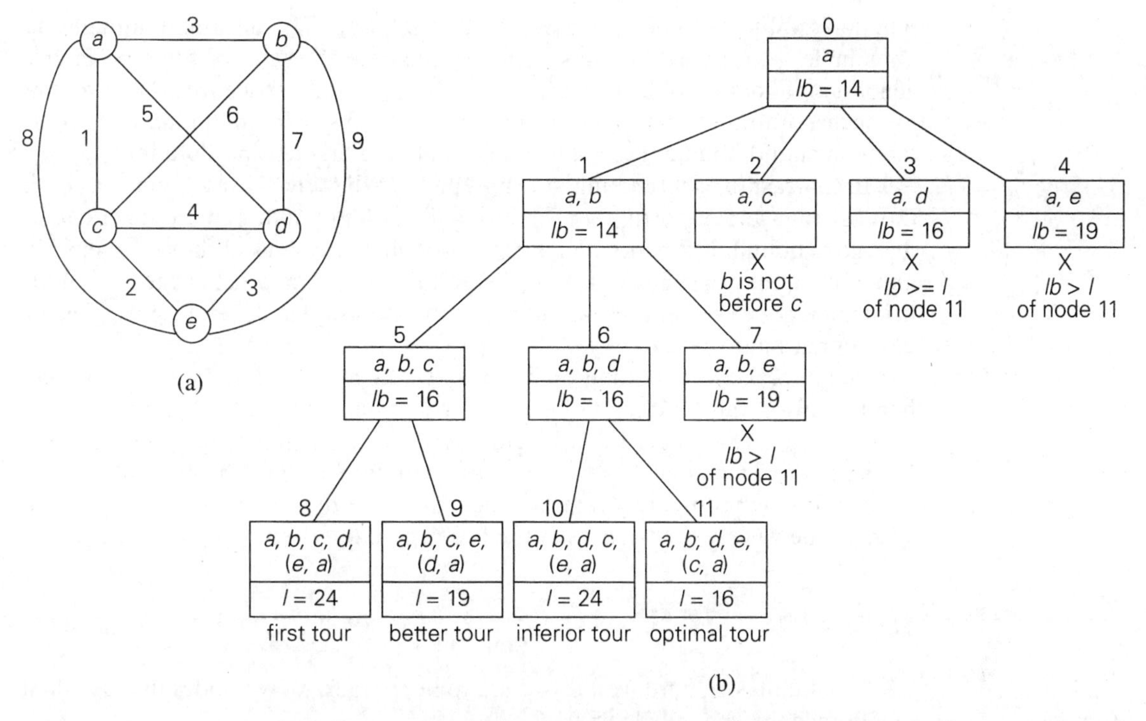

FIGURE 12.9 (a) Weighted graph. (b) State-space tree of the branch-and-bound algorithm to find a shortest Hamiltonian circuit in this graph. The list of vertices in a node specifies a beginning part of the Hamiltonian circuits represented by the node.

Figure 12.9a. To reduce the amount of potential work, we take advantage of two observations made in Section 3.4. First, without loss of generality, we can consider only tours that start at a. Second, because our graph is undirected, we can generate only tours in which b is visited before c. In addition, after visiting $n - 1 = 4$ cities, a tour has no choice but to visit the remaining unvisited city and return to the starting one. The state-space tree tracing the algorithm's application is given in Figure 12.9b.

The comments we made at the end of the preceding section about the strengths and weaknesses of backtracking are applicable to branch-and-bound as well. To reiterate the main point: these state-space tree techniques enable us to solve many large instances of difficult combinatorial problems. As a rule, however, it is virtually impossible to predict which instances will be solvable in a realistic amount of time and which will not.

Incorporation of additional information, such as a symmetry of a game's board, can widen the range of solvable instances. Along this line, a branch-and-bound algorithm can be sometimes accelerated by a knowledge of the objective function's value of some nontrivial feasible solution. The information might be obtainable—say, by exploiting specifics of the data or even, for some problems, generated randomly—before we start developing a state-space tree. Then we can use such a solution immediately as the best one seen so far rather than waiting for the branch-and-bound processing to lead us to the first feasible solution.

In contrast to backtracking, solving a problem by branch-and-bound has both the challenge and opportunity of choosing the order of node generation and finding a good bounding function. Though the best-first rule we used above is a sensible approach, it may or may not lead to a solution faster than other strategies. (Artificial intelligence researchers are particularly interested in different strategies for developing state-space trees.)

Finding a good bounding function is usually not a simple task. On the one hand, we want this function to be easy to compute. On the other hand, it cannot be too simplistic—otherwise, it would fail in its principal task to prune as many branches of a state-space tree as soon as possible. Striking a proper balance between these two competing requirements may require intensive experimentation with a wide variety of instances of the problem in question.

Exercises 12.2

1. What data structure would you use to keep track of live nodes in a best-first branch-and-bound algorithm?
2. Solve the same instance of the assignment problem as the one solved in the section by the best-first branch-and-bound algorithm with the bounding function based on matrix columns rather than rows.
3. **a.** Give an example of the best-case input for the branch-and-bound algorithm for the assignment problem.

   b. In the best case, how many nodes will be in the state-space tree of the branch-and-bound algorithm for the assignment problem?
4. Write a program for solving the assignment problem by the branch-and-bound algorithm. Experiment with your program to determine the average size of the cost matrices for which the problem is solved in a given amount of time, say, 1 minute on your computer.
5. Solve the following instance of the knapsack problem by the branch-and-bound algorithm:

item	weight	value	
1	10	\$100	
2	7	\$63	$W = 16$
3	8	\$56	
4	4	\$12	

6. a. Suggest a more sophisticated bounding function for solving the knapsack problem than the one used in the section.

b. Use your bounding function in the branch-and-bound algorithm applied to the instance of Problem 5.

7. Write a program to solve the knapsack problem with the branch-and-bound algorithm.

8. a. Prove the validity of the lower bound given by formula (12.2) for instances of the traveling salesman problem with symmetric matrices of integer intercity distances.

b. How would you modify lower bound (12.2) for nonsymmetric distance matrices?

9. Apply the branch-and-bound algorithm to solve the traveling salesman problem for the following graph:

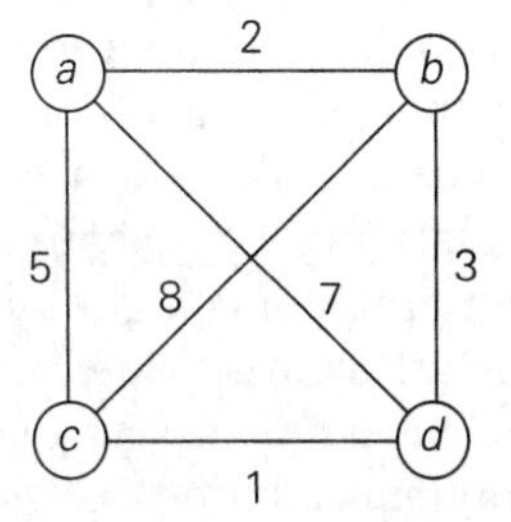

(We solved this problem by exhaustive search in Section 3.4.)

10. As a research project, write a report on how state-space trees are used for programming such games as chess, checkers, and tic-tac-toe. The two principal algorithms you should read about are the minimax algorithm and alpha-beta pruning.

12.3 Approximation Algorithms for *NP*-Hard Problems

In this section, we discuss a different approach to handling difficult problems of combinatorial optimization, such as the traveling salesman problem and the knapsack problem. As we pointed out in Section 11.3, the decision versions of these problems are *NP*-complete. Their optimization versions fall in the class of ***NP-hard problems***—problems that are at least as hard as *NP*-complete problems.[2] Hence, there are no known polynomial-time algorithms for these problems, and there are serious theoretical reasons to believe that such algorithms do not exist. What then are our options for handling such problems, many of which are of significant practical importance?

2. The notion of an *NP*-hard problem can be defined more formally by extending the notion of polynomial reducibility to problems that are not necessarily in class *NP*, including optimization problems of the type discussed in this section (see [Gar79, Chapter 5]).

If an instance of the problem in question is very small, we might be able to solve it by an exhaustive-search algorithm (Section 3.4). Some such problems can be solved by the dynamic programming technique we demonstrated in Section 8.2. But even when this approach works in principle, its practicality is limited by dependence on the instance parameters being relatively small. The discovery of the branch-and-bound technique has proved to be an important breakthrough, because this technique makes it possible to solve many large instances of difficult optimization problems in an acceptable amount of time. However, such good performance cannot usually be guaranteed.

There is a radically different way of dealing with difficult optimization problems: solve them approximately by a fast algorithm. This approach is particularly appealing for applications where a good but not necessarily optimal solution will suffice. Besides, in real-life applications, we often have to operate with inaccurate data to begin with. Under such circumstances, going for an approximate solution can be a particularly sensible choice.

Although approximation algorithms run a gamut in level of sophistication, most of them are based on some problem-specific heuristic. A ***heuristic*** is a common-sense rule drawn from experience rather than from a mathematically proved assertion. For example, going to the nearest unvisited city in the traveling salesman problem is a good illustration of this notion. We discuss an algorithm based on this heuristic later in this section.

Of course, if we use an algorithm whose output is just an approximation of the actual optimal solution, we would like to know how accurate this approximation is. We can quantify the accuracy of an approximate solution s_a to a problem of minimizing some function f by the size of the relative error of this approximation,

$$re(s_a) = \frac{f(s_a) - f(s^*)}{f(s^*)},$$

where s^* is an exact solution to the problem. Alternatively, since $re(s_a) = f(s_a)/f(s^*) - 1$, we can simply use the ***accuracy ratio***

$$r(s_a) = \frac{f(s_a)}{f(s^*)}$$

as a measure of accuracy of s_a. Note that for the sake of scale uniformity, the accuracy ratio of approximate solutions to maximization problems is usually computed as

$$r(s_a) = \frac{f(s^*)}{f(s_a)}$$

to make this ratio greater than or equal to 1, as it is for minimization problems.

Obviously, the closer $r(s_a)$ is to 1, the better the approximate solution is. For most instances, however, we cannot compute the accuracy ratio, because we typically do not know $f(s^*)$, the true optimal value of the objective function. Therefore, our hope should lie in obtaining a good upper bound on the values of $r(s_a)$. This leads to the following definitions.

DEFINITION A polynomial-time approximation algorithm is said to be a ***c*-approximation algorithm**, where $c \geq 1$, if the accuracy ratio of the approximation it produces does not exceed c for any instance of the problem in question:

$$r(s_a) \leq c. \tag{12.3}$$

The best (i.e., the smallest) value of c for which inequality (12.3) holds for all instances of the problem is called the ***performance ratio*** of the algorithm and denoted R_A.

The performance ratio serves as the principal metric indicating the quality of the approximation algorithm. We would like to have approximation algorithms with R_A as close to 1 as possible. Unfortunately, as we shall see, some approximation algorithms have infinitely large performance ratios ($R_A = \infty$). This does not necessarily rule out using such algorithms, but it does call for a cautious treatment of their outputs.

There are two important facts about difficult combinatorial optimization problems worth keeping in mind. First, although the difficulty level of solving most such problems exactly is the same to within a polynomial-time transformation of one problem to another, this equivalence does not translate into the realm of approximation algorithms. Finding good approximate solutions is much easier for some of these problems than for others. Second, some of the problems have special classes of instances that are both particularly important for real-life applications and easier to solve than their general counterparts. The traveling salesman problem is a prime example of this situation.

Approximation Algorithms for the Traveling Salesman Problem

We solved the traveling salesman problem by exhaustive search in Section 3.4, mentioned its decision version as one of the most well-known *NP*-complete problems in Section 11.3, and saw how its instances can be solved by a branch-and-bound algorithm in Section 12.2. Here, we consider several approximation algorithms, a small sample of dozens of such algorithms suggested over the years for this famous problem. (For a much more detailed discussion of the topic, see [Law85], [Hoc97], [App07], and [Gut07].)

But first let us answer the question of whether we should hope to find a polynomial-time approximation algorithm with a finite performance ratio on all instances of the traveling salesman problem. As the following theorem [Sah76] shows, the answer turns out to be no, unless $P = NP$.

THEOREM 1 If $P \neq NP$, there exists no c-approximation algorithm for the traveling salesman problem, i.e., there exists no polynomial-time approximation algorithm for this problem so that for all instances

$$f(s_a) \leq cf(s^*)$$

for some constant c.

PROOF By way of contradiction, suppose that such an approximation algorithm A and a constant c exist. (Without loss of generality, we can assume that c is a positive integer.) We will show that this algorithm could then be used for solving the Hamiltonian circuit problem in polynomial time. We will take advantage of a variation of the transformation used in Section 11.3 to reduce the Hamiltonian circuit problem to the traveling salesman problem. Let G be an arbitrary graph with n vertices. We map G to a complete weighted graph G' by assigning weight 1 to each edge in G and adding an edge of weight $cn + 1$ between each pair of vertices not adjacent in G. If G has a Hamiltonian circuit, its length in G' is n; hence, it is the exact solution s^* to the traveling salesman problem for G'. Note that if s_a is an approximate solution obtained for G' by algorithm A, then $f(s_a) \leq cn$ by the assumption. If G does not have a Hamiltonian circuit in G, the shortest tour in G' will contain at least one edge of weight $cn + 1$, and hence $f(s_a) \geq f(s^*) > cn$. Taking into account the two derived inequalities, we could solve the Hamiltonian circuit problem for graph G in polynomial time by mapping G to G', applying algorithm A to get tour s_a in G', and comparing its length with cn. Since the Hamiltonian circuit problem is NP-complete, we have a contradiction unless $P = NP$. ■

Greedy Algorithms for the TSP The simplest approximation algorithms for the traveling salesman problem are based on the greedy technique. We will discuss here two such algorithms.

Nearest-neighbor algorithm

The following well-known greedy algorithm is based on the ***nearest-neighbor*** heuristic: always go next to the nearest unvisited city.

Step 1 Choose an arbitrary city as the start.

Step 2 Repeat the following operation until all the cities have been visited: go to the unvisited city nearest the one visited last (ties can be broken arbitrarily).

Step 3 Return to the starting city.

EXAMPLE 1 For the instance represented by the graph in Figure 12.10, with a as the starting vertex, the nearest-neighbor algorithm yields the tour (Hamiltonian circuit) s_a: $a - b - c - d - a$ of length 10.

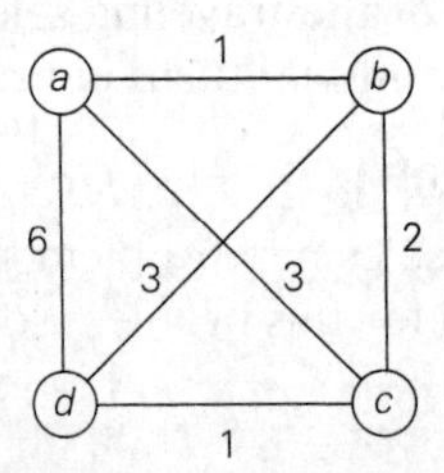

FIGURE 12.10 Instance of the traveling salesman problem.

The optimal solution, as can be easily checked by exhaustive search, is the tour s^*: $a - b - d - c - a$ of length 8. Thus, the accuracy ratio of this approximation is

$$r(s_a) = \frac{f(s_a)}{f(s^*)} = \frac{10}{8} = 1.25$$

(i.e., tour s_a is 25% longer than the optimal tour s^*). ■

Unfortunately, except for its simplicity, not many good things can be said about the nearest-neighbor algorithm. In particular, nothing can be said in general about the accuracy of solutions obtained by this algorithm because it can force us to traverse a very long edge on the last leg of the tour. Indeed, if we change the weight of edge (a, d) from 6 to an arbitrary large number $w \geq 6$ in Example 1, the algorithm will still yield the tour $a - b - c - d - a$ of length $4 + w$, and the optimal solution will still be $a - b - d - c - a$ of length 8. Hence,

$$r(s_a) = \frac{f(s_a)}{f(s^*)} = \frac{4 + w}{8},$$

which can be made as large as we wish by choosing an appropriately large value of w. Hence, $R_A = \infty$ for this algorithm (as it should be according to Theorem 1).

Multifragment-heuristic algorithm

Another natural greedy algorithm for the traveling salesman problem considers it as the problem of finding a minimum-weight collection of edges in a given complete weighted graph so that all the vertices have degree 2. (With this emphasis on edges rather than vertices, what other greedy algorithm does it remind you of?) An application of the greedy technique to this problem leads to the following algorithm [Ben90].

Step 1 Sort the edges in increasing order of their weights. (Ties can be broken arbitrarily.) Initialize the set of tour edges to be constructed to the empty set.

Step 2 Repeat this step n times, where n is the number of cities in the instance being solved: add the next edge on the sorted edge list to the set of tour edges, provided this addition does not create a vertex of degree 3 or a cycle of length less than n; otherwise, skip the edge.

Step 3 Return the set of tour edges.

As an example, applying the algorithm to the graph in Figure 12.10 yields $\{(a, b), (c, d), (b, c), (a, d)\}$. This set of edges forms the same tour as the one produced by the nearest-neighbor algorithm. In general, the multifragment-heuristic algorithm tends to produce significantly better tours than the nearest-neighbor algorithm, as we are going to see from the experimental data quoted at the end of this section. But the performance ratio of the multifragment-heuristic algorithm is also unbounded, of course.

There is, however, a very important subset of instances, called ***Euclidean***, for which we can make a nontrivial assertion about the accuracy of both the nearest-neighbor and multifragment-heuristic algorithms. These are the instances in which intercity distances satisfy the following natural conditions:

- *triangle inequality* $d[i, j] \leq d[i, k] + d[k, j]$ for any triple of cities i, j, and k (the distance between cities i and j cannot exceed the length of a two-leg path from i to some intermediate city k to j)
- *symmetry* $d[i, j] = d[j, i]$ for any pair of cities i and j (the distance from i to j is the same as the distance from j to i)

A substantial majority of practical applications of the traveling salesman problem are its Euclidean instances. They include, in particular, geometric ones, where cities correspond to points in the plane and distances are computed by the standard Euclidean formula. Although the performance ratios of the nearest-neighbor and multifragment-heuristic algorithms remain unbounded for Euclidean instances, their accuracy ratios satisfy the following inequality for any such instance with $n \geq 2$ cities:

$$\frac{f(s_a)}{f(s^*)} \leq \frac{1}{2}(\lceil \log_2 n \rceil + 1),$$

where $f(s_a)$ and $f(s^*)$ are the lengths of the heuristic tour and shortest tour, respectively (see [Ros77] and [Ong84]).

Minimum-Spanning-Tree–Based Algorithms There are approximation algorithms for the traveling salesman problem that exploit a connection between Hamiltonian circuits and spanning trees of the same graph. Since removing an edge from a Hamiltonian circuit yields a spanning tree, we can expect that the structure of a minimum spanning tree provides a good basis for constructing a shortest tour approximation. Here is an algorithm that implements this idea in a rather straightforward fashion.

Twice-around-the-tree algorithm

Step 1 Construct a minimum spanning tree of the graph corresponding to a given instance of the traveling salesman problem.

Step 2 Starting at an arbitrary vertex, perform a walk around the minimum spanning tree recording all the vertices passed by. (This can be done by a DFS traversal.)

Step 3 Scan the vertex list obtained in Step 2 and eliminate from it all repeated occurrences of the same vertex except the starting one at the end of the list. (This step is equivalent to making shortcuts in the walk.) The vertices remaining on the list will form a Hamiltonian circuit, which is the output of the algorithm.

EXAMPLE 2 Let us apply this algorithm to the graph in Figure 12.11a. The minimum spanning tree of this graph is made up of edges (a, b), (b, c), (b, d), and (d, e) (Figure 12.11b). A twice-around-the-tree walk that starts and ends at a is

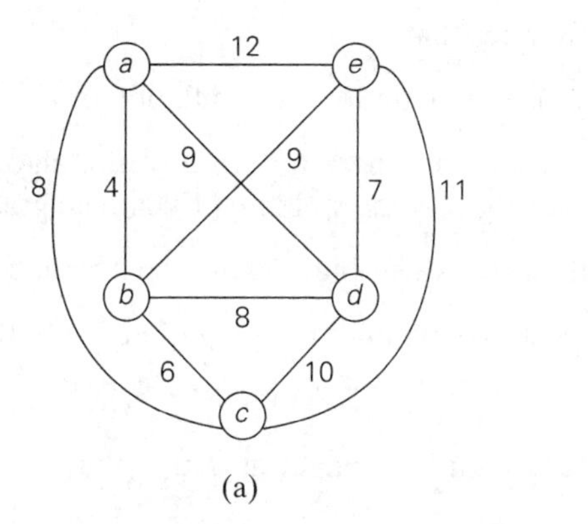

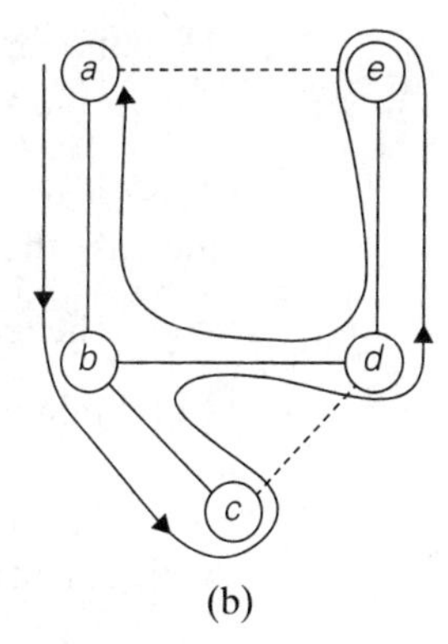

FIGURE 12.11 Illustration of the twice-around-the-tree algorithm. (a) Graph. (b) Walk around the minimum spanning tree with the shortcuts.

$$a, \ b, \ c, \ b, \ d, \ e, \ d, \ b, \ a.$$

Eliminating the second b (a shortcut from c to d), the second d, and the third b (a shortcut from e to a) yields the Hamiltonian circuit

$$a, \ b, \ c, \ d, \ e, \ a$$

of length 39. ■

The tour obtained in Example 2 is not optimal. Although that instance is small enough to find an optimal solution by either exhaustive search or branch-and-bound, we refrained from doing so to reiterate a general point. As a rule, we do not know what the length of an optimal tour actually is, and therefore we cannot compute the accuracy ratio $f(s_a)/f(s^*)$. For the twice-around-the-tree algorithm, we can at least estimate it above, provided the graph is Euclidean.

THEOREM 2 The twice-around-the-tree algorithm is a 2-approximation algorithm for the traveling salesman problem with Euclidean distances.

PROOF Obviously, the twice-around-the-tree algorithm is polynomial time if we use a reasonable algorithm such as Prim's or Kruskal's in Step 1. We need to show that for any Euclidean instance of the traveling salesman problem, the length of a tour s_a obtained by the twice-around-the-tree algorithm is at most twice the length of the optimal tour s^*, i.e.,

$$f(s_a) \leq 2f(s^*).$$

Since removing any edge from s^* yields a spanning tree T of weight $w(T)$, which must be greater than or equal to the weight of the graph's minimum spanning tree $w(T^*)$, we get the inequality

$$f(s^*) > w(T) \geq w(T^*).$$

This inequality implies that

$$2f(s^*) > 2w(T^*) = \text{the length of the walk obtained in Step 2 of the algorithm.}$$

The possible shortcuts outlined in Step 3 of the algorithm to obtain s_a cannot increase the total length of the walk in a Euclidean graph, i.e.,

$$\text{the length of the walk obtained in Step 2} \geq \text{the length of the tour } s_a.$$

Combining the last two inequalities, we get the inequality

$$2f(s^*) > f(s_a),$$

which is, in fact, a slightly stronger assertion than the one we needed to prove. ■

Christofides Algorithm There is an approximation algorithm with a better performance ratio for the Euclidean traveling salesman problem—the well-known ***Christofides algorithm*** [Chr76]. It also uses a minimum spanning tree but does this in a more sophisticated way than the twice-around-the-tree algorithm. Note that a twice-around-the-tree walk generated by the latter algorithm is an Eulerian circuit in the multigraph obtained by doubling every edge in the graph given. Recall that an Eulerian circuit exists in a connected multigraph if and only if all its vertices have even degrees. The Christofides algorithm obtains such a multigraph by adding to the graph the edges of a minimum-weight matching of all the odd-degree vertices in its minimum spanning tree. (The number of such vertices is always even and hence this can always be done.) Then the algorithm finds an Eulerian circuit in the multigraph and transforms it into a Hamiltonian circuit by shortcuts, exactly the same way it is done in the last step of the twice-around-the-tree algorithm.

EXAMPLE 3 Let us trace the Christofides algorithm in Figure 12.12 on the same instance (Figure 12.12a) used for tracing the twice-around-the-tree algorithm in Figure 12.11. The graph's minimum spanning tree is shown in Figure 12.12b. It has four odd-degree vertices: a, b, c, and e. The minimum-weight matching of these four vertices consists of edges (a, b) and (c, e). (For this tiny instance, it can be found easily by comparing the total weights of just three alternatives: (a, b) and (c, e), (a, c) and (b, e), (a, e) and (b, c).) The traversal of the multigraph, starting at vertex a, produces the Eulerian circuit $a - b - c - e - d - b - a$, which, after one shortcut, yields the tour $a - b - c - e - d - a$ of length 37. ■

The performance ratio of the Christofides algorithm on Euclidean instances is 1.5 (see, e.g., [Pap82]). It tends to produce significantly better approximations to optimal tours than the twice-around-the-tree algorithm does in empirical tests. (We quote some results of such tests at the end of this subsection.) The quality of a tour obtained by this heuristic can be further improved by optimizing shortcuts made on the last step of the algorithm as follows: examine the multiply-visited cities in some arbitrary order and for each make the best possible shortcut. This

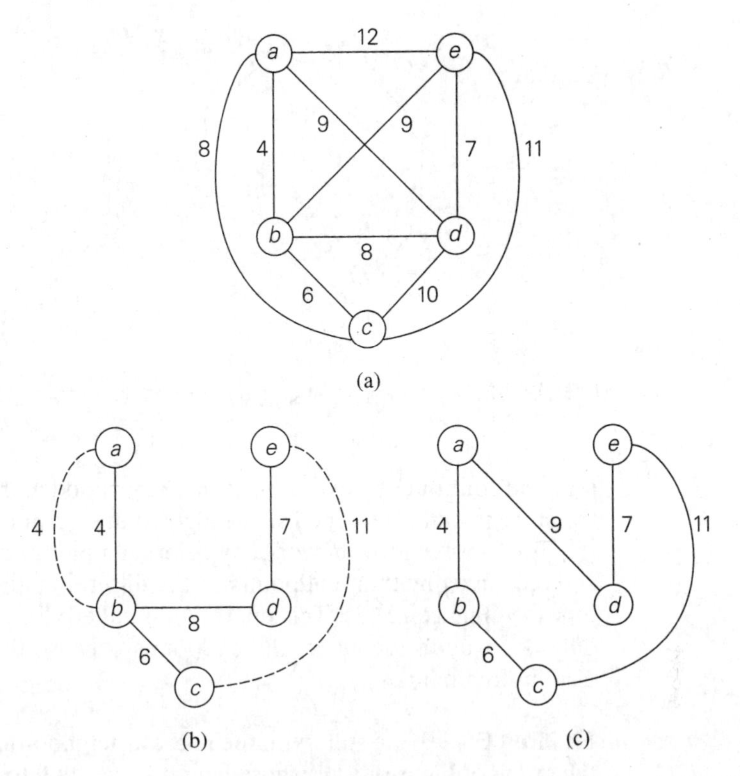

FIGURE 12.12 Application of the Christofides algorithm. (a) Graph. (b) Minimum spanning tree with added edges (in dash) of a minimum-weight matching of all odd-degree vertices. (c) Hamiltonian circuit obtained.

enhancement would have not improved the tour $a - b - c - e - d - a$ obtained in Example 3 from $a - b - c - e - d - b - a$ because shortcutting the second occurrence of b happens to be better than shortcutting its first occurrence. In general, however, this enhancement tends to decrease the gap between the heuristic and optimal tour lengths from about 15% to about 10%, at least for randomly generated Euclidean instances [Joh07a].

Local Search Heuristics For Euclidean instances, surprisingly good approximations to optimal tours can be obtained by iterative-improvement algorithms, which are also called ***local search*** heuristics. The best-known of these are the ***2-opt***, ***3-opt***, and ***Lin-Kernighan*** algorithms. These algorithms start with some initial tour, e.g., constructed randomly or by some simpler approximation algorithm such as the nearest-neighbor. On each iteration, the algorithm explores a neighborhood around the current tour by replacing a few edges in the current tour by other edges. If the changes produce a shorter tour, the algorithm makes it the current

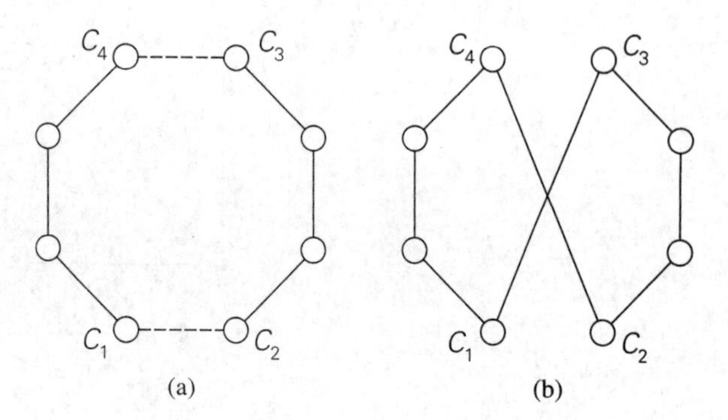

FIGURE 12.13 2-change: (a) Original tour. (b) New tour.

tour and continues by exploring its neighborhood in the same manner; otherwise, the current tour is returned as the algorithm's output and the algorithm stops.

The 2-opt algorithm works by deleting a pair of nonadjacent edges in a tour and reconnecting their endpoints by the different pair of edges to obtain another tour (see Figure 12.13). This operation is called the ***2-change***. Note that there is only one way to reconnect the endpoints because the alternative produces two disjoint fragments.

EXAMPLE 4 If we start with the nearest-neighbor tour $a - b - c - d - e - a$ in the graph of Figure 12.11, whose length l_{nn} is equal to 39, the 2-opt algorithm will move to the next tour as shown in Figure 12.14. ■

To generalize the notion of the 2-change, one can consider the ***k-change*** for any $k \geq 2$. This operation replaces up to k edges in a current tour. In addition to 2-changes, only the 3-changes have proved to be of practical interest. The two principal possibilities of 3-changes are shown in Figure 12.15.

There are several other local search algorithms for the traveling salesman problem. The most prominent of them is the ***Lin-Kernighan*** algorithm [Lin73], which for two decades after its publication in 1973 was considered the best algorithm to obtain high-quality approximations of optimal tours. The Lin-Kernighan algorithm is a variable-opt algorithm: its move can be viewed as a 3-opt move followed by a sequence of 2-opt moves. Because of its complexity, we have to refrain from discussing this algorithm here. The excellent survey by Johnson and McGeoch [Joh07a] contains an outline of the algorithm and its modern extensions as well as methods for its efficient implementation. This survey also contain results from the important empirical studies about performance of many heuristics for the traveling salesman problem, including of course, the Lin-Kernighan algorithm. We conclude our discussion by quoting some of these data.

Empirical Results The traveling salesman problem has been the subject of intense study for the last 50 years. This interest was driven by a combination of pure

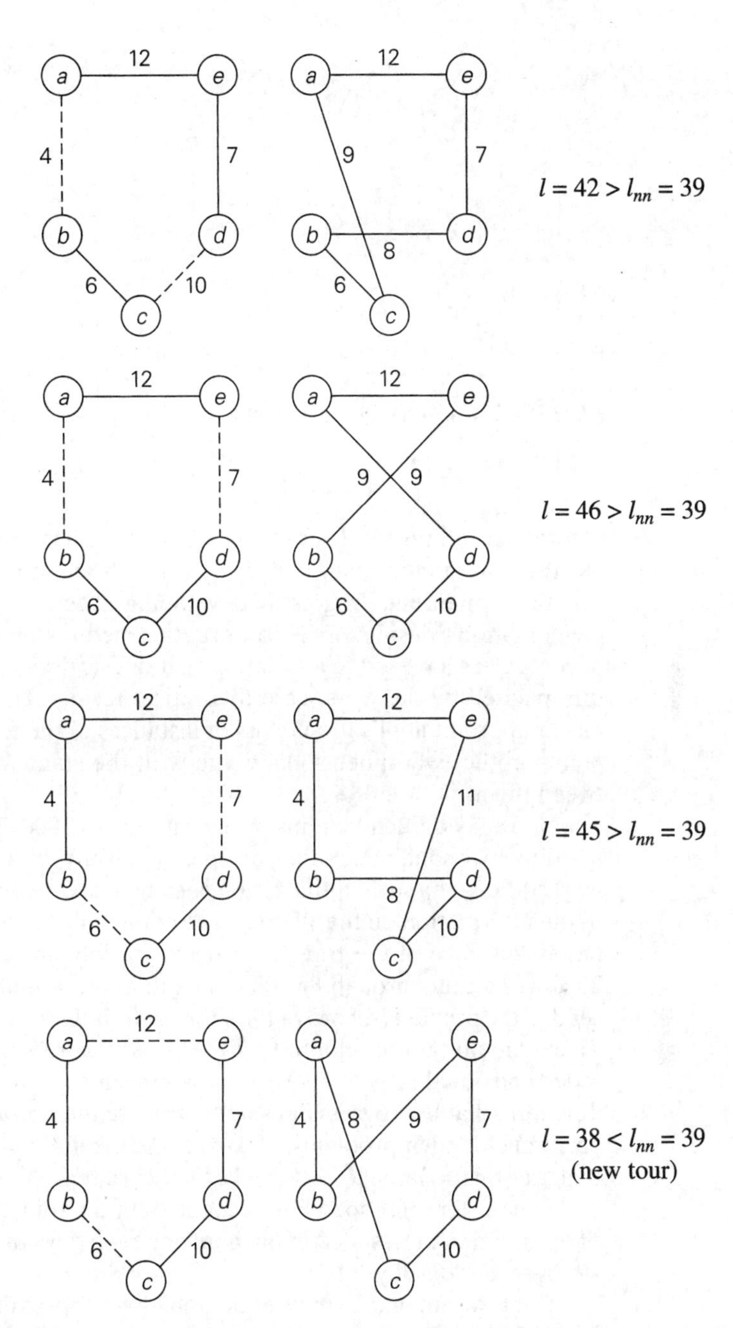

FIGURE 12.14 2-changes from the nearest-neighbor tour of the graph in Figure 12.11.

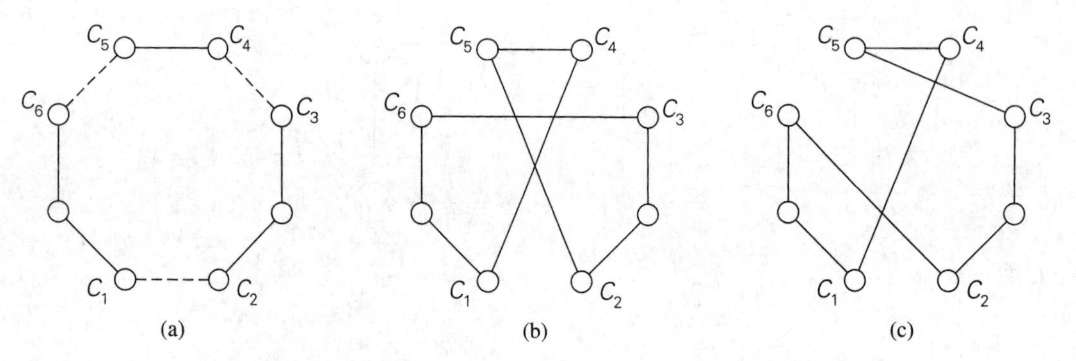

FIGURE 12.15 3-change: (a) Original tour. (b), (c) New tours.

theoretical interest and serious practical needs stemming from such newer applications as circuit-board and VLSI-chip fabrication, X-ray crystallography, and genetic engineering. Progress in developing effective heuristics, their efficient implementation by using sophisticated data structures, and the ever-increasing power of computers have led to a situation that differs drastically from a pessimistic picture painted by the worst-case theoretical results. This is especially true for the most important applications class of instances of the traveling salesman problem: points in the two-dimensional plane with the standard Euclidean distances between them.

Nowadays, Euclidean instances with up to 1000 cities can be solved exactly in quite a reasonable amount of time—typically, in minutes or faster on a good workstation—by such optimization packages as *Concord* [App]. In fact, according to the information on the Web site maintained by the authors of that package, the largest instance of the traveling salesman problem solved exactly as of January 2010 was a tour through 85,900 points in a VLSI application. It significantly exceeded the previous record of the shortest tour through all 24,978 cities in Sweden. There should be little doubt that the latest record will also be eventually superseded and our ability to solve ever larger instances exactly will continue to expand. This remarkable progress does not eliminate the usefulness of approximation algorithms for such problems, however. First, some applications lead to instances that are still too large to be solved exactly in a reasonable amount of time. Second, one may well prefer spending seconds to find a tour that is within a few percent of optimum than to spend many hours or even days of computing time to find the shortest tour exactly.

But how can one tell how good or bad the approximate solution is if we do not know the length of an optimal tour? A convenient way to overcome this difficulty is to solve the linear programming problem describing the instance in question by ignoring the integrality constraints. This provides a lower bound—called the ***Held-Karp bound***—on the length of the shortest tour. The Held-Karp bound is typically very close (less than 1%) to the length of an optimal tour, and this bound can be computed in seconds or minutes unless the instance is truly huge. Thus, for a tour

TABLE 12.1 Average tour quality and running times for various heuristics on the 10,000-city random uniform Euclidean instances [Joh07a]

Heuristic	% excess over the Held-Karp bound	Running time (seconds)
nearest neighbor	24.79	0.28
multifragment	16.42	0.20
Christofides	9.81	1.04
2-opt	4.70	1.41
3-opt	2.88	1.50
Lin-Kernighan	2.00	2.06

s_a obtained by some heuristic, we estimate the accuracy ratio $r(s_a) = f(s_a)/f(s^*)$ from *above* by the ratio $f(s_a)/HK(s^*)$, where $f(s_a)$ is the length of the heuristic tour s_a and $HK(s^*)$ is the Held-Karp lower bound on the shortest-tour length.

The results (see Table 12.1) from a large empirical study [Joh07a] indicate the average tour quality and running times for the discussed heuristics.[3] The instances in the reported sample have 10,000 cities generated randomly and uniformly as integral-coordinate points in the plane, with the Euclidean distances rounded to the nearest integer. The quality of tours generated by the heuristics remain about the same for much larger instances (up to a million cities) as long as they belong to the same type of instances. The running times quoted are for expert implementations run on a Compaq ES40 with 500 Mhz Alpha processors and 2 gigabytes of main memory or its equivalents.

Asymmetric instances of the traveling salesman problem—i.e., those with a nonsymmetic matrix of intercity distances—have proved to be significantly harder to solve, both exactly and approximately, than Euclidean instances. In particular, exact optimal solutions for many 316-city asymmetric instances remained unknown at the time of the state-of-the-art survey by Johnson et al. [Joh07b].

Approximation Algorithms for the Knapsack Problem

The knapsack problem, another well-known *NP*-hard problem, was also introduced in Section 3.4: given n items of known weights $w_1, \ldots, w_n$ and values $v_1, \ldots, v_n$ and a knapsack of weight capacity W, find the most valuable subset of the items that fits into the knapsack. We saw how this problem can be solved by exhaustive search (Section 3.4), dynamic programming (Section 8.2),

3. We did not include the results for the twice-around-the-tree heuristic because of the inferior quality of its approximations with the average excess of about 40%. Nor did we quote the results for the most sophisticated local search heuristics with the average excess over optimum of less than a fraction of 1%.

and branch-and-bound (Section 12.2). Now we will solve this problem by approximation algorithms.

Greedy Algorithms for the Knapsack Problem We can think of several greedy approaches to this problem. One is to select the items in decreasing order of their weights; however, heavier items may not be the most valuable in the set. Alternatively, if we pick up the items in decreasing order of their value, there is no guarantee that the knapsack's capacity will be used efficiently. Can we find a greedy strategy that takes into account both the weights and values? Yes, we can, by computing the value-to-weight ratios v_i/w_i, $i = 1, 2, \dots, n$, and selecting the items in decreasing order of these ratios. (In fact, we already used this approach in designing the branch-and-bound algorithm for the problem in Section 12.2.) Here is the algorithm based on this greedy heuristic.

Greedy algorithm for the discrete knapsack problem

Step 1 Compute the value-to-weight ratios $r_i = v_i/w_i$, $i = 1, \dots, n$, for the items given.

Step 2 Sort the items in nonincreasing order of the ratios computed in Step 1. (Ties can be broken arbitrarily.)

Step 3 Repeat the following operation until no item is left in the sorted list: if the current item on the list fits into the knapsack, place it in the knapsack and proceed to the next item; otherwise, just proceed to the next item.

EXAMPLE 5 Let us consider the instance of the knapsack problem with the knapsack capacity 10 and the item information as follows:

item	weight	value
1	7	\$42
2	3	\$12
3	4	\$40
4	5	\$25

Computing the value-to-weight ratios and sorting the items in nonincreasing order of these efficiency ratios yields

item	weight	value	value/weight
1	4	\$40	10
2	7	\$42	6
3	5	\$25	5
4	3	\$12	4

The greedy algorithm will select the first item of weight 4, skip the next item of weight 7, select the next item of weight 5, and skip the last item of weight 3. The solution obtained happens to be optimal for this instance (see Section 12.2, where we solved the same instance by the branch-and-bound algorithm). ■

Does this greedy algorithm always yield an optimal solution? The answer, of course, is no: if it did, we would have a polynomial-time algorithm for the *NP*-hard problem. In fact, the following example shows that no finite upper bound on the accuracy of its approximate solutions can be given either.

EXAMPLE 6

item	weight	value	value/weight
1	1	2	2
2	W	W	1

The knapsack capacity is $W > 2$.

Since the items are already ordered as required, the algorithm takes the first item and skips the second one; the value of this subset is 2. The optimal selection consists of item 2 whose value is W. Hence, the accuracy ratio $r(s_a)$ of this approximate solution is $W/2$, which is unbounded above. ■

It is surprisingly easy to tweak this greedy algorithm to get an approximation algorithm with a finite performance ratio. All it takes is to choose the better of two alternatives: the one obtained by the greedy algorithm or the one consisting of a single item of the largest value that fits into the knapsack. (Note that for the instance of the preceding example, the second alternative is better than the first one.) It is not difficult to prove that the performance ratio of this ***enhanced greedy algorithm*** is 2. That is, the value of an optimal subset s^* will never be more than twice as large as the value of the subset s_a obtained by this enhanced greedy algorithm, and 2 is the smallest multiple for which such an assertion can be made.

It is instructive to consider the continuous version of the knapsack problem as well. In this version, we are permitted to take arbitrary fractions of the items given. For this version of the problem, it is natural to modify the greedy algorithm as follows.

Greedy algorithm for the continuous knapsack problem

Step 1 Compute the value-to-weight ratios v_i/w_i, $i = 1, \ldots, n$, for the items given.

Step 2 Sort the items in nonincreasing order of the ratios computed in Step 1. (Ties can be broken arbitrarily.)

Step 3 Repeat the following operation until the knapsack is filled to its full capacity or no item is left in the sorted list: if the current item on the list fits into the knapsack in its entirety, take it and proceed to the next item; otherwise, take its largest fraction to fill the knapsack to its full capacity and stop.

For example, for the four-item instance used in Example 5 to illustrate the greedy algorithm for the discrete version, the algorithm will take the first item of weight 4 and then 6/7 of the next item on the sorted list to fill the knapsack to its full capacity.

It should come as no surprise that this algorithm always yields an optimal solution to the continuous knapsack problem. Indeed, the items are ordered according to their efficiency in using the knapsack's capacity. If the first item on the sorted list has weight w_1 and value v_1, no solution can use w_1 units of capacity with a higher payoff than v_1. If we cannot fill the knapsack with the first item or its fraction, we should continue by taking as much as we can of the second-most efficient item, and so on. A formal rendering of this proof idea is somewhat involved, and we will leave it for the exercises.

Note also that the optimal value of the solution to an instance of the continuous knapsack problem can serve as an upper bound on the optimal value of the discrete version of the same instance. This observation provides a more sophisticated way of computing upper bounds for solving the discrete knapsack problem by the branch-and-bound method than the one used in Section 12.2.

Approximation Schemes We now return to the discrete version of the knapsack problem. For this problem, unlike the traveling salesman problem, there exist polynomial-time ***approximation schemes***, which are parametric families of algorithms that allow us to get approximations $s_a^{(k)}$ with any predefined accuracy level:

$$\frac{f(s^*)}{f(s_a^{(k)})} \le 1 + 1/k \quad \text{for any instance of size } n,$$

where k is an integer parameter in the range $0 \le k < n$. The first approximation scheme was suggested by S. Sahni in 1975 [Sah75]. This algorithm generates all subsets of k items or less, and for each one that fits into the knapsack it adds the remaining items as the greedy algorithm would do (i.e., in nonincreasing order of their value-to-weight ratios). The subset of the highest value obtained in this fashion is returned as the algorithm's output.

EXAMPLE 7 A small example of an approximation scheme with $k = 2$ is provided in Figure 12.16. The algorithm yields {1, 3, 4}, which is the optimal solution for this instance. ■

You can be excused for not being overly impressed by this example. And, indeed, the importance of this scheme is mostly theoretical rather than practical. It lies in the fact that, in addition to approximating the optimal solution with any predefined accuracy level, the time efficiency of this algorithm is polynomial in n Indeed, the total number of subsets the algorithm generates before adding extra elements is

$$\sum_{j=0}^{k} \binom{n}{j} = \sum_{j=0}^{k} \frac{n(n-1)\cdots(n-j+1)}{j!} \le \sum_{j=0}^{k} n^j \le \sum_{j=0}^{k} n^k = (k+1)n^k.$$

item	weight	value	value/weight
1	4	$40	10
2	7	$42	6
3	5	$25	5
4	1	$ 4	4

capacity $W = 10$

(a)

subset	added items	value
∅	1, 3, 4	$69
{1}	3, 4	$69
{2}	4	$46
{3}	1, 4	$69
{4}	1, 3	$69
{1, 2}	not feasible	
{1, 3}	4	$69
{1, 4}	3	$69
{2, 3}	not feasible	
{2, 4}		$46
{3, 4}	1	$69

(b)

FIGURE 12.16 Example of applying Sahni's approximation scheme for $k = 2$. (a) Instance. (b) Subsets generated by the algorithm.

For each of those subsets, it needs $O(n)$ time to determine the subset's possible extension. Thus, the algorithm's efficiency is in $O(kn^{k+1})$. Note that although it is polynomial in n, the time efficiency of Sahni's scheme is exponential in k. More sophisticated approximation schemes, called ***fully polynomial schemes***, do not have this shortcoming. Among several books that discuss such algorithms, the monographs [Mar90] and [Kel04] are especially recommended for their wealth of other material about the knapsack problem.

Exercises 12.3

1. **a.** Apply the nearest-neighbor algorithm to the instance defined by the intercity distance matrix below. Start the algorithm at the first city, assuming that the cities are numbered from 1 to 5.

$$\begin{bmatrix} 0 & 14 & 4 & 10 & \infty \\ 14 & 0 & 5 & 8 & 7 \\ 4 & 5 & 0 & 9 & 16 \\ 10 & 8 & 9 & 0 & 32 \\ \infty & 7 & 16 & 32 & 0 \end{bmatrix}$$

   b. Compute the accuracy ratio of this approximate solution.

2. **a.** Write pseudocode for the nearest-neighbor algorithm. Assume that its input is given by an $n \times n$ intercity distance matrix.

   b. What is the time efficiency of the nearest-neighbor algorithm?

3. Apply the twice-around-the-tree algorithm to the graph in Figure 12.11a with a walk around the minimum spanning tree that starts at the same vertex a but differs from the walk in Figure 12.11b. Is the length of the obtained tour the same as the length of the tour in Figure 12.11b?

4. Prove that making a shortcut of the kind used by the twice-around-the-tree algorithm cannot increase the tour's length in a Euclidean graph.

5. What is the time efficiency class of the greedy algorithm for the knapsack problem?

6. Prove that the performance ratio R_A of the enhanced greedy algorithm for the knapsack problem is equal to 2.

7. Consider the greedy algorithm for the bin-packing problem, which is called the ***first-fit*** (*FF*) ***algorithm***: place each of the items in the order given into the first bin the item fits in; when there are no such bins, place the item in a new bin and add this bin to the end of the bin list.

   a. Apply *FF* to the instance

   $$s_1 = 0.4, \quad s_2 = 0.7, \quad s_3 = 0.2, \quad s_4 = 0.1, \quad s_5 = 0.5$$

   and determine whether the solution obtained is optimal.

   b. Determine the worst-case time efficiency of *FF*.

   c. Prove that *FF* is a 2-approximation algorithm.

8. The ***first-fit decreasing*** (*FFD*) approximation algorithm for the bin-packing problem starts by sorting the items in nonincreasing order of their sizes and then acts as the first-fit algorithm.

   a. Apply *FFD* to the instance

   $$s_1 = 0.4, \quad s_2 = 0.7, \quad s_3 = 0.2, \quad s_4 = 0.1, \quad s_5 = 0.5$$

   and determine whether the solution obtained is optimal.

   b. Does *FFD* always yield an optimal solution? Justify your answer.

   c. Prove that *FFD* is a 1.5-approximation algorithm.

   d. Run an experiment to determine which of the two algorithms—*FF* or *FFD*—yields more accurate approximations on a random sample of the problem's instances.

9. **a.** Design a simple 2-approximation algorithm for finding a ***minimum vertex cover*** (a vertex cover with the smallest number of vertices) in a given graph.

   b. Consider the following approximation algorithm for finding a ***maximum independent set*** (an independent set with the largest number of vertices) in a given graph. Apply the 2-approximation algorithm of part (a) and output all the vertices that are not in the obtained vertex cover. Can we claim that this algorithm is a 2-approximation algorithm, too?

10. a. Design a polynomial-time greedy algorithm for the graph-coloring problem.

b. Show that the performance ratio of your approximation algorithm is infinitely large.

12.4 Algorithms for Solving Nonlinear Equations

In this section, we discuss several algorithms for solving nonlinear equations in one unknown,

$$f(x) = 0. \qquad \textbf{(12.4)}$$

There are several reasons for this choice among subareas of numerical analysis. First of all, this is an extremely important problem from both a practical and theoretical point of view. It arises as a mathematical model of numerous phenomena in the sciences and engineering, both directly and indirectly. (Recall, for example, that the standard calculus technique for finding extremum points of a function $f(x)$ is based on finding its critical points, which are the roots of the equation $f'(x) = 0$.) Second, it represents the most accessible topic in numerical analysis and, at the same time, exhibits its typical tools and concerns. Third, some methods for solving equations closely parallel algorithms for array searching and hence provide examples of applying general algorithm design techniques to problems of continuous mathematics.

Let us start with dispelling a misconception you might have about solving equations. Your experience with equation solving from middle school to calculus courses might have led you to believe that we can solve equations by "factoring" or by applying a readily available formula. Sorry to break it to you, but you have been deceived (with the best of educational intentions, of course): you were able to solve all those equations only because they had been carefully selected to make it possible. In general, we cannot solve equations exactly and need approximation algorithms to do so.

This is true even for solving the quadratic equation

$$ax^2 + bx + c = 0$$

because the standard formula for its roots

$$x_{1,2} = \frac{-b \pm \sqrt{b^2 - 4ac}}{2a}$$

requires computing the square root, which can be done only approximately for most positive numbers. In addition, as we discussed in Section 11.4, this canonical formula needs to be modified to avoid the possibility of low-accuracy solutions.

What about formulas for roots of polynomials of degrees higher than two? Such formulas for third- and fourth-degree polynomials exist, but they are too cumbersome to be of practical value. For polynomials of degrees higher than four, there can be no general formula for their roots that would involve only the polynomial's coefficients, arithmetical operations, and radicals (taking roots). This remarkable result was published first by the Italian mathematician and physician Paolo Ruffini (1765–1822) in 1799 and rediscovered a quarter century later by the Norwegian mathematician Niels Abel (1802–1829); it was developed further by the French mathematician Evariste Galois (1811–1832).[4]

The impossibility of such a formula can hardly be considered a great disappointment. As the great German mathematician Carl Friedrich Gauss (1777–1855) put it in his thesis of 1801, the algebraic solution of an equation was no better than devising a symbol for the root of the equation and then saying that the equation had a root equal to the symbol [OCo98].

We can interpret solutions to equation (12.4) as points at which the graph of the function $f(x)$ intersects with the x-axis. The three algorithms we discuss in this section take advantage of this interpretation. Of course, the graph of $f(x)$ may intersect the x-axis at a single point (e.g., $x^3 = 0$), at multiple or even infinitely many points ($\sin x = 0$), or at no point ($e^x + 1 = 0$). Equation (12.4) would then have a single root, several roots, and no roots, respectively. It is a good idea to sketch a graph of the function before starting to approximate its roots. It can help to determine the number of roots and their approximate locations. In general, it is a good idea to isolate roots, i.e., to identify intervals containing a single root of the equation in question.

Bisection Method

This algorithm is based on an observation that the graph of a continuous function must intersect with the x-axis between two points a and b at least once if the function's values have opposite signs at these two points (Figure 12.17).

The validity of this observation is proved as a theorem in calculus courses, and we take it for granted here. It serves as the basis of the following algorithm, called the ***bisection method***, for solving equation (12.4). Starting with an interval $[a, b]$ at whose endpoints $f(x)$ has opposite signs, the algorithm computes the value of $f(x)$ at the middle point $x_{mid} = (a + b)/2$. If $f(x_{mid}) = 0$, a root was found and the algorithm stops. Otherwise, it continues the search for a root either on $[a, x_{mid}]$ or on $[x_{mid}, b]$, depending on which of the two halves the values of $f(x)$ have opposite signs at the endpoints of the new interval.

Since we cannot expect the bisection algorithm to stumble on the exact value of the equation's root and stop, we need a different criterion for stopping the algo-

4. Ruffini's discovery was completely ignored by almost all prominent mathematicians of that time. Abel died young after a difficult life of poverty. Galois was killed in a duel when he was only 21 years old. Their results on the solution of higher-degree equations are now considered to be among the crowning achievements in the history of mathematics.

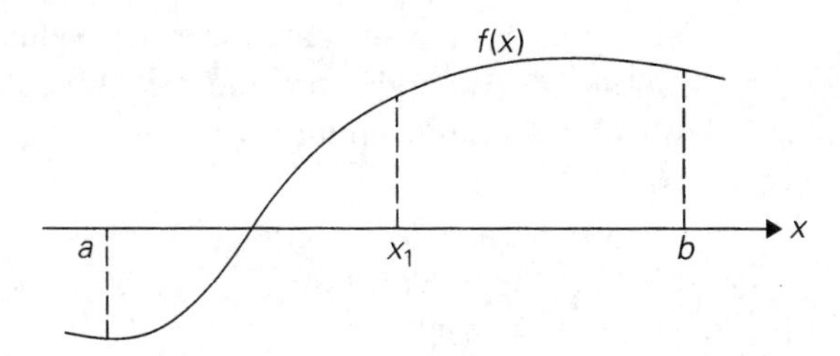

FIGURE 12.17 First iteration of the bisection method: x_1 is the middle point of interval $[a, b]$.

rithm. We can stop the algorithm after the interval $[a_n, b_n]$ bracketing some root x^* becomes so small that we can guarantee that the absolute error of approximating x^* by x_n, the middle point of this interval, is smaller than some small preselected number $\varepsilon > 0$. Since x_n is the middle point of $[a_n, b_n]$ and x^* lies within this interval as well, we have

$$|x_n - x^*| \le \frac{b_n - a_n}{2}. \tag{12.5}$$

Hence, we can stop the algorithm as soon as $(b_n - a_n)/2 < \varepsilon$ or, equivalently,

$$x_n - a_n < \varepsilon. \tag{12.6}$$

It is not difficult to prove that

$$|x_n - x^*| \le \frac{b_1 - a_1}{2^n} \quad \text{for } n = 1, 2, \ldots. \tag{12.7}$$

This inequality implies that the sequence of approximations $\{x_n\}$ can be made as close to root x^* as we wish by choosing n large enough. In other words, we can say that $\{x_n\}$ ***converges*** to root x^*. Note, however, that because any digital computer represents extremely small values by zero (Section 11.4), the convergence assertion is true in theory but not necessarily in practice. In fact, if we choose ε below a certain machine-dependent threshold, the algorithm may never stop! Another source of potential complications is round-off errors in computing values of the function in question. Therefore, it is a good practice to include in a program implementing the bisection method a limit on the number of iterations the algorithm is allowed to run.

Here is pseudocode of the bisection method.

ALGORITHM *Bisection*$(f(x), a, b, eps, N)$

//Implements the bisection method for finding a root of $f(x) = 0$
//Input: Two real numbers a and b, $a < b$,
// a continuous function $f(x)$ on $[a, b]$, $f(a)f(b) < 0$,
// an upper bound on the absolute error $eps > 0$,
// an upper bound on the number of iterations N

```
//Output: An approximate (or exact) value x of a root in (a, b)
//or an interval bracketing the root if the iteration number limit is reached
n ← 1  //iteration count
while n ≤ N do
    x ← (a + b)/2
    if x − a < eps return x
    fval ← f(x)
    if fval = 0 return x
    if fval * f(a) < 0
        b ← x
    else a ← x
    n ← n + 1
return "iteration limit", a, b
```

Note that we can use inequality (12.7) to find in advance the number of iterations that should suffice, at least in theory, to achieve a preselected accuracy level. Indeed, choosing the number of iterations n large enough to satisfy $(b_1 - a_1)/2^n < \varepsilon$, i.e.,

$$n > \log_2 \frac{b_1 - a_1}{\varepsilon}, \tag{12.8}$$

does the trick.

EXAMPLE 1 Let us consider equation

$$x^3 - x - 1 = 0. \tag{12.9}$$

It has one real root. (See Figure 12.18 for the graph of $f(x) = x^3 - x - 1$.) Since $f(0) < 0$ and $f(2) > 0$, the root must lie within interval [0, 2]. If we choose the error tolerance level as $\varepsilon = 10^{-2}$, inequality (12.8) would require $n > \log_2(2/10^{-2})$ or $n \geq 8$ iterations.

Figure 12.19 contains a trace of the first eight iterations of the bisection method applied to equation (12.9).

Thus, we obtained $x_8 = 1.3203125$ as an approximate value for the root x^* of equation (12.9), and we can guarantee that

$$|1.3203125 - x^*| < 10^{-2}.$$

Moreover, if we take into account the signs of the function $f(x)$ at a_8, b_8, and x_8, we can assert that the root lies between 1.3203125 and 1.328125. ■

The principal weakness of the bisection method as a general algorithm for solving equations is its slow rate of convergence compared with other known methods. It is for this reason that the method is rarely used. Also, it cannot be extended to solving more general equations and systems of equations. But it does have several strong points. It always converges to a root whenever we start with an

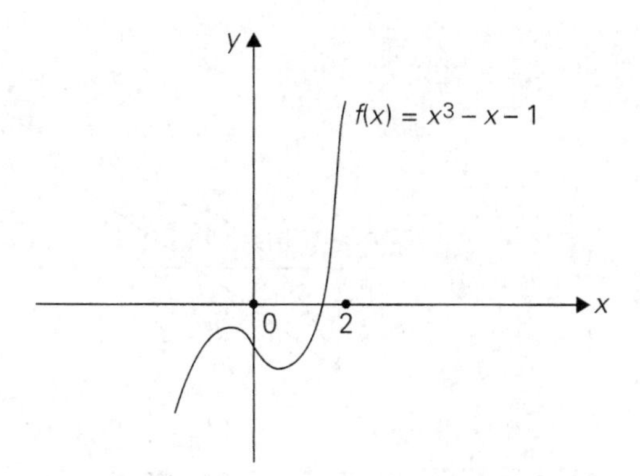

FIGURE 12.18 Graph of function $f(x) = x^3 - x - 1$.

n	a_n	b_n	x_n	$f(x_n)$
1	0.0−	2.0+	1.0	−1.0
2	1.0−	2.0+	1.5	0.875
3	1.0−	1.5+	1.25	−0.296875
4	1.25−	1.5+	1.375	0.224609
5	1.25−	1.375+	1.3125	−0.051514
6	1.3125−	1.375+	1.34375	0.082611
7	1.3125−	1.34375+	1.328125	0.014576
8	1.3125−	1.328125+	1.3203125	−0.018711

FIGURE 12.19 Trace of the bisection method for solving equation (12.8). The signs after the numbers in the second and third columns indicate the sign of $f(x) = x^3 - x - 1$ at the corresponding endpoints of the intervals.

interval whose properties are very easy to check. And it does not use derivatives of the function $f(x)$ as some faster methods do.

What important algorithm does the method of bisection remind you of? If you have found it to closely resemble binary search, you are correct. Both of them solve variations of the searching problem, and they are both divide-by-half algorithms. The principal difference lies in the problem's domain: discrete for binary search and continuous for the bisection method. Also note that while binary search requires its input array to be sorted, the bisection method does not require its function to be nondecreasing or nonincreasing. Finally, whereas binary search is very fast, the bisection method is relatively slow.

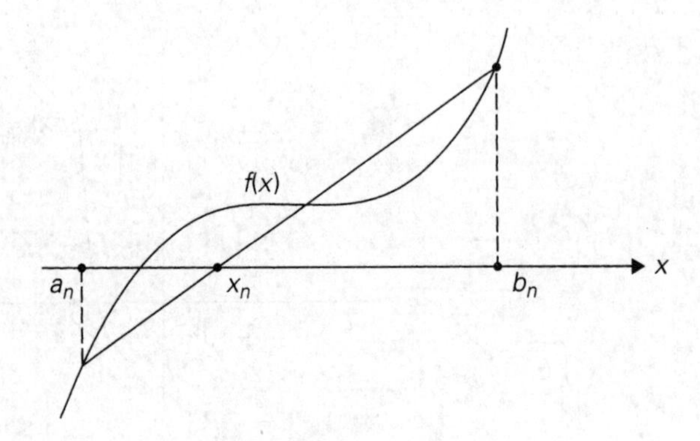

FIGURE 12.20 Iteration of the method of false position.

Method of False Position

The ***method of false position*** (also known by its name in Latin, ***regula falsi***) is to interpolation search as the bisection method is to binary search. Like the bisection method, it has, on each iteration, some interval $[a_n, b_n]$ bracketing a root of a continuous function $f(x)$ that has opposite-sign values at a_n and b_n. Unlike the bisection method, however, it computes the next root approximation not as the middle of $[a_n, b_n]$ but as the x-intercept of the straight line through the points $(a_n, f(a_n))$ and $(b_n, f(b_n))$ (Figure 12.20).

You are asked in the exercises to show that the formula for this x-intercept can be written as

$$x_n = \frac{a_n f(b_n) - b_n f(a_n)}{f(b_n) - f(a_n)}. \tag{12.10}$$

EXAMPLE 2 Figure 12.21 contains the results of the first eight iterations of this method for solving equation (12.9).

Although for this example the method of false position does not perform as well as the bisection method, for many instances it yields a faster converging sequence. ■

Newton's Method

Newton's method, also called the ***Newton-Raphson method***, is one of the most important general algorithms for solving equations. When applied to equation (12.4) in one unknown, it can be illustrated by Figure 12.22: the next element x_{n+1} of the method's approximation sequence is obtained as the x-intercept of the tangent line to the graph of function $f(x)$ at x_n.

The analytical formula for the elements of the approximation sequence turns out to be

$$x_{n+1} = x_n - \frac{f(x_n)}{f'(x_n)} \quad \text{for } n = 0, 1, \ldots. \tag{12.11}$$

	a_n	b_n	x_n	$f(x_n)$
1	0.0−	2.0+	0.333333	−1.296296
2	0.333333−	2.0+	0.676471	−1.366909
3	0.676471−	2.0+	0.960619	−1.074171
4	0.960619−	2.0+	1.144425	−0.645561
5	1.144425−	2.0+	1.242259	−0.325196
6	1.242259−	2.0+	1.288532	−0.149163
7	1.288532−	2.0+	1.309142	−0.065464
8	1.309142−	2.0+	1.318071	−0.028173

FIGURE 12.21 Trace of the method of false position for equation (12.9). The signs after the numbers in the second and third columns indicate the sign of $f(x) = x^3 - x - 1$ at the corresponding endpoints of the intervals.

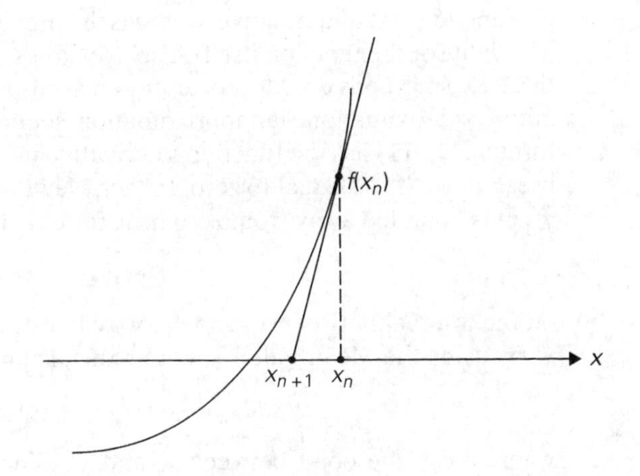

FIGURE 12.22 Iteration of Newton's method.

In most cases, Newton's algorithm guarantees convergence of sequence (12.11) if an initial approximation x_0 is chosen "close enough" to the root. (Precisely defined prescriptions for choosing x_0 can be found in numerical analysis textbooks.) It may converge for initial approximations far from the root as well, but this is not always true.

EXAMPLE 3 Computing $\sqrt{a}$ for $a \geq 0$ can be done by finding a nonnegative root of equation $x^2 - a = 0$. If we use formula (12.11) for this case of $f(x) = x^2 - a$ and $f'(x) = 2x$, we obtain

$$x_{n+1} = x_n - \frac{f(x_n)}{f'(x_n)} = x_n - \frac{x_n^2 - a}{2x_n} = \frac{x_n^2 + a}{2x_n} = \frac{1}{2}(x_n + \frac{a}{x_n}),$$

which is exactly the formula we used in Section 11.4 for computing approximate values of square roots. ■

EXAMPLE 4 Let us apply Newton's method to equation (12.9), which we previously solved with the bisection method and the method of false position. Formula (12.11) for this case becomes

$$x_{n+1} = x_n - \frac{x_n^3 - x_n - 1}{3x_n^2 - 1}.$$

As an initial element of the approximation sequence, we take, say, $x_0 = 2$. Figure 12.23 contains the results of the first five iterations of Newton's method. ■

You cannot fail to notice how much faster Newton's approximation sequence converges to the root than the approximation sequences of both the bisection method and the method of false position. This very fast convergence is typical of Newton's method if an initial approximation is close to the equation's root. Note, however, that on each iteration of this method we need to evaluate new values of the function and its derivative, whereas the previous two methods require only one new value of the function itself. Also, Newton's method does not bracket a root as these two methods do. Moreover, for an arbitrary function and arbitrarily chosen initial approximation, its approximation sequence may diverge. And, because formula (12.11) has the function's derivative in the denominator, the method may break down if it is equal to zero. In fact, Newton's method is most effective when $f'(x)$ is bounded away from zero near root x^*. In particular, if

$$|f'(x)| \geq m_1 > 0$$

on the interval between x_n and x^*, we can estimate the distance between x_n and x^* by using the Mean Value Theorem of calculus as follows:

$$f(x_n) - f(x^*) = f'(c)(x_n - x^*),$$

where c is some point between x_n and x^*. Since $f(x^*) = 0$ and $|f'(c)| \geq m_1$, we obtain

n	x_n	x_{n+1}	$f(x_{n+1})$
0	2.0	1.545455	1.145755
1	1.545455	1.359615	0.153705
2	1.359615	1.325801	0.004625
3	1.325801	1.324719	$4.7 \cdot 10^{-6}$
4	1.324719	1.324718	$5 \cdot 10^{-12}$

FIGURE 12.23 Trace of Newton's method for equation (12.9).

$$|x_n - x^*| \le \frac{|f(x_n)|}{m_1}. \tag{12.12}$$

Formula (12.12) can be used as a criterion for stopping Newton's algorithm when its right-hand side becomes smaller than a preselected accuracy level ε. Other possible stopping criteria are

$$|x_n - x_{n-1}| < \varepsilon$$

and

$$|f(x_n)| < \varepsilon,$$

where ε is a small positive number. Since the last two criteria do not necessarily imply closeness of x_n to root x^*, they should be considered inferior to the one based on (12.12).

The shortcomings of Newton's method should not overshadow its principal strengths: fast convergence for an appropriately chosen initial approximation and applicability to much more general types of equations and systems of equations.

Exercises 12.4

1. **a.** Find on the Internet or in your library a procedure for finding a real root of the general cubic equation $ax^3 + bx^2 + cx + d = 0$ with real coefficients.
   b. What general algorithm design technique is it based on?
2. Indicate how many roots each of the following equations has:
   a. $xe^x - 1 = 0$ **b.** $x - \ln x = 0$ **c.** $x \sin x - 1 = 0$
3. **a.** Prove that if $p(x)$ is a polynomial of an odd degree, then it must have at least one real root.
   b. Prove that if x_0 is a root of an n-degree polynomial $p(x)$, the polynomial can be factored into
   $$p(x) = (x - x_0)q(x),$$
   where $q(x)$ is a polynomial of degree $n - 1$. Explain what significance this theorem has for finding the roots of a polynomial.
   c. Prove that if x_0 is a root of an n-degree polynomial $p(x)$, then
   $$p'(x_0) = q(x_0),$$
   where $q(x)$ is the quotient of the division of $p(x)$ by $x - x_0$.
4. Prove inequality (12.7).
5. Apply the bisection method to find the root of the equation
   $$x^3 + x - 1 = 0$$
   with an absolute error smaller than 10^{-2}.

6. Derive formula (12.10) underlying the method of false position.
7. Apply the method of false position to find the root of the equation

$$x^3 + x - 1 = 0$$

with an absolute error smaller than 10^{-2}.
8. Derive formula (12.11) underlying Newton's method.
9. Apply Newton's method to find the root of the equation

$$x^3 + x - 1 = 0$$

with an absolute error smaller than 10^{-2}.
10. Give an example that shows that the approximation sequence of Newton's method may diverge.

11. *Gobbling goat* There is a grassy field in the shape of a circle with a radius of 100 feet. A goat is attached by a rope to a hook at a fixed point on the field's border. How long should the rope be to let the goat reach only half of the grass in the field?

SUMMARY

- *Backtracking* and *branch-and-bound* are two algorithm design techniques for solving problems in which the number of choices grows at least exponentially with their instance size. Both techniques construct a solution one component at a time, trying to terminate the process as soon as one can ascertain that no solution can be obtained as a result of the choices already made. This approach makes it possible to solve many large instances of *NP*-hard problems in an acceptable amount of time.

- Both backtracking and branch-and-bound employ, as their principal mechanism, a *state-space tree*—a rooted tree whose nodes represent partially constructed solutions to the problem in question. Both techniques terminate a node as soon as it can be guaranteed that no solution to the problem can be obtained by considering choices that correspond to the node's descendants.

- *Backtracking* constructs its state-space tree in the depth-first-search fashion in the majority of its applications. If the sequence of choices represented by a current node of the state-space tree can be developed further without violating the problem's constraints, it is done by considering the first remaining legitimate option for the next component. Otherwise, the method backtracks by undoing the last component of the partially built solution and replaces it by the next alternative.

- *Branch-and-bound* is an algorithm design technique that enhances the idea of generating a state-space tree with the idea of estimating the best value obtainable from a current node of the decision tree: if such an estimate is not superior to the best solution seen up to that point in the processing, the node is eliminated from further consideration.

- Approximation algorithms are often used to find approximate solutions to difficult problems of combinatorial optimization. The *performance ratio* is the principal metric for measuring the accuracy of such approximation algorithms.

- The *nearest-neighbor* and *multifragment heuristic* are two simple greedy algorithms for approximating a solution to the traveling salesman problem. The performance ratios of these algorithms are unbounded above, even for the important subset of *Euclidean graphs*.

- The *twice-around-the-tree* and *Christofides* algorithms exploit the graph's minimum spanning tree to construct an Eulerian circuit and then transform it into a Hamiltonian circuit (an approximate solution to the TSP) by shortcuts. For Euclidean graphs, the performance ratios of these algorithms are 2 and 1.5, respectively.

- *Local search heuristics*—the *2-opt, 3-opt*, and *Lin-Kernighan* algorithms—work by replacing a few edges in the current tour to find a shorter one until no such replacement can be found. These algorithms are capable of finding in seconds a tour that is within a few percent of optimum for large Euclidean instances of the traveling salesman problem.

- A sensible greedy algorithm for the knapsack problem is based on processing an input's items in descending order of their value-to-weight ratios. For the continuous version of the problem, this algorithm always yields an exact optimal solution.

- *Polynomial approximation schemes* for the knapsack problem are polynomial-time parametric algorithms that approximate solutions with any predefined accuracy level.

- Solving nonlinear equations is one of the most important areas of numerical analysis. Although there are no formulas for roots of nonlinear equations (with a few exceptions), several algorithms can solve them approximately.

- The *bisection method* and the *method of false position* are continuous analogues of binary search and interpolation search, respectively. Their principal advantage lies in bracketing a root on each iteration of the algorithm.

- *Newton's method* generates a sequence of root approximations that are x-intercepts of tangent lines to the function's graph. With a good initial approximation, it typically requires just a few iterations to obtain a high-accuracy approximation to the equation's root.

Epilogue

Science is nothing but trained and organized common sense.

—Thomas H. Huxley (1825–1895), English biologist and educator

Well, we have arrived. It was a long road. Not as long as it took humanity to travel from Euclid's algorithm, which opens this book, to the latest developments in algorithmics, but it was long enough. So let us now take a last look at what we have learned during the journey.

We started with the well-accepted proposition that the notion of an algorithm constitutes the cornerstone of computer science. And since computer programs are just implementations of algorithms on particular machines, algorithms lie at the heart of practical computing, too.

Like any science, computer science is concerned with classifying its principal subject. Although algorithms can be classified in numerous ways, two of them are particularly important. We can classify algorithms by their underlying design technique and by their efficiency. These two principal dimensions reflect the needs of computing practice as well: we need design techniques as a guide for developing a new algorithm, and we need a framework to ascertain the efficiency of a given algorithm.

We discussed 10 general design techniques in this book:

brute force	*dynamic programming*
decrease-and-conquer	*greedy technique*
divide-and-conquer	*iterative improvement*
transform-and-conquer	*backtracking*
space-time trade-offs	*branch-and-bound*

We showed how these techniques apply to a variety of important problems in computer science: sorting, searching, string processing, graphs, and some geometric and numerical problems. Although these basic techniques are not all applicable to every problem, taken collectively they provide a powerful tool kit for designing new algorithms and classifying existing ones. Moreover, these techniques can be thought of as general problem-solving approaches not limited to the computing domain. The puzzles included in the book make this point.

The analysis framework classifies algorithms by the order of growth of their running time as a function of input size. It does so by investigating the number of times the algorithm's basic operation is executed. The main tools are summation formulas and recurrence relations for nonrecursive and recursive algorithms, respectively. We saw that a surprisingly large number of algorithms fall into one of the few classes on the following list.

class	**notation**	**important examples**
constant time	$\Theta(1)$	hashing (on average)
logarithmic	$\Theta(\log n)$	binary search (worst and average cases)
linear	$\Theta(n)$	sequential search (worst and average cases)
linearithmic	$\Theta(n \log n)$	advanced sorting algorithms
quadratic	$\Theta(n^2)$	elementary sorting algorithms
cubic	$\Theta(n^3)$	Gaussian elimination
exponential	$\Omega(a^n)$	combinatorial problems

For some algorithms, we must distinguish between the worst-, best-, and average-case efficiencies. The average case is particularly difficult to investigate, and we discussed how one can do this empirically.

We touched on the limitations of algorithms. We saw that there are two major reasons for such limitations: the intrinsic complexity of a problem and the necessity of dealing with rounded-off numbers for most numerical problems. We also discussed approaches for dealing with such limitations.

It should come as no surprise, however, that there are areas of algorithmics that were not covered in this book. The most important of them are randomized and parallel algorithms. A ***randomized algorithm*** is an algorithm that makes random choices during its execution. For example, we can randomly select an array's element to serve as a pivot in sorting the array by quicksort. Unlike a deterministic algorithm, a randomized algorithm behaves differently on different runs with the same input and may even yield different results. For many applications, this variability can be to our advantage, yielding randomized algorithms that are faster or simpler (or both) than their deterministic counterparts.

One of the most impressive randomized algorithms discovered to date is the Miller-Rabin algorithm for primality testing of integers (e.g., [Cor09]). This randomized algorithm solves the problem in an acceptable amount of time for thousand-digit numbers with the probability of yielding an erroneous answer smaller than the probability of hardware malfunction. It is much faster than the best known deterministic algorithms for solving this problem, which is crucial for modern cryptology. If you want to learn more about randomized algorithms, the monograph by R. Motwani and P. Raghavan [Mot95] and the excellent survey by R. M. Karp [Kar91] are particularly recommended.

The vast majority of computers in use today still closely resemble the machine outlined more than half a century ago by John von Newmann. The central assumption of this architecture is that instructions are executed one after another, one operation at a time. Accordingly, algorithms designed to be executed on such machines are called ***sequential algorithms***. It is this kind of algorithm that we discussed in this book. The central assumption of the von Neumann model does not hold for some newer computers that can execute operations concurrently, i.e., in parallel. Algorithms that take advantage of this capability are called ***parallel algorithms***.

Consider, as an example, the problem of computing the sum of n numbers stored in an array $A[0..n-1]$. One can prove that any sequential algorithm that uses only multiplications, additions, and subtractions requires at least $n-1$ steps to solve this problem. However, if we can pair and find the sum of elements $A[0]$ and $A[1]$, $A[2]$ and $A[3]$, and so on, in parallel, the size of the problem will be reduced by half. Repeating this operation until the entire sum is computed yields an algorithm that requires just $\lceil \log_2 n \rceil$ steps.

There is a wealth of books devoted to parallel algorithms. Several general-purpose algorithm textbooks include separate chapters on parallel algorithms ([Hor07] providing particularly extensive coverage of them) or discuss them together with sequential algorithms (see [Ber05], [Mil05]).

The juggernaut of technological innovations has also produced some promising breakthroughs—such as quantum computing and DNA computing—that might have a dramatic impact on the computational capabilities and algorithms of the future. ***Quantum computing*** (see, e.g., [Yan08]) seeks to exploit a quantum physics phenomenon of having a subatomic particle in two states simultaneously. Hence, at least theoretically, a system of n such particles, called "qubits," can simultaneously contain 2^n bits of information. In 1994, Peter Shor of AT&T Research Labs presented an algorithm for factoring an integer that took advantage of this theoretical possibility [Sho94]. This algorithm requires only $O(b^3)$ time and $O(b)$ space on b-bit number inputs. Moreover, IBM researchers were able to build a 7-qubit computer that actually implemented Shor's algorithm and successfully factored the number 15 into 3 and 5. Although technological problems of scaling this and similar approaches to larger problems are formidable and may still prove to be insurmountable, quantum computing has the potential to change our current ideas about the difficulty of some computational problems. But it is worth pointing out that integer factoring (more accurately, its decision version), though certainly difficult, is believed not to be NP-complete. Therefore, solving it efficiently on a quantum computer would not imply existence of polynomial-time quantum algorithms for all intractable problems. In fact, it is believed that the class of problems that can be solved in polynomial time on a quantum computer and the class of *NP*-complete problems are disjoint subsets of class *NP*.

If quantum computing seeks to harness the power of quantum physics to solve difficult computational problems, ***DNA computing*** seeks to accomplish the same goal via exploiting the mechanism of gene selection. The most famous example of this approach was provided in the same year, 1994, by Len Adleman [Adl94], a U.S. computer scientist well-known for his participation in the development of the very important RSA encryption algorithm. He showed how the problem of finding a Hamiltonian path in a directed graph can in principle be solved by generating DNA strands representing paths in the graph and discarding those of them that did not satisfy the definition of such a path. The existence of a Hamiltonian path is known to be an *NP*-complete problem, and Adleman's approach is similar to exhaustive search. But a huge number of biochemical processes are occurring in parallel, leaving the hope of yielding a solution in an acceptable amount of time. Adleman was able to solve the Hamiltonian path problem for a small graph of seven vertices, although he had to repeat parts of the procedure several times to purify the DNA solution.

Scaling Adleman's approach to larger graphs would require an exponentially fast growing number of the nucleotides needed by this procedure. Although the true potential of DNA computing remains unclear, several research teams around the world continue working on it. Among a few reported successes, NASA announced in 2002 that a team led by Adleman developed a DNA computer that solved a problem requiring evaluation of 1 million alternatives for their ability to satisfy 24 different criteria. The same year, researchers from the Weizmann Institute of Science in Israel unveiled a programmable molecular computing machine composed of enzymes and DNA molecules instead of silicon microchips. While that computer was limited to solving decision problems only, researchers at the California Institute of Technology announced in 2011 the most complex biochemical circuit to date, which was able to compute the square root rounded down to the nearest integer of a number up to 15.

So, whichever direction you take in your future journey through the land of algorithms in your studies and your career, the road ahead is as exciting as it has ever been. Not many areas of science and engineering can make this claim with the same assurance that algorithmics can. Have a great trip!

APPENDIX A

Useful Formulas for the Analysis of Algorithms

This appendix contains a list of useful formulas and rules that are helpful in the mathematical analysis of algorithms. More advanced material can be found in [Gra94], [Gre07], [Pur04], and [Sed96].

Properties of Logarithms

All logarithm bases are assumed to be greater than 1 in the formulas below; $\lg x$ denotes the logarithm base 2, $\ln x$ denotes the logarithm base $e = 2.71828\ldots$; x, y are arbitrary positive numbers.

1. $\log_a 1 = 0$
2. $\log_a a = 1$
3. $\log_a x^y = y \log_a x$
4. $\log_a xy = \log_a x + \log_a y$
5. $\log_a \frac{x}{y} = \log_a x - \log_a y$
6. $a^{\log_b x} = x^{\log_b a}$
7. $\log_a x = \frac{\log_b x}{\log_b a} = \log_a b \log_b x$

Combinatorics

1. Number of permutations of an n-element set: $P(n) = n!$
2. Number of k-combinations of an n-element set: $C(n, k) = \frac{n!}{k!(n-k)!}$
3. Number of subsets of an n-element set: 2^n

Important Summation Formulas

1. $\sum_{i=l}^{u} 1 = \underbrace{1+1+\cdots+1}_{u-l+1 \text{ times}} = u - l + 1$ (l, u are integer limits, $l \le u$); $\sum_{i=1}^{n} 1 = n$

2. $\sum_{i=1}^{n} i = 1+2+\cdots+n = \frac{n(n+1)}{2} \approx \frac{1}{2}n^2$

3. $\sum_{i=1}^{n} i^2 = 1^2+2^2+\cdots+n^2 = \frac{n(n+1)(2n+1)}{6} \approx \frac{1}{3}n^3$

4. $\sum_{i=1}^{n} i^k = 1^k+2^k+\cdots+n^k \approx \frac{1}{k+1}n^{k+1}$

5. $\sum_{i=0}^{n} a^i = 1+a+\cdots+a^n = \frac{a^{n+1}-1}{a-1}$ $(a \ne 1)$; $\sum_{i=0}^{n} 2^i = 2^{n+1}-1$

6. $\sum_{i=1}^{n} i2^i = 1\cdot 2+2\cdot 2^2+\cdots+n2^n = (n-1)2^{n+1}+2$

7. $\sum_{i=1}^{n} \frac{1}{i} = 1+\frac{1}{2}+\cdots+\frac{1}{n} \approx \ln n + \gamma$, where $\gamma \approx 0.5772\ldots$ (Euler's constant)

8. $\sum_{i=1}^{n} \lg i \approx n \lg n$

Sum Manipulation Rules

1. $\sum_{i=l}^{u} ca_i = c\sum_{i=l}^{u} a_i$

2. $\sum_{i=l}^{u} (a_i \pm b_i) = \sum_{i=l}^{u} a_i \pm \sum_{i=l}^{u} b_i$

3. $\sum_{i=l}^{u} a_i = \sum_{i=l}^{m} a_i + \sum_{i=m+1}^{u} a_i$, where $l \le m < u$

4. $\sum_{i=l}^{u} (a_i - a_{i-1}) = a_u - a_{l-1}$

Approximation of a Sum by a Definite Integral

$$\int_{l-1}^{u} f(x)dx \le \sum_{i=l}^{u} f(i) \le \int_{l}^{u+1} f(x)dx \quad \text{for a nondecreasing } f(x)$$

$$\int_{l}^{u+1} f(x)dx \le \sum_{i=l}^{u} f(i) \le \int_{l-1}^{u} f(x)dx \quad \text{for a nonincreasing } f(x)$$

Floor and Ceiling Formulas

The *floor* of a real number x, denoted $\lfloor x \rfloor$, is defined as the greatest integer not larger than x (e.g., $\lfloor 3.8 \rfloor = 3$, $\lfloor -3.8 \rfloor = -4$, $\lfloor 3 \rfloor = 3$). The *ceiling* of a real number x, denoted $\lceil x \rceil$, is defined as the smallest integer not smaller than x (e.g., $\lceil 3.8 \rceil = 4$, $\lceil -3.8 \rceil = -3$, $\lceil 3 \rceil = 3$).

1. $x - 1 < \lfloor x \rfloor \le x \le \lceil x \rceil < x + 1$
2. $\lfloor x + n \rfloor = \lfloor x \rfloor + n$ and $\lceil x + n \rceil = \lceil x \rceil + n$ for real x and integer n
3. $\lfloor n/2 \rfloor + \lceil n/2 \rceil = n$
4. $\lceil \lg(n + 1) \rceil = \lfloor \lg n \rfloor + 1$

Miscellaneous

1. $n! \approx \sqrt{2\pi n}\left(\frac{n}{e}\right)^n$ as $n \to \infty$ (Stirling's formula)
2. Modular arithmetic (n, m are integers, p is a positive integer)

$$(n + m) \bmod p = (n \bmod p + m \bmod p) \bmod p$$

$$(nm) \bmod p = ((n \bmod p)(m \bmod p)) \bmod p$$

APPENDIX B

Short Tutorial on Recurrence Relations

Sequences and Recurrence Relations

DEFINITION A (numerical) ***sequence*** is an ordered list of numbers.

Examples: 2, 4, 6, 8, 10, 12, ... (positive even integers)
0, 1, 1, 2, 3, 5, 8, ... (the Fibonacci numbers)
0, 1, 3, 6, 10, 15, ... (numbers of key comparisons in selection sort)

A sequence is usually denoted by a letter (such as x or a) with a subindex (such as n or i) written in curly brackets, e.g., $\{x_n\}$. We use the alternative notation $x(n)$. This notation stresses the fact that a sequence is a function: its argument n indicates a position of a number in the list, while the function's value $x(n)$ stands for that number itself. $x(n)$ is called the ***generic term*** of the sequence.

There are two principal ways to define a sequence:

- by an explicit formula expressing its generic term as a function of n, e.g., $x(n) = 2n$ for $n \geq 0$
- by an equation relating its generic term to one or more other terms of the sequence, combined with one or more explicit values for the first term(s), e.g.,

$$x(n) = x(n-1) + n \quad \text{for } n > 0, \tag{B.1}$$

$$x(0) = 0. \tag{B.2}$$

It is the latter method that is particularly important for analysis of recursive algorithms (see Section 2.4 for a detailed discussion of this topic).

An equation such as (B.1) is called a ***recurrence equation*** or ***recurrence relation*** (or simply a ***recurrence***), and an equation such as (B.2) is called its ***initial condition***. An initial condition can be given for a value of n other than 0 (e.g., for $n = 1$) and for some recurrences (e.g., for the recurrence $F(n) = F(n-1) + F(n-2)$

defining the Fibonacci numbers—see Section 2.5), more than one value needs to be specified by initial conditions.

To solve a given recurrence subject to a given initial condition means to find an explicit formula for the generic term of the sequence that satisfies both the recurrence equation and the initial condition or to prove that such a sequence does not exist. For example, the solution to recurrence (B.1) subject to initial condition (B.2) is

$$x(n) = \frac{n(n+1)}{2} \quad \text{for } n \geq 0. \tag{B.3}$$

It can be verified by substituting this formula into (B.1) to check that the equality holds for every $n > 0$, i.e., that

$$\frac{n(n+1)}{2} = \frac{(n-1)(n-1+1)}{2} + n$$

and into (B.2) to check that $x(0) = 0$, i.e., that

$$\frac{0(0+1)}{2} = 0.$$

Sometimes it is convenient to distinguish between a general solution and a particular solution to a recurrence. Recurrence equations typically have an infinite number of sequences that satisfy them. A ***general solution*** to a recurrence equation is a formula that specifies all such sequences. Typically, a general solution involves one or more arbitrary constants. For example, for recurrence (B.1), the general solution can be specified by the formula

$$x(n) = c + \frac{n(n+1)}{2}, \tag{B.4}$$

where c is such an arbitrary constant. By assigning different values to c, we can get all the solutions to equation (B.1) and only these solutions.

A ***particular solution*** is a specific sequence that satisfies a given recurrence equation. Usually we are interested in a particular solution that satisfies a given initial condition. For example, sequence (B.3) is a particular solution to (B.1)–(B.2).

Methods for Solving Recurrence Relations

No universal method exists that would enable us to solve every recurrence relation. (This is not surprising, because we do not have such a method even for solving much simpler equations in one unknown $f(x) = 0$ for an arbitrary function $f(x)$.) There are several techniques, however, some more powerful than others, that can solve a variety of recurrences.

Method of Forward Substitutions Starting with the initial term (or terms) of the sequence given by the initial condition(s), we can use the recurrence equation to generate the few first terms of its solution in the hope of seeing a pattern that can be expressed by a closed-end formula. If such a formula is found, its validity should be either checked by direct substitution into the recurrence equation and the initial condition (as we did for (B.1)–(B.2)) or proved by mathematical induction.

For example, consider the recurrence

$$x(n) = 2x(n-1) + 1 \quad \text{for } n > 1, \tag{B.5}$$

$$x(1) = 1. \tag{B.6}$$

We obtain the few first terms as follows:

$$\begin{aligned} x(1) &= 1, \\ x(2) &= 2x(1) + 1 = 2 \cdot 1 + 1 = 3, \\ x(3) &= 2x(2) + 1 = 2 \cdot 3 + 1 = 7, \\ x(4) &= 2x(3) + 1 = 2 \cdot 7 + 1 = 15. \end{aligned}$$

It is not difficult to notice that these numbers are one less than consecutive powers of 2:

$$x(n) = 2^n - 1 \quad \text{for } n = 1, 2, 3, \text{ and } 4.$$

We can prove the hypothesis that this formula yields the generic term of the solution to (B.5)–(B.6) either by direct substitution of the formula into (B.5) and (B.6) or by mathematical induction.

As a practical matter, the method of forward substitutions works in a very limited number of cases because it is usually very difficult to recognize the pattern in the first few terms of the sequence.

Method of Backward Substitutions This method of solving recurrence relations works exactly as its name implies: using the recurrence relation in question, we express $x(n-1)$ as a function of $x(n-2)$ and substitute the result into the original equation to get $x(n)$ as a function of $x(n-2)$. Repeating this step for $x(n-2)$ yields an expression of $x(n)$ as a function of $x(n-3)$. For many recurrence relations, we will then be able to see a pattern and express $x(n)$ as a function of $x(n-i)$ for an arbitrary $i = 1, 2, \ldots$. Selecting i to make $n-i$ reach the initial condition and using one of the standard summation formulas often leads to a closed-end formula for the solution to the recurrence.

As an example, let us apply the method of backward substitutions to recurrence (B.1)–(B.2). Thus, we have the recurrence equation

$$x(n) = x(n-1) + n.$$

Replacing n by $n-1$ in the equation yields $x(n-1) = x(n-2) + n - 1$; after substituting this expression for $x(n-1)$ in the initial equation, we obtain

$$x(n) = [x(n-2) + n - 1] + n = x(n-2) + (n-1) + n.$$

Replacing n by $n-2$ in the initial equation yields $x(n-2) = x(n-3) + n - 2$; after substituting this expression for $x(n-2)$, we obtain

$$x(n) = [x(n-3) + n - 2] + (n-1) + n = x(n-3) + (n-2) + (n-1) + n.$$

Comparing the three formulas for $x(n)$, we can see the pattern arising after i such substitutions:[1]

$$x(n) = x(n-i) + (n-i+1) + (n-i+2) + \cdots + n.$$

Since initial condition (B.2) is specified for $n = 0$, we need $n - i = 0$, i.e., $i = n$, to reach it:

$$x(n) = x(0) + 1 + 2 + \cdots + n = 0 + 1 + 2 + \cdots + n = n(n+1)/2.$$

The method of backward substitutions works surprisingly well for a wide variety of simple recurrence relations. You can find many examples of its successful applications throughout this book (see, in particular, Section 2.4 and its exercises).

Linear Second-Order Recurrences with Constant Coefficients An important class of recurrences that can be solved by neither forward nor backward substitutions are recurrences of the type

$$ax(n) + bx(n-1) + cx(n-2) = f(n), \qquad \textbf{(B.7)}$$

where a, b, and c are real numbers, $a \neq 0$. Such a recurrence is called ***second-order linear recurrence with constant coefficients***. It is ***second-order*** because elements $x(n)$ and $x(n-2)$ are two positions apart in the unknown sequence in question; it is ***linear*** because the left-hand side is a linear combination of the unknown terms of the sequence; it has ***constant coefficients*** because of the assumption that a, b, and c are some fixed numbers. If $f(n) = 0$ for every n, the recurrence is said to be ***homogeneous***; otherwise, it is called ***inhomogeneous***.

Let us consider first the homogeneous case:

$$ax(n) + bx(n-1) + cx(n-2) = 0. \qquad \textbf{(B.8)}$$

Except for the degenerate situation of $b = c = 0$, equation (B.8) has infinitely many solutions. All these solutions, which make up the general solution to (B.8), can be obtained by one of the three formulas that follow. Which of the three formulas applies to a particular case depends on the roots of the quadratic equation with the same coefficients as recurrence (B.8):

$$ar^2 + br + c = 0. \qquad \textbf{(B.9)}$$

Quadratic equation (B.9) is called the ***characteristic equation*** for recurrence equation (B.8).

THEOREM 1 Let r_1, r_2 be two roots of characteristic equation (B.9) for recurrence relation (B.8).

---

1. Strictly speaking, the validity of the pattern's formula needs to be proved by mathematical induction on i. It is often easier, however, to get the solution first and then verify it (e.g., as we did for $x(n) = n(n+1)/2$).

Case 1 If r_1 and r_2 are real and distinct, the general solution to recurrence (B.8) is obtained by the formula

$$x(n) = \alpha r_1^n + \beta r_2^n,$$

where α and β are two arbitrary real constants.

Case 2 If r_1 and r_2 are equal to each other, the general solution to recurrence (B.8) is obtained by the formula

$$x(n) = \alpha r^n + \beta n r^n,$$

where $r = r_1 = r_2$ and α and β are two arbitrary real constants.

Case 3 If $r_{1,2} = u \pm iv$ are two distinct complex numbers, the general solution to recurrence (B.8) is obtained as

$$x(n) = \gamma^n[\alpha \cos n\theta + \beta \sin n\theta],$$

where $\gamma = \sqrt{u^2 + v^2}$, $\theta = \arctan v/u$, and α and β are two arbitrary real constants.

Case 1 of this theorem arises, in particular, in deriving the explicit formula for the nth Fibonacci number (Section 2.5). First, we need to rewrite the recurrence defining this sequence as

$$F(n) - F(n-1) - F(n-2) = 0.$$

Its characteristic equation is

$$r^2 - r - 1 = 0,$$

with the roots

$$r_{1,2} = \frac{1 \pm \sqrt{1 - 4(-1)}}{2} = \frac{1 \pm \sqrt{5}}{2}.$$

Since this characteristic equation has two distinct real roots, we have to use the formula indicated in Case 1 of Theorem 1:

$$F(n) = \alpha \left(\frac{1+\sqrt{5}}{2}\right)^n + \beta \left(\frac{1-\sqrt{5}}{2}\right)^n.$$

So far, we have ignored initial conditions $F(0) = 0$ and $F(1) = 1$. Now, we take advantage of them to find specific values of constants α and β. We do this by substituting 0 and 1—the values of n for which the initial conditions are given—into the last formula and equating the results to 0 and 1, respectively:

$$F(0) = \alpha \left(\frac{1+\sqrt{5}}{2}\right)^0 + \beta \left(\frac{1-\sqrt{5}}{2}\right)^0 = 0,$$

$$F(1) = \alpha \left(\frac{1+\sqrt{5}}{2}\right)^1 + \beta \left(\frac{1-\sqrt{5}}{2}\right)^1 = 1.$$

After some standard algebraic simplifications, we get the following system of two linear equations in two unknowns α and β:

$$\alpha + \qquad \beta = 0$$

$$\left(\frac{1+\sqrt{5}}{2}\right)\alpha + \left(\frac{1-\sqrt{5}}{2}\right)\beta = 1.$$

Solving the system (e.g., by substituting $\beta = -\alpha$ into the second equation and solving the equation obtained for α), we get the values $\alpha = 1/\sqrt{5}$ and $\beta = -1/\sqrt{5}$ for the unknowns. Thus,

$$F(n) = \frac{1}{\sqrt{5}}\left(\frac{1+\sqrt{5}}{2}\right)^n - \frac{1}{\sqrt{5}}\left(\frac{1-\sqrt{5}}{2}\right)^n = \frac{1}{\sqrt{5}}(\phi^n - \hat{\phi}^n),$$

where $\phi = (1+\sqrt{5})/2 \approx 1.61803$ and $\hat{\phi} = -1/\phi \approx -0.61803$.

As another example, let us solve the recurrence

$$x(n) - 6x(n-1) + 9x(n-2) = 0.$$

Its characteristic equation

$$r^2 - 6r + 9 = 0$$

has two equal roots $r_1 = r_2 = 3$. Hence, according to Case 2 of Theorem 1, its general solution is given by the formula

$$x(n) = \alpha 3^n + \beta n 3^n.$$

If we want to find its particular solution for which, say, $x(0) = 0$ and $x(1) = 3$, we substitute $n = 0$ and $n = 1$ into the last equation to get a system of two linear equations in two unknowns. Its solution is $\alpha = 0$ and $\beta = 1$, and hence the particular solution is

$$x(n) = n3^n.$$

Let us now turn to the case of inhomogeneous linear second-order recurrences with constant coefficients.

THEOREM 2 The general solution to inhomogeneous equation (B.7) can be obtained as the sum of the general solution to the corresponding homogeneous equation (B.8) and a particular solution to inhomogeneous equation (B.7).

Since Theorem 1 gives a complete recipe for finding the general solution to a homogeneous second-order linear equation with constant coefficients, Theorem 2 reduces the task of finding all solutions to equation (B.7) to finding just one particular solution to it. For an arbitrary function $f(n)$ in the right-hand side of equation (B.7), it is still a difficult task with no general help available. For a few simple classes of functions, however, a particular solution can be found. Specifically, if $f(n)$ is a nonzero constant, we can look for a particular solution that is a constant as well.

As an example, let us find the general solution to the inhomogeneous recurrence

$$x(n) - 6x(n-1) + 9x(n-2) = 4.$$

If $x(n) = c$ is its particular solution, constant c must satisfy the equation

$$c - 6c + 9c = 4,$$

which yields $c = 1$. Since we have already found above the general solution to the corresponding homogeneous equation

$$x(n) - 6x(n-1) + 9x(n-2) = 0,$$

the general solution to $x(n) - 6x(n-1) + 9x(n-2) = 4$ is obtained by the formula

$$x(n) = \alpha 3^n + \beta n 3^n + 1.$$

Before leaving this topic, we should note that the results analogous to those of Theorems 1 and 2 hold for the general ***linear kth degree recurrence with constant coefficients***,

$$a_k x(n) + a_{k-1} x(n-1) + \cdots + a_0 x(n-k) = f(n). \quad \textbf{(B.10)}$$

The practicality of this generalization is limited, however, by the necessity of finding roots of the kth degree polynomial

$$a_k r^k + a_{k-1} r^{k-1} + \cdots + a_0 = 0, \quad \textbf{(B.11)}$$

which is the characteristic equation for recurrence (B.10).

Finally, there are several other, more sophisticated techniques for solving recurrence relations. Purdom and Brown [Pur04] provide a particularly thorough discussion of this topic from the analysis of algorithms perspective.

Common Recurrence Types in Algorithm Analysis

There are a few recurrence types that arise in the analysis of algorithms with remarkable regularity. This happens because they reflect one of the fundamental design techniques.

Decrease-by-One A decrease-by-one algorithm solves a problem by exploiting a relationship between a given instance of size n and a smaller instance of size $n - 1$. Specific examples include recursive evaluation of $n!$ (Section 2.4) and insertion sort (Section 4.1). The recurrence equation for investigating the time efficiency of such algorithms typically has the following form:

$$T(n) = T(n-1) + f(n), \quad \textbf{(B.12)}$$

where function $f(n)$ accounts for the time needed to reduce an instance to a smaller one and to extend the solution of the smaller instance to a solution of the larger instance. Applying backward substitutions to (B.12) yields

$$\begin{aligned} T(n) &= T(n-1) + f(n) \\ &= T(n-2) + f(n-1) + f(n) \\ &= \dots \\ &= T(0) + \sum_{j=1}^{n} f(j). \end{aligned}$$

For a specific function $f(x)$, the sum $\sum_{j=1}^{n} f(j)$ can usually be either computed exactly or its order of growth ascertained. For example, if $f(n) = 1$, $\sum_{j=1}^{n} f(j) = n$; if $f(n) = \log n$, $\sum_{j=1}^{n} f(j) \in \Theta(n \log n)$; if $f(n) = n^k$, $\sum_{j=1}^{n} f(j) \in \Theta(n^{k+1})$. The sum $\sum_{j=1}^{n} f(j)$ can also be approximated by formulas involving integrals (see, in particular, the appropriate formulas in Appendix A).

Decrease-by-a-Constant-Factor A decrease-by-a-constant-factor algorithm solves a problem by reducing its instance of size n to an instance of size n/b ($b = 2$ for most but not all such algorithms), solving the smaller instance recursively, and then, if necessary, extending the solution of the smaller instance to a solution of the given instance. The most important example is binary search; other examples include exponentiation by squaring (introduction to Chapter 4), Russian peasant multiplication, and the fake-coin problem (Section 4.4).

The recurrence equation for investigating the time efficiency of such algorithms typically has the form

$$T(n) = T(n/b) + f(n), \qquad \textbf{(B.13)}$$

where $b > 1$ and function $f(n)$ accounts for the time needed to reduce an instance to a smaller one and to extend the solution of the smaller instance to a solution of the larger instance. Strictly speaking, equation (B.13) is valid only for $n = b^k$, $k = 0, 1, \dots$. For values of n that are not powers of b, there is typically some round-off, usually involving the floor and/or ceiling functions. The standard approach to such equations is to solve them for $n = b^k$ first. Afterward, either the solution is tweaked to make it valid for all n's (see, for example, Problem 7 in Exercises 2.4), or the order of growth of the solution is established based on the ***smoothness rule*** (Theorem 4 in this appendix).

By considering $n = b^k$, $k = 0, 1, \dots$, and applying backward substitutions to (B.13), we obtain the following:

$$\begin{aligned} T(b^k) &= T(b^{k-1}) + f(b^k) \\ &= T(b^{k-2}) + f(b^{k-1}) + f(b^k) \\ &= \cdots \\ &= T(1) + \sum_{j=1}^{k} f(b^j). \end{aligned}$$

For a specific function $f(x)$, the sum $\sum_{j=1}^{k} f(b^j)$ can usually be either computed exactly or its order of growth ascertained. For example, if $f(n) = 1$,

$$\sum_{j=1}^{k} f(b^j) = k = \log_b n.$$

If $f(n) = n$, to give another example,

$$\sum_{j=1}^{k} f(b^j) = \sum_{j=1}^{k} b^j = b\frac{b^k - 1}{b - 1} = b\frac{n - 1}{b - 1}.$$

Also, recurrence (B.13) is a special case of recurrence (B.14) covered by the ***Master Theorem*** (Theorem 5 in this appendix). According to this theorem, in particular, if $f(n) \in \Omega(n^d)$ where $d > 0$, then $T(n) \in \Omega(n^d)$ as well.

Divide-and-Conquer A divide-and-conquer algorithm solves a problem by dividing its given instance into several smaller instances, solving each of them recursively, and then, if necessary, combining the solutions to the smaller instances into a solution to the given instance. Assuming that all smaller instances have the same size n/b, with a of them being actually solved, we get the following recurrence valid for $n = b^k$, $k = 1, 2, \ldots$:

$$T(n) = aT(n/b) + f(n), \quad \textbf{(B.14)}$$

where $a \geq 1$, $b \geq 2$, and $f(n)$ is a function that accounts for the time spent on dividing the problem into smaller ones and combining their solutions. Recurrence (B.14) is called the ***general divide-and-conquer recurrence.***[2]

Applying backward substitutions to (B.14) yields the following:

2. In our terminology, for $a = 1$, it covers decrease-by-a-constant-factor, not divide-and-conquer, algorithms.

$$\begin{aligned}T(b^k) &= aT(b^{k-1}) + f(b^k)\\ &= a[aT(b^{k-2}) + f(b^{k-1})] + f(b^k) = a^2T(b^{k-2}) + af(b^{k-1}) + f(b^k)\\ &= a^2[aT(b^{k-3}) + f(b^{k-2})] + af(b^{k-1}) + f(b^k)\\ &= a^3T(b^{k-3}) + a^2f(b^{k-2}) + af(b^{k-1}) + f(b^k)\\ &= \cdots\\ &= a^kT(1) + a^{k-1}f(b^1) + a^{k-2}f(b^2) + \cdots + a^0f(b^k)\\ &= a^k[T(1) + \sum_{j=1}^{k} f(b^j)/a^j].\end{aligned}$$

Since $a^k = a^{\log_b n} = n^{\log_b a}$, we get the following formula for the solution to recurrence (B.14) for $n = b^k$:

$$T(n) = n^{\log_b a}[T(1) + \sum_{j=1}^{\log_b n} f(b^j)/a^j]. \qquad \textbf{(B.15)}$$

Obviously, the order of growth of solution $T(n)$ depends on the values of the constants a and b and the order of growth of the function $f(n)$. Under certain assumptions about $f(n)$ discussed in the next section, we can simplify formula (B.15) and get explicit results about the order of growth of $T(n)$.

Smoothness Rule and the Master Theorem We mentioned earlier that the time efficiency of decrease-by-a-constant-factor and divide-and-conquer algorithms is usually investigated first for n's that are powers of b. (Most often $b = 2$, as it is in binary search and mergesort; sometimes $b = 3$, as it is in the better algorithm for the fake-coin problem of Section 4.4, but it can be any integer greater than or equal to 2.) The question we are going to address now is when the order of growth observed for n's that are powers of b can be extended to all its values.

DEFINITION Let $f(n)$ be a nonnegative function defined on the set of natural numbers. $f(n)$ is called ***eventually nondecreasing*** if there exists some nonnegative integer n_0 so that $f(n)$ is nondecreasing on the interval $[n_0, \infty)$, i.e.,

$$f(n_1) \le f(n_2) \quad \text{for all } n_2 > n_1 \ge n_0.$$

For example, the function $(n - 100)^2$ is eventually nondecreasing, although it is decreasing on the interval [0, 100], and the function $\sin^2 \frac{\pi n}{2}$ is a function that is not eventually nondecreasing. The vast majority of functions we encounter in the analysis of algorithms *are* eventually nondecreasing. Most of them are, in fact, nondecreasing on their entire domains.

DEFINITION Let $f(n)$ be a nonnegative function defined on the set of natural numbers. $f(n)$ is called ***smooth*** if it is eventually nondecreasing and

$$f(2n) \in \Theta(f(n)).$$

It is easy to check that functions which do not grow too fast, including $\log n$, n, $n \log n$, and n^α where $\alpha \geq 0$, are smooth. For example, $f(n) = n \log n$ is smooth because

$$f(2n) = 2n \log 2n = 2n(\log 2 + \log n) = (2 \log 2)n + 2n \log n \in \Theta(n \log n).$$

Fast-growing functions, such as a^n where $a > 1$ and $n!$, are not smooth. For example, $f(n) = 2^n$ is not smooth because

$$f(2n) = 2^{2n} = 4^n \notin \Theta(2^n).$$

THEOREM 3 Let $f(n)$ be a smooth function as just defined. Then, for any fixed integer $b \geq 2$,

$$f(bn) \in \Theta(f(n)),$$

i.e., there exist positive constants c_b and d_b and a nonnegative integer n_0 such that

$$d_b f(n) \leq f(bn) \leq c_b f(n) \quad \text{for } n \geq n_0.$$

(The same assertion, with obvious changes, holds for the O and Ω notations.)

PROOF We will prove the theorem for the O notation only; the proof of the Ω part is the same. First, it is easy to check by induction that if $f(2n) \leq c_2 f(n)$ for $n \geq n_0$, then

$$f(2^k n) \leq c_2^k f(n) \quad \text{for } k = 1, 2, \ldots \text{ and } n \geq n_0.$$

The induction basis for $k = 1$ checks out trivially. For the general case, assuming that $f(2^{k-1}n) \leq c_2^{k-1} f(n)$ for $n \geq n_0$, we obtain

$$f(2^k n) = f(2 \cdot 2^{k-1} n) \leq c_2 f(2^{k-1} n) \leq c_2 c_2^{k-1} f(n) = c_2^k f(n).$$

This proves the theorem for $b = 2^k$.

Consider now an arbitrary integer $b \geq 2$. Let k be a positive integer such that $2^{k-1} \leq b < 2^k$. We can estimate $f(bn)$ above by assuming without loss of generality that $f(n)$ is nondecreasing for $n \geq n_0$:

$$f(bn) \leq f(2^k n) \leq c_2^k f(n).$$

Hence, we can use c_2^k as a required constant for this value of b to complete the proof. ■

The importance of the notions introduced above stems from the following theorem.

THEOREM 4 ***(Smoothness Rule)*** Let $T(n)$ be an eventually nondecreasing function and $f(n)$ be a smooth function. If

$$T(n) \in \Theta(f(n)) \quad \text{for values of } n \text{ that are powers of } b,$$

where $b \geq 2$, then

$$T(n) \in \Theta(f(n)).$$

(The analogous results hold for the cases of O and Ω as well.)

PROOF We will prove just the O part; the Ω part can be proved by the analogous argument. By the theorem's assumption, there exist a positive constant c and a positive integer $n_0 = b^{k_0}$ such that

$$T(b^k) \leq cf(b^k) \quad \text{for } b^k \geq n_0,$$

$T(n)$ is nondecreasing for $n \geq n_0$, and $f(bn) \leq c_b f(n)$ for $n \geq n_0$ by Theorem 3. Consider an arbitrary value of $n, n \geq n_0$. It is bracketed by two consecutive powers of b: $n_0 \leq b^k \leq n < b^{k+1}$. Therefore,

$$T(n) \leq T(b^{k+1}) \leq cf(b^{k+1}) = cf(bb^k) \leq cc_b f(b^k) \leq cc_b f(n).$$

Hence, we can use the product cc_b as a constant required by the $O(f(n))$ definition to complete the O part of the theorem's proof. ■

Theorem 4 allows us to expand the information about the order of growth established for $T(n)$ on a convenient subset of values (powers of b) to its entire domain. Here is one of the most useful assertions of this kind.

THEOREM 5 ***(Master Theorem)*** Let $T(n)$ be an eventually nondecreasing function that satisfies the recurrence

$$T(n) = aT(n/b) + f(n) \quad \text{for } n = b^k, \; k = 1, 2, \ldots$$
$$T(1) = c,$$

where $a \geq 1$, $b \geq 2$, $c > 0$. If $f(n) \in \Theta(n^d)$ where $d \geq 0$, then

$$T(n) \in \begin{cases} \Theta(n^d) & \text{if } a < b^d, \\ \Theta(n^d \log n) & \text{if } a = b^d, \\ \Theta(n^{\log_b a}) & \text{if } a > b^d. \end{cases}$$

(Similar results hold for the O and Ω notations, too.)

PROOF We will prove the theorem for the principal special case of $f(n) = n^d$. (A proof of the general case is a minor technical extension of the same argument—see, e.g., [Cor09].) If $f(n) = n^d$, equality (B.15) yields for $n = b^k$, $k = 0, 1, \ldots,$

$$T(n) = n^{\log_b a}[T(1) + \sum_{j=1}^{\log_b n} b^{jd}/a^j] = n^{\log_b a}[T(1) + \sum_{j=1}^{\log_b n} (b^d/a)^j].$$

The sum in this formula is that of a geometric series, and therefore

$$\sum_{j=1}^{\log_b n} (b^d/a)^j = (b^d/a)\frac{(b^d/a)^{\log_b n} - 1}{(b^d/a) - 1} \quad \text{if } b^d \neq a$$

and

$$\sum_{j=1}^{\log_b n} (b^d/a)^j = \log_b n \quad \text{if } b^d = a.$$

If $a < b^d$, then $b^d/a > 1$, and therefore

$$\sum_{j=1}^{\log_b n} (b^d/a)^j = (b^d/a)\frac{(b^d/a)^{\log_b n} - 1}{(b^d/a) - 1} \in \Theta((b^d/a)^{\log_b n}).$$

Hence, in this case,

$$\begin{aligned} T(n) &= n^{\log_b a}[T(1) + \sum_{j=1}^{\log_b n} (b^d/a)^j] \in n^{\log_b a}\Theta((b^d/a)^{\log_b n}) \\ &= \Theta(n^{\log_b a}(b^d/a)^{\log_b n}) = \Theta(a^{\log_b n}(b^d/a)^{\log_b n}) \\ &= \Theta(b^{d \log_b n}) = \Theta(b^{\log_b n^d}) = \Theta(n^d). \end{aligned}$$

If $a > b^d$, then $b^d/a < 1$, and therefore

$$\sum_{j=1}^{\log_b n} (b^d/a)^j = (b^d/a)\frac{(b^d/a)^{\log_b n} - 1}{(b^d/a) - 1} \in \Theta(1).$$

Hence, in this case,

$$T(n) = n^{\log_b a}[T(1) + \sum_{j=1}^{\log_b n} (b^d/a)^j] \in \Theta(n^{\log_b a}).$$

If $a = b^d$, then $b^d/a = 1$, and therefore

$$\begin{aligned} T(n) &= n^{\log_b a}[T(1) + \sum_{j=1}^{\log_b n} (b^d/a)^j] = n^{\log_b a}[T(1) + \log_b n] \\ &\in \Theta(n^{\log_b a} \log_b n) = \Theta(n^{\log_b b^d} \log_b n) = \Theta(n^d \log_b n). \end{aligned}$$

Since $f(n) = n^d$ is a smooth function for any $d \geq 0$, a reference to Theorem 4 completes the proof. ■

Theorem 5 provides a very convenient tool for a quick efficiency analysis of divide-and-conquer and decrease-by-a-constant-factor algorithms. You can find examples of such applications throughout the book.

References

[Ade62] Adelson-Velsky, G.M. and Landis, E.M. An algorithm for organization of information. *Soviet Mathematics Doklady*, vol. 3, 1962, 1259–1263.

[Adl94] Adleman, L.M. Molecular computation of solutions to combinatorial problems. *Science*, vol. 266, 1994, 1021–1024.

[Agr04] Agrawal, M., Kayal, N., and Saxena, N. PRIMES is in *P. Annals of Mathematics*, vol. 160, no. 2, 2004, 781–793.

[Aho74] Aho, A.V., Hopcroft, J.E., and Ullman, J.D. *The Design and Analysis of Computer Algorithms*. Addison-Wesley, 1974.

[Aho83] Aho, A.V., Hopcroft, J.E., and Ullman, J.D. *Data Structures and Algorithms*. Addison-Wesley, 1983.

[Ahu93] Ahuja, R.K., Magnanti, T.L., and Orlin, J.B. *Network Flows: Theory, Algorithms, and Applications*. Prentice Hall, 1993.

[App] Applegate, D.L., Bixby, R.E., Chvátal, V., and Cook, W.J. *The Traveling Salesman Problems*. www.tsp.gatech.edu/index.html.

[App07] Applegate, D.L., Bixby, R.E., Chvátal, V., and Cook, W.J. *The Traveling Salesman Problem: A Computational Study*. Princeton University Press, 2007.

[Arb93] Arbel, A. *Exploring Interior-Point Linear Programming: Algorithms and Software (Foundations of Computing)*. MIT Press, 1993.

[Aro09] Arora, S. and Barak, B. *Computational Complexity: A Modern Approach*. Cambridge University Press, 2009.

[Ata09] Atallah, M.J. and Blanton, M., eds. *Algorithms and Theory of Computation Handbook*, 2nd ed. (two-volume set), Chapman and Hall/CRC, 2009.

[Avi07] Avidan, S. and Shamir, A. Seam carving for content-aware image resizing. *ACM Transactions on Graphics*, vol. 26, no. 3, article 10, July 2007, 9 pages.

[Azi10] Aziz, A. and Prakash, A. *Algorithms for Interviews*. algorithmsforinterviews.com, 2010.

[Baa00] Baase, S. and Van Gelder, A. *Computer Algorithms: Introduction to Design and Analysis*, 3rd ed. Addison-Wesley, 2000.

[Bae81] Baecker, R. (with assistance of D. Sherman) *Sorting out Sorting*. 30-minute color sound film. Dynamic Graphics Project, University of Toronto, 1981. video.google.com/videoplay?docid=3970523862559-774879#docid=-4110947752111188923.

[Bae98] Baecker, R. Sorting out sorting: a case study of software visualization for teaching computer science. In *Software Visualization: Programming as a Multimedia Experience,* edited by J. Stasko, J. Domingue, M.C. Brown, and B.A. Price. MIT Press, 1998, 369–381.

[BaY95] Baeza-Yates, R.A. Teaching algorithms. *ACM SIGACT News,* vol. 26, no. 4, Dec. 1995, 51–59.

[Bay72] Bayer, R. and McGreight, E.M. Organization and maintenance of large ordered indices. *Acta Informatica*, vol. 1, no. 3, 1972, 173–189.

[Bel09] Bell, J., and Stevens, B. A survey of known results and research areas for *n*-queens. *Discrete Mathematics*, vol. 309, issue 1, Jan. 2009, 1–31.

[Bel57] Bellman, R.E. *Dynamic Programming*. Princeton University Press, 1957.

[Ben00] Bentley, J. *Programming Pearls*, 2nd ed. Addison-Wesley, 2000.

[Ben90] Bentley, J.L. Experiments on traveling salesman heuristics. In *Proceedings of the First Annual ACM-SIAM Symposium on Discrete Algorithms*, 1990, 91–99.

[Ben93] Bentley, J.L. and McIlroy, M.D. Engineering a sort function. *Software—Practice and Experience*, vol. 23, no. 11, 1993, 1249–1265.

[Ber03] Berlekamp, E.R., Conway, J.H., and Guy, R.K. *Winning Ways for Your Mathematical Plays*, 2nd ed., volumes 1–4. A K Peters, 2003.

[Ber00] Berlinski, D. *The Advent of the Algorithm: The Idea That Rules the World*. Harcourt, 2000.

[Ber05] Berman, K.A. and Paul, J.L. *Algorithms: Sequential, Parallel, and Distributed*. Course Technology, 2005.

[Ber01] Berstekas, D.P. *Dynamic Programming and Optimal Control: 2nd Edition (Volumes 1 and 2)*. Athena Scientific, 2001.

[Blo73] Bloom, M., Floyd, R.W., Pratt, V., Rivest, R.L., and Tarjan, R.E. Time bounds for selection. *Journal of Computer and System Sciences,* vol. 7, no. 4, 1973, 448–461.

[Bog] Bogomolny, A. *Interactive Mathematics Miscellany and Puzzles*. www.cut-the-knot.org.

[Boy77] Boyer, R.S. and Moore, J.S. A fast string searching algorithm. *Communications of the ACM,* vol. 21, no. 10, 1977, 762–772.

[Bra96] Brassard, G. and Bratley, P. *Fundamentals of Algorithmics.* Prentice Hall, 1996.

[Car79] Carmony, L. Odd pie fights. *Mathematics Teacher*, vol. 72, no. 1, 1979, 61–64.

[Cha98] Chabert, Jean-Luc, ed. *A History of Algorithms: From the Pebble to the Microchip.* Translated by Chris Weeks. Springer, 1998.

[Cha00] Chandler, J.P. Patent protection of computer programs. *Minnesota Intellectual Property Review*, vol. 1, no. 1, 2000, 33–46.

[Chr76] Christofides, N. *Worst-Case Analysis of a New Heuristic for the Traveling Salesman Problem.* Technical Report, GSIA, Carnegie-Mellon University, 1976.

[Chv83] Chvátal, V. *Linear Programming*. W.H. Freeman, 1983.

[Com79] Comer, D. The ubiquitous B-tree. *ACM Computing Surveys*, vol. 11, no. 2, 1979, 121–137.

[Coo71] Cook, S.A. The complexity of theorem-proving procedures. In *Proceeding of the Third Annual ACM Symposium on the Theory of Computing,* 1971, 151–158.

[Coo87] Coopersmith, D. and Winograd, S. Matrix multiplication via arithmetic progressions. In *Proceedings of Nineteenth Annual ACM Symposium on the Theory of Computing,* 1987, 1–6.

[Cor09] Cormen, T.H., Leiserson, C.E., Rivest, R.L., and Stein, C. *Introduction to Algorithms*, 3rd ed. MIT Press, 2009.

[Cra07] Crack, T.F. *Heard on the Street: Quantitative Questions from Wall Street Job Interviews*, 10th ed., self-published, 2007.

[Dan63] Dantzig, G.B. *Linear Programming and Extensions.* Princeton University Press, 1963.

[deB10] de Berg, M., Cheong, O., van Kreveld, M., and Overmars, M. *Computational Geometry: Algorithms and Applications*, 3rd ed. Springer, 2010.

[Dew93] Dewdney, A.K. *The (New) Turing Omnibus.* Computer Science Press, 1993.

[Dij59] Dijkstra, E.W. A note on two problems in connection with graphs. *Numerische Mathematik,* vol. 1, 1959, 269–271.

[Dij76] Dijkstra, E.W. *A Discipline of Programming*. Prentice-Hall, 1976.

[Dud70] Dudley, U. The first recreational mathematics book. *Journal of Recreational Mathematics*, 1970, 164–169.

[Eas10] Easley, D., and Kleinberg, J. *Networks, Crowds, and Markets: Reasoning About a Highly Connected World*. Cambridge University Press, 2010.

[Edm65] Edmonds, J. Paths, trees, and flowers. *Canadian Journal of Mathematics*, vol. 17, 1965, 449–467.

[Edm72] Edmonds, J. and Karp, R.M. Theoretical improvements in algorithmic efficiency for network flow problems. *Journal of the ACM*, vol. 19, no. 2, 1972, 248–264.

[Flo62] Floyd, R.W. Algorithm 97: shortest path. *Communications of the ACM*, vol. 5, no. 6, 1962, 345.

[For57] Ford, L.R., Jr., and Fulkerson, D.R. A simple algorithm for finding maximal network flows and an application to the Hitchcock problem. *Canadian Journal of Mathematics*, vol. 9, no. 2, 1957, 210–218.

[For62] Ford, L.R., Jr., and Fulkerson, D.R. *Flows in Networks.* Princeton University Press, 1962.

[For68] Forsythe, G.E. What to do till the computer scientist comes. *American Mathematical Monthly,* vol. 75, no. 5, 1968, 454–462.

[For69] Forsythe, G.E. Solving a quadratic equation on a computer. In *The Mathematical Sciences,* edited by COSRIMS and George Boehm, MIT Press, 1969, 138–152.

[Gal62] Gale, D. and Shapley, L.S. College admissions and the stability of marriage. *American Mathematical Monthly*, vol. 69, Jan. 1962, 9–15.

[Gal86] Galil, Z. Efficient algorithms for finding maximum matching in graphs. *Computing Surveys*, vol. 18, no. 1, March 1986, 23–38.

[Gar99] Gardiner, A. *Mathematical Puzzling*. Dover, 1999.

[Gar78] Gardner, M. *aha! Insight*. Scientific American/W.H. Freeman, 1978.

[Gar88] Gardner, M. *Hexaflexagons and Other Mathematical Diversions*: *The First Scientific American Book of Puzzles and Games*. University of Chicago Press, 1988.

[Gar94] Gardner, M. *My Best Mathematical and Logic Puzzles.* Dover, 1994.

[Gar79] Garey, M.R. and Johnson, D.S. *Computers and Intractability: A Guide to the Theory of NP-Completeness*. W.H. Freeman, 1979.

[Ger03] Gerald, C.F. and Wheatley, P.O. *Applied Numerical Analysis*, 7th ed. Addison-Wesley, 2003.

[Gin04] Ginat, D. Embedding instructive assertions in program design. In *Proceedings of ITiCSE'04*, June 28–30, 2004, Leeds, UK, 62–66.

[Gol94] Golomb, S.W. *Polyominoes: Puzzles, Patterns, Problems, and Packings*. Revised and expanded second edition. Princeton University Press, 1994.

[Gon91] Gonnet, G.H. and Baeza-Yates, R. *Handbook of Algorithms and Data Structures in Pascal and C*, 2nd ed. Addison-Wesley, 1991.

[Goo02] Goodrich, M.T. and Tamassia, R. *Algorithm Design: Foundations, Analysis, and Internet Examples*. John Wiley & Sons, 2002.

[Gra94] Graham, R.L., Knuth, D.E., and Patashnik, O. *Concrete Mathematics: A Foundation for Computer Science*, 2nd ed. Addison-Wesley, 1994.

[Gre07] Green, D.H. and Knuth, D.E. *Mathematics for Analysis of Algorithms*, 3rd edition. Birkhäuser, 2007.

[Gri81] Gries, D. *The Science of Programming*. Springer, 1981.

[Gus89] Gusfield, D. and Irwing, R.W. *The Stable Marriage Problem: Structure and Algorithms*. MIT Press, 1989.

[Gut07] Gutin, G. and Punnen, A.P., eds. *Traveling Salesman Problem and Its Variations*. Springer, 2007.

[Har92] Harel, D. *Algorithmics: The Spirit of Computing*, 2nd ed. Addison-Wesley, 1992.

[Har65] Hartmanis, J. and Stearns, R.E. On the computational complexity of algorithms. *Transactions of the American Mathematical Society,* vol. 117, May 1965, 285–306.

[Hea63] Heap, B.R. Permutations by interchanges. *Computer Journal*, vol. 6, 1963, 293–294.

[Het10] Hetland, M.L. *Python Algorithms: Mastering Basic Algorithms in the Python Language.* Apress, 2010.

[Hig93] Higham, N.J. The accuracy of floating point summation. *SIAM Journal on Scientific Computing*, vol. 14, no. 4, July 1993, 783–799.

[Hoa96] Hoare, C.A.R. Quicksort. In *Great Papers in Computer Science,* Phillip Laplante, ed. West Publishing Company, 1996, 31–39.

[Hoc97] Hochbaum, D.S., ed. *Approximation Algorithms for NP-Hard Problems*. PWS Publishing, 1997.

[Hop87] Hopcroft, J.E. Computer science: the emergence of a discipline. *Communications of the ACM,* vol. 30, no. 3, March 1987, 198–202.

[Hop73] Hopcroft, J.E. and Karp, R.M. An $n^{5/2}$ algorithm for maximum matchings in bipartite graphs. *SIAM Journal on Computing*, vol. 2, 1973, 225–231.

[Hor07] Horowitz, E., Sahni, S., and Rajasekaran, S. *Computer Algorithms*, 2nd ed. Silicon Press, 2007.

[Hor80] Horspool, R.N. Practical fast searching in strings. *Software—Practice and Experience,* vol. 10, 1980, 501–506.

[Hu02] Hu, T.C. and Shing, M.T. *Combinatorial Algorithms: Enlarged Second edition*. Dover, 2002.

[Huf52] Huffman, D.A. A method for the construction of minimum redundancy codes. *Proceedings of the IRE,* vol. 40, 1952, 1098–1101.

[Joh07a] Johnson, D.S. and McGeoch, L.A. Experimental analysis of heuristics for the STSP. In *The Traveling Salesman Problem and Its Variations,* edited by G. Gutin and A.P. Punnen, Springer, 2007, 369–443.

[Joh07b] Johnson, D.S., Gutin, G., McGeoch, L.A., Yeo, A., Zhang, W., and Zverovitch, A. Experimental analysis of heuristics for the ATSP. In *The Traveling Salesman Problem and Its Variations,* edited by G. Gutin and A.P. Punnen, Springer, 2007, 445–487.

[Joh04] Johnsonbaugh, R. and Schaefer, M. *Algorithms*. Pearson Education, 2004.

[Kar84] Karmarkar, N. A new polynomial-time algorithm for linear programming. *Combinatorica*, vol. 4, no. 4, 1984, 373–395.

[Kar72] Karp, R.M. Reducibility among combinatorial problems. In *Complexity of Computer Communications*, edited by R.E. Miller and J.W. Thatcher. Plenum Press, 1972, 85–103.

[Kar86] Karp, R.M. Combinatorics, complexity, and randomness. *Communications of the ACM*, vol. 29, no. 2, Feb. 1986, 89–109.

[Kar91] Karp, R.M. An introduction to randomized algorithms. *Discrete Applied Mathematics,* vol. 34, Nov. 1991, 165–201.

[Kel04] Kellerer, H., Pferschy, U., and Pisinger, D. *Knapsack Problems*. Springer, 2004.

[Ker99] Kernighan, B.W. and Pike. R. *The Practice of Programming*. Addison-Wesley, 1999.

[Kha79] Khachian, L.G. A polynomial algorithm in linear programming. *Soviet Mathematics Doklady*, vol. 20, 1979, 191–194.

[Kle06] Kleinberg, J. and Tardos, É. *Algorithm Design.* Pearson, 2006.

[Knu75] Knuth, D.E. Estimating the efficiency of backtrack programs. *Mathematics of Computation,* vol. 29, Jan. 1975, 121–136.

[Knu76] Knuth, D.E. Big omicron and big omega and big theta. *ACM SIGACT News*, vol. 8, no. 2, 1976, 18–23.

[Knu96] Knuth, D.E. *Selected Papers on Computer Science*. CSLI Publications and Cambridge University Press, 1996.

[KnuI] Knuth, D.E. *The Art of Computer Programming, Volume 1: Fundamental Algorithms*, 3rd ed. Addison-Wesley, 1997.

[KnuII] Knuth, D.E. *The Art of Computer Programming, Volume 2: Seminumerical Algorithms*, 3rd ed. Addison-Wesley, 1998.

[KnuIII] Knuth, D.E. *The Art of Computer Programming, Volume 3: Sorting and Searching*, 2nd ed. Addison-Wesley, 1998.

[KnuIV] Knuth, D.E. *The Art of Computer Programming, Volume 4A, Combinatorial Algorithms, Part 1*. Addison-Wesley, 2011.

[Knu77] Knuth, D.E., Morris, Jr., J.H., and Pratt, V.R. Fast pattern matching in strings. *SIAM Journal on Computing*, vol. 5, no. 2, 1977, 323–350.

[Kol95] Kolman, B. and Beck, R.E. *Elementary Linear Programming with Applications*, 2nd ed. Academic Press, 1995.

[Kor92] Kordemsky, B.A. *The Moscow Puzzles*. Dover, 1992.

[Kor05] Kordemsky, B.A. *Mathematical Charmers*. Oniks, 2005 (in Russian).

[Kru56] Kruskal, J.B. On the shortest spanning subtree of a graph and the traveling salesman problem. *Proceedings of the American Mathematical Society*, vol. 7, 1956, 48–50.

[Laa10] Laakmann, G. *Cracking the Coding Interview*, 4th ed. CareerCup, 2010.

[Law85] Lawler, E.L., Lenstra, J.K., Rinnooy Kan, A.H.G., and Shmoys, D.B., eds. *The Traveling Salesman Problem*. John Wiley, 1985.

[Lev73] Levin, L.A. Universal sorting problems. *Problemy Peredachi Informatsii,* vol. 9, no. 3, 1973, 115–116 (in Russian). English translation in *Problems of Information Transmission*, vol. 9, 265–266.

[Lev99] Levitin, A. Do we teach the right algorithm design techniques? In *Proceedings of SIGCSE'99*, New Orleans, LA, 1999, 179–183.

[Lev11] Levitin, A. and Levitin, M. *Algorithmic Puzzles*. Oxford University Press, 2011.

[Lin73] Lin, S. and Kernighan, B.W. An effective heuristic algorithm for the traveling-salesman problem. *Operations Research*, vol. 21, 1973, 498–516.

[Man89] Manber, U. *Introduction to Algorithms: A Creative Approach.* Addison-Wesley, 1989.

[Mar90] Martello, S. and Toth, P. *Knapsack Problems: Algorithms and Computer Implementations*. John Wiley, 1990.

[Mic10] Michalewicz, Z. and Fogel, D.B. *How to Solve It: Modern Heuristics*, second, revised and extended edition. Springer, 2010.

[Mil05] Miller, R. and Boxer, L. *Algorithms Sequential and Parallel: A Unified Approach*, 2nd ed. Charles River Media, 2005.

[Mor91] Moret, B.M.E. and Shapiro, H.D. *Algorithms from P to NP. Volume I: Design and Efficiency.* Benjamin Cummings, 1991.

[Mot95] Motwani, R. and Raghavan, P. *Randomized Algorithms.* Cambridge University Press, 1995.

[Nea09] Neapolitan, R. and Naimipour, K. *Foundations of Algorithms, Fourth Edition*. Jones and Bartlett, 2009.

[Nem89] Nemhauser, G.L., Rinnooy Kan, A.H.G., and Todd, M.J., eds. *Optimization*. North-Holland, Amsterdam, 1989.

[OCo98] O'Connor, J.J. and Robertson, E.F. *The MacTutor History of Mathematics archive,* June 1998, www-history.mcs.st-andrews.ac.uk/history/Mathematicians/Abel.html.

[Ong84] Ong, H.L. and Moore, J.B. Worst-case analysis of two traveling salesman heuristics. *Operations Research Letters*, vol. 2, 1984, 273–277.

[ORo98] O'Rourke, J. *Computational Geometry in C*, 2nd ed. Cambridge University Press, 1998.

[Ove80] Overmars, M.H. and van Leeuwen, J. Further comments on Bykat's convex hull algorithm. *Information Processing Letters,* vol. 10, no. 4/5, 1980, 209–212.

[Pan78] Pan, V.Y. Strassen's algorithm is not optimal. *Proceedings of Nineteenth Annual IEEE Symposium on the Foundations of Computer Science*, 1978, 166–176.

[Pap82] Papadimitriou, C.H. and Steiglitz, K. *Combinatorial Optimization: Algorithms and Complexity*. Prentice-Hall, 1982.

[Par95] Parberry, I. *Problems on Algorithms*. Prentice-Hall, 1995.

[Pol57] Pólya, G. *How to Solve It: A New Aspect of Mathematical Method*, 2nd ed. Doubleday, 1957.

[Pre85] Preparata, F.P. and Shamos, M.I. *Computational Geometry: An Introduction.* Springer, 1985.

[Pri57] Prim, R.C. Shortest connection networks and some generalizations. *Bell System Technical Journal,* vol. 36, no. 1, 1957, 1389–1401.

[Pur04] Purdom, P.W., Jr., and Brown, C. *The Analysis of Algorithms.* Oxford University Press, 2004.

[Raw91] Rawlins, G.J.E. *Compared to What? An Introduction to the Analysis of Algorithms*. Computer Science Press, 1991.

[Rei77] Reingold, E.M., Nievergelt, J., and Deo, N. *Combinatorial Algorithms: Theory and Practice.* Prentice-Hall, 1977.

[Riv78] Rivest, R.L., Shamir, A., and Adleman, L.M. A method for obtaining digital signatures and public-key cryptosystems. *Communications of the ACM,* vol. 21, no. 2, Feb. 1978, 120–126.

[Ros07] Rosen, K. *Discreet Mathematics and Its Applications*, 6th ed., McGraw-Hill, 2007.

[Ros77] Rosenkrantz, D.J., Stearns, R.E., and Lewis, P.M. An analysis of several heuristics for the traveling salesman problem. *SIAM Journal of Computing*, vol. 6, 1977, 563–581.

[Roy59] Roy, B. Transitivité et connexité. *Comptes rendus de l'Académie des Sciences*, vol. 249, 216–218, 1959.

[Sah75] Sahni, S. Approximation algorithms for the 0/1 knapsack problem. *Journal of the ACM*, vol. 22, no. 1, Jan. 1975, 115–124.

[Sah76] Sahni, S. and Gonzalez, T. *P*-complete approximation problems. *Journal of the ACM*, vol. 23, no. 3, July 1976, 555–565.

[Say05] Sayood, K. *Introduction to Data Compression*, 3rd ed. Morgan Kaufmann Publishers, 2005.

[Sed02] Sedgewick, R. *Algorithms in C/C++/Java, Parts 1–5: Fundamentals, Data Structures, Sorting, Searching, and Graph Algorithms*, 3rd ed. Addison-Wesley Professional, 2002.

[Sed96] Sedgewick, R. and Flajolet, P. *An Introduction to the Analysis of Algorithms.* Addison-Wesley Professional, 1996.

[Sed11] Sedgewick, R. and Wayne, K. *Algorithms, Fourth Edition.* Pearson Education, 2011.

[Sha07] Shaffer, C.A., Cooper, M., and Edwards, S.H. Algorithm visualization: a report on the state of the field. *ACM SIGCSE Bulletin,* vol. 39, no. 1, March 2007, 150–154.

[Sha98] Shasha, D. and Lazere, C. *Out of Their Minds: The Lives and Discoveries of 15 Great Computer Scientists.* Copernicus, 1998.

[She59] Shell, D.L. A high-speed sorting procedure. *Communications of the ACM*, vol. 2, no. 7, July 1959, 30–32.

[Sho94] Shor, P.W. Algorithms for quantum computation: discrete algorithms and factoring. *Proceedings 35th Annual Symposium on Foundations of Computer Science* (Shafi Goldwasser, ed.). IEEE Computer Society Press, 1994, 124–134.

[Sip05] Sipser, M. *Introduction to the Theory of Computation*, 2nd ed. Course Technology, 2005.

[Ski10] Skiena, S.S. *Algorithm Design Manual*, 2nd ed. Springer, 2010.

[Str69] Strassen, V. Gaussian elimination is not optimal. *Numerische Mathematik,* vol. 13, no. 4, 1969, 354–356.

[Tar83] Tarjan, R.E. *Data Structures and Network Algorithms.* Society for Industrial and Applied Mathematics, 1983.

[Tar85] Tarjan, R.E. Amortized computational complexity. *SIAM Journal on Algebraic and Discrete Methods,* vol. 6, no. 2, Apr. 1985, 306–318.

[Tar87] Tarjan, R.E. Algorithm design. *Communications of the ACM,* vol. 30, no. 3, March 1987, 204–212.

[Tar84] Tarjan, R.E. and van Leeuwen, J. Worst-case analysis of set union algorithms. *Journal of the ACM*, vol. 31, no. 2, Apr. 1984, 245–281.

[War62] Warshall, S. A theorem on boolean matrices. *Journal of the ACM,* vol. 9, no. 1, Jan. 1962, 11–12.

[Wei77] Weide, B. A survey of analysis techniques for discrete algorithms. *Computing Surveys*, vol. 9, no. 4, 1977, 291–313.

[Wil64] Williams, J.W.J. Algorithm 232 (heapsort). *Communications of the ACM,* vol. 7, no. 6, 1964, 347–348.

[Wir76] Wirth, N. *Algorithms + Data Structures = Programs.* Prentice-Hall, Englewood Cliffs, NJ, 1976.

[Yan08] Yanofsky, N.S. and Mannucci, M.A. *Quantum Computing for Computer Scientists*. Cambridge University Press, 2008.

[Yao82] Yao, F. Speed-up in dynamic programming. *SIAM Journal on Algebraic and Discrete Methods,* vol. 3, no. 4, 1982, 532–540.

Hints to Exercises

CHAPTER 1

Exercises 1.1

1. It is probably faster to do this by searching the Web, but your library should be able to help, too.

2. One can find arguments supporting either view. There is a well-established principle pertinent to the matter, though: scientific facts or mathematical expressions of them are not patentable. (Why do you think this is the case?) But should this preclude granting patents for all algorithms?

3. You may assume that you are writing your algorithms for a human rather than a machine. Still, make sure that your descriptions do not contain obvious ambiguities. Knuth provides an interesting comparison between cooking recipes and algorithms [KnuI, p. 6].

4. There is a quite straightforward algorithm for this problem based on the definition of $\lfloor\sqrt{n}\rfloor$.

5. Try to design an algorithm that always makes less than mn comparisons.

6. **a.** Just follow Euclid's algorithm as described in the text.

   b. Compare the number of divisions made by the two algorithms.

7. Prove that if d divides both m and n (i.e., $m = sd$ and $n = td$ for some positive integers s and t), then it also divides both n and $r = m \bmod n$ and vice versa. Use the formula $m = qn + r$ $(0 \le r < n)$ and the fact that if d divides two integers u and v, it also divides $u + v$ and $u - v$ (why?).

8. Perform one iteration of the algorithm for two arbitrarily chosen integers $m < n$.

9. The answer to part (a) can be given immediately, the answer to part (b) can be given by checking the algorithm's performance on all pairs $1 < m < n \le 10$.

10. a. Use the equality

$$\gcd(m, n) = \gcd(m - n, n) \quad \text{for } m \geq n > 0.$$

b. The key is to figure out the total number of distinct integers that can be written on the board, starting with an initial pair m, n where $m > n \geq 1$. You should exploit a connection of this question to the question of part (a). Considering small examples, especially those with $n = 1$ and $n = 2$, should help, too.

11. Of course, for some coefficients, the equation will have no solutions.

12. Tracing the algorithm by hand for, say, $n = 10$ and studying its outcome should help answering both questions.

Exercises 1.2

1. The farmer would have to make several trips across the river, starting with the only one possible.

2. Unlike the Old World puzzle of Problem 1, the first move solving this puzzle is not obvious.

3. The principal issue here is a possible ambiguity.

4. Your algorithm should work correctly for all possible values of the coefficients, including zeros.

5. You almost certainly learned this algorithm in one of your introductory programming courses. If this assumption is not true, you have a choice between designing such an algorithm on your own or looking it up.

6. You may need to make a field trip to refresh your memory.

7. Question (a) is difficult, though the answer to it—discovered in the 1760s by the German mathematician Johann Lambert—is well-known. By comparison, question (b) is incomparably simpler.

8. You probably know two or more different algorithms for sorting an array of numbers.

9. You can: decrease the number of times the inner loop is executed, make that loop run faster (at least for some inputs), or, more significantly, design a faster algorithm from scratch.

Exercises 1.3

1. Trace the algorithm on the input given. Use the definitions of stability and being in-place that were introduced in the section.

2. If you do not recall any searching algorithms, you should design a simple searching algorithm (without succumbing to the temptation to find one in the latter chapters of the book).

3. This algorithm is introduced later in the book, but you should have no trouble designing it on your own.

4. If you have not encountered this problem in your previous courses, you may look up the answers on the Web or in a discrete structures textbook. The answers are, in fact, surprisingly simple.

5. No efficient algorithm for solving this problem for an arbitrary graph is known. This particular graph does have Hamiltonian circuits that are not difficult to find. (You need to find just one of them.)

6. **a.** Put yourself (mentally) in a passenger's place and ask yourself what criterion for the "best" route you would use. Then think of people that may have different needs.

   b. The representation of the problem by a graph is straightforward. Give some thoughts, though, to stations where trains can be changed.

7. **a.** What are tours in the traveling salesman problem?

   b. It would be natural to consider vertices colored the same color as elements of the same subset.

8. Create a graph whose vertices represent the map's regions. You will have to decide on the edges on your own.

9. Assume that the circumference in question exists and find its center first. Also, do not forget to give a special answer for $n \leq 2$.

10. Be careful not to miss some special cases of the problem.

Exercises 1.4

1. **a.** Take advantage of the fact that the array is not sorted.

   b. We used this trick in implementing one of the algorithms in Section 1.1.

2. **a.** For a sorted array, there is a spectacularly efficient algorithm you almost certainly have heard about.

   b. Unsuccessful searches can be made faster.

3. **a.** *Push(x)* puts x on the top of the stack; *pop* deletes the item from the top of the stack.

   b. *Enqueue(x)* adds x to the rear of the queue; *dequeue* deletes the item from the front of the queue.

4. Just use the definitions of the graph properties in question and data structures involved.

5. There are two well-known algorithms that can solve this problem. The first uses a stack; the second uses a queue. Although these algorithms are discussed later in the book, do not miss this chance to discover them by yourself!

6. The inequality $h \leq n - 1$ follows immediately from the height's definition. The lower bound inequality follows from the inequality $2^{h+1} - 1 \geq n$, which can be

proved by considering the largest number of vertices a binary tree of height h can have.

7. You need to indicate how each of the three operations of the priority queue will be implemented.

8. Because of insertions and deletions, using an array of the dictionary's elements (sorted or unsorted) is not the best implementation possible.

9. You need to know about postfix notation in order to answer one of these questions. (If you are not familiar with it, find the information on the Internet.)

10. There are several algorithms for this problem. Keep in mind that the words may contain multiple occurrences of the same letter.

CHAPTER 2

Exercises 2.1

1. The questions are indeed as straightforward as they appear, though some of them may have alternative answers. Also, keep in mind the caveat about measuring an integer's size.

2. a. The sum of two matrices is defined as the matrix whose elements are the sums of the corresponding elements of the matrices given.

b. Matrix multiplication requires two operations: multiplication and addition. Which of the two would you consider basic and why?

3. Will the algorithm's efficiency vary on different inputs of the same size?

4. a. Gloves are not socks: they can be right-handed and left-handed.

b. You have only two qualitatively different outcomes possible. Find the number of ways to get each of the two.

5. a. First, prove that if a positive decimal integer n has b digits in its binary representation, then

$$2^{b-1} \le n < 2^b.$$

Then, take binary logarithms of the terms in these inequalities.

b. The proof is similar to the proof of formula (2.1).

c. The formulas will be the same, with just one small adjustment to account for the different radix.

d. How can we switch from one logarithm base to another?

6. Insert a verification of whether the problem is already solved.

7. A similar question was investigated in the section.

8. Use either the difference between or the ratio of $f(4n)$ and $f(n)$, whichever is more convenient for getting a compact answer. If it is possible, try to get an answer that does not depend on n.

9. If necessary, simplify the functions in question to single out terms defining their orders of growth to within a constant multiple. (We discuss formal methods for answering such questions in the next section; however, the questions can be answered without knowledge of such methods.)

10. a. Use the formula $\sum_{i=0}^{n} 2^i = 2^{n+1} - 1$.

b. Use the formula for the sum of the first n odd numbers or the formula for the sum of arithmetic progression.

Exercises 2.2

1. Use the corresponding counts of the algorithm's basic operation (see Section 2.1) and the definitions of O, Θ, and Ω.

2. Establish the order of growth of $n(n+1)/2$ first and then use the informal definitions of O, Θ, and Ω. (Similar examples were given in the section.)

3. Simplify the functions given to single out the terms defining their orders of growth.

4. a. Check carefully the pertinent definitions.

b. Compute the ratio limits of every pair of consecutive functions on the list.

5. First, simplify some of the functions. Then, use the list of functions in Table 2.2 to "anchor" each of the functions given. Prove their final placement by computing appropriate limits.

6. a. You can prove this assertion either by computing an appropriate limit or by applying mathematical induction.

b. Compute $\lim_{n\to\infty} a_1^n/a_2^n$.

7. Prove the correctness of (a), (b), and (c) by using the appropriate definitions; construct a counterexample for (d) (e.g., by constructing two functions behaving differently for odd and even values of their arguments).

8. The proof of part (a) is similar to the one given for the theorem's assertion in Section 2.2. Of course, different inequalities need to be used to bound the sum from below.

9. Follow the analysis plan used in the text when the algorithm was mentioned for the first time.

10. You may use straightforward algorithms for all the four questions asked. Use the O notation for the time efficiency class of one of them, and the Θ notation for the three others.

11. The problem can be solved in two weighings.

12. You should walk intermittently left and right from your initial position until the door is reached.

Exercises 2.3

1. Use the common summation formulas and rules listed in Appendix A. You may need to perform some simple algebraic operations before applying them.
2. Find a sum among those in Appendix A that looks similar to the sum in question and try to transform the latter to the former. Note that you do not have to get a closed-form expression for a sum before establishing its order of growth.
3. Just follow the formulas in question.
4. **a.** Tracing the algorithm to get its output for a few small values of n (e.g., $n = 1$, 2, and 3) should help if you need it.

   b. We faced the same question for the examples discussed in this section. One of them is particularly pertinent here.

   c. Follow the plan outlined in the section.

   d. As a function of n, the answer should follow immediately from your answer to part (c). You may also want to give an answer as a function of the number of bits in the n's representation (why?).

   e. Have you not encountered this sum somewhere?
5. **a.** Tracing the algorithm to get its output for a few small values of n (e.g., $n = 1$, 2, and 3) should help if you need it.

   b. We faced the same question for the examples discussed in the section. One of them is particularly pertinent here.

   c. You can either follow the section's plan by setting up and computing a sum or answer the question directly. (Try to do both.)

   d. Your answer will immediately follow from the answer to part (c).

   e. Does the algorithm always have to make two comparisons on each iteration? This idea can be developed further to get a more significant improvement than the obvious one—try to do it for a four-element array and then generalize the insight. But can we hope to find an algorithm with a better than linear efficiency?
6. **a.** Elements $A[i, j]$ and $A[j, i]$ are symmetric with respect to the main diagonal of the matrix.

   b. There is just one candidate here.

   c. You may investigate the worst case only.

   d. Your answer will immediately follow from the answer to part (c).

   e. Compare the problem the algorithm solves with the way it does this.
7. Computing a sum of n numbers can be done with $n - 1$ additions. How many does the algorithm make in computing each element of the product matrix?
8. Set up a sum for the number of times all the doors are toggled and find its asymptotic order of growth by using some formulas from Appendix A.

9. For the general step of the proof by induction, use the formula

$$\sum_{i=1}^{n+1} i = \sum_{i=1}^{n} i + (n+1).$$

The young Gauss computed the sum $1+2+\cdots+99+100$ by noticing that it can be computed as the sum of 50 pairs, each with the same sum.

10. There are at least two different ways to solve this problem, which comes from a collection of Wall Street interview questions.

11. a. Setting up a sum should pose no difficulties. Using the standard summation formulas and rules will require more effort than in the previous examples, however.

b. Optimize the algorithm's innermost loop.

12. Set up a sum for the number of squares after n iterations of the algorithm and then simplify it to get a closed-form answer.

13. To derive a formula expressing the total number of digits as a function of the number of pages n, where $1 \leq n \leq 1000$, it is convenient to partition the function's domain into several natural intervals.

Exercises 2.4

1. Each of these recurrences can be solved by the method of backward substitutions.

2. The recurrence relation in question is almost identical to the recurrence relation for the number of multiplications, which was set up and solved in the section.

3. a. The question is similar to that about the efficiency of the recursive algorithm for computing $n!$.

b. Write pseudocode for the nonrecursive algorithm and determine its efficiency.

4. a. Note that you are asked here about a recurrence for the function's values, not about a recurrence for the number of times its operation is executed. Just follow the pseudocode to set it up. It is easier to solve this recurrence by *forward* substitutions (see Appendix B).

b. This question is very similar to one we have already discussed.

c. You may want to include the subtractions needed to decrease n.

5. a. Use the formula for the number of disk moves derived in the section.

b. Solve the problem for three disks to investigate the number of moves made by each of the disks. Then generalize the observations and prove their validity for the general case of n disks.

6. The required algorithm and the method of its analysis are similar to those of the classic version of the puzzle. Because of the additional constraint, more than two smaller instances of the puzzle need to be solved here.

7. a. Consider separately the cases of even and odd values of n and show that for both of them $\lfloor \log_2 n \rfloor$ satisfies the recurrence relation and its initial condition.

b. Just follow the algorithm's pseudocode.

8. a. Use the formula $2^n = 2^{n-1} + 2^{n-1}$ without simplifying it; do not forget to provide a condition for stopping your recursive calls.

b. A similar algorithm was investigated in the section.

c. A similar question was investigated in the section.

d. A bad efficiency class of an algorithm by itself does not mean that the algorithm is bad. For example, the classic algorithm for the Tower of Hanoi puzzle is optimal despite its exponential-time efficiency. Therefore, a claim that a particular algorithm is not good requires a reference to a better one.

9. a. Tracing the algorithm for $n = 1$ and $n = 2$ should help.

b. It is very similar to one of the examples discussed in the section.

10. Get the basic operation count either by solving a recurrence relation or by computing directly the number of the adjacency matrix elements the algorithm checks in the worst case.

11. a. Use the definition's formula to get the recurrence relation for the number of multiplications made by the algorithm.

b. Investigate the right-hand side of the recurrence relation. Computing the first few values of $M(n)$ may be helpful, too.

12. You might want to use the neighborhood's symmetry to obtain a simple formula for the number of squares added to the neighborhood on the nth iteration of the algorithm.

13. The minimum amount of time needed to fry three hamburgers is smaller than 4 minutes.

14. Solve first a simpler version in which a celebrity must be present.

Exercises 2.5

1. Use a search engine.

2. Set up an equation expressing the number of rabbits after n months in terms of the number of rabbits in some previous months.

3. There are several ways to solve this problem. The most elegant of them makes it possible to put the problem in this section.

4. Writing down the first, say, ten Fibonacci numbers makes the pattern obvious.

5. It is easier to substitute ϕ^n and $\hat{\phi}^n$ into the recurrence equation separately. Why will this suffice?

6. Use an approximate formula for $F(n)$ to find the smallest values of n to exceed the numbers given.

7. Set up the recurrence relations for $C(n)$ and $Z(n)$, with appropriate initial conditions, of course.

8. All the information needed on each iteration of the algorithm is the values of the last two consecutive Fibonacci numbers. Modify the algorithm $Fib(n)$ to take advantage of this fact.

9. Prove it by mathematical induction.

10. Consider first a small example such as computing gcd(13, 8).

11. Take advantage of the special nature of the rectangle's dimensions.

12. The last k digits of an integer N can be obtained by computing $N \bmod 10^k$. Performing all operations of your algorithms modulo 10^k (see Appendix A) will enable you to circumvent the exponential growth of the Fibonacci numbers. Also note that Section 2.6 is devoted to a general discussion of the empirical analysis of algorithms.

Exercises 2.6

1. Does it return a correct comparison count for every array of size 2?

2. Debug your comparison counting and random input generating for small array sizes first.

3. On a reasonably fast desktop, you may well get zero time, at least for smaller sizes in your sample. Section 2.6 mentions a trick for overcoming this difficulty.

4. Check how fast the count values grow with doubling the input size.

5. A similar question was discussed in the section.

6. Compare the values of the functions $\lg \lg n$ and $\lg n$ for $n = 2^k$.

7. Insert the division counter in a program implementing the algorithm and run it for the input pairs in the range indicated.

8. Get the empirical data for random values of n in a range of, say, between 10^2 and 10^4 or 10^5 and plot the data obtained. (You may want to use different scales for the axes of your coordinate system.)

CHAPTER 3

Exercises 3.1

1. a. Think of algorithms that have impressed you with their efficiency and/or sophistication. Neither characteristic is indicative of a brute-force algorithm.

b. Surprisingly, it is not a very easy question to answer. Mathematical problems (including those you've studied in your secondary school and college courses) are a good source of such examples.

2. a. The first question was all but answered in the section. Expressing the answer as a function of the number of bits can be done by using the formula relating the two metrics.

b. How can we compute $(ab) \bmod m$?

3. It helps to have done the exercises in question.

4. a. The most straightforward algorithm, which is based on substituting x_0 into the formula, is quadratic.

b. Analyzing what unnecessary computations the quadratic algorithm does should lead you to a better (linear) algorithm.

c. How many coefficients does a polynomial of degree n have? Can one compute its value at an arbitrary point without processing all of them?

5. For each of the three network topologies, what properties of the matrix should the algorithm check?

6. The answer to four of the questions is yes.

7. a. Just apply the brute-force thinking to the problem in question.

b. The problem can be solved in one weighing.

8. Just trace the algorithm on the input given. (It was done for another input in the section.)

9. Although the majority of elementary sorting algorithms are stable, do not rush with your answer. A general remark about stability made in Section 1.3, where the notion of stability is introduced, could be helpful, too.

10. Generally speaking, implementing an algorithm for a linked list poses problems if the algorithm requires accessing the list's elements not in sequential order.

11. Just trace the algorithm on the input given. (See an example in the section.)

12. a. A list is sorted if and only if all its adjacent elements are in a correct order. Why?

b. Add a boolean flag to register the presence or absence of switches.

c. Identify worst-case inputs first.

13. Can bubblesort change the order of two equal elements in its input?

14. Thinking about the puzzle as a sorting-like problem may or may not lead you to the most simple and efficient solution.

Exercises 3.2

1. Modify the analysis of the algorithm's version in Section 2.1.

2. As a function of p, what kind of function is C_{avg}?

3. Solve a simpler problem with a single gadget first. Then design a better than linear algorithm for the problem with two gadgets.

4. The content of this quote from Mahatma Gandhi is more thought provoking than this drill.

5. For each input, one iteration of the algorithm yields all the information you need to answer the question.

6. It will suffice to limit your search for an example to binary texts and patterns.

7. The answer, surprisingly, is yes.

8. a. For a given occurrence of A in the text, what are the substrings you need to count?

b. For a given occurrence of B in the text, what are the substrings you need to count?

9. You may use either bit strings or a natural-language text for the visualization program. It would be a good idea to implement, as an option, a search for all occurrences of a given pattern in a given text.

10. Test your program thoroughly. Be especially careful about the possibility of words read diagonally with wrapping around the table's border.

11. A (very) brute-force algorithm can simply shoot at adjacent feasible cells starting at, say, one of the corners of the board. Can you suggest a better strategy? (You can investigate relative efficiencies of different strategies by making two programs implementing them play each other.) Is your strategy better than the one that shoots at randomly generated cells of the opponent's board?

Exercises 3.3

1. You may want to consider two versions of the answer: without taking into account the comparison and assignments in the algorithm's innermost loop and with them.

2. Sorting n real numbers can be done in $O(n \log n)$ time.

3. a. Solving the problem for $n = 2$ and $n = 3$ should lead you to the critical insight.

b. Where would you put the post office if it did not have to be at one of the village locations?

4. a. Check requirements (i)–(iii) by using basic properties of absolute values.

b. For the Manhattan distance, the points in question are defined by the equation $|x - 0| + |y - 0| = 1$. You can start by sketching the points in the positive quadrant of the coordinate system (i.e., the points for which $x, y \geq 0$) and then sketch the rest by using the symmetries.

c. The assertion is false. You can choose, say, $p_1(0, 0)$ and $p_2(1, 0)$ and find p_3 to complete a counterexample.

5. **a.** Prove that the Hamming distance does satisfy the three axioms of a distance metric.

   b. Your answer should include two parameters.

6. True; prove it by mathematical induction.

7. Your answer should be a function of two parameters: n and k. A special case of this problem (for $k = 2$) is solved in the text.

8. Review the examples given in the section.

9. Some of the extreme points of a convex hull are easier to find than others.

10. If there are other points of a given set on the straight line through p_i and p_j, which of all these points need to be preserved for further processing?

11. Your program should work for any set of n distinct points, including sets with many collinear points.

12. **a.** The set of points satisfying inequality $ax + by \le c$ is the half-plane of the points on one side of the straight line $ax + by = c$, including all the points on the line itself. Sketch such a half-plane for each of the inequalities and find their intersection.

    b. The extreme points are the vertices of the polygon obtained in part (a).

    c. Compute and compare the values of the objective function at the extreme points.

Exercises 3.4

1. **a.** Identify the algorithm's basic operation and count the number of times it will be executed.

   b. For each of the time amounts given, find the largest value of n for which this limit won't be exceeded.

2. How different is the traveling salesman problem from the problem of finding a Hamiltonian circuit?

3. Your algorithm should check the well-known condition that is both necessary and sufficient for the existence of an Eulerian circuit in a connected graph.

4. Generate the remaining $4! - 6 = 18$ possible assignments, compute their costs, and find the one with the minimal cost.

5. Make the size of your counterexample as small as possible.

6. Rephrase the problem so that the sum of elements in one subset, rather than two, needs to be checked on each try of a possible partition.

7. Follow the definitions of a clique and of an exhaustive-search algorithm.

8. Try all possible orderings of the elements given.

9. Use common formulas of elementary combinatorics.

10. **a.** Add all the elements in the magic square in two different ways.

    b. What combinatorial objects do you have to generate here?

11. a. For testing, you may use alphametic collections available on the Internet.

b. Given the absence of electronic computers in 1924, you must refrain here from using the Internet.

Exercises 3.5

1. a. Use the definitions of the adjacency matrix and adjacency lists given in Section 1.4.

b. Perform the DFS traversal the same way it is done for another graph in the text (see Figure 3.10).

2. Compare the efficiency classes of the two versions of DFS for sparse graphs.

3. a. What is the number of such trees equal to?

b. Answer this question for connected graphs first.

4. Perform the BFS traversal the same way it is done in the text (see Figure 3.11).

5. You may use the fact that the level of a vertex in a BFS tree indicates the number of edges in the shortest (minimum-edge) path from the root to that vertex.

6. a. What property of a BFS forest indicates a cycle's presence? (The answer is similar to the one for a DFS forest.)

b. The answer is no. Find two examples supporting this answer.

7. Given the fact that both traversals can reach a new vertex if and only if it is adjacent to one of the previously visited vertices, which vertices will be visited by the time either traversal halts (i.e., its stack or queue becomes empty)?

8. Use a DFS forest and a BFS forest for parts (a) and (b), respectively.

9. Use either DFS or BFS.

10. a. Follow the instructions of the problem's statement.

b. Trying both traversals should lead you to a correct answer very fast.

11. You can apply BFS without an explicit sketch of a graph representing the states of the puzzle.

CHAPTER 4

Exercises 4.1

1. Solve the problem for $n = 1$.

2. You may consider pouring soda from a filled glass into an empty glass as one move.

3. It's easier to use the bottom-up approach.

4. Use the fact that all the subsets of an n-element set $S = \{a_1, \ldots, a_n\}$ can be divided into two groups: those that contain a_n and those that do not.

5. The answer is no.

6. Use the same idea that underlies insertion sort.
7. Trace the algorithm as we did in the text for another input (see Figure 4.4).
8. a. The sentinel should stop the smallest element from moving beyond the first position in the array.
   b. Repeat the analysis performed in the text for the sentinel version.
9. Recall that one can access elements of a singly linked list only sequentially.
10. Compare the running times of the algorithm's inner loop.
11. a. Answering the questions for an array of three elements should lead to the general answers.
    b. Assume for simplicity that all elements are distinct and that inserting $A[i]$ in each of the $i+1$ possible positions among its predecessors is equally likely. Analyze the sentinel version of the algorithm first.
12. a. Note that it's more convenient to sort sublists in parallel, i.e., compare $A[0]$ with $A[h_i]$, then $A[1]$ with $A[1+h_i]$, and so on.
    b. Recall that, generally speaking, sorting algorithms that can exchange elements far apart are not stable.

Exercises 4.2

1. Trace the algorithm as it is done in the text for another digraph (see Figure 4.7).
2. a. You need to prove two assertions: (i) if a digraph has a directed cycle, then the topological sorting problem does not have a solution; (ii) if a digraph has no directed cycles, then the problem has a solution.
   b. Consider an extreme type of a digraph.
3. a. How does it relate to the time efficiency of DFS?
   b. Do you know the length of the list to be generated by the algorithm? Where should you put, say, the first vertex being popped off a DFS traversal stack for the vertex to be in its final position?
4. Try to do this for a small example or two.
5. Trace the algorithm on the instances given as it is done in the section (see Figure 4.8).
6. a. Use a proof by contradiction.
   b. If you have difficulty answering the question, consider an example of a digraph with a vertex with no incoming edges and write down its adjacency matrix.
   c. The answer follows from the definitions of the source and adjacency lists.
7. For each vertex, store the number of edges entering the vertex in the remaining subgraph. Maintain a queue of the source vertices.
9. a. Trace the algorithm on the input given by following the steps of the algorithm as indicated.

b. Determine the efficiency for each of the three principal steps of the algorithm and then determine the overall efficiency. Of course, the answers depend on whether a digraph is represented by its adjacency matrix or by its adjacency lists.

10. Take advantage of topological sorting and the graph's symmetry.

Exercises 4.3

1. Use standard formulas for the numbers of these combinatorial objects. For the sake of simplicity, you may assume that generating one combinatorial object takes the same time as, say, one assignment.

2. We traced the algorithms on smaller instances in the section.

3. See an outline of this algorithm in the section.

4. a. Trace the algorithm for $n = 2$; take advantage of this trace in tracing the algorithm for $n = 3$ and then use the latter for $n = 4$.

b. Show that the algorithm generates $n!$ permutations and that all of them are distinct. Use mathematical induction.

c. Set up a recurrence relation for the number of swaps made by the algorithm. Find its solution and the solution's order of growth. You may need the formula: $e \approx \sum_{i=0}^{n} \frac{1}{i!}$ for large values of n.

5. We traced both algorithms on smaller instances in the section.

6. Tricks become boring after they have been given away.

7. This is not a difficult exercise because of the obvious way of getting bit strings of length n from bit strings of length $n - 1$.

8. You may still mimic the binary addition without using it explicitly.

9. Just trace the algorithms for $n = 4$.

10. There are several decrease-and–conquer algorithms for this problem. They are more subtle than one might expect. Generating combinations in a predefined order (increasing, decreasing, lexicographic) helps with both a design and a correctness proof. The following simple property is very helpful in that regard. Assuming with no loss of generality that the underlying set is $\{1, 2, \ldots, n\}$, there are $\binom{n-i}{k-1}$ k-subsets whose smallest element is i, $i = 1, 2, \ldots, n - k + 1$.

11. Represent the disk movements by flipping bits in a binary n-tuple.

12. Thinking about the switches as bits of a bit string could be helpful but not necessary.

Exercises 4.4

1. Take care of the length of the longest piece present.

2. If the instance of size n is to compute $\lfloor \log_2 n \rfloor$, what is the instance of size $n/2$? What is the relationship between the two?

3. For part (a), take advantage of the formula that gives the immediate answer. The most efficient prop for answering questions (b)–(d) is a binary search tree that mirrors the algorithm's operations in searching for an arbitrary search key.

4. Estimate the ratio of the average number of key comparisons made by sequential search to the average number made by binary search in successful searches.

5. How would you reach the middle element in a linked list?

6. **a.** Use the comparison $K \le A[m]$ where $m \leftarrow \lfloor (l + r)/2 \rfloor$ until $l = r$. Then check whether the search is successful or not.

   b. The analysis is almost identical to that of the text's version of binary search.

7. Number the pictures and use this numbering in your questions.

8. The algorithm is quite similar to binary search, of course. In the worst case, how many key comparisons does it make on each iteration and what fraction of the array remains to be processed?

9. Start by comparing the middle element $A[m]$ with $m + 1$.

10. It is obvious how one needs to proceed if $n \bmod 3 = 0$ or $n \bmod 3 = 1$; it is somewhat less so if $n \bmod 3 = 2$.

11. **a.** Trace the algorithm for the numbers given as it is done in the text for another input (see Figure 4.14b).

    b. How many iterations does the algorithm perform?

12. You may implement the algorithm either recursively or nonrecursively.

13. The fastest way to answer the question is to use the formula that exploits the binary representation of n, which is mentioned at the end of the section.

14. Use the binary representation of n.

15. **a.** Use forward substitutions (see Appendix B) into the recurrence equations given in the text.

    b. On observing the pattern in the first 15 values of n obtained in part (a), express it analytically. Then prove its validity by mathematical induction.

    c. Start with the binary representation of n and translate into binary the formula for $J(n)$ obtained in part (b).

Exercises 4.5

1. **a.** The answer follows immediately from the formula underlying Euclid's algorithm.

   b. Let $r = m \bmod n$. Investigate two cases of r's value relative to n's value.

2. Trace the algorithm on the input given, as is done in the section for another input.

3. The nonrecursive version of the algorithm was applied to a particular instance in the section's example.

4. Write an equation of the straight line through the points $(l, A[l])$ and $(r, A[r])$ and find the x coordinate of the point on this line whose y coordinate is v.

5. Construct an array for which interpolation search decreases the remaining subarray by one element on each iteration.

6. a. Solve the inequality $\log_2 \log_2 n + 1 > 6$.

b. Compute $\lim_{n\to\infty} \frac{\log \log n}{\log n}$. Note that to within a constant multiple, one can consider the logarithms to be natural, i.e., base e.

7. a. The definition of the binary search tree suggests such an algorithm.

b. What is the worst-case input for your algorithm? How many key comparisons does it make on such an input?

8. a. Consider separately three cases, (i) the key's node is a leaf, (ii) the key's node has one child, (iii) the key's node has two children.

b. Assume that you know a location of the key to be deleted.

9. Starting at an arbitrary vertex of the graph, traverse a sequence of its untraversed edges until either all the edges are traversed or no untraversed edge is available.

10. Follow the plan used in the section for analyzing the normal version of the game.

11. Play several rounds of the game on the graph paper to become comfortable with the problem. Considering special cases of the spoiled square's location should help you to solve it.

12. Do yourself a favor: try to design an algorithm on your own. It does not have to be optimal, but it should be reasonably efficient.

13. Start by comparing the search number with the last element in the first row.

CHAPTER 5

Exercises 5.1

1. In more than one respect, this question is similar to the divide-and-conquer computation of the sum of n numbers.

2. Unlike Problem 1, a divide-and-conquer algorithm for this problem can be more efficient by a constant factor than the brute-force algorithm.

3. How would you compute a^8 by solving two exponentiation problems of size 4? How about a^9?

4. Look at the notations used in the theorem's statement.

5. Apply the Master Theorem.

6. Trace the algorithm as it was done for another input in the section.

7. How can mergesort reverse a relative ordering of two elements?

8. a. Use backward substitutions, as usual.

b. What inputs minimize the number of key comparisons made by mergesort? How many comparisons are made by mergesort on such inputs during the merging stage?

c. Do not forget to include key moves made both before the split and during the merging.

9. Modify mergesort to solve the problem.

11. A divide-and-conquer algorithm works by reducing a problem's instance to several smaller instances of the *same* problem.

Exercises 5.2

1. We traced the algorithm on another instance in the section.

2. Use the rules for stopping the scans.

3. The definition of stability of a sorting algorithm was given in Section 1.3. Your example does not have to be large.

4. Trace the algorithm to see on which inputs index i gets out of bounds.

5. Study what the section's version of quicksort does on such arrays. You should base your answers on the number of key comparisons, of course.

6. Where will splits occur on the inputs in question?

7. a. Computing the ratio $n^2/(n \log_2 n)$ for $n = 10^6$ is incorrect.

b. Think the best-case and worst-case inputs.

8. Use the partition idea.

9. a. You may want to first solve the two-color flag problem, i.e., rearrange efficiently an array of R's and B's. (A similar problem is Problem 8 in this section's exercises.)

b. Extend the definition of a partition.

11. Use the partition idea.

Exercises 5.3

1. The problem is almost identical to the one discussed in the section.

2. Trace the algorithm on a small input.

3. This can be done by an algorithm discussed in an earlier chapter of the book.

4. Use strong induction on the number of internal nodes.

5. This is a standard exercise that you have probably done in your data structures course. With the traversal definitions given at the end of the section, you should be able to trace them even if you have never encountered these algorithms before.

6. Your pseudocode can simply mirror the traversal definition.

7. If you do not know the answer to this important question, you may want to check the results of the traversals on a small binary search tree. For a proof, answer this question: What can be said about two nodes with keys k_1 and k_2 if $k_1 < k_2$?

8. Find the root's label of the binary tree first, and then identify the labels of the nodes in its left and right subtrees.

9. Use strong induction on the number of internal nodes.

11. Breaking the chocolate bar can be represented by a binary tree.

Exercises 5.4

1. You might want to answer the question for $n = 2$ first and then generalize it.

2. Trace the algorithm on the input given. You will have to use it again in order to compute the products of two-digit numbers as well.

3. **a.** Take logarithms of both sides of the equality.

   b. What did we use the closed-form formula for?

4. **a.** How do we multiply by powers of 10?

   b. Try to repeat the argument for, say, $98 * 76$.

5. Counting the number of one-digit additions made by the pen-and-pencil algorithm in multiplying, say, two four-digit numbers, should help answer the general question.

6. Check the formulas by simple algebraic manipulations.

7. Trace Strassen's algorithm on the input given. (It takes some work, but it would have been much more of it if you were asked to stop the recursion when $n = 1$.) It is a good idea to check your answer by multiplying the matrices by the brute-force (definition-based) algorithm, too.

8. Use the method of backward substitutions to solve the recurrence given in the text.

9. The recurrence for the number of multiplications in Pan's algorithm is similar to that for Strassen's algorithm. Use the Master Theorem to find the order of growth of its solution.

Exercises 5.5

1. **a.** How many points need to be considered in the combining-solutions stage of the algorithm?

b. Design a simpler algorithm in the same efficiency class.

2. Divide the rectangle in Figure 5.7b into eight congruent rectangles and show that each of these rectangles can contain no more than one point of interest.

3. Recall (see Section 5.1) that the number of comparisons made by mergesort in the worst case is $C_{worst}(n) = n \log_2 n - n + 1$ (for $n = 2^k$). You may use just the highest-order term of this formula in the recurrence you need to set up.

6. The answer to part (a) comes directly from a textbook on plane geometry.

7. Use the formula relating the value of a determinant with the area of a triangle.

8. It must be in $\Omega(n)$, of course. (Why?)

9. Design a sequence of n points for which the algorithm decreases the problem's size just by 1 on each of its recursive calls.

11. Apply an idea used in this section to construct a decagon with its vertices at ten given points.

12. The path cannot cross inside the fenced area, but it can go along the fence.

CHAPTER 6

Exercises 6.1

1. This problem is similar to one of the examples in the section.

2. a. Compare every element in one set with all the elements in the other.

b. In fact, you can use presorting in three different ways: sort elements of just one of the sets, sort elements of each of the sets separately, and sort elements of the two sets together.

3. a. How do we find the smallest and largest elements in a sorted list?

b. The brute-force algorithm and the divide-and-conquer algorithm are both linear.

4. Use the known results about the average-case comparison numbers of the algorithms in this question.

5. a. The problem is similar to one of the preceding problems in these exercises.

b. How would you solve this problem if the student information were written on index cards? Better yet, think how somebody else, who has never taken a course on algorithms but possesses a good dose of common sense, would solve this problem.

6. a. Many problems of this kind have exceptions for one particular configuration of points. As to the question about a solution's uniqueness, you can get the answer by considering a few small "random" instances of the problem.

b. Construct a polygon for a few small "random" instances of the problem. Try to construct polygons in some systematic fashion.

7. It helps to think about real numbers as ordered points on the real line. Considering the special case of $s = 0$, with a given array containing both negative and positive numbers, might be helpful, too.

8. After sorting the a_i's and b_i's, the problem can be solved in linear time.

9. Start by sorting the number list given.

10. **a.** Sort the points in nondecreasing order of their x coordinates and then scan them right to left.

    b. Think of choice problems with two desirable characteristics to take into account.

11. Use the presorting idea twice.

Exercises 6.2

1. Trace the algorithm as we did in solving another system in the section.

2. **a.** Use the Gaussian elimination results as explained in the text.

   b. It is one of the varieties of the transform-and-conquer technique. Which one?

3. To find the inverse, you can either solve the system with three simultaneous right-hand side vectors representing the columns of the 3×3 identity matrix or use the LU decomposition of the system's coefficient matrix found in Problem 2.

4. Though the final answer is correct, its derivation contains an error you have to find.

5. Pseudocode of this algorithm is quite straightforward. If you are in doubt, see the section's example tracing the algorithm. The order of growth of the algorithm's running time can be found by following the standard plan for the analysis of nonrecursive algorithms.

6. Estimate the ratio of the algorithm running times by using the approximate formulas for the number of divisions and the number of multiplications in both algorithms.

7. **a.** This is a "normal" case: one of the two equations should not be proportional to the other.

   b. The coefficients of one equation should be the same or proportional to the corresponding coefficients of the other equation, whereas the right-hand sides should not.

   c. The two equations should be either the same or proportional to each other (including the right-hand sides).

8. **a.** Manipulate the matrix rows above a pivot row the same way the rows below the pivot row are changed.

   b. Are the Gauss-Jordan method and Gaussian elimination based on the same algorithm design technique or on different ones?

c. Derive a formula for the number of multiplications in the Gauss-Jordan method in the same manner this was done for Gaussian elimination in the section.

9. How long will it take to compute the determinant compared to the time needed to apply Gaussian elimination to the system?

10. a. Apply Cramer's rule to the system given.

b. How many distinct determinants are there in the Cramer's rule formulas?

11. a. If x_{ij} is the number of times the panel in the ith row and jth column needs to be toggled in a solution, what can be said about x_{ij}? After you answer this question, show that the binary matrix representing an initial state of the board can be represented as a linear combination (in modulo-2 arithmetic) of n^2 binary matrices each representing the effect of toggling an individual panel.

b. Set up a system of four equations in four unknowns (see part (a)) and solve it by Gaussian elimination, performing all operations in modulo-2 arithmetic.

c. If you believe that a system of nine equations in nine unknowns is too large to solve by hand, write a program to solve the problem.

Exercises 6.3

1. Use the definition of AVL trees. Do not forget that an AVL tree is a special case of a binary search tree.

2. For both questions, it is easier to construct the required trees bottom up, i.e., for smaller values of n first.

3. The single L-rotation and the double RL-rotation are the mirror images of the single R-rotation and the double LR-rotation, whose diagrams can be found in the section.

4. Insert the keys one after another doing appropriate rotations the way it was done in the section's example.

5. a. An efficient algorithm immediately follows from the definition of the binary search tree of which the AVL tree is a special case.

b. The correct answer is opposite to the one that immediately comes to mind.

7. a. Trace the algorithm for the input given (see Figure 6.8 for an example).

b. Keep in mind that the number of key comparisons made in searching for a key in a 2-3 tree depends not only on its node's depth but also on whether the key is the first or second one in the node.

8. False; find a simple counterexample.

9. Where will the smallest and largest keys be located?

Exercises 6.4

1. **a.** Trace the algorithm outlined in the text on the input given.
   b. Trace the algorithm outlined in the text on the input given.
   c. A mathematical fact may not be established by checking its validity on a single example.
2. For a heap represented by an array, only the parental dominance requirement needs to be checked.
3. **a.** What structure does a complete tree of height h with the largest number of nodes have? What about a complete tree with the smallest number of nodes?
   b. Use the results established in part (a).
4. First, express the right-hand side as a function of h. Then, prove the obtained equality by either using the formula for the sum $\sum i2^i$ given in Appendix A or by mathematical induction on h.
5. **a.** Where in a heap should one look for its smallest element?
   b. Deleting an arbitrary element of a heap can be done by generalizing the algorithm for deleting its root.
6. Fill in a table with the time efficiency classes of efficient implementations of the three operations: finding the largest element, finding and deleting the largest element, and adding a new element.
7. Trace the algorithm on the inputs given (see Figure 6.14 for an example).
8. As a rule, sorting algorithms that can exchange far-apart elements are not stable.
9. One can claim that the answers are different for the two principal representations of a heap.
10. This algorithm is less efficient than heapsort because it uses the array rather than the heap to implement the priority queue.
12. Pick the spaghetti rods up in a bundle and place them end down (i.e., vertically) onto a tabletop.

Exercises 6.5

1. Set up sums and simplify them by using the standard formulas and rules for sum manipulation. Do not forget to include the multiplications outside the inner loop.
2. Take advantage of the fact that the value of x^i can be easily computed from the previously computed x^{i-1}.
3. **a.** Use the formulas for the number of multiplications (and additions) for both algorithms.
   b. Does Horner's rule use any extra memory?

4. Apply Horner's rule to the instance given the same way it is applied to another one in the section.

5. Compute $p(2)$ where $p(x) = x^8 + x^7 + x^5 + x^2 + 1$.

6. If you implement the algorithm for long division by $x - c$ efficiently, the answer might surprise you.

7. **a.** Trace the left-to-right binary exponentiation algorithm on the instance given the same way it is done for another instance in the section.

   b. The answer is yes: the algorithm can be extended to work for the zero exponent as well. How?

8. Trace the right-to-left binary exponentiation algorithm on the instance given the same way it is done for another instance in the section.

9. Compute and use the binary digits of n "on the fly."

10. Use a formula for the sum of the terms of this special kind of a polynomial.

11. Compare the number of operations needed to implement the task in question.

12. Although there exists exactly one such polynomial, there are several different ways to represent it. You may want to generalize *Lagrange's interpolation formula* for $n = 2$:

$$p(x) = y_1 \frac{x - x_2}{x_1 - x_2} + y_2 \frac{x - x_1}{x_2 - x_1}$$

Exercises 6.6

1. **a.** Use the rules for computing $\text{lcm}(m, n)$ and $\gcd(m, n)$ from the prime factors of m and n.

   b. The answer immediately follows from the formula for computing lcm (m, n).

2. Use a relationship between minimization and maximization problems.

3. Prove the assertion by induction on k.

4. **a.** Base your algorithm on the following observation: a graph contains a cycle of length 3 if and only if it has two adjacent vertices i and j that are also connected by a path of length 2.

   b. Do not jump to a conclusion in answering this question.

5. An easier solution is to reduce the problem to another one with a known algorithm. Since we did not discuss many geometric algorithms in the book, it should not be difficult to figure out to which one this problem needs to be reduced.

6. Express this problem as a maximization problem of a function in one variable.

7. Introduce double-indexed variables x_{ij} to indicate an assignment of the ith person to the jth job.

8. Take advantage of the specific features of this instance to reduce the problem to one with fewer variables.

9. Create a new graph.

10. Solve first the one-dimensional version of the post office location problem (Problem 3(a) in Exercises 3.3).

11. a. Create a state-space graph for the problem as it is done for the river-crossing puzzle in the section.

b. Create a state-space graph for the problem.

c. Look at the state obtained after the first six river crossings in the solution to part (b).

12. The problem can be solved by reduction to a well-known problem about a graph traversal.

CHAPTER 7

Exercises 7.1

1. Yes, it is possible. How?

2. Check the algorithm's pseudocode to see what it does upon encountering equal values.

3. Trace the algorithm on the input given (see Figure 7.2 for an example).

4. Check whether the algorithm can reverse a relative ordering of equal elements.

5. Where will $A[i]$ be in the sorted array?

6. Take advantage of the standard traversals of such trees.

7. a. Follow the definitions of the arrays B and C in the description of the method.

b. Find, say, $B[C[3]]$ for the example in part (a).

8. Start by finding the target positions for all the statures.

9. a. Use linked lists to hold nonzero elements of the matrices.

b. Represent each of the given polynomials by a linked list with nodes containing exponent i and coefficient a_i for each nonzero term $a_i x^i$.

10. You may use a search of the literature/Internet to answer this question.

Exercises 7.2

1. Trace the algorithm in the same way it is done in the section for another instance of the string-matching problem.

2. A special alphabet notwithstanding, this application is not different than applications to natural-language strings.

3. For each pattern, fill in its shift table and then determine the number of character comparisons (both successful and unsuccessful) on each trial and the total number of trials.
4. Find an example of a binary string of length m and a binary string of length n ($n \geq m$) so that Horspool's algorithm makes
   a. the largest possible number of character comparisons before making the smallest possible shift.
   b. the smallest possible number of character comparisons.
5. It is logical to try a worst-case input for Horspool's algorithm.
6. Can the algorithm shift the pattern by more than one position without the possibility of missing another matching substring?
7. For each pattern, fill in the two shift tables and then determine the number of character comparisons (both successful and unsuccessful) on each trial and the total number of trials.
8. Check the description of the Boyer-Moore algorithm.
9. Check the descriptions of the algorithms.
11. **a.** A brute-force algorithm fits the bill here.
    b. Enhance the input before a search.

Exercises 7.3

1. Apply the open hashing (separate chaining) scheme to the input given, as is done in the text for another input (see Figure 7.5). Then compute the largest number and average number of comparisons for successful searches in the constructed table.
2. Apply the closed hashing (open addressing) scheme to the input given as it is done in the text for another input (see Figure 7.6). Then compute the largest number and average number of comparisons for successful searches in the constructed table.
3. How many different addresses can such a hash function produce? Would it distribute keys evenly?
4. The question is quite similar to computing the probability of having the same result in n throws of a fair die.
5. Find the probability that n people have different birthdays. As to the hashing connection, what hashing phenomenon deals with coincidences?
6. **a.** There is no need to insert a new key at the end of the linked list it is hashed to.
   b. Which operations are faster in a sorted linked list and why? For sorting, do we have to copy all elements in the nonempty lists in an array and then apply a general purpose sorting algorithm, or is there a way to take advantage of the sorted order in each of the nonempty linked lists?

7. A direct application of hashing solves the problem.

8. Consider this question as a mini-review: the answers are in Section 7.3 for hashing and in the appropriate sections of the book for the others. Of course, you should use the best algorithms available.

9. If you need to refresh your memory, check the book's table of contents.

Exercises 7.4

1. Thinking about searching for information should lead to a variety of examples.

2. **a.** Use the standard rules of sum manipulation and, in particular, the geometric series formula.

   b. You will need to take logarithms base $\lceil m/2 \rceil$ in your derivation.

3. Find this value from the inequality in the text that provides the upper-bound of the B-tree's height.

4. Follow the insertion algorithm outlined in the section.

5. The algorithm is suggested by the definition of the B-tree.

6. **a.** Just follow the description of the algorithm given in the statement of the problem. Note that a new key is always inserted in a leaf and that full nodes are always split on the way down, even though the leaf for the new key may have a room for it.

   b. Can a split of a full node cause a cascade of splits through the chain of its ancestors? Can we get a taller search tree than necessary?

CHAPTER 8

Exercises 8.1

1. Compare the definitions of the two techniques.

2. Use the table generated by the dynamic programming algorithm in solving the problem's instance in Example 1 of the section.

3. **a.** The analysis is similar to that of the top-down recursive computation of the nth Fibonacci number in Section 2.5.

   b. Set up and solve a recurrence for the number of candidate solutions that need to be processed by the exhaustive search algorithm.

4. Apply the dynamic programming algorithm to the instance given as it is done in Example 2 of the section. Note that there are two optimal coin combinations here.

5. Adjust formula (8.5) for inadmissible cells and their immediate neighbors.

6. The problem is similar to the change-making problem discussed in the section.

7. **a.** Relate the number of the rook's shortest paths to the square in the ith row and the jth column of the chessboard to the numbers of the shortest paths to the adjacent squares.

   b. Consider one shortest path as 14 consecutive moves to adjacent squares.

8. One can solve the problem in quadratic time.

9. Use a well-known formula from elementary combinatorics relating $C(n, k)$ to smaller binomial coefficients.

10. **a.** Topologically sort the dag's vertices first.

    b. Create a dag with $n + 1$ vertices: one vertex to start and the others to represent the coins given.

11. Let $F(i, j)$ be the order of the largest all-zero submatrix of a given matrix with its low right corner at (i, j). Set up a recurrence relating $F(i, j)$ to $F(i - 1, j)$, $F(i, j - 1)$, and $F(i - 1, j - 1)$.

12. **a.** In the situation where teams A and B need i and j games, respectively, to win the series, consider the result of team A winning the game and the result of team A losing the game.

    b. Set up a table with five rows ($0 \leq i \leq 4$) and five columns ($0 \leq j \leq 4$) and fill it by using the recurrence derived in part (a).

    c. Your pseudocode should be guided by the recurrence set up in part (a). The efficiency answers follow immediately from the table's size and the time spent on computing each of its entries.

Exercises 8.2

1. **a.** Use formulas (8.6)–(8.7) to fill in the appropriate table, as is done for another instance of the problem in the section.

   b., c. What would the equality of the two terms in

   $$\max\{F(i - 1, j), v_i + F(i - 1, j - w_i)\}$$

   mean?

2. **a.** Write pseudocode to fill the table in Figure 8.4 (say, row by row) by using formulas (8.6)–(8.7).

   b. An algorithm for identifying an optimal subset is outlined in the section via an example.

3. How many values does the algorithm compute? How long does it take to compute one value? How many table cells need to be traversed to identify the composition of an optimal subset?

4. Use the definition of $F(i, j)$ to check whether it is always true that

   a. $F(i, j - 1) \leq F(i, j)$ for $1 \leq j \leq W$.

   b. $F(i - 1, j) \leq F(i, j)$ for $1 \leq i \leq n$.

5. The problem is similar to one of the problems discussed in Section 8.1.

6. Trace the calls of the function *MemoryKnapsack*(i, j) on the instance in question. (An application to another instance can be found in the section.)
7. The algorithm applies formula (8.6) to fill *some* of the table's cells. Why can we still assert that its efficiencies are in $\Theta(nW)$?
8. One of the reasons deals with the time efficiency; the other deals with the space efficiency.
9. You may want to include algorithm visualizations in your report.

Exercises 8.3

1. Continue applying formula (8.8) as prescribed by the algorithm.
2. **a.** The algorithm's time efficiency can be investigated by following the standard plan of analyzing the time efficiency of a nonrecursive algorithm.

   b. How much space do the two tables generated by the algorithm use?
3. $k = R[1, n]$ indicates that the root of an optimal tree is the kth key in the list of ordered keys $a_1, \ldots, a_n$. The roots of its left and right subtrees are specified by $R[1, k-1]$ and $R[k+1, n]$, respectively.
4. Use a space-for-time trade-off.
5. If the assertion were true, would we not have a simpler algorithm for constructing an optimal binary search tree?
6. The structure of the tree should simply minimize the average depth of its nodes. Do not forget to indicate a way to distribute the keys among the nodes of the tree.
7. **a.** Since there is a one-to-one correspondence between binary search trees for a given set of n orderable keys and binary trees with n nodes (why?), you can count the latter. Consider all the possibilities of partitioning the nodes between the left and right subtrees.

   b. Compute the values in question using the two formulas.

   c. Use the formula for the nth Catalan number and Stirling's formula for $n!$.
8. Change the bounds of the innermost loop of algorithm *OptimalBST* by exploiting the monotonicity of the root table mentioned at the end of the section.
9. Assume that $a_1, \ldots, a_n$ are distinct keys ordered from the smallest to the largest, $p_1, \ldots, p_n$ are the probabilities of searching for them, and $q_0, q_1, \ldots, q_n$ are probabilities of unsuccessful searches for keys in intervals $(-\infty, a_1)$, $(a_1, a_2), \ldots, (a_n, \infty)$, respectively; $(p_1 + \cdots + p_n) + (q_0 + \cdots + q_n) = 1$. Set up a recurrence relation similar to recurrence (8.8) for the expected number of key comparisons that takes into account both successful and unsuccessful searches.
10. See the memory function solution for the knapsack problem in Section 8.2.
11. **a.** It is easier to find a general formula for the number of multiplications needed for computing $(A_1 \cdot A_2) \cdot A_3$ and $A_1 \cdot (A_2 \cdot A_3)$ for matrices A_1 with

dimensions $d_0 \times d_1$, A_2 with dimensions $d_1 \times d_2$, and A_3 with dimensions $d_2 \times d_3$ and then choose some specific values for the dimensions to get a required example.

b. You can get the answer by following the approach used for counting binary trees.

c. The recurrence relation for the optimal number of multiplications in computing $A_i \cdot \ldots \cdot A_j$ is very similar to the recurrence relation for the optimal number of comparisons in searching a binary search tree composed of keys $a_i, \ldots, a_j$.

Exercises 8.4

1. Apply the algorithm to the adjacency matrix given, as is done in the section for another matrix.

2. a. The answer can be obtained either by considering how many values the algorithm computes or by following the standard plan for analyzing the efficiency of a nonrecursive algorithm (i.e., by setting up a sum to count its basic operation's executions).

b. What is the efficiency class of the traversal-based algorithm for sparse graphs represented by their adjacency lists?

3. Show that one can simply overwrite elements of $R^{(k-1)}$ with elements of $R^{(k)}$ without any other changes in the algorithm.

4. What happens if $R^{(k-1)}[i, k] = 0$?

5. Show first that formula (8.11) (from which the superscripts can be eliminated according to the solution to Problem 3)

$$r_{ij} = r_{ij} \textbf{ or } (r_{ik} \textbf{ and } r_{kj})$$

is equivalent to

$$\textbf{if } r_{ik} \; r_{ij} \leftarrow (r_{ij} \textbf{ or } r_{kj}).$$

6. a. What property of the transitive closure indicates a presence of a directed cycle? Is there a better algorithm for checking this?

b. Which elements of the transitive closure of an undirected graph are equal to 1? Can you find such elements with a faster algorithm?

7. See an example of applying the algorithm to another instance in the section.

8. What elements of matrix $D^{(k-1)}$ does $d_{ij}^{(k)}$, the element in the ith row and the jth column of matrix $D^{(k)}$, depend on? Can these values be changed by the overwriting?

9. Your counterexample must contain a cycle of a negative length.

10. It will suffice to store, in a single matrix P, indices of intermediate vertices k used in updates of the distance matrices. This matrix can be initialized with all its elements equal to, say, -1.

CHAPTER 9

Exercises 9.1

1. You may use integer divisions in your algorithm.
2. You can apply the greedy approach either to each of its rows (or columns) or to the entire cost matrix.
3. Considering the case of two jobs might help. Of course, after forming a hypothesis, you will have to prove the algorithm's optimality for an arbitrary input or find a specific counterexample showing that it is not the case.
4. Only the earliest-finish-first algorithm always yields an optimal solution.
5. Simply apply the greedy approach to the situation at hand. You may assume that $t_1 \le t_2 \le \cdots \le t_n$.
6. Think the minimum positive amount of water among all the vessels in their current state.
7. The minimum number of messages for $n = 4$ is six.
8. For both versions of the problem, it is not difficult to get to a hypothesis about the solution's form after considering the cases of $n = 1$, 2, and 3. It is proving the solutions' optimality that is at the heart of this problem.
9. **a.** Trace the algorithm for the graph given. An example can be found in the text.

   b. After the next fringe vertex is added to the tree, add all the unseen vertices adjacent to it to the priority queue of fringe vertices.
10. Applying Prim's algorithm to a weighted graph that is not connected should help in answering this question.
11. Check whether the proof of the algorithm's correctness is valid for negative edge weights.
12. The answer is no. Give a counterexample.
13. Since Prim's algorithm needs weights on a graph's edges, some weights have to be assigned. As to the second question, think of other algorithms that can solve this problem.
14. Strictly speaking, the wording of the question asks you to prove two things: the fact that at least one minimum spanning tree exists for any weighted connected graph and the fact that a minimum spanning tree is unique if all the weights are distinct numbers. The proof of the former stems from the obvious observation about finiteness of the number of spanning trees for a weighted connected

graph. The proof of the latter can be obtained by repeating the correctness proof of Prim's algorithm with a minor adjustment at the end.

15. Consider two cases: the key's value was decreased (this is the case needed for Prim's algorithm) and the key's value was increased.

Exercises 9.2

1. Trace the algorithm for the given graphs the same way it is done for another input in the section.

2. Two of the four assertions are true; the other two are false.

3. Applying Kruskal's algorithm to a disconnected graph should help to answer the question.

4. One way to answer the question is to transform a graph with negative weights to one with all positive weights.

5. Is the general trick of transforming maximization problems to their minimization counterparts (see Section 6.6) applicable here?

6. Substitute the three operations of the disjoint subsets' ADT—$makeset(x)$, $find(x)$, and $union(x, y)$—in the appropriate places of the algorithm's pseudocode given in the section.

7. Follow the plan used in Section 9.1 to prove the correctness of Prim's algorithm.

8. The argument is very similar to the one made in the section for the union-by-size version of quick find.

11. The question is not trivial, because introducing extra points (called *Steiner points*) may make the total length of the network smaller than that of a minimum spanning tree of the square. Solving first the problem for three equidistant points might give you an indication of what a solution to the problem in question might look like.

Exercises 9.3

1. One of the questions requires no changes in either the algorithm or the graph; the others require simple adjustments.

2. Just trace the algorithm on the input graphs the same way it was done for an example in the section.

3. Your counterexample can be a graph with just three vertices.

4. Only one of the assertions is correct. Find a small counterexample for the other.

5. Simplify the pseudocode given in the section by implementing the priority queue as an unordered array and eliminating the parental labeling of vertices.

6. Prove it by induction on the number of vertices included in the tree constructed by the algorithm.

7. Topologically sort the dag's vertices first.

8. To get a graph, connect numbers on adjacent levels that can be components of a sum from the apex to the base. Then figure out how to deal with the fact that the weights are assigned to vertices rather than edges.

9. Take advantage of the ways of thinking used in geometry and physics.

10. Before you embark on implementing a shortest-path algorithm, you would have to decide what criterion determines the "best route." Of course, it would be highly desirable to have a program asking the user which of several possible criteria s/he wants to be applied.

Exercises 9.4

1. See the example given in the section.

2. After combining the two nodes with the lowest probabilities, resolve the tie arising on the next iteration in two different ways. For each of the two Huffman codes obtained, compute the mean and variance of the codeword length.

3. You may base your answers on the way Huffman's algorithm works or on the fact that Huffman codes are known to be optimal prefix codes.

4. The maximal length of a codeword relates to the height of Huffman's coding tree in an obvious fashion. Try to find a set of n specific frequencies for an alphabet of size n for which the tree has the shape yielding the longest codeword possible.

5. **a.** What is the most appropriate data structure for an algorithm whose principal operation is finding the two smallest elements in a given set and replacing them by their sum?

   b. Identify the principal operations of the algorithm, the number of times they are executed, and their efficiencies for the data structure used.

6. Maintain two queues: one for given frequencies, the other for weights of new trees.

7. It would be natural to use one of the standard traversal algorithms.

8. Generate the codewords right to left.

10. A similar example was discussed at the end of Section 9.4. Construct Huffman's tree and then come up with specific questions that would yield that tree. (You are allowed to ask questions such as: Is this card the ace, or a seven, or an eight?)

CHAPTER 10

Exercises 10.1

1. Start at an arbitrary integer point x and investigate whether a neighboring point is a better location for the post office than x is.

2. Sketch the feasible region of the problem in question. Follow this up by either applying the Extreme-Point Theorem or by inspecting level lines, whichever is more appropriate. Both methods were illustrated in the text.
3. Sketch the feasible region of the problem. Then choose values of the parameters c_1 and c_2 to obtain a desired behavior of the objective function's level lines.
4. What is the principal difference between maximizing a linear function, say, $f(x) = 2x$, on a closed vs. semi-open interval, e.g., $0 \le x \le 1$ vs. $0 \le x < 1$?
5. Trace the simplex method on the instances given, as was done for an example in the text.
6. When solving the problem by hand, you might want to start by getting rid of fractional coefficients in the problem's statement. Also, note that the problem's specifics make it possible to replace its equality constraint by one inequality constraint. You were asked to solve this problem directly in Problem 8 of Exercises 6.6.
7. The specifics of the problem make it possible to see the optimal solution at once. Sketching its feasible region for $n = 2$ or $n = 3$, though not necessary, may help to see both this solution and the number of iterations needed by the simplex method to find it.
8. Consider separately two versions of the problem: continuous and 0-1 (see Example 2 in Section 6.6).
9. If $x' = (x'_1, x'_2, \ldots, x'_n)$ and $x'' = (x''_1, x''_2, \ldots, x''_n)$ are two distinct optimal solutions to the same linear programming problem, what can we say about any point of the line segment with the endpoints at x' and x''? Note that any such point x can be expressed as $x = tx' + (1-t)x'' = (tx'_1 + (1-t)x''_1, tx'_2 + (1-t)x''_2, \ldots, tx'_n + (1-t)x''_n)$, where $0 \le t \le 1$.
10. **a.** You will need to use the notion of a matrix transpose, defined as the matrix whose rows are the columns of the given matrix.

    b. Apply the general definition to the specific problem given. Note the change from maximization to minimization, the change of the roles played by the objective function's coefficients and the constraints' right-hand sides, the transposition of the constraints, and the reversal of their signs.

    c. You may use either the simplex method or the geometric approach.

Exercises 10.2

1. What properties of the elements of the modified adjacency matrix stem from the source and sink definitions, respectively?
2. See the algorithm and an example illustrating it in the text.
3. Of course, the value (capacity) of an optimal flow (cut) is the same for any optimal solution. The question is whether distinct flows (cuts) can yield the same optimal value.

4. **a.** Add extra vertices and edges to the network given.
   b. If an intermediate vertex has a constraint on the flow amount that can flow through it, split the vertex in two.

5. Take advantage of the recursive structure of a rooted tree.

6. **a.** Sum the equations expressing the flow-conservation requirements.
   b. Sum the equations defining the flow value and flow-conservation requirements for the vertices in set X inducing the cut.

7. **a.** Use template (10.11) given in the text.
   b. Use either an add-on tool of your spreadsheet or some software available on the Internet.

10. Use edge capacities to impose the problem's constraints. Also, take advantage of the solution to Problem 4(a).

Exercises 10.3

1. You may (but do not have to) use the algorithm described in the section.

2. See an application of this algorithm to another bipartite graph in the section.

3. The definition of a matching and its cardinality should lead you to the answers to these questions with no difficulty.

4. **a.** You do not have to check the inequality for each subset S of V if you can point out a subset for which the inequality does not hold. Otherwise, fill in a table for all the subsets S of the indicated set V with columns for S, $R(S)$, and $|R(S)| \geq |S|$.
   b. Think time efficiency.

5. Reduce the problem to finding a maximum matching in a bipartite graph.

6. Transform a given bipartite graph into a network by making vertices of the former be intermediate vertices of the latter.

7. Since this greedy algorithm is arguably simpler than the augmenting-path algorithm given in the section, should we expect a positive or negative answer? Of course, this point cannot be substituted for a more specific argument or a counterexample.

8. Start by presenting a tree given as a BFS tree.

9. For pointers regarding an efficient implementation of the algorithm, see [Pap82, Section 10.2].

10. Although not necessary, thinking about the problem as one dealing with matching squares of a chessboard might lead you to a short and elegant proof that this well-known puzzle has no solution.

Exercises 10.4

1. A marriage matching is obtained by selecting three matrix cells, one cell from each row and column. To determine the stability of a given marriage matching, check each of the remaining matrix cells for a blocking pair.
2. It suffices to consider each member of one sex (say, the men) as a potential member of a blocking pair.
3. An application of the men-proposing version to another instance is given in the section. For the women-proposing version, reverse the roles of the sexes.
4. You may use either the men-proposing or women-proposing version of the algorithm.
5. The time efficiency is clearly defined by the number of proposals made. You may (but are not required to) provide the exact number of proposals in the worst and best cases, respectively; an appropriate Θ class will suffice.
6. Prove it by contradiction.
7. Prove it by contradiction.
8. Choose data structures so that the innermost loop of the algorithm can run in constant time.
9. The principal references are [Gal62] and [Gus89].
10. Consider four boys, three of whom rate the fourth boy as the least desired roommate. Complete these rankings to obtain an instance with no stable pairing.

CHAPTER 11

Exercises 11.1

1. Is it possible to solve the puzzle by making fewer moves than the brute-force algorithm? Why?
2. Since you know that the number of disk moves made by the classic algorithm is $2^n - 1$, you can simply prove (e.g., by mathematical induction) that for any algorithm solving this problem, the number of disk moves $M(n)$ made by the algorithm is greater than or equal to $2^n - 1$. Alternatively, you can show that if $M^*(n)$ is the minimum needed number of disk moves, then $M^*(n)$ satisfies the recurrence relation

$$M^*(n) = 2M^*(n-1) + 1 \quad \text{for } n > 1 \text{ and } M^*(1) = 1,$$

whose solution is $2^n - 1$.

3. All these questions have straightforward answers. If a trivial lower bound is tight, don't forget to mention a specific algorithm that proves its tightness.
4. Reviewing Section 4.4, where the fake-coin problem was introduced, should help in answering the question.

5. Pay attention to comparison losers.
6. Think inversions.
7. Divide the set of vertices of an input graph into two disjoint subsets U and W having $\lfloor n/2 \rfloor$ and $\lceil n/2 \rceil$ vertices, respectively, and show that any algorithm will have to check for an edge between every pair of vertices (u, w), where $u \in U$ and $w \in W$, before the graph's connectivity can be established.
8. The question and the answer are quite similar to the case of two n-element sorted lists discussed in the section. So is the proof of the lower bound.
9. Simply follow the transformation formula suggested in the section.
10. **a.** Check whether the formulas hold for two arbitrary square matrices.
    b. Use a formula similar to the one showing that multiplication of arbitrary square matrices can be reduced to multiplication of symmetric matrices.
11. What problem with a known lower bound is most similar to the one in question? After finding an appropriate reduction, do not forget to indicate an algorithm that makes the lower bound tight.
12. Use the problem reduction method.

Exercises 11.2

1. **a.** Prove first that $2^h \geq l$ by induction on h.
   b. Prove first that $3^h \geq l$ by induction on h.
2. **a.** How many outcomes does the problem have?
   b. Of course, there are many ways to solve this simple problem.
   c. Thinking about a, b, and c as points on the real line should help.
3. This is a straightforward question. You may assume that the three elements to be sorted are distinct. (If you need help, see decision trees for the three-element selection sort and three-element insertion sort in the section.)
4. Compute a nontrivial lower bound for sorting a four-element array and then identify a sorting algorithm whose number of comparisons in the worst case matches the lower bound.
5. This is not an easy task. None of the standard sorting algorithms can do this. Try to design a special algorithm that squeezes as much information as possible from each of its comparisons.
6. This is a very straightforward question. Use the obvious observation that sequential search in a sorted list can be stopped as soon as an element larger than the search key is encountered.
7. **a.** Start by transforming the logarithms to the same base.
   b. The easiest way is to prove that

$$\lim_{n\to\infty} \frac{\lceil \log_2(n+1) \rceil}{\lceil \log_3(2n+1) \rceil} > 1.$$

To get rid of the ceiling functions, you can use

$$\frac{f(n)-1}{g(n)+1} < \frac{\lceil f(n) \rceil}{\lceil g(n) \rceil} < \frac{f(n)+1}{g(n)-1}$$

where $f(n) = \log_2(n+1)$ and $g(n) = \log_3(2n+1)$ and show that

$$\lim_{n\to\infty} \frac{f(n)-1}{g(n)+1} = \lim_{n\to\infty} \frac{f(n)+1}{g(n)-1} > 1.$$

8. The answer to the first question follows directly from inequality (11.1). The answer to the second is no (why?).

9. a. Think losers.

b. Think the height of the tournament tree or, alternatively, the number of steps needed to reduce an n-element set to a one-element set by halving.

c. After the winner has been determined, which player can be the second best?

10. a. How many outcomes does this problem have?

b. Draw a ternary decision tree that solves the problem.

c. Show that each of the two cases—weighing two coins (one on each cup of the scale) or four coins (two on each cup of the scale)—yields at least one situation with more than three outcomes still possible. The latter cannot be resolved uniquely with a single weighing.[1]

d. Decide first whether you should start with weighing two coins. Do not forget that you can take advantage of the extra coin known to be genuine.

e. This is a famous puzzle. The principal insight is that of the solution to part (d).

11. If you want to solve the problem in the spirit of the section, represent the process of assembling the puzzle by a binary tree.

Exercises 11.3

1. Check the definition of a decidable decision problem.

2. First, determine whether $n^{\log_2 n}$ is a polynomial function. Then, read carefully the definitions of tractable and intractable problems.

3. All four combinations are possible, and none of the examples needs to be large.

4. Simply use the definition of the chromatic number. Solving Problem 5 first might be helpful but not necessary.

5. This problem should be already familiar to you.

---

1. This approach of using information-theoretic reasoning for the problem was suggested by Brassard and Bratley [Bra96].

6. What is a proper measure of an input's size for this problem?
7. See the formulation of the decision version of graph coloring and the verification algorithm for the Hamiltonian circuit problem given in the section.
8. You may start by expressing the partition problem as a linear equation with 0-1 variables x_i, $i = 1, \dots, n$.
9. If you are not familiar with the notions of a clique, vertex cover, and independent set, it would be a good idea to start by finding a maximum-size clique, a minimum-size vertex cover, and a maximum-size independent set for a few simple graphs such as those in Problem 4. As far as Problem 9 is concerned, try to find a relationship between these three notions. You will find it useful to consider the *complement* of your graph, which is the graph with the same vertices and the edges between vertices that are *not* adjacent in the graph itself.
10. The same problem in a different wording can be found in the section.
11. Just two of them do not contradict the current state of our knowledge about the complexity classes.
12. The problem you need was mentioned explicitly in the section.

Exercises 11.4

1. As the given definition of the number of significant digits requires, compute the relative errors of the approximations. One of the answers doesn't agree with our intuitive idea of this notion.
2. Use the definitions of the absolute and relative errors and the properties of the absolute value.
3. Compute the value of $\sum_{i=0}^{5} \frac{0.5^i}{i!}$ and the magnitude of the difference between it and $\sqrt{e} = 1.648721\dots$.
4. Apply the formula for the area of a trapezoid to each of the n approximating trapezoid strips and sum them up.
5. Apply formulas (11.7) and (11.9) to the integrals given.
6. Find an upper bound for the second derivative of $e^{\sin x}$ and use formula (11.9) to find a value of n guaranteeing the truncation error smaller than the given error limit.
7. A similar problem is discussed in the section.
8. Consider all possible values for the coefficients a, b, and c. Keep in mind that solving an equation means finding all its roots or proving that no roots exist.
9. **a.** Prove that every element x_n of the sequence is (i) positive, (ii) greater than $\sqrt{D}$ (by computing $x_{n+1} - \sqrt{D}$), and (iii) decreasing (by computing $x_{n+1} - x_n$). Then take the limit of both sides of equality (11.15) as n goes to infinity.

b. Use the equality

$$x_{n+1} - \sqrt{D} = \frac{(x_n - \sqrt{D})^2}{2x_n}.$$

10. It is done for $\sqrt{2}$ in the section.

CHAPTER 12

Exercises 12.1

1. a. Resume the algorithm by backtracking from the first solution's leaf.

b. How can you get the second solution from the first one by exploiting a symmetry of the board?

2. Think backtracking applied backward.

3. a. Take advantage of the general template for backtracking algorithms. You will have to figure out how to check whether no two queens attack each other in a given placement of the queens.

To make your comparison with an exhaustive-search algorithm easier, you may consider the version that finds all the solutions to the problem without taking advantage of the symmetries of the board. Also note that an exhaustive-search algorithm can try either all placements of n queens on n distinct squares of the $n \times n$ board, or only placements of the queens in different rows, or only placements in different rows and different columns.

b. Although it is interesting to see how accurate such an estimate is for a single random path, you would want to compute the average of several of them to get a reasonably accurate estimate of the tree size.

4. Consider separately six cases of different remainders of the division of n by 6. The cases of $n \bmod 6 = 2$ and $n \bmod 6 = 3$ are harder than the others and require an adjustment of a greedy placement of the queens.

5. Another instance of this problem is solved in the section.

6. Note that without loss of generality, one can assume that vertex a is colored with color 1 and hence associate this information with the root of the state-space tree.

7. This application of backtracking is quite straightforward.

8. a. Another instance of this problem is solved in the section.

b. Some of the nodes will be deemed promising when, in fact, they are not.

9. A minor change in the template given does the job.

11. Make sure that your program does not duplicate tree nodes for the same board position. And, of course, if a given instance of the puzzle does not have a solution, your program should issue a message to that effect.

Exercises 12.2

1. What operations does a best-first branch-and-bound algorithm perform on the live nodes of its state-space tree?
2. Use the smallest numbers selected from the columns of the cost matrix to compute the lower bounds. With this bounding function, it's more logical to consider four ways to assign job 1 for the nodes on the first level of the tree.
3. **a.** Your answer should be an $n \times n$ matrix with a simple structure making the algorithm work the fastest.

   b. Sketch the structure of the state-space tree for your answer to part (a).
5. A similar problem is solved in the section.
6. Take into account more than a single item from those not included in the subset under consideration.
8. A Hamiltonian circuit must have exactly two edges incident to each vertex of the graph.
9. A similar problem is solved in the section.

Exercises 12.3

1. **a.** Start by marking the first column of the matrix and finding the smallest element in the first row and an unmarked column.

   b. You will have to find an optimal solution by exhaustive search or by a branch-and-bound algorithm or by some other method.
2. **a.** The simplest approach is to mark matrix columns that correspond to visited cities. Alternatively, you can maintain a linked list of unvisited cities.

   b. Following the standard plan for analyzing algorithm efficiency should pose no difficulty (and yield the same result for either of the two options mentioned in the hint to part (a)).
3. Do the walk in the clockwise direction.
4. Extend the triangle inequality to the case of $k \geq 1$ intermediate vertices and prove its validity by mathematical induction.
5. First, determine the time efficiency of each of the three steps of the algorithm.
6. You will have to prove two facts:

   i. $f(s^*) \leq 2f(s_a)$ for any instance of the knapsack problem, where $f(s_a)$ is the value of the approximate solution obtained by the enhanced greedy algorithm and $f(s^*)$ is the optimal value of the exact solution to the same instance.

   ii. The smallest constant for which such an assertion is true is 2.

   In order to prove (i), use the value of the optimal solution to the continuous version of the problem and its relationship to the value of the approximate solution. In order to prove (ii), find a family of three-item instances that prove the point (two of them can be of weight $W/2$ and the third one can be of a weight slightly more than $W/2$).

7. a. Trace the algorithm on the instance given and then answer the question whether you can put the same items in fewer bins.

b. What is the basic operation of this algorithm? What inputs make the algorithm run the longest?

c. Prove first the inequality

$$B_{FF} < 2\sum_{i=1}^{n} s_i \quad \text{for any instance with } B_{FF} > 1$$

where B_{FF} is the number of bins obtained by applying the first-fit (*FF*) algorithm to an instance with sizes $s_1, s_2, \ldots, s_n$. To prove it, take advantage of the fact that there can be no more than one bin that is half full or less.

8. a. Trace the algorithm on the instance given and then answer the question whether you can put the same items in fewer bins.

b. You can answer the question either with a theoretical argument or by providing a counterexample.

c. Take advantage of the two following properties:

i. All the items placed by *FFD* in extra bins, i.e., bins after the first B^* ones, have size at most 1/3.

ii. The total number of items placed in extra bins is at most $B^* - 1$.

(B^* is the optimal number of bins.)

d. This task has two versions of dramatically different levels of difficulty. What are they?

9. a. One such algorithm is based on the idea similar to that of the source removal algorithm for the transitive closure except that it starts with an arbitrary edge of the graph.

b. Recall the warning that polynomial-time equivalence of solving *NP*-hard problems exactly does not imply the same for their approximate solving.

10. a. Color the vertices without introducing new colors unnecessarily.

b. Find a sequence of graphs G_n for which the ratio

$$\frac{\chi_a(G_n)}{\chi^*(G_n)}$$

(where $\chi_a(G_n)$ and $\chi^*(G_n)$ are the number of colors obtained by the greedy algorithm and the minimum number of colors, respectively) can be made as large as one wishes.

Exercises 12.4

1. It might help your search to know that the solution was first published by Italian Renaissance mathematician Girolamo Cardano.

2. You can answer these questions without using calculus or a sophisticated calculator by representing equations in the form $f_1(x) = f_2(x)$ and graphing functions $f_1(x)$ and $f_2(x)$.

3. a. Use the property underlying the bisection method.

b. Use the definition of division of polynomial $p(x)$ by $x - x_0$, i.e., the equality

$$p(x) = q(x)(x - x_0) + r,$$

where x_0 is a root of $p(x)$, $q(x)$ and r are the quotient and remainder of this division, respectively.

c. Differentiate both sides of the equality given in part (b) and substitute x_0 in the result.

4. Use the fact that $|x_n - x^*|$ is the distance between x_n, the middle of interval $[a_n, b_n]$, and root x^*.

5. Sketch the graph to determine a general location of the root and choose an initial interval bracketing it. Use an appropriate inequality given in Section 12.4 to determine the smallest number of iterations required. Perform the iterations of the algorithm, as is done for the example in the section.

6. Write an equation of the line through the points $(a_n, f(a_n))$ and $(b_n, f(b_n))$ and find its x-intercept.

7. See the example given in the section. As a stopping criterion, you may use either the length of interval $[a_n, b_n]$ or inequality (12.12).

8. Write an equation of the tangent line to the graph of the function at $(x_n, f(x_n))$ and find its x-intercept.

9. See the example given in the section. Of course, you may start with a different x_0 than the one used in that example.

10. Consider, for example, $f(x) = \sqrt[3]{x}$.

11. Derive an equation for the area in question and then solve it by using one of the methods discussed in the section.

Index

The index covers the main text, the exercises, the epilogue, and the appendices. The following indicators are used after page numbers: "ex" for exercises, "fig" for figures, "n" for footnotes, "sum" for summaries.

E

M

V

W